ALSO BY KEACH HAGEY

*The King of Content:
Sumner Redstone's Battle for Viacom, CBS, and
Everlasting Control of His Media Empire*

# THE OPTIMIST

*Sam Altman, OpenAI, and the Race to Invent the Future*

# KEACH HAGEY

*Independent Publishers Since 1923*

Copyright © 2025 by Keach Hagey

All rights reserved
Printed in the United States of America
First Edition

For information about permission to reproduce selections from this book, write to Permissions, W. W. Norton & Company, Inc., 500 Fifth Avenue, New York, NY 10110

For information about special discounts for bulk purchases, please contact W. W. Norton Special Sales at specialsales@wwnorton.com or 800-233-4830

Manufacturing by Sheridan
Book design by Lovedog Studio
Production manager: Julia Druskin

ISBN 978-1-324-07596-7

W. W. Norton & Company, Inc.
500 Fifth Avenue, New York, NY 10110
www.wwnorton.com

W. W. Norton & Company Ltd.
15 Carlisle Street, London W1D 3BS

10 9 8 7 6 5 4 3 2 1

For Wesley

# CONTENTS

*PROLOGUE* .................................................................. 1

## PART I  1985–2005

CHAPTER 1   **CHICAGO** ............................................... 21

CHAPTER 2   **ST. LOUIS** ............................................. 39

CHAPTER 3   **"WHERE ARE YOU?"** ............................... 55

CHAPTER 4   **AMONG THE "NERD'S NERDS"** ............... 67

## PART II  2005–2012

CHAPTER 5   **"STOPPING OUT"** ................................... 85

CHAPTER 6   **"WHERE YOU AT?"** ................................ 99

CHAPTER 7   **FROM "WEAK" TO "COOL"** ................... 110

CHAPTER 8   **THE DOUCHEBAG BADGE** ..................... 118

## PART III  2012–2019

CHAPTER 9   **"A RIDE ON A ROCKET"** ........................ 131

CHAPTER 10  **"SAM ALTMAN FOR PRESIDENT"** ........... 149

CHAPTER 11  **"A MANHATTAN PROJECT FOR AI"** ........ 167

CHAPTER 12  **ALTRUISTS** ......................................... 189

CHAPTER 13  **PIVOTING TO PROFIT** ......................... 215

## PART IV  2019-2024

CHAPTER 14  **PRODUCTS** ........................................... 241

CHAPTER 15  **CHATGPT** ........................................... 254

CHAPTER 16  **THE BLIP** .......................................... 276

CHAPTER 17  **PROMETHEUS UNBOUND** ..................... 295

EPILOGUE ......................................................... 311

*ACKNOWLEDGMENTS* ............................................... 317

*NOTES* ........................................................... 321

*INDEX* ........................................................... 341

# PROLOGUE

**ON A BALMY MID-NOVEMBER EVENING IN 2023, PETER THIEL,** the famed venture capitalist, threw a birthday party for his husband, Matt Danzeisen, at YESS, an avant-garde Japanese restaurant located in a century-old converted bank building in Los Angeles's Arts District. Seated next to him in the cavernous, temple-like space was his friend Sam Altman.[1] Thiel had backed Altman's first venture fund more than a decade before, and had remained a mentor to the younger investor as the latter became the face of the artificial intelligence revolution as the CEO of OpenAI. OpenAI's launch of ChatGPT a year earlier had propelled tech stocks out of a slump and to one of their best years in decades. Yet Thiel was worried.

Years before he met Altman, Thiel had taken another AI-obsessed prodigy under his wing named Eliezer Yudkowsky. Thiel funded his institute, which was based on the idea that humans had to figure out how to make AI friendly before they built one that was smarter than they were. But Thiel felt that Yudkowsky had become "extremely black-pilled and Luddite," boiling his doomer argument down to: "All you can do is go to Burning Man and do lots of drugs while you were waiting for the AI to come along and kill you." In March, Yudkowsky had published an op-ed in *Time* magazine arguing that unless the current wave of generative AI research was halted, "literally everyone on Earth will die."[2]

"You don't understand how Eliezer has programmed half the people in your company to believe this stuff," Thiel warned Altman. "You need to take this more seriously."

Altman picked at his vegetarian dish and tried not to roll his eyes. This was not the first dinner where Thiel had warned him that the company had been taken over by "the EAs," by which he meant people who subscribed to the philosophy of effective altruism, a data-driven cousin of utilitarianism. The EAs had lately pivoted from trying to end global poverty to trying to prevent runaway AI from murdering humanity. Thiel had repeatedly predicted that "the AI safety people" would "destroy" OpenAI. Thiel had backed the company from the beginning, first with a personal donation when it was just a small nonprofit research lab in 2015, and then again earlier in 2023 through his Founders Fund, after OpenAI had sprouted a for-profit subsidiary designed to gobble up billions from Microsoft and other investors. But he was also a well-known catastrophist who, as the joke went in Silicon Valley, had correctly predicted seventeen of the last two financial crises.

"Well, it was kind of true of Elon, but we got rid of Elon," Altman responded, referring to the messy 2018 split with his co-founder, Elon Musk, who once referred to the attempt to create artificial intelligence as "summoning the demon."[3] "And then there were the Anthropic people," Altman continued, meaning the more than one dozen OpenAI employees who left at the end of 2020 to set up their own rival lab after losing trust in Altman. "But we separated from them." The more than seven hundred employees who were left had been riding a rocket ship and were about to have the chance to buy beachfront second homes with the imminent close of a tender offer valuing OpenAI at more than $80 billion. There was no need to panic.

**ALTMAN HAD** long made optimism the core of his personal brand. But anyone in his position would have reason to feel as positive as he did in that moment. The elfin thirty-eight-year-old was wrapping up the best year of a charmed career, a year when he became a household name, had senators eating out of his hand, met with presidents and prime ministers around the world, and—most important within the value system of Silicon Valley—delivered a new technology that seemed like it was very possibly going to change everything. When OpenAI launched its uncannily humanlike chatbot, ChatGPT (short for generative pre-trained transformer) the previous November, it was an instant smash, reaching 100 million users in less than three months, the fastest-growing app in the world to date.[4] When OpenAI, only a few months later, unveiled a more formidable successor, GPT-4—it could pass the bar exam and ace the AP biology test—the dizzying rate of progress suggested that the company's audacious mission to safely create the world's first artificial general intelligence, or AGI, might indeed be within reach. Even the most determined AI skeptics—including one Stanford computer science professor who memorably dismissed the original ChatGPT to me as a "dancing dog"—felt their doubts soften. For a few giddy months, as every company in America frantically spun up AI task forces and tried to estimate the productivity gains that AI would bring, it felt as if we were all collectively stepping into a science fiction short story, and Altman was the author.

Altman was not actually writing the code. He was, instead, the visionary, the evangelizer, and the dealmaker; in the nineteenth century, he would have been called the "promoter." His specialty, honed over years of advising and then running the prestigious startup accelerator Y Combinator, was to take the nearly impossible, convince others that it was in fact possible, and then raise so much money that it actually became possible. "He's the only person I've ever met in my life who only wants to work on things that could change the world, even if there's only a one percent chance of them working," said Ali Rowghani, who ran a fund at Y Combinator during Altman's tenure as the accelerator's president.

Perhaps more than anyone else, Altman embodies the "add a zero" ethos of Silicon Valley, a mindset he learned at the knee of his original mentor, the hacker, entrepreneur, and essayist Paul Graham, who co-founded Y Combinator. Graham's advice to his startups was nearly always to stop thinking too small, to think about how to elevate their business model so that the "millions" in the slide deck of revenue projections could be replaced with "billions." As he took the reins at OpenAI in 2019, Altman blogged about his own personal philosophy of success: "It's useful to focus on adding another zero to whatever you define as your success metric—money, status, impact on the world, whatever."[5] Thiel said he was drawn to Altman because he was "just at the absolute center of the Silicon Valley zeitgeist," a millennial born in 1985 who entered the tech world during the sweet spot between dot-com crash and the financial crisis, when startup optimism was once again possible but tech had not yet become what Thiel derided as a calcified "track." Riding along with Altman's investments in the mid-2010s, as the talk of a tech bubble got louder, Thiel had to override his contrarian tendencies, and was glad he did. "Sam was extremely optimistic, which was an important trait to invest in these things because they all looked fully valued," he said. As it turned out, YC-minted unicorns like Stripe and Airbnb still had a lot of room to grow. For nearly all of his adult life, with the brief exception of the aftershocks of the financial crisis of 2008 and a global pandemic that began in 2020, Altman has seen tech markets do nothing but rise.

But this time, Thiel's pessimism was on the mark. As the two investing partners celebrated beneath the exposed rafters of LA's hottest new restaurant, four members of OpenAI's six-person board—including two with direct ties to the EA community—were holding secret video meetings about firing Altman. While Yudkowsky himself had nothing to do with it, his influential blog, *LessWrong*, had helped place fear of the existential risk from AI at the heart of the EA movement.

This fear had influenced the founding of OpenAI itself, whose stated goal was "to advance digital intelligence in a way that is most likely to benefit humanity as a whole, unconstrained by the need to generate financial return." The fears appeared even more clearly in OpenAI's unusual 2018 charter, which declares that because "we are concerned about late-stage AGI development becoming a competitive race without time for adequate safety precautions," the company is committed to "stop competing with and start assisting" any "value-aligned" projects that reach AGI before it does. The fear manifested most bizarrely in the company's governance structure, in which a for-profit subsidiary was controlled by a nonprofit board with fiduciary duty to no investor but rather "to humanity."[6] Investors were warned that their money could be wiped out if the board determined that that's what was necessary to fulfill this all-important mission.

The charter had flowed in part from "AI Principles" developed at a 2017 conference organized by the Future of Humanity Institute, a nonprofit organization focused on mitigating the existential risk from AI that had been bankrolled by billionaires including Musk and Skype founder Jaan Tallinn. Altman had attended the conference and signed on to the "principles." Earlier, in 2015, the same year he co-founded OpenAI, Altman wrote on his blog that AGI was "probably the greatest threat to the continued existence of humanity," recommending the book *Superintelligence: Paths, Dangers, Strategies* by Nick Bostrom, a philosopher at Oxford University who had been a frequent guest at the conferences organized by Yudkowsky's institute over the years.[7] The AI safety concerns popularized by Bostrom—most notably the parable of the paperclip-making AI who destroys humanity not out of spite, but because people got in the way of its programmed need to turn all matter in the universe into paperclips, a fable cribbed and bastardized from Yudkowsky—were fundamental to OpenAI's initial ability to recruit the world's top AI research scientists, not least because Musk shared those concerns and lent his fortune to the effort. And even as Altman took his post-ChatGPT victory lap, he was always careful to hint at potential doom in his ecstatic visions of the future, asking US senators during a televised hearing to regulate AI because "if this technology goes wrong, it can go quite wrong."[8]

Many in the industry saw this as little more than savvy marketing. Indeed, Altman's focus in 2023 had largely been on what he did best: striking deals for more investment, whipping up fascination in the press, and serving as the global prophet of an unimaginably prosperous future. He may in fact have been worried about where everything was heading, but his foot was placed firmly on the gas.

However, contrary to Thiel's dark suggestions, the members of OpenAI's board were not meeting that night because they feared that OpenAI was hurtling too fast toward AGI. In fact, their reasons for wanting to oust Altman had little to do with effective altruism or existential risk. Instead, they had to do with something like Thiel's ultimate praise of Altman as the pure distillation of "the Silicon Valley zeitgeist."

According to the early twenty-first-century Silicon Valley zeitgeist that Y Combinator cultivated and Altman exemplifies, founders are kings, emperors, gods. The venture capitalists who fund them trip over themselves to prove how "founder-friendly" they are, which in practice means how unlikely they would be to replace a founding CEO or otherwise give the founders a hard time. Thiel took this imperative to its logical conclusion by naming his venture firm the "Founders Fund" and declaring that it would never fire a founder.[9] In the end, the success of a startup is usually regarded as secondary to the relationship between the venture capitalist and the founder. After all, if this business doesn't work out, the founder can always take another round of funding and start something new, something that, this time, might reach a value denomination that begins with a *b*.

Founders had won so much leverage by the time Altman took the reins at Y Combinator in 2014 that he wrote a blog gently pushing back against the trend of startups' founders cashing investors' checks without giving them board seats, arguing that some experienced venture capitalists might actually have some expertise to offer their companies. But, lest any of the YC startups worry Altman had gone too far, he concluded the post by saying "it's a good idea to keep enough control so that investors can't fire you."[10]

OpenAI was intended to be a break from this dynamic. Not only did Altman not own the kind of super-voting shares that ensured that someone like Mark Zuckerberg would be a king for life, he owned almost no equity at all. He had agreed to this unprecedented situation early in OpenAI's development, first because it was originally a nonprofit, and then as a means of staying on the board while satisfying

PROLOGUE

the charter's requirement that the majority of directors be independent, that is, without equity in the company. He argued that his lack of power was an essential form of accountability.

Yet the board had found that Altman, used to maneuvering quickly in the opaque world of venture capital, had so much de facto power that they felt they could not do their jobs. Five days after eating Japanese food with Thiel, Altman was fired by the company he had co-founded. The reason was that he was "not consistently candid in his communications with the board."

**I MET** Altman eight months earlier, during the initial wave of AI fever. *The Wall Street Journal* had sent me out to San Francisco to interview him at OpenAI's headquarters in the city's Mission District, the funky enclave where I'd spent a summer during college working as a barista. Back then, during the first dot-com bubble, the neighborhood was plastered with homemade flyers telling the tech people to go home— *Die Yuppie Scum*—and leave the punks and freaks in peace. In the intervening years, as the next tech boom overwhelmed the Mission, it had become something like Brooklyn, but with better burritos.

A new kind of euphoria was in the air. It was mid-March, the gray season in New York when everyone has just about stopped believing in the sun, but San Francisco was luminous. After five days of rain, the air had been scrubbed clean and the sun had come out. Every billboard along Highway 101 hawked some kind of AI. The evening news led with tales of GPT-4's passing the LSAT. The Mission was full of tasteful sprigs of wildflowers, coffee with notes of blueberry, and homeless people.

OpenAI's office is located in a nondescript former mayonnaise factory on the formerly industrial edge of the Mission near Potrero Hill. It lacks any kind of sign identifying what lies within—a choice that seemed silly at the time but came to seem more prudent later that month when Yudkowsky called for the bombing of data centers to stop rogue AI. When I approached, I encountered a bewildered-looking investor with a lanyard doing the same thing I was doing: walking back and forth between the unmarked door and the unmarked freight bay around the corner, unable to believe that either one of these could be the entrance to Silicon Valley's most exciting and terrifying company. It did not help that when we asked the security guard near the freight bay if we were in the right place, he refused to say.

Eventually, we both made it into the lobby, which felt like a cross between a greenhouse and a day spa. Succulents and ferns hovered at all levels. The sound of water trickling from stone fountains mixed with the murmured chatter of venture capitalists who had come for some kind of networking event to learn about the AI investing ecosystem.

After a while, Altman bounded into the room in blindingly white sneakers, smiling gamely, his dimples making him seem even younger than his thirty-seven years. The first thing that anyone notices about Altman—and the fact that so many of the early stories about him revolve around—is that he's small, a rail-thin five feet, eight inches. The second thing is the intensity of his green-eyed gaze, which he levels directly at you, as though he is speaking to the most important person in the world. He apologized that his last meeting, which was still listed on the conference room scheduling screen as "AI Manhattan Project," had run long.

"It was about how we can think about doing more new efforts on alignment," Altman explained, when asked about that meeting's ominous title. "As capabilities are advancing, can we have more of a coordinated effort with other groups on how we address the AGI safety problem in particular. Which I think we have exciting ideas for."

We were talking two days after the release of GPT-4. Altman was leading an organization of such ostensibly epochal significance that mere products and profits were far from his mind. He seemed to revel in the weirdness of this.

"I was extremely fortunate that early in my career, I made more money than I could ever need," he said. "I want to work on what I think is going to be interesting and important and useful and impactful and critical to get right, but I don't need more money. And also,

I think we will make some decisions here that are," and he paused, groping for the right word, "*strange* over time."

He outlined a future where the globe's population would vote on how AI should turn out. "Very deeply we would like this technology to be governed by, and the benefits of it shared with, everybody," he said. "If you can't do it as a government project, and I think there are many reasons why that just may not be a good idea or practical, a nonprofit is a reasonable approach." And his definition of safe AGI was quite broad. Asked what safety meant to him, he spoke of a future in which "the vast majority of people in the world are much better off than they would have been in a world where AGI just didn't happen." For many, that would mean a different line of work because, he said, "I don't think most people love their jobs."

At only one moment during our two-hour interview did the altruistic mask slip to reveal the fierce competitor beneath. Over the previous month, both Google and Anthropic had announced the impending releases of their own generative-AI chatbots, and it appeared that the industry was entering into the exact kind of competitive AI arms race that the OpenAI charter openly fretted about. But when asked about the competition, Altman would only say: "Um, well, they've raced to release press releases," he said, adding, "Obviously they're behind now."

This brief instance of chest-beating notwithstanding, the OpenAI offices that he led us through pulsed with cultish do-goodery, like a prestigious private school with an honor code. There was a cafeteria, a 1980s-style living room filled with board games, and a replica of a university library, complete with book ladder and glowing desk lamps, inspired by the Bender Room in Stanford University's Green Library. At the end of the shelves was a stack of vinyl records topped, that day, by the soundtrack to *Blade Runner*.

But Altman's obvious pride and joy was the feature that had been the hardest to build: the winding central staircase that he had designed so that all of the then–four hundred employees would have to pass each other each day. Standing at the top of the staircase, geeking out about the engineering feats it had required, Altman did seem more than a little bit like a priest in a temple. And then something he had said during our interview rushed back into my inner ear.

"It wasn't that long ago that almost no one believed in AGI," he said. "And still, maybe most people don't. But I think more people are willing to entertain it now. And I think a lot of the world is going through a process that most of the people here have gone through in previous years, which is, like, really grappling with this. And it is hard. It is exciting. It is terrifying. It's a *lot*. And so I expect that process to unfold in the world over the next few years, and we'll try to be a voice of some guidance along the way."

Altman was here, not so much to sell technology as to sell *belief*. And in that he had already succeeded beyond anyone's wildest imaginings. When asked about Altman's idealism for the *Journal* profile that was the purpose of my visit to OpenAI's office, Thiel said, "We should treat him as more of a messiah figure."

**A LITTLE** more than a year later, in April 2024, I walked into the Baccarat Hotel in Midtown Manhattan to find Altman curled up in one of the opulent lobby's overstuffed leather chairs. I was early, but he had gotten there even earlier, discreetly installed his security detail in the corner, and—I later discovered after it was too late to do anything about it—arranged to pick up the check. When he saw me, he leaped up, threw his arms wide open, and greeted me with a hug. This was Altman: warm, charming, thoughtful, kind.

He was wearing his typical uniform of a cornflower-blue long-sleeved T-shirt, dark indigo hipster jeans, and immaculate gray New Balance sneakers. In a few days, he would turn thirty-nine, and a few flecks of gray were creeping into his mop of brown hair. He had ordered an espresso—one of the two he typically allows himself per day, the first in lieu of breakfast—and seemed genuinely happy, enlivened by an extended visit to New York.

This was surprising, because he had made it quite clear over the course of months of negotiations that he did not want this book to be written. The *Wall Street Journal* profile of him that my colleague Berber Jin and I had written a year earlier, after the interview at OpenAI's offices, had led to a book deal. When initially informed about it, he argued that it was too soon, and too focused on him. After considering it for several months, he told me he would not be participating. Since the subject of my last book had been in something close to a vegetative state during the entire time I was reporting his biography, I was not particularly daunted by this, and kept making calls. Then, a few months before our meeting in New York, Altman had had a change of heart. He'd help, a little, but requested I make clear how odious he found the project.

"I really object to the warping of history that comes when one person gets imputed onto a company, or a movement, or a tech revolution, because that's just not how the world works and is unfair to the exceptional work of other people," he said, a flicker of anger in his voice. "I think people should not encourage that viewpoint."

While noble, this view had become somewhat harder to support since Altman had retaken the CEO's seat a mere five days after his firing. Nearly all of the company's 770 employees had signed a petition threatening to quit and go work at Microsoft if he weren't reinstated. Clearly, to OpenAI's employees and investors, Altman was indispensable. And hardly a day had passed in the last eighteen months that the business press had not run a story about the AI revolution illustrated with his photograph. Whether he liked it or not, it was his face, not the OpenAI logo, that was the symbol of what many were calling the biggest technological advance of our lifetimes, if not human history.

His other objection was less modest, more plausible, and way more Sam.

"There's something about the superstition of not prematurely celebrating, and I think this is a decade or two premature," he objected. "OpenAI has a long way to go."

Altman has a skill for convincing people that he can see into the future. When he talks publicly about OpenAI's technology, he often disses its current products—he had recently told a prominent podcaster that GPT-4, the company's most advanced product, "kind of sucks"—and invites the audience to focus on what the company's current rate of improvement implies about what is to come.[11] This is pure investor thinking. In my first years at the *Journal*, when I would often get to the office early to cover companies' earnings reports, I always marveled at how nothing in the companies' press releases ever seemed to move the stock; investors were instead focused on the forward-looking "guidance" that the company's CFO would give in mumbly jargon during the call with analysts. In the world of venture capital–backed private startups that Altman grew up in, this near-shamanistic ability to lead a boardroom full of VCs in a group visualization of a startup's revenue going up and to the right was the whole game. And no one was better at it than Altman.

In Altman's vision of the future, AGI will inevitably become an "extension of our wills" without which "we just don't feel like ourselves." Students will get free or very cheap AI tutors that will make them "smarter and better prepared than anybody today." The price of goods and services will plummet as AI handles much of the work of lawyers, graphic designers, and computer programmers, leaving people free to spend more time in the "flow state" of whatever creative work they want to pursue. Wages of those affected will be replaced with a universal basic income, or UBI, which itself might be derived from some share of the wealth that all these new robot slaves produce. Governments will co-invest with private industry to build vast data centers, powered by cheap nuclear energy made possible by unlocking the secrets of the nuclear fusion reaction that powers the sun. AI will flow like electricity, allowing us to cure cancer and solve the mysteries of physics that have kept us bound to this planet. We will live longer as diseases fall away one by one. Humanity will step into a new age of health and abundance.

Altman has done much more than just talk about this vision. Through his portfolio of more than four hundred startup investments that have made him a billionaire several times over, he has placed large personal bets on companies that may help realize it.[12] He has invested at least $375 million in Helion, a YC-backed startup that is trying to make nuclear fusion a source of clean, renewable energy, and he has backed and taken public the nuclear fission microreactor startup Oklo. He has co-founded the company behind Worldcoin, a for-profit cryptocurrency project that travels the world with a bowling ball–sized orb, scanning people's irises in exchange for cryptocurrency that could one day be used to distribute a kind of global UBI. And he personally funded long-running studies into the efficacy of UBI.

His other investments include moonshots that would extend the human lifespan by ten years, treat Parkinson's disease with stem cells, bring back supersonic commercial flight, and create an implant to connect a brain to a computer. One gets the sense that he would never personally invest in anything as square as enterprise software (never mind that that's largely what OpenAI and its partner Microsoft have to date actually produced). As Paul Graham once said of Altman's portfolio of wildly ambitious projects, "I think his goal is to make the whole future."[13]

Altman says he didn't set out to create an interlocking series of investments that promise to remake every facet of human existence. It just kind of happened that way. "I have for a very long time believed that energy and intelligence are the two most important things," he said. "I didn't realize how much they fit together. I got totally lucky on that."

In these ways, Altman is little different from many of his peers, even if he is a better and more ambitious investor than most. One thing that sets him apart, however, is his interest in politics. He doesn't just want to create a new technology and give it to the world. He has always aspired to be, in the words of Patrick Chung, the investor who first discovered him, a "great man of history." He has told friends and associates in 2016 and 2017 that he was mulling a run for president, and flirted with a run for California governor in 2017, drafting a national platform later that year after deciding to back other candidates rather than run himself. After ChatGPT exploded into global consciousness, Altman made the first of many trips to the White House, and then set off on a global tour that put him face-to-face with world leaders from French president Emmanuel Macron to Indian prime minister Narendra Modi.

Altman gamely indulged my questions for another two and a half hours, occasionally taking out a little spiral-bound notebook from his pocket to jot down a reminder to check with one or another of the people I'd interviewed about him. It was both endearing—as though he were at my service—and unsettling. Altman sits at the center of Silicon Valley's economy of favor-trading, and I could only imagine the reputational repercussions that would ripple out from this project. As Graham once memorably put it, "Sam is extremely good at becoming powerful."

Yet his power since his firing and return—referred to within OpenAI as "the blip"—had changed, becoming both more expansive and more conventional. He was warier, more guarded. He had stopped bragging about the company's strange governance structure. His excitement about UBI had morphed into wanting to give people free (or cheap) access to ChatGPT instead of money. His fame meant he no longer had time for hobbies. They had come for him, and likely would again. And as successful as he had been at batting down each successive threat—the investigation following the firing had found that the board had not uncovered any clearly fireable offence so much as lost trust and made a judgment—new ones kept arising seemingly every week. Beneath them all was the question, first whispered, then murmured, then popping up in elaborate online essays from the company's defectors: Can we trust this person to lead us to AGI?

**SINCE RETURNING** to OpenAI, Altman has been trying to reshape the company into a more typical for-profit in which he may have as much as $10 billion in equity after all. He is now much more than the face of the AI revolution—he is its undisputed leader and controller. He has attracted powerful enemies, most notably Musk, who has sued Altman and OpenAI, accusing them of betraying the original nonprofit mission. (OpenAI says Musk's suit is baseless, and Musk's critics point out that he is now a competitor after starting his own AI company, xAI.) And Altman's stature has made the question of who he really is more urgent than ever.

To write this book, I conducted more than 250 interviews with Altman's family, friends, teachers, mentors, co-founders, colleagues, investors, and portfolio companies, in addition to hours of interviews with Altman himself. The person who emerged is a brilliant dealmaker with a need for speed and a love of risk, who believes in technological progress with an almost religious conviction—yet who sometimes moves too fast for the people around him and whose aversion to confrontation has occasionally allowed bigger conflicts to fester. Yet every time Altman has been knocked down, he has managed to come back with more power. As Graham famously wrote of him in 2008, "You could parachute him into an island full of cannibals and come back in 5 years and he'd be the king."[14]

To understand Altman requires understanding his family. The story that follows begins with his father, Jerry Altman, who was far more than the "real estate developer" he is sometimes described as (if he is described at all). Jerry Altman's political activism and creative dealmaking had a lasting impact on affordable housing policy, and he passed his passion for both on to his son. Sam's mother, Connie Gibstine, bequeathed her son both her family's "science brain" and her own ferocious work ethic. Jerry and Connie's daily affirmations that Sam and his siblings could be whatever they wanted to be helped fuel Sam's confidence and optimism. At the same time, within their ultimately unhappy marriage lie many of the seeds of Altman's own self-professed history of anxiety and the rift that led his sister, Annie, to break off contact with the rest of her family after Jerry died in 2018.

Growing up as the firstborn son of Jewish professionals in suburban St. Louis, Altman demonstrated early on that he was special, and was treated as such. His childhood was shaped by progressive institutions, from the social justice–focused Central Reform Congregation synagogue to the rigorous John Burroughs School, which both taught that he had a moral duty to improve the world. Through computers, he discovered both an intellectual calling and, through AOL Instant Messenger chats, a way to break through the awkwardness of being a gay teenager in the Midwest in the late 1990s. At Burroughs, Altman's decision to speak out against what he saw as displays of intolerance of his sexuality would teach him about the potentially life-changing value of taking big risks.

As an undergraduate at Stanford, Altman met the co-founders of his first startup, Loopt, a location-based social network for the flip-phone era. It was the only company Altman ran before OpenAI, and Loopt's story offers a preview of what was to come, from the relative ease it had raising capital from prestigious venture capital firms like Sequoia Capital to the staff mutinies that Altman faced as a young CEO leading a struggling startup.

The most important thing that Loopt did, in the end, was introduce Altman to Paul Graham and Y Combinator. In Altman, Graham saw a distillation of everything it took to succeed with a startup. Loopt was sold for parts in 2012, but Altman remained close to YC, advising its startups even as he oversaw his own investment fund backed by Thiel. And when Graham decided to retire, he chose Altman as his successor. This put Altman at the very center of power in Silicon Valley. Under his leadership, YC went from incubating dozens to hundreds of startups a year, pushed into the hard sciences, and created an arm for moonshots that would end up spawning a nonprofit research lab called OpenAI. With Altman still busy leading YC, the job of rounding up talent to staff the lab was left to his friend Greg Brockman, the former CTO of YC-funded payments company Stripe.

This book presents new information not only on Altman's early life and career, but on his time so far at OpenAI. It tells the story of how Altman and Musk bonded over weekly dinners discussing the dangers and promise of AI technology, and then how Altman won the power struggle against the older, richer entrepreneur thanks to his alliance with Brockman. It reveals how Altman, who longtime Sequoia Capital chief Michael Moritz calls a "mercantile spirit," oversaw the creation of the first commercial product to ever use a large language model, which up until then had been the purview of academic research. And it shows how, with the release of ChatGPT and GPT-4, Altman used his YC-honed mastery of telling startup stories to tell one of the greatest startup stories of all time.

**ALTMAN IS** wary of discussing it with journalists, but friends including Thiel say he is sympathetic to the notion, common in Silicon Valley, that AGI has already been invented and we are living in a computer simulation that it has created. "Sam lands on the simulation side and I land on the non-simulation side," Thiel told the *Journal* for our profile. "There are some ways in which you can say: How is the AI different from God?" When asked about this, Altman brushes it off as "freshman dorm" talk, but not before acknowledging, like Descartes before him, that "you can't be certain of anything other than your own awareness," not even that you exist. "It's very similar to what a lot of the Eastern religions would say about 'we only exist in consciousness,'" he said during our first interview.

Altman is a seeker. He does not believe in God, but meditates regularly and has embraced elements of Hinduism's Advaita Vedanta philosophy. Shortly after ChatGPT was released, he tweeted that one thing he believes that few others do is the "absolute equivalence of brahman and atman."[15] Advaita, which roughly means nondualism, holds that there is no difference between Brahman (the eternal consciousness that is the fabric of all reality) and Atman (the individual soul or self), and that the world that we experience is an illusory manifestation of Brahman. "I'm certainly willing to believe that consciousness is somehow the fundamental substrate and we're all just in the dream or the simulation or whatever," Altman told podcaster Lex Fridman. "I think it's interesting how much the Silicon Valley religion of the simulation has gotten close to Brahman and how little space there is between them."[16]

In other words: This is all a dream. And in the dream, anything is possible.

# PART I

# 1985–2005

CHAPTER 1

# CHICAGO

**A BRISK WIND WHIPPED IN FROM LAKE MICHIGAN ON APRIL 29,** 1983, as Chicago's politicos packed into an auditorium on Navy Pier for the inauguration of the city's first Black mayor. A band played, and Harold Washington, in a dark suit and silver tie that picked up the gray in his mustache and hair, marched down the aisle alongside his fiancée, a large fuchsia corsage on her shoulder. As he climbed the dais, the crowd erupted, chanting "We want Harold! We want Harold!" For a city that for decades had been controlled by Mayor Richard J. Daley and the Democratic party machine, Washington was something entirely new.[1]

He had stunned the country by beating both the incumbent mayor and the late Daley's son in the Democratic primary, and then overcame the racism of a general election marked by the Republican candidate's dog whistle of a slogan, "Epton for Mayor, Before It's Too Late." Someone scrawled the n-word on a Catholic church during a campaign stop.[2] (As Washington was fond of saying, and as Barack Obama—the young community organizer who tried to work for Washington and ended up harnessing the coalition he built to launch his own political career—liked to repeat: "Politics ain't beanbag.")[3] In the end, a surge in Black voter registration, along with a coalition of Latinos and white "lakefront liberals," put Washington over the top.[4] "My election was the result of the greatest grassroots effort in the history of the City of Chicago," he said from the podium.[5]

Looking on proudly from the audience that day was Jerry Altman—

a quiet, conservatively dressed man in his early thirties who had been a member of that grassroots effort. Jerry had jet black hair he wore parted on the side, eyes that crinkled into slits when he smiled, and a wild laugh that sometimes drifted into a giggle. He had spent the last year traveling with Washington's campaign as a volunteer by day and working as an affordable housing consultant at night. Now he would serve on the mayor's transition team for housing. Washington also tapped Altman's consultancy for a task force to stop landlords from walking away from their buildings in poor parts of town.[6]

A failed romance had brought Jerry to Chicago, but there he found a home in the country's mecca of community organizing. This was the city that had given rise to activists like Gale Cincotta, a mother of six from a neighborhood beset by white flight. The horror she felt over her children's underfunded schools led to a national crusade against redlining and then the 1977 Community Reinvestment Act.[7] Jerry forged alliances with Cincotta and her organizing partner, Shel Trapp, who were famous for antics like nailing a rat to an alderman's door to protest the city's slow response to its rodent problem. Jerry also worked alongside activists trained by the legendary Saul Alinsky, whose confrontational approach made Chicago a beacon for activists—and made Alinsky a lasting bogeyman of the Right.

Jerry, however, was not an organizer, and certainly not a rat-nailer. Rather, he was a financial specialist with a passion for affordable housing, who believed that, through creative dealmaking, he could convince the business community it was in their interest to invest in homes for the poor. Within a few years, he would do just that, by pioneering a method of funding affordable housing through a quirk of the tax code. Jerry's innovation would inspire the federal Low-Income Housing Tax Credit program, the primary way affordable housing gets built to this day.

"It was a movement," said Leroy Kennedy, who worked with Jerry on Washington's task force. "There was a whole lot of enthusiasm, motivation, and, frankly, happiness that somebody actually cared about what was going on here."

**JERRY ALTMAN** grew up comfortable, in the affluent St. Louis suburb of Clayton, the youngest son of a shoe factory owner. In many ways, the Altmans' history is a familiar tale of Jewish immigration and entrepreneurship. Jerry's grandfather Harry Altman—Sam Altman's great-grandfather—was born in Płock, a Jewish community near Warsaw, then part of the Russian Empire. He fled westward in Europe with his wife, Birdie, rather than fight in Siberia in the Russo-Japanese War of 1904. Harry then came to the United States alone. After stopping in New York, he began working for the Pitzitz family, which owned stores throughout the South, including one in the small town of Nicholls, Georgia. After two years, he earned enough money to send for Birdie, who went to work in the Pitzitz's store, which the Altmans eventually bought.[8]

The couple had five children. A daughter, Minnie, died in childhood of the Spanish flu. The survivors were Sam, Jack, Sol, and Reba. The Altmans were not religious, but it nonetheless bothered Birdie that her children had little exposure to their Orthodox Jewish heritage in the small Southern town. So she arranged to sell the family's business in Nicholls and purchase a commercial building in Atlanta. The deal went south, and Birdie sued, eventually winning a victory in the Georgia Supreme Court.[9] As Sunny Altman, Sol's granddaughter, said: "My great-grandmother was the businesswoman, much more than my great-grandfather."

With Birdie's dreams of setting the family up in Atlanta dashed, the Altmans moved instead to the port city of Brunswick. They started over in real estate, buying up stores that were going out of business, selling the merchandise, then selling the property. One of these deals was a shoe store in North Carolina that had burned down, but its inventory was intact. At the request of their son Jack, then in his late teens, they kept the inventory and opened a store on the town's main drag of Newcastle Street. They named it Altman's Shoes.[10]

The business thrived. Birdie sent for more family from Europe, and they established a store in Brunswick. When the Altmans' eldest son, Sam, finished college, he opened a store, Altman's Feminine Apparel, on the same street. Aside from their youngest brother, Sol, who became a lawyer, the Altman clan's fortunes revolved around shoes. Reba married a man named Phil Salkin and opened another shoe store in Brunswick called Salkin's. Jack attended the University of Georgia and returned to Brunswick, where he learned to make shoes of his own design that he sold to wealthy patrons at The Cloister hotel on nearby Sea Island during the Depression.

When World War II broke out, Jack Altman was sent to the Pacific, where he flew in a division of the Army Air Corps known as the "Aleutian Tigers." He was wounded and shipped to a St. Louis hospital in 1942. There he met and married a beautiful, dark-haired volunteer named Sylvia Harris. Sylvia was from a good Jewish family from University City, a suburb. She was the daughter of a Metropolitan Life Insurance executive. The service was held at Temple Israel, one of the oldest synagogues in Reform Judaism.

A shoe designer could not have landed in a better place. In the early twentieth century, St. Louis was the center of the American shoe industry, home to behemoths like International Shoe Company, Brown Shoe, and Hamilton-Brown that together made more than half the shoes purchased by the US military during World War I.[11] "First in booze, first in shoes, last in the American League," was the chestnut that St. Louis natives repeated about their city and its abysmal Browns baseball team. Jack started cobbling with a partner and named their two-man operation Joy's Shoemakers. His designs caught the eye of an Englishman named Sam Wolfe, who owned a handful of shoe companies. The two went into business together under the name Deb Shoe Company, making fashionable women's shoes from designs Jack often copied during scouting trips to Europe. He would buy shoes in Italy in the sizes of the women in the family, and return to shower his female relatives with Ferragamos. The formula was successful enough for Deb to expand to three factories in nearby Washington, Missouri, where unions were less powerful and production costs cheaper.

The Altmans settled in the affluent St. Louis suburb of Clayton and raised three children, Gail, Jack, and Jerold, who everyone called Jerry. They spent summers vacationing with their cousins in Georgia and attended Clayton's well-regarded public schools. When Jerry was seven, his grandmother Birdie had a stroke. Jerry's mom, Sylvia, traveled to Brunswick in January of 1958 to care for her mother-in-law. One day, lugging groceries in front of a shoe store on Newcastle Street, she slipped on wet pavement and broke her hip. She was flown, in traction, to St. Louis. She never again got out of bed. After a six-month ordeal, she died from an infection at the age of thirty-six.[12]

Sylvia's death caused fissures in the family that would never heal. Two years later, her widower, Jack Altman, a robust, broad-chested man with a hairline receding above a round, somewhat severe face, married his secretary, Thelma Noerper, a delicately built blonde barely clearing five feet. Noerper—divorced—brought her grown daughter, Sally, into the family. Jerry was quick to accept Thelma as a mother. His older siblings, Gail, thirteen, and Jack, ten, were not. "Jerry was young," cousin Richard Altman said. "It was easier for a child who was seven or eight years old. He became close to Thelma, but the rest of them never did."

Jerry's dad was a workaholic, constantly on the road with his company's sales team. When he was home, he was exacting. "Jack was a very intense man, a very anxious man," said Bob Nawrocki, Sally's son. "He was pretty controlling of his kids. I think Thelma became the sounding board for Jerry."

Jack Jr. idolized his father and worked at the shoe factory during summers and vacations. On Shabbat, they would attend Temple Israel, just the two of them. When Jack took the family to Europe for three weeks in 1965 on a business trip, Jack Jr. spent days with his father sneaking around with a Minolta spy camera, snapping photos of shoe designs that soon would appear in storefront windows in the Midwest.

Jerry had a very different personality and disliked his father's relentless focus on business. "Jerry probably has one of the highest EQs I ever met," Nawrocki said. Though Jerry never said a word against his father in Nawrocki's presence, he did once say, after Nawrocki mentioned what a good grandfather Jack had been to him, "Well, my experience was very different."

While all the children started in public school, Jack and Thelma realized the boys were just coasting and sent them to the all-boys Saint Louis Country Day School, which Jerry started in the fifth grade. The hundred-acre campus in the posh suburb of Ladue was favored by old St. Louis families. Jerry was a bit of a loner there, but well-liked. "Jerry was quiet, had a great sense of humor, and loved baseball," recalled his classmate Ed Hall. "He didn't have a mean bone in his body." He was more studious than athletic. He loved Russian class, playing one of the tsars in *Ivan Tsarevich and the Gray Wolf*.[13] By junior year, Jerry had a "mission beige" 1966 Pontiac Tempest that he kept immaculately clean, inside and out. Many years later, Sam Altman would also adore cars—much, much faster ones.

As Jerry headed for high school graduation in 1969, St. Louis was roiling with racial discontent, particularly over housing. As in many American cities in the years after World War II, middle-class white families had left St. Louis for surrounding suburbs, bringing their tax dollars with them. But St. Louis, which entered the country as part of a slave state but never went through Reconstruction, has a distinctively odious history of segregation. You can see it today by driving across Delmar Boulevard, the so-called "Delmar divide" that separates the white, wealthy south from the Black, bombed-out north. But the true divide is between St. Louis City and the surrounding, suburban St. Louis County.

Racially restrictive covenants, stating that only white people could live in a house, were attached to 80 percent of the homes built in St. Louis County before 1950.[14] (Although many cities had such arrangements, it was a case originating in St. Louis, *Shelley v. Kraemer*, that went to the Supreme Court and rendered such covenants illegal in 1948.) Meanwhile, the city's planners systematically tore down poor, Black neighborhoods like Mill Creek Valley in the name of slum clearance, and herded their residents into thirty-three eleven-story buildings whose name by the end of the 1960s was synonymous with failed housing policy—Pruitt-Igoe.[15]

Completed in 1954, the housing project was originally intended to have two complexes on its fifty-five-acre site: one for Blacks, named after the Tuskegee Airman Wendell O. Pruitt, and one for whites, named after the US congressman William Igoe. But a 1955 court decision ended segregation in federal housing, and the project became almost entirely Black. The federal government built Pruitt-Igoe but left it to St. Louis to maintain, which the city failed to do. The police referred to it as "Korea" and "Fort Apache" and rarely ventured there without dogs.* Then, in 1969, the city's cash-strapped housing authority raised rents in the complex by six times. In response, more than one thousand Pruitt-Igoe residents staged a rent strike. To make their point, they protested at the home of St. Louis's housing director, who happened to live in Clayton, threatening to build a tent city there. An activist group's press release promised to "bring any evicted public housing tenant that volunteers to dramatize the human suffering and poverty of the urban public-housing concentration camps to [live in] the city of Clayton."[16] The housing authority agreed to the protesters' demands. Three years later, it would demolish Pruitt-Igoe on national television.

Jerry knew Pruitt-Igoe well, having spent a summer driving a Good Humor ice cream truck through the projects. He made decent money, but never lingered after dark. As the rent strike revved up during the spring of his senior year, Country Day remained a world apart. "We were really isolated from reality in that school," said Jerry's classmate and friend Walker Igleheart. But Jerry seemed more sensitive than the others. "He's the kind of guy who wanted to do good in the world," said Joe Rechter, another Country Day alum. But he hadn't yet figured out how.

---

* As if all this were not enough, the US Army secretly conducted tests of aerosolized radiological weapons sprayed from the roofs of the buildings of Pruitt-Igoe, chosen for its similarity to the Kolpino housing complex in Leningrad.

It was in college, while majoring in economics at the Wharton School at the University of Pennsylvania, that Jerry began to develop an idea. "He was very interested in housing policy and making housing fairer," said Leah Bird, who befriended Jerry midway through college. They had long conversations about politics and policy, particularly about Philadelphia's controversial mayor, Frank Rizzo. "He knew that legislation needed to be changed," Bird said. Jerry dabbled in Philly's local campaigns and worked at a shoe store there, but had no intention of going into the family business.

Instead, Jerry went into city government, moving and becoming a planning administrator in Hartford, Connecticut, where he convinced the city council to consider wild schemes like spending $25,000 to study "the possibility of setting up a community-based corporation to produce solar energy products." His pitch had elements that would become classic (Sam) Altmanisms: it was built around the hope of attaining federal grant money, delivered in a tone of authority, professed to be free from profit motives, and the pitchman seemed most excited by the novelty of creating new forms of nonprofit corporations. "A community-based firm with a lesser emphasis on profits could undertake employment and job-training programs which private businesses might be reluctant to try," Jerry argued to the *Hartford Courant*.[17]

Soon, Jerry was appointed assistant city manager and began dating Megan O'Neill, the daughter of a Hartford judge and state representative. Megan was in law school and working as a researcher at the Hartford Institute of Criminal and Social Justice. "Everybody had left Hartford," said Frank Hartmann, who led the institute and knew both Jerry and Megan. "So the issue was, how do you make Hartford a decent place to live again?" Megan carried out a five-year study on neighborhood safety that argued for the kind of collaboration between police and residents that would become known as community policing.[18]

"She was a piece of work," Hartmann recalled fondly. "Very bright. Very outspoken. She was a lot to handle." Megan would eventually go on to serve as an assistant state attorney general. Jerry was also "clearly bright," Hartmann recalled. "He was in some ways larger than Hartford. So I think nobody expected him to hang around."

Jerry and Megan were married in the yard behind the O'Neills' white Cotswold cottage in West Hartford in 1977. A *New York Times* announcement noted the bride planned to keep her maiden name. Bird remembers attending a party at the house and finding it unsettlingly WASPy. "There wasn't a person of color in sight," she recalled.

After Megan graduated from law school in 1979, she landed a position at a firm in Chicago. Jerry went with her. But her job didn't work out, and Megan wanted to return to Hartford. Jerry did not. The marriage ended. It hadn't lasted two years.

**LIKE JERRY'S** grandfather before him, marriage had brought him by chance to the best possible place to ply his craft. He used his Hartford contacts to land Aetna Life & Casualty, Hartford's largest employer, as a client that he could work for remotely while remaining in Chicago, still an epicenter for progressive causes, due to Alinsky's acolytes. The insurance industry had taken one look at the punishment administered to the banks for their redlining, thanks to activists like Gale Cincotta, and resolved to make good with their local communities before they were exposed for their own decades of discrimination. Jerry's specialty was putting together financial packages to fund low-income housing, with a mixture of government money and corporate investment. "Jerry was trying to bridge the gap between boardrooms and neighborhoods," said Richard Manson, who met Jerry while consulting for Aetna in 1981.

The work took Jerry all over the country. In an area of Brooklyn's Park Slope that the *Daily News* described as "impoverished," "crime-ridden" and "a wasteland of vacant lots," Jerry helped Aetna partner with local and federal resources to redevelop rowhouses that had been gutted by arson.[19] Today, these homes, across the street from a hipster vintage clothing store and an upscale restaurant, are worth between $2 and $3 million each.

In Pennsylvania, Jerry worked with community organizer Mike Eichler, who had met Alinsky and been trained in the confrontational tactics of his organization, in the Mon Valley, an area around Pittsburgh that was depleted by the steel industry's collapse. Eichler had been retained by a group of corporations headquartered in Pittsburgh, and in turn hired Jerry, who he had heard was "technically competent at putting together real estate deals." They hit it off. Eichler's unusual plan to train the unemployed steelworkers to form their own nonprofit organization and raise the money to invest in their town's economic development made perfect sense to Jerry.

"It was a highly controversial strategy because all throughout the country, but especially in Pittsburgh, there's a stereotype of people who worked and lived in these towns as being uneducated, 'strong back, no mind,' not able to understand the changes in the economy, not able to be retrained, et cetera," Eichler said. Neither he nor Jerry believed that. "That was highly unusual, because most technically competent people did the same stereotyping of the local residents."

Eichler would identify willing locals, and Jerry would train them. Sometimes that meant hanging out at an auto parts store as he was interrupted by customers coming in for mufflers and taillights, until the owner understood the finer points of pitching a bank for financing a piece of commercial real estate. Together, the outfit developed a four-story building, rented it out, and used the revenue to build low-income housing for unemployed steelworkers. Eichler was struck by Jerry's relentless optimism. "It wasn't like, 'I don't know if this is going to work. Should I really be talking like this?' It was like, 'Of course that's going to work!'"

**BACK IN** Chicago, a friend of Jerry's set him up with a woman a few years his junior with whom, at first glance, he had an almost comical amount in common. They were both from Clayton, both liberal, both from upper-middle-class Reform Jewish families that found their way in large measure through real estate.

Connie Gibstine was a smart, ambitious medical student with a direct gaze, vivacious personality, and fierce competitive streak. After attending medical school at the University of Missouri, she moved to Chicago for her internship and then residency in dermatology at Northwestern University.

Connie's grandfather, Herman Gibstine, immigrated as a child to St. Louis from Aleksandria, then part of the Russian Empire, in 1893 along with his parents and four siblings. His father, Morris Gibstein—the spelling of the name changed when they arrived in the United States—had been a prosperous lumber merchant in the old country, but took up work as a tailor in St. Louis. His children went into the millinery business. While still a teenager around the turn of the century, Samuel Gibstine, the middle child, launched Samuel Gibstine & Company, a millinery successful enough to inspire his older sister to open a shop specializing in Easter hats trimmed with French flowers and hire their younger brother, Herman, to work in it.[20] By 1905, at the age of twenty, Samuel had sold his hat shop and gone into real estate. The Samuel Gibstine Real Estate Company would become one of the city's largest buyers and sellers of property.[21]

Herman's transition from hats to real estate was a bit slower, and a lot more tumultuous. By the early 1920s he was married, working as treasurer of the American Millinery Company, and active in Freemasonry.[22] The couple had two children, Eleanor and Marvin, and a tidy brick home in University City. One August in 1925, someone left an iron at the hat factory on overnight, triggering the sprinkler system, ruining the inventory, and causing up to $50,000 in damages.[23] The firm went bankrupt. As creditors were lining up to pick the bones, Gibstine claimed his home had been burgled and someone stole $1,500 he had stuffed under his mattress, the proceeds from selling "166 and a half dozen" hats.[24] The federal judge handling the case demanded he turn over the money or be charged with contempt. He fled to Seattle, where his sister had settled, and was arrested there and indicted for fraud. The charges were dropped after Herman's relatives paid off the creditors.[25]

Herman returned to St. Louis and co-founded Haffner & Gibstine Real Estate on the bustling commercial corridor of Easton Avenue, since renamed Martin Luther King Drive, selling bungalows in neighborhoods like Richmond Heights, Hamilton Heights, and The Ville. He changed partners over the years, but continued to trade in districts that, decades later, white flight would hollow out and his grandson-in-law, Jerry Altman, would try to revitalize.

**MARVIN, HERMAN'S** son, had a brain for math and science. He attended Soldan High—a St. Louis public school with a largely wealthy and Jewish student body—enrolled in Washington University in 1940, and went on to Washington University's school of medicine, obtaining his medical degree in 1946.[26] That year, he married Peggy Francis, the daughter of a real estate executive from Clayton, when she was midway through the University of Illinois. Their engagement announcement was triumphantly titled, "Miss Peggy Francis Engaged to Doctor."[27] They were married at a hotel overlooking St. Louis's Forest Park, and, after a stint in Germany with the Army, settled in a yellow brick house on a pleasant, but not fancy, street in Clayton. They had four children.[28]

Marvin was both a tinkerer and a worrier. In the early 1960s, he built a bomb shelter in their home's basement, complete with two three-story bunk beds. "We had cans of water, and then also, like, trash cans that I guess were going to be toilets?" recalled Connie, their third-born. There had been bomb drills in school—children crouching against the wall or under their desks in preparation for a weapon that would, through basic geopolitical logic, annihilate all life on earth. "I remember going into my parents' room in the night, just scared, thinking, 'What if they don't let me walk home to the bomb shelter when the bomb drops?'" When the threat seemed to pass, Marvin turned the basement into his ham radio studio, where he broadcast under the call letters WN0BVQ. "He would come upstairs all excited that he had connected with somebody via Morse code—on his ham radio, which he'd actually built—in Australia," Connie said. "Had he been born a generation or two later, he would have been a computer person."

Connie was the only of four kids who followed her father into medicine. She went to Clayton High School, where she worked on the yearbook, served in student government, and was co-captain of the JV cheerleading squad. She attended the University of Missouri for undergrad, where she majored in biology and minored in physics—but "only because I had enough classes in it; that's not my brain at all"—and then stayed at Columbia for med school. Three years in, her father died of cancer at age fifty-six, leaving her mother a widow at fifty-two.

Connie moved to Chicago in 1980 to become a dermatologist—a specialty she chose in part so she could control her hours and one day have a family. As a resident, she was serious and practical. "Connie was always the voice of reason," said Dr. Amy Paller, who was a resident with her. "She efficiently cuts to the business." She was also ambitious, co-authoring academic papers on psoriasis and other ailments. They were residents during the early years of the AIDS epidemic. "People would come in and if we saw something that looked like capillary sarcoma, it was a death sentence at that point," Paller said. The experience deeply affected Connie's view on the physical risks of being a gay man, which, years later, would influence the reaction she had to her son coming out. "We spent whole clinics treating these nice, thirty-year-old guys who were just wasting away," she said. "It was awful. It was *awful*."

Connie and Jerry were married back in St. Louis on the Wash U campus, with a great deal less fanfare than Jerry's first wedding. Both were eager to return to Chicago. The couple moved into a high-rise just south of the Gold Coast. Connie continued her residency, while Jerry spent a year doing his consulting work at night while volunteering for Harold Washington during the day, "just for free," Connie said, "because he wanted to see a Black man" become mayor.

**WASHINGTON'S CAMPAIGN** was about more than giving representation to the 40 percent of Chicagoans who were African American, though this was the base of his support. It was about smashing "the machine"—the patronage system that determined who got jobs and political favors, which Mayor Daley had administered since the 1950s. Daley's reign, with its emphasis on "slum clearance," had been a disaster for the city's poor residents.[29] With Washington in office, Jerry's consultancy, the Housing Agenda, tried to shore up struggling neighborhoods' existing housing stock by using the city's criminal code to force landlords to maintain their buildings. The work became a national blueprint, submitted by Washington to Congress as part of a 1984 hearing on the state of the country's cities.[30]

That same year, Jerry came up with a way to get corporations to fund affordable housing that would inspire federal legislation. He had been hired at an affiliate of Local Initiatives Support Corporation, or LISC, a national nonprofit that pooled corporate money to invest in urban development and affordable housing. In this role, he came up with a novel mechanism: use a portion of the tax code that allowed the builders of affordable housing to speed up the rate at which buildings "depreciated," and thus enjoyed a lower tax burden, and pass those savings on to the project's investors. It was not quite a tax credit, but it was close. "Jerry was the idea guy," said Andy Ditton, who had hired him at LISC. "He was extremely adept at thinking outside the box."

They used this trick to lure funding from Continental Bank, Chicago's second largest, for the rehabilitation of a six-unit apartment building in the Bucktown neighborhood, and invited Chicago's politicos to the ribbon cutting, including Congressman Dan Rostenkowski, the powerful chairman of the House Ways and Means Committee. Impressed, "Rosty" stopped Ditton afterwards and asked him how he did it. Ditton's team then spent months working with Rostenkowski's staff to craft provisions in a 1986 federal bill that set aside $10 billion for Low-Income Housing Tax Credits.

Jerry continued to spend a lot of time on the road as well, though he increasingly applied his Chicago connections—such as his relationship with rat-nailing organizers Gale Cincotta and Shel Trapp—to this work. In Lowell, Massachusetts, where a working-class, largely Latino neighborhood called The Acre had been slated to be bulldozed, Jerry worked alongside LISC consultant Bill Traynor to keep it intact. Cincotta had introduced Jerry to a local organization called Coalition for a Better Acre, and Jerry suggested they incorporate as a nonprofit and present a proposal for a development project that Aetna could fund. "Aetna was willing to stand behind CBA," Jerry told public policy professor Ross Gittell, for his book *Renewing Cities*. "We knew they were controversial, we chose them on purpose, we thought they were truly representative of the neighborhood's interest and that they had a lot of potential."[31]

Traynor was dazzled by Jerry's skill at structuring deals, a knack Jerry would one day pass on to Sam. "I saw him come into our world with humility and good listening skills, but also with a really, really smart, savvy sense of how you do things," Traynor said. The group began with a modest $2.8 million project to build twenty-four homes on sixteen vacant lots. Before that was remotely completed, they turned their eyes to a crumbling 200-unit property called Cement City. Traynor recalled Jerry's audacity at a breakfast with the business community, selling it as a $20 million HUD redevelopment project, despite having no experience with anything of that scale. "We didn't even have site control of the property!" Traynor said, laughing.

Connie was not especially impressed by any of this. "I felt like he was kind of not directed," she said. "And he wasn't." Although she already had a medical degree, she nudged Jerry to attend law school and, for fun, decided to join him there. They enrolled in night classes at Loyola University in 1984. Classmates remember them sitting next to each other and sharing healthy dinners they brought from home. "In my mind, it was to focus him somewhat," she said. "And I kind of just always liked school, so I did it too." Connie never ended up practicing law, but that did not suppress her competitiveness when it came to law school. Before classes began, Connie learned she was pregnant—and would be twice more before getting her degree. "We started together," Connie said. "I took one year longer than he did because I had all these kids. I mean, I took the Illinois bar when I was seven months pregnant with a third kid. That delayed me a little bit."

**SAMUEL HARRIS** Altman was born on April 22, 1985. He was an easy baby—"You could vacuum right by his crib and he'd stay asleep," as Connie put it. But it wasn't until the Altmans had two more sons, Max and Jack, over the next three years, followed by a girl, Annie, a few years later, that they realized just how unusual Sam was. At age two, he would entertain himself by inserting *Sesame Street* videotapes into the family's VCR and playing them on his own. "He could just do it," Connie said. "But then Max comes along, who's great also, but had no idea how to work those kinds of things at similar ages." When Sam was three, as the family was trying to call Peggy in St. Louis one day, Connie realized he grasped the concept of area codes. "He said, 'What numbers do you have to dial before I call grandma?'" Connie recalled. "And my other kids probably understood that when they were, like, ten." To her, it seemed that "he was kind of born an adult."

From an early age Sam was also inclined to retreat into his mind, displaying none of the typical little boy's need for constant motion. By then, the Altmans had moved to a house in Chicago's gentrifying Lincoln Park neighborhood, shaded with trees and dotted with pocket parks. One day while Connie was in her final, uncomfortable days of pregnancy with Max, she took Sam to a playground to get their blood flowing. But Sam, then two, had no interest in the slide or swings. "No, Mommy, I'm going to sit with you here, and let's just sit together and watch the babies play," she remembers him telling her. "And the 'babies' were older than he was," she said. The episode has stuck with her as evidence of his uncanny maturity and cerebral nature. He's "so not physical," she says. But it was also evidence of a profound filial devotion that one day would break his family apart.

As Sam neared kindergarten, Connie was disappointed by their educational options. Getting into the right public magnet school felt unappealingly political. Private schools seemed out of reach for a family with three kids. "It was not just the tuition," Connie said, "but the families were, like, going skiing at their family places in Switzerland over Thanksgiving break, stuff which was not part of our life at all."

Through all the pregnancies, Connie worked as a dermatologist, scheduling appointments right up until her due dates. She loved dermatology. "Some of my friends, who are, like, thoracic surgeons, will say, 'How can you stand treating acne? It's so boring, so pedestrian.' But it's life-changing for these kids," she said, referring to her many teenaged patients, "and it's very satisfying for me to do." After her clinic was bought by the city's largest hospital and HMOs became more dominant, though, she chafed against the insurance model's strictures, having grown up in a household where her father's patients paid mostly in cash. "I was looking to leave the group because they had all these rules about who we share medical assistance with," she said.

Jerry's law degree had failed to "focus" him, and he was traveling more than ever, consulting on projects across the country. In 1987, a few months into Harold Washington's second term and just hours after he dedicated a new housing project, the mayor suffered a heart attack and died slumped over his desk. Washington had welcomed the "do-gooders" who came knocking on the city's door.[32] Now that he was gone, it was not clear what place they would have in government.

"It didn't really matter where he worked," Connie said about Jerry. They began to think about returning to St. Louis. Connie could support her widowed mother, and benefit from the childcare that a young, energetic grandmother could provide. Jerry's parents had passed, and his brother was in the area, though they didn't talk much. The

Altmans' Chicago property taxes were about to go up with a major reassessment, and Connie had had it with big-city inconveniences like being unable to walk her grocery carts to her car because the stores had erected posts to keep people from stealing them. "It was just really a harder life than it had to be," she said.

So when Sam was four, Max was two, and Jack was just ten weeks old, the Altmans returned to Clayton.

CHAPTER 2

# ST. LOUIS

**THE ALTMANS SETTLED IN A BROWN BRICK TUDOR IN CLAY**-ton's Hillcrest neighborhood, a prewar subdivision of charming, architecturally distinct homes on the city's edge that was not very different from the one in which Connie grew up. Each house seemed plucked from a different Grimm's fairy tale. Yards were flush with dogwoods and azaleas. The Altman children grew up riding their bikes down the block to Forest Park, an expanse of woods and golf greens bigger than Central Park, and slipping through a hole in the fence behind their house onto the campus of the highly rated local elementary school. The orderly, walkable community with the public school at its center and the public park at its margins gave Sam the sense that not only were humans capable of organizing themselves around the public good; it was their default state. Years later, while advising a batch of startups in the Bay Area after a rainstorm, far from the wild cicada drone of the Missouri suburbs, he revealed how deeply its landscape was lodged in his heart. "This smell, after it rains for the first time. You know what it's called?" he asked his charges. "Petrichor. It's my favorite smell. You only get to smell this once or twice a year, because it has to not rain for a while, and then rain. It's the smell of summer in St. Louis."[1]

The Altmans told their children daily that they could do anything. But there were tensions in their household from almost the moment they moved to St. Louis. Since Chicago, Connie had been dreaming of starting her own practice, but the prospect now seemed overwhelming with so many young children at home. Instead, she joined the dermatology practice of Dr. Lawrence Samuels, a St. Louis native who specialized in hair transplants, "so I could just go and work and see patients and leave and go home to my kids and not have to run it," she said. But Samuels saw no signs of an employee on the mommy track. "She worked as hard as any male that I've ever worked with," he said. After giving birth to their fourth child, Annie, "she was back at work in two weeks."

Jerry struggled to find a foothold in St. Louis. He continued to consult in far-flung places like New Orleans and Palm Beach County, where he helped build housing for a poor, Black community a half hour from Mar-a-Lago that had never had public water or sewers. "He would be in Las Vegas one day, and the Mon Valley another, and then Little Rock, Arkansas," said Mike Eichler, who worked for LISC with Jerry on these projects, referring to the Monongahela Valley of Pennsylvania. As they crisscrossed the country in rental cars, they brainstormed about a new, less combative form of community organizing that would be more accepting of the capitalist reality around them. *What if, instead of embarrassing the president of the local bank for its decades of redlining, we approached him as a partner who could expand his business and be seen as a leader in the community by investing in affordable housing?* Jerry ultimately joined the board of the organization that Eichler founded as a retort to the confrontational tactics of his mentor, Saul Alinksy, called the Consensus Organizing Institute. "Jerry helped me develop a whole new method of community organizing," Eichler said.

Unfortunately, St. Louis was not fertile ground for their approach. While many cities enticed corporations to invest in affordable housing by offering bridge financing from a federal grant program designed to help poor people and stop "blight," St. Louis's mayor proclaimed that he was philosophically opposed to spending the city's $20 million in federal block grant money on affordable housing.[2] "I just don't believe the city of St. Louis should become the region's final repository for all the poor, underemployed, and undereducated," said Mayor Vincent C. Schoemehl Jr. Instead, according to a 1991 *St. Louis Post-Dispatch* investigation, he burned most of the money on administrative patronage jobs and steered what was left over to the city's affluent neighborhoods. In his typical understated tone, Jerry, then leading a group of community development associations, deadpanned to the paper, "There is an imbalance currently between resources and commitment to the downtown as compared to the neighborhoods, particularly poor neighborhoods."[3] In part for this reason, St. Louis had little of the culture of public-private partnership that Jerry specialized in. That same year, he complained to the local paper that the city "lacks the effective corporate commitment to community development evident in cities such as Pittsburgh, Cleveland, Kansas City, and Baltimore," and "lacks the foundation commitment evident in Philadelphia, New Orleans, and Washington."[4] For a man who rarely raised his voice, you can almost hear him scream.

Jerry had set out to create a life in contrast to his workaholic father, but had ended up on airplanes nearly as often. "He watched his father be gone from home a lot, building his shoe-manufacturing business, and he wanted to be not focused on making money, but focused on his family and doing good," Connie said. "And he was focused on doing good, but he was never around." She resented Jerry's travel. "It was a source of tension because he would be gone evenings and weekends, doing community organizing things while I was home with all these little kids supporting everybody," she said.

This anger persisted for years. "I know she got very upset about him not being present when the boys were little, and going on a lot of work trips," said their daughter, Annie. "By the time I came along, the dynamic was him putting his tail between his legs and going to do the dishes after dinner every night, and constantly getting passive aggression from her and almost feeling like, acting like, he deserved it and it was OK."

After Annie was born, Jerry stopped traveling and took a local job leading a nonprofit that developed real estate in poorer neighborhoods, often saving architectural jewels from the city's more prosperous past from the wrecking ball in the process. Connie, for her part, decided it was time to run her "own show," founding her own clinic with a colleague from Dr. Samuels's office. The partnership did not last long. "With Connie's drive and work ethic, they just weren't compatible," Samuels said. So Connie finally struck out on her own, founding her own dermatology practice at Missouri Baptist Hospital. She could arrange her schedule to work four days a week and avoid the kind of cosmetic procedures that attracted more demanding patients. "I just had this little old-fashioned show, which worked great," she said. But she still struggled with keeping track of her kids' orthodontist appointments and birthday parties, which—a generation after the women's movement—still fell to her, despite her being the primary breadwinner. "We were told we could have it all," she said. "It's not true."

Connie insisted that the family eat dinner together every night, which she cooked herself. At the table, the kids played math games and Twenty Questions to guess what was for dessert.[5] The rest of the time, it was Ping-Pong, video games, and board games.[6] Sam always won the card game Samurai, Jack later recalled, jokingly, "because he always declared himself the Samurai leader. 'I have to win, and I'm in charge of everything.'"[7] Every Thanksgiving break, the family played touch football. When it came to sports, the results were different. Both Sam and Jack played water polo in high school. "I was dramatically better than him," Jack told the *Journal*. "You can put that in there." But the real competition, as the siblings later acknowledged on Annie's podcast, was to be "Mom's favorite." It was such a joke inside the family that Connie at one point had shirts printed up with that phrase for each of them. "I come from a family that loves to rank things in order to make meaning from them," Annie later wrote.[8]

The mantle of Connie's favorite rotated among the siblings, though it lingered longest on Sam. Max excelled at math, but the rest of them excelled at everything. Jack, the most well-rounded, "can sing and dance and write poems and got eight hundred on every single standardized test he ever took," Connie said. Annie, the baby, nine years younger than Sam, was both "very, very brilliant" and "a little bit spoiled," according to Steve Roberts, the local real estate developer who coached her softball team with Jerry. She was a daddy's girl. Jerry's college friend Leah Bird recalled her as a "smart, vivacious, funny, slightly acerbic person who adored her father." Sam, then, was less special in the family for his intelligence than for his strange, precocious comfort in the world of adults and his refusal to cross Connie. "I always used to say you could drop him in New York City at age ten and he could have figured out his life and figured out how the city worked," Connie said. She and Jerry prided themselves on never nagging their children to do their homework or bugging them about what they wore. "My only rule with my kids was there was no talking back," she said. "Sam never tried, but the others tried once, and I shut that down."

The one area where Sam did not follow Connie was religion. If it had been up to Connie, the family would have probably skipped joining a synagogue altogether. "I grew up with a super-science father whose belief was man made God, God didn't make man," she said. "That is my belief, too." But Jerry attended services regularly and was bar mitzvahed as an adult. So the family met somewhere in the middle, and joined the most progressive synagogue in the city, Central Reform Congregation.

Established in 1984 with the express purpose of remaining within the city limits when the rest of the city's Jewish congregations were leaving for the surrounding county, CRC saw its mission as social justice, according to Rabbi Susan Talve, its founding rabbi. "If we were going to fight racism and all the other systems of caste that were causing so many disparities, we needed to be on the front line," she said. At the height of the AIDS crisis, the synagogue was a welcoming place for LGBTQ people, going out of its way to hire gay and lesbian religious school teachers. It declared itself proudly pro-choice and feminist, a haven for what Talve calls both "Jews by choice" and "Jews of color." And when, after years of renting space from a Unitarian church, it put up its own building, the circular mosaic in its front lobby featured a zodiac calendar woven with ancient Jewish imagery. During one recent visit, Rabbi Talve pointed to a blue figure in the mosaic and said, "We made our messiah a trans person with tattoos."

Sam grew up attending CRC's Hebrew school on Saturday mornings, and was bar mitzvahed in the congregation like all his siblings. "He grew up with the language that we're here to repair the world," Talve said. "*Tikkun olam*, we call it." The synagogue keeps a self-serve food pantry that it calls its "Tikkun closet." Talve recalls Sam because he was the same age as one of her children. "I remember always loving Sam," she said. "He was kind and quiet, really cute. I remember him being really smart and asking great questions." At one point, the Altman boys wrote to members of the congregation and asked them to make donations so that the synagogue could buy a second torah. Family members describe Judaism as more important to Sam than it was to those in his parents' generation.

**SAM'S PARENTS** bought him his first computer, a boxy, beige Mac LC II, for his eighth birthday. It was technically a computer for the whole family, but for a few years, before his father started using it for work, it resided in Sam's room. "It was very exciting and cool," Sam said. He had taught himself to program in his elementary school's computer lab. "We had these [Apple] IIGSes that could boot up into a BASIC prompt and you could program things like 'Print every prime number from one to a million,'" he recalled. But he quickly found the simple language tiresome. "I spent a lot of time thinking about what would happen if we could ever make the computer learn to think and not just have to sit here and program it one line of BASIC at a time," he said. One of his clearest memories is sitting up late one night in his desk chair, alongside his lofted bed, envisioning this future. "I just remember thinking that someday the computer was going to learn to think."

Word of his computer proficiency traveled around Ralph M. Captain Elementary School. "The teachers would go find Sam in third grade and tap him on the shoulder and be like, 'We need him for a minute. We don't know how to do this,'" Connie said. "And he would go to their classroom or the library and show them how to do what they didn't know how to do." His elementary school teachers began giving him separate, more advanced lessons.

But Clayton's junior high school did not believe in this kind of tracking. "They didn't want to hurt anybody's feelings," Connie said. "Everybody got participation trophies and that whole thing." She and Jerry thought Sam needed more rigorous schooling and asked him if he wanted to apply to John Burroughs School, which had surpassed Country Day's coed successor to become the top private school in the greater St. Louis area. (A flavor of the rivalry between the two schools can be found in one Country Day alumnus's description of Burroughs's culture as "designer homeless.") Sam declined. "My dad really didn't believe in private school," Sam recalled. "And I did kind of have this anti–private school bent myself," he said. "I was excited to go to junior high with my friends."

But in sixth grade he found a chaotic, underfunded public school where the teachers didn't always have control of their classrooms. "Sixth grade was not good," he said. "Seventh grade was worse." One day in seventh-grade science class, one of his favorite subjects, the teacher could not get through a lesson on electricity because a student was so disruptive he had to be physically removed from the classroom with the help of the vice principal and the gym teacher. By that point, one of Sam's friends, Matt Mendelsohn—who today runs the endowment at Yale—had left for Burroughs and was reporting good things. "And then I was like, OK, yeah, I'd like to go," Sam said.

John Burroughs School was founded in 1923 on an eighteen-acre campus in Ladue by a group of parents inspired by the progressive ideas of the philosopher John Dewey. They believed that the school's purpose was to develop each child's "latent possibilities of power" so that "in adulthood he shall make his contribution to the improvement of human society."[9] They named the school after the nature essayist John Burroughs, not because he had anything to do with St. Louis, but more as a philosophical North Star. Though largely forgotten today,

Burroughs was once one of the most famous writers in America, a transcendentalist friend of Walt Whitman's who spun his observations of the birds and flowers of his native Catskill Mountains into poetic, scientifically precise, and essentially atheistic essays in magazines like *The Atlantic*. He had died the year before the school was founded, leaving behind a cultish following that had at one point included President Theodore Roosevelt and Thomas Edison, and more than a dozen schools were named for him around the country. Today, the most visible tie at the St. Louis school to his legacy is a statue near the athletic fields of the bearded old man gazing up at the universe, his face slack with wonder.

Sam arrived a year late to the school that began in seventh grade and cultivated a family-like atmosphere. Lunch was served family-style around tables with assigned seating, headed by a teacher at each end, and stretched a full forty-minute period to encourage conversation. Classes were usually limited to twelve to fourteen students. Participation in sports was mandatory. There was no dress code. "The student body is made up of kids that are kind," said Meridith Thorpe, director of admissions and tuition aid, during a recent visit. "That's a nonnegotiable for us." Sam was shocked to discover that no one locked their lockers. "He loved the fact that you could leave your backpack on the floor and nobody stole it," Connie said.

Amid all the touchy-feely exhortations to "leave no part of yourself at the door" was ferocious, and often crushing, pressure to perform academically and get into a good college. All sophomores were required to take the PSAT as the beginning of their journey toward performing well on the SAT. All-nighters were not uncommon. "When we graduated, they basically said, 'Our goal was to give you more stuff to do than time to do it,'" said Brian Jump, a classmate of Sam's. "People didn't really sleep," recalled another friend and classmate, Kurt Roedinger. Many Burroughs alumni describe the moment when they got to college and realized that the hardest thing they would ever do was behind them.

Sam was at home in this environment. "He was definitely the

smartest person in our class," Roedinger said. Despite taking all the most advanced classes, he still had time for his own interests, including reading science fiction authors like Isaac Asimov and Arthur C. Clarke, and computer programming that was way beyond anything in the school curriculum. "He was the only person I knew at that age that had a C++ programming book on his bookshelf," Jump said. He bonded with the school's computer science teacher, Georgeann Kepchar, showing up early to her second-period AP Computer Science class to talk to her about AI at a time when it was still largely considered a science fictional notion. "Ms. Kepchar was like this unbelievably important force in my life," he said at a recent alumni event.[10] He became an upperclassman just as the first 3G cellular data networks appeared in the United States, allowing internet browsing on a phone. Sam was an early tech adopter. "I used the internet on my phone and I was probably the only kid to do that," he said.

Yet he did not have the typical techie personality, said Andy Abbott, the head of school who got to know Sam when Abbott served as the principal of grades 11 and 12. Sam edited the yearbook, co-edited the school's literary magazine, and charmed teachers with his wit and humor. "His personality kind of reminded me of Malcolm Gladwell," Abbott said. "He can talk about anything and it's really interesting. It's that way about computers. And it's that way about Faulkner. And it's that way about politics. And it's that way about human rights. It just seemed like he had read everything and had an interesting take on it." Abbott remembers half-jokingly trying to talk him out of a tech career. "Oh, don't go in that direction, Sam," he recalls saying. "You're so personable!"

Although he was always, as his brother Jack described it, a "voracious reader," sometimes Sam seemed to have detailed knowledge of obscure things that book-learning could not quite explain. Jump, who went on to join the merchant marine, remembers telling Sam about his plan to ride a Jet Ski from Miami to the Bahamas. Instead of trying to convince his friend to abandon what Jump later admitted was "an incredibly terrible idea," Sam offered both encouragement and uncannily detailed technical advice. "He knew about the exact Jet Ski that I was taking. It was one of the first four-stroke Jet Skis so it had increased reliability. We talked about the range on it," Jump recalled. Sam had fully formed opinions on the various models of VHF radios and the chances their antennas would be able to reach the Coast Guard should Jump get into trouble. "Eighteen-year-olds don't know this stuff," Jump said.

In fact, Sam had good reasons for knowing about this stuff. After a summer spent at YMCA camp during which the counselors tried to entice him to accept Jesus Christ as his Lord and Savior, the Altmans had sent him to a series of summer programs well beyond the Bible Belt, including one aboard a boat in the Caribbean called Broadreach, where he discovered a love of scuba diving. He had also somehow carved out time to follow in his late grandfather Marvin's footsteps and join a local ham radio club, a move so unexpected that one of the club's leaders told Connie at pickup that they had never had a teenager come to a meeting before without a father or grandfather. Ham radios use VHF, short for very high frequency, radio waves, the same kind used by marine radios. Even at this early point, Sam showed an ability to marshal life experiences to connect via a shared niche interest with almost anyone he came in contact with—particularly if that interest had anything to do with technology.

**BY THE** time Sam was twelve, he knew he was gay. Until he came out to his mother at age sixteen, she thought he was just a bit antisocial. "To not be part of the junior high male group who's making boob jokes and stuff, I think Sam had a hard time," she said. "He used to not go to social events and I thought, 'Oh, you are unhappy. Maybe you should let me take you to this party.' And he said, 'No, no, I'd rather stay home.' And I didn't realize why he wasn't going, because I just missed it."

When he finally did sit her down to tell her, after researching on AOL how to come out to your parents, her first reaction was: "Are you sure?" She could only think about her AIDS patients back in Chicago. "I was just so fearful for his physical safety, his health, and that his life would be difficult," she said. "But he took it as I was unhappy that he was gay." She asked how he had known at twelve, given that she could not recall thinking about sexuality one way or another at that age. "Well, Mom, that's because you were normal," he told her. "It wouldn't have come to mind because you weren't different. When you're different, you think about it. You recognize it." Sam would later tell *The New Yorker* that the computer in his room was a crucial tool for navigating this challenging time. "Growing up gay in the Midwest in the two-thousands was not the most awesome thing," he said. "And finding AOL chat rooms was transformative. Secrets are bad when you're eleven or twelve."[11]

Though friendly with many, Sam did not have a large group of friends. His best friend was Sally Che, who would go on to Harvard and a career as a doctor. They had met in eighth grade, during a class trip to Burroughs's camp in the Ozarks, Drey Land. As they were floating down a river on inner tubes, Sam, who was still the new kid in a class who had mostly arrived together as seventh graders, paddled over to Che and asked her name. "Oh, my name is Sally," she had replied. Sam, wearing a tie-dyed T-shirt, replied by playfully flipping over her raft, leaving her floundering. "That's how he introduced himself," Che said, laughing, during a recent interview near her home in McClean, Virginia. Che, who described herself as "a pretty good girl in high school" and a "classic Asian" who took her studies and parents' directives seriously, was drawn to Sam's unusually expanded sense of the possible. "He's somebody who colored within the box, not getting me in trouble, but he sort of colored wildly in the box—like *colors*."

She teased him incessantly about his "horrific" fashion choices—which included a strange devotion to tie-dyed shirts and hair dyed blond with Sun-In, inspired by the frosted tips of Justin Timberlake from NSYNC—and about being a messy eater. "He was an embarrassment to eat out with," she said. But she loved that he didn't care.

"He was born with his level of confidence. I don't think he super cared what other people thought." It just seemed innate. "You never knew where it came from."

One weekend, they took Connie's green Acura on a road trip to Chicago, blasting Blink-182's angsty, bratty anthem "What's My Age Again" at top volume. (*My friends say I should act my age. What's my age again?*) "He was a crazy driver, and I was very cautious," she said. On the way home, just to mess with her, he insisted they switch seats while the car was hurtling down the highway at ninety miles an hour. "Wildly dangerous," she recalled.

By high school, their main hangout was a coffee shop called Coffee Cartel, which was popular with LGBTQ people of all ages. Sometimes Sam would have to nudge Che to get her away from her studies. One Saturday night, he called her. "Let's go to Coffee Cartel," he said. She looked at her watch. "But it's almost nine," she replied. Sam makes fun of her for it to this day. Needless to say, they went out. "He brought out this restlessness in people. It was this adventurousness about him. He was so spontaneous."

It was Che who set Sam up with his boyfriend, Nathan Watters, who also went on to become a doctor. At first Watters had not liked Sam at all. Watters had been a close friend of Che's in seventh grade, having met her while sharing a keyboard with her in band, until Sam moved in in eighth grade and "stole her away," as Watters put it. Watters found Sam's tie-dyed shirts and bleached hair "obnoxious," and was put off by his outspokenness. "Something about him rubbed me the wrong way," he said.

But as they entered high school, their relationship started to change. "We had this animosity in person during the day, but then at night, with the advent of AIM, we would stay up all night, doing our homework, just chatting back and forth, being super nice to each other, maybe poking fun, but actually building a friendship," Watters said. Sam invited Watters to come to a youth group that would meet once a week in the Central West End. "It was for queer kids from all over St. Louis to have a safe place to get together," he said. Afterwards, a group of them would go to Coffee Cartel, where Sam would order actual coffee—a grown-up move that Watters found pretentious.

By senior year, Sam, Watters, and Che were in an Honors English seminar together. Watters and Che had dinner plans. Watters asked Che if they could push dinner back to 7:30, because he was going to be late. Sam, overhearing, was confused. He also had dinner plans with Che that night. "It's not possible," Watters replied, just before class started. In retrospect, it was obvious she was setting them up. "She knew that if she could just get us together under the right circumstances that maybe we would enjoy each other's company," Watters said. After dinner, they went back to Watters's house and hung out until Sam had to go home. After that, they were dating.

Sam didn't drink or do drugs, Watters said, but he did have a weakness for driving fast. "He loved to speed," he said. "He knew these tricks. There was this road on the way to the coffee shop where he would be on this little boulevard, and then you had to go four lanes over, and he would just slam" across the lanes to scare Watters. "He was very into cars, very into the showiness of it all, the speed of it all."

Sam was so obsessed with the book *The Picture of Dorian Gray* by Oscar Wilde that Watters bought him a leather-bound, hand-illustrated edition decorated with gold leaf for his birthday. In the book, considered one of the greatest homoerotic novels of all time (though there is no overt homosexuality in it), a beautiful young man makes a deal with the devil that makes his portrait age instead of him. "He chases youth a lot," Watters said of Sam, pointing to his boyfriends later in life, who were often around a decade younger. "I think he always has chased youth."

Several of the rest of Sam's small crew, including Roedinger and a talented musician named Rick Pernikoff, were also gay. "Those guys, obviously going through high school gay, dealing with that, I think had probably forged a lifelong bond," said Jump, who was Pernikoff's best friend. Their social life mostly consisted of going out to Steak 'n Shake, Starbucks, or Coffee Cartel, drinking coffee late into the night and studying. For a while, Sam had an after-school job at J.Crew, and they'd go visit him at the mall. Every so often, Pernikoff would drag him to a party. Mostly, though, they would just drive around the suburbs in his mom's Acura blasting Third Eye Blind.

By senior year, he "basically willed into existence" a chapter of the Gay Straight Alliance, Roedinger said. "He's always been a born leader—not just in charge of things, but could get people to go along." But in St. Louis in the early 2000s, sexuality remained a largely taboo topic. Gay couples did not publicly attend dances together or walk around holding hands.

That year, the Gay Straight Alliance organized an assembly to educate the student body about many of their fellow students who identified as gay. When some members of a Christian student group called KLIFE learned of the assembly, their families requested that they be excused from attending, and Burroughs's administration agreed. Sam was incensed, and decided he would stand up during the school's morning assembly and say so, taking advantage of a school tradition known as a "sound off." He had trouble sleeping the night before. But when he stepped up to the podium, he exuded pure confidence. His friends might have known he was gay, but the larger student body did not, and he used this bombshell for maximum rhetorical impact.

"People don't have to watch any assembly, that's fine with me," Sam summarized his message for the *Journal* years later. "But the message this sends and the way they went about it was horrible." The speech made waves. In its wake, a tall, blond, conservative student named Barbara Ann Smith, who went by Barbie, wore a "straight pride" shirt to school and was asked to remove it. Smith, who would go on to a notable legal career after clerking for Supreme Court Justice Samuel Alito Jr. and appear frequently on Fox Business, said wearing the shirt "is one of the things that when I think about my high school experience I regret the most because I realize how hurtful that must have been to my classmates and friends and others." She said Sam was widely admired by the student body at the time, while she was and continues to be politically conservative, she also admired him and never intended to hurt him. "Even in a group of very smart people, he stood out as someone who was going to change the world," Roedinger said.

Burroughs had a veritable army of college counselors—roughly six for one hundred students—and Sam, like his siblings, was a gifted standardized test taker. The result was that when it came time to apply to college, Sam was prepared. He set his sights on three options: Harvard, Stanford, and the prestigious Morehead-Cain Scholarship at the University of North Carolina at Chapel Hill, the oldest merit scholarship in the country, which gives the 3 percent of the applicants who manage to qualify for it a free ride. When he got into all three, although Stanford had been his dream since childhood, he offered to go to UNC. "He said, 'Mom, I know you have three more kids to put through college, I can do this. You don't need to spend the money,'" Connie recalled.

Although the kids all had college funds, money had been a source of stress for the family. As Sam entered high school, Jerry left his nonprofit job, telling his executive assistant, Kathy Kingsbury, that "he wanted to build wealth for his family." He took a position at a Kansas City–based property management company called Cohen-Esrey, which specialized in syndicating the kind of affordable housing tax credits that were his specialty to investors, but it didn't last long. Then one day, he was coaching his kids' softball practice alongside local developer Steve Roberts, when the topic of work came up.

Roberts, a former alderman, was, along with his brother, Michael, on his way to becoming one of the city's most prolific property developers, as part of a business empire that would include the only Black-owned wireless affiliate of Sprint PCS. Jerry signed on to work for the Roberts brothers, who were intent on reviving the dilapidated blocks north of Delmar near where they had grown up watching the white people flee within months of them and other middle-class Black families moving in. But it would be several years before these plans would bear financial fruit. In the meantime, putting four kids through college would strain the family finances. "In those days, Connie was just getting her practice going as a dermatologist," Roberts said. "It would have been a stretch for them to have four children and send them off to private school and still pay the mortgage on a very nice home in Clayton, Missouri."

But Connie blanched envisioning her gay, Jewish son at a big Southern state school that was just starting its computer science program. "I felt like that was punishing him for being so good, and so diligent," she said. She and Jerry told him to follow his dream.

CHAPTER 3

# "WHERE ARE YOU?"

**IT DOESN'T USUALLY RAIN IN APRIL ON THE STANFORD CAM**-pus. By then, the Peninsula's brief rainy season has passed, leaving sprawling lawns of emerald green and the sweet scent of star jasmine in the air. But on Saturday, April 9, 2005, clouds had piled up against the foothills overlooking the 8,000-acre campus and a chilly drizzle was soaking its palms and roses, perfect weather for sleeping off the hangovers that much of the student populace had inflicted upon themselves in the course of the previous evening's keggers on Greek Row.

But in one dorm room crammed with tech gear, at the western edge of campus, in party-free housing for nerds known in Stanford parlance as FroSoCo (short for Freshman Sophomore College), eighteen-year-old sophomore Sam Altman woke up, clear-headed, in the bed of his boyfriend, Nick Sivo, and began what would turn out to be the most consequential thirty-six hours of his life.

The couple had met almost immediately upon arriving on campus the previous school year. Altman was sitting in the auditorium-style lecture hall of CS106X, Stanford's accelerated single-quarter combination of its introductory computer science courses, CS106A and CS106B, when a shy fellow freshman with large brown eyes and the translucently pale skin of someone deeply committed to hours-long coding sprints sat down next to him. Sivo was an Eagle Scout and valedictorian of his Texas high school who had been raised in the Church of Jesus Christ of Latter-day Saints. Like Altman's, Sivo's mother was a doctor—a radiologist. His father was a CPA who worked as a CFO for a software consultancy, so he grew up in a house filled with computers, and was first drawn to them by his desire to play, and then make, video games. His parents tolerated his tinkering even when he shorted out the family computer or wrecked the installation of its operating system. But it was boredom that truly drove him into programming. Math came easily, and by high school he was so frustrated with the busywork of having to show all the steps while solving quadratic equations that he wrote software to automate these and other repetitive tasks. He had taught himself Microsoft QBasic from the MS-DOS 6 documentation, and later, Visual Basic, another programming language developed by Microsoft, from a book titled *Teach Yourself Visual Basic in 24 Hours*.

In the lecture hall, Sivo asked Altman about the weird device he had with him—a Compaq TC1000 "laplet," part laptop, part tablet, with a snap-on keyboard that folded into six shapes. They struck up a conversation but did not exchange names.

For most of freshman year, they were just acquaintances. Altman was dating someone else, and had thrown himself into campus life with a frenzied intensity. He founded a student organization with national ambitions to fight for gay marriage rights, penning an editorial in *The Stanford Daily* calling an anti–gay marriage article in the conservative *Stanford Review*, founded in 1987 by Peter Thiel, "juvenile" and "having no rational basis."[1] He threw himself into his computer science courses with similar abandon. Blake Ross, a fellow computer science major who lived with Altman in Donner, a modernist freshman dorm with a piano in the lobby, went to check on Altman's progress on a computer science assignment and found him wrestling with an error message from the software program. "Thirty minutes later, drop by Sam's room again, it's a crime scene," Ross told *Insider*. "He's dismembered the compiler itself. Kid's knee-deep in a swamp of low-level code—like rewiring your espresso machine because the leaf on your latte isn't arboreally sound." Altman explained that he had found a bug in the assignment.[2]

Altman was driven by a combination of impatience and almost unimaginable ambition. At the start of the school year, as he sat with the course catalog and an Excel spreadsheet, finding himself strangely enthralled by the jigsaw puzzle of plotting the next four years in a way that knocked out prerequisites and satisfied the required courses as quickly as possible, he paused and realized he should first decide what he actually liked to work on. Grabbing a loose piece of paper, he scribbled a list. Topping it, in order, were AI, nuclear energy, and education.

It was not until that spring, as Altman and Sivo were headed to summer fellowships working for Andrew Ng, a young assistant professor of computer science specializing in AI who was building autonomous helicopters, that they became friends. "Soldering" was the subject of the email that Ng wrote formally introducing Sivo to Altman. Altman had been "working on the Stanford electric vehicle project (and some psychiatry research, and some computer security research)," Ng explained before asking them to meet up to solder some circuits. Although they worked on largely different projects that summer spent in Stanford's Summer Research College program, they both became immersed in how location and GPS worked from the helicopter project, and began hacking on projects together after class. By December 2004, they were an item.

That fall, they had drawn into housing about as far apart as it was possible to be at Stanford, a sprawling campus build on the footprint of a gigantic horse farm: Altman in a four-person triple in Mirrielees, quiet, apartment-style housing on the eastern side of campus, and Sivo in FroSoCo, so far to the largely uninhabited west that one had to look out for mountain lions wandering down from the foothills. Altman found Mirrielees socially isolating and depressing. "Mirrielees was tough," he said in an interview with *The Wall Street Journal* years later. "I definitely didn't feel the same connection I did freshman year."

He spent more time off-campus, playing poker at Bay 101, a run-down cardroom in San Jose, where he profited from a boom in the game's popularity that lowered the average skill level considerably. "There were a bunch of terrible players," he said. "At the time, if you went to grind low-limit poker, it was a very profitable thing to do, and, I felt, super fun." The game taught him risk management, but "that feels like the least interesting thing I learned about it," he said. "All these layers of different strategies—it's a very complex game when you get good at it. Like, shockingly complex. There are always more things to learn and more ways to look at it, new things to try." Whenever he had a windfall, he would buy some Apple stock. Friends from college remember him tracking the stock price obsessively. "I was a huge Apple fanboy," Altman said. "They were my computers and I had a very strong relationship with them."

Back on campus, he continued to hang out with the friends he met freshman year, most notably Alok Deshpande, a lanky, earnest computer science major with a wide smile who had lived in the dorm next to him. The son of parents who had met in engineering school in India and raised him and his sisters in Atlanta, Deshpande had grown up attending private school and excelling in math competitions. "He was always good at math and science and physics," said his mother, Sheila Deshpande. "Sometimes a teacher would be sick, and the teacher would ask him to conduct the class." He and Altman became friends almost instantly, and their families had dinner together a few months into freshman year. Sheila Deshpande remembers Altman as friendly, if sometimes a little distant. "I sometimes felt that he was there, but not there," she said.

Altman and Deshpande had begun talking about doing something with location and mobile phones. Mobile phone makers had originally resisted putting any kind of tracking device in phones, but in 1999 Congress passed a law requiring them to figure out how to point rescue workers to the location of an incoming 911 call. Initially, carriers like Cingular complied by triangulating a signal between cell towers, the same process they used to place a call. But other carriers, notably Sprint and Nextel, which merged in 2005, opted to install global positioning system (GPS) chips in their phones, a more expensive but far more accurate option that could locate users within fifty feet, instead of five hundred. Wireless carriers were slow to follow the new law, because of the cost, even as they desperately tried to spin up businesses of "location-based services" to justify the investment that the Federal Communications Commission was forcing them to make. Meanwhile, privacy advocates like the Electronic Frontier Foundation cried foul. "You've built a surveillance system," said a staff attorney at the EFF in 2000. "This is just another way the government can track people's moments, not necessarily with their knowledge."[3] The FCC pushed back the deadline several times, but eventually settled on 2005 as the year that all cell phones were required to be able to beam back their location.

As that deadline loomed, Altman—whose father would spend 2005 helping the Roberts brothers close the sale of Sprint PCS affiliate Alamosa Holdings to Sprint—climbed on stage at a campus entrepreneurship event and held his flip phone in the air dramatically. He declared that he'd just learned that all phones would soon be location-enabled, and thought it would be cool to build some kind of service around this new capability. He invited anyone who wanted to join him to get in touch. During the networking portion of the event, Altman exchanged numbers with Peter Deming, a senior majoring in business who was hanging around campus so he could finish the sailing season before going on to a job at Goldman Sachs.

All sailors love maps, and Deming was no exception. He found the idea of location on a phone intoxicating. He had launched a startup before and had experience raising money. While he couldn't program, he could help with things like market research, market sizing, and putting together a pitch deck. A few weeks later, Altman, Deshpande, and Deming met for dinner at PF Chang's in Palo Alto to brainstorm. *What was the most useful thing you could do with mobile location data?* The obvious answer was something like Google Maps, but that app had gone live weeks before, and the team decided that was too hard anyway. They all agreed there should be some community component. Ultimately, Altman had the clearest idea: a social network based on a location service that would help your friends find you.

Sivo resisted joining at first. He and Altman had just started dating, and he didn't want to complicate things. But they kept working on projects together and eventually slid into collaborating. Sivo joined subsequent dinners at PF Chang's and the four became co-founders. On Deshpande's suggestion, they named their product Viendo, after the Spanish word for seeing, a riff on the idea of seeing friends.

Silicon Valley was just beginning to shake off the trauma of the dot-com bubble's bursting, which had vaporized some $5 trillion from public markets and sent tech investors fleeing to the safety of proven business models and demonstrable revenue for half a decade. Because startups themselves and the venture capitalists who stuffed them with funding regardless of their fundamentals were seen as the culprits of the crash, some classmates laughed at Altman when he mentioned wanting to work on a startup freshman year.[4]

But in the time since the crash, as the Enron and WorldCom accounting scandals further eroded Americans' faith in corporate America, a quiet revolution was happening in technology. Everything was getting faster, cheaper, and better connected, such that a handful of Harvard students were able to create TheFacebook in their dorm room.

Altman and his co-founders decided to enter an undergraduate business plan competition organized by Stanford BASES (Business Association of Stanford Entrepreneurial Students), one the largest student-run entrepreneurship organizations in the world, which offered $2,000 for its top prize. Deming spent the subsequent weeks in the business school library looking up data that could support their projections and making the presentation look like a real pitch deck for a venture capitalist. There would be no demo, just a slide deck and Altman's pitching skills.

When the competition arrived, the co-founders made their way through the spring drizzle to the buildings in the Old Quad topped with red tile roofs. Altman was calm. While everyone who met him was taken by his intelligence, it was his "preternatural confidence," in the words of Mark Jacobstein, who would join the company a few months later, that truly set him apart. "He had that reality distortion field." Altman idolized the original wielder of the "RDF," Steve Jobs. For the pitch, he had honed his message to almost Apple-like simplicity: The most commonly asked question on a cell phone was "Where are you?"

**ALTMAN AND** Deshpande took the stage in a wood-paneled classroom in the Old Quad and pitched Viendo, with Altman taking the lead. Viendo would answer the "Where are you?" question for its users by providing a map displaying the live location of your friends. You would automatically be able to see if a friend was too far away to bother reaching out to for lunch if you didn't happen to have your bicycle with you that day, or if someone was close enough to quickly ping for a cup of coffee by the library. In retrospect, it was the solution to a problem experienced only by a fairly exclusive subset of the population—that is, young people with few responsibilities who live on an improbably gargantuan campus and are unhindered by even a shred of paranoia that someone might know their exact location at all times.

Nevertheless, Viendo won the competition and, even more importantly, convinced one of the judges that there might just be a real business there. Patrick Chung was only a few months into his career as an associate at New Enterprise Associates, America's largest venture capital firm, which sponsored the competition to get an early look at talent and investment opportunities. He wasn't much older than the teenagers on stage.

Raised in Toronto by a nuclear physicist and finance executive who had both immigrated from Taiwan, he had attended the strict British-style all-boys prep school Upper Canada College and maintained that era's slightly formal accent that set him apart from his surf-speaking California compatriots. After college at Harvard and a master's at Oxford, he was recruited to join McKinsey. He soon found that he detested the work. He left after a little more than a year to try the startup life with a friend from high school and fellow McKinsey refugee. Called Zefer, their consulting company pulled in $100 million in revenues building "e-businesses" for clients like Burger King and Gillette, and tried to go public in 2000. But when the bubble popped, their IPO was pulled, leading to a modestly profitable sale and Chung's return to school—this time to Harvard for a JD/MBA—while he figured out what he wanted to do with his life. When NEA recruited him to its Menlo Park office, he was skeptical at first, having never dreamed of moving to the West Coast. "The idea of California was unappealing to me—a bunch of movie stars and kombucha and convertibles," he recalled. But he was pleasantly surprised to find NEA full of "some of the nicest, most accomplished people I had ever met," including a Nobel laureate, and took the job.

After the competition, Chung approached the co-founders. "Do you want to do this for real?" he asked.

"What do you mean, 'for real?'" Altman replied.

Chung explained that, in order to get access to the location data they wanted, they would have to sign a deal with each mobile carrier, like Cingular (soon to be AT&T), Verizon, and Sprint. Luckily, he already had a relationship with the CTO of Sprint and offered to fly out to Kansas City to introduce Altman to the CTO in person, even before writing a check to fund the company. There was just one thing: if Sprint was interested and NEA's partners decided to invest, the idea would not wait. NEA's partners would not invest in a company run by students still fully enrolled in school. They would have to drop out.

It was a lot to take in, but Altman didn't have time to worry about it right then. As it turned out, the contest was only the second most important thing that he had scheduled that day. The most important was the red-eye to Boston he had booked that evening out of San Francisco International Airport for his interview to be part of the inaugural Summer Founders Program organized by a new kind of investment firm called Y Combinator. Named for a mathematical term for a program that runs programs, Y Combinator was the brainchild of the startup guru Paul Graham; his girlfriend, Jessica

Livingston; and Graham's co-founders from his first startup, Robert Tappan Morris and Trevor Blackwell.[5] Their vision was a three-month program for startups, modeled on graduate school, that would fit into the summer internship slot and offer both a modest investment—$6,000 per founder for roughly 6 percent of the company, a number Graham picked after hearing it was what MIT paid its grad students over the summer—and advice in the intense environment of a Cambridge residency.

Graham was a celebrity among the geek set thanks to the essays he posted on his personal website on topics ranging from the programming language Lisp to the nature of wealth. In the early days of Google, he was the first "Paul" it surfaced, ahead of the Beatle and the apostle; to this day, his essays are near the top if one searches the word "essays." "Pretty much everything he does is featured on *Slashdot*, which is sort of like *The New York Times* of the computer world—everybody reads it," Aaron Swartz told *The Stanford Daily* that spring.[6] Like Altman, Swartz was a nineteen-year-old founder from Stanford bound for the program, except Swartz was already tech-world famous for having helped create the popular RSS-feed format at age fourteen. The investment aside, mentorship by Graham would be priceless.

Altman had learned about the opportunity from his friend Blake Ross, who had posted something to Facebook about it a day before the deadline for applications. Ross had met Graham the previous fall at Foo Camp, an annual gathering of hackers in Northern California convened by Tim O'Reilly of O'Reilly Media, the influential tech publisher and conference organizer ("Foo" stands for Friends of O'Reilly, as well as being the placeholder name for a variable in computer programming), and had maintained a correspondence with him. A tech wunderkind who had helped create the Firefox web browser while still a teenager, Ross had also been interviewed by Livingston for her book on startup founders, *Founders at Work*. Altman had no such personal connection, but later described himself as a "big fanboy" of Graham's essays.[7]

But the timing was not ideal.[8] Viendo's co-founders already had summer gigs lined up: Deshpande at a Stanford program in India, Sivo at his father's tech company, and Altman at a tech job at Goldman Sachs. "They had the reputation for the hardest technical interview," Altman explained. "It was actually very brilliant marketing to the CS [Computer Science] department: 'this is the hardest tech interview.'" Deming, the sailor, had a real job starting at Goldman a few weeks after graduation.

Still, Graham's pitch was hard to resist. "The SFP is like a summer job, except that instead of salary we give you seed funding to start your own company with your friends," Graham wrote. "If that sounds more exciting than spending the summer working in a cube farm, I encourage you to apply."[9] Altman realized he was having much more fun working on this project than he was going to have working at an investment bank.[10]

Unable to convince his co-founders to ditch their internships, Altman filled out the thirty-two-question application himself, with some emailed help from his co-founders. When he got word back from Graham that they were one of twenty startups invited to Cambridge for interviews, the timing proved even worse: the interview was scheduled for the same weekend as the business plan competition. Altman tried to push it back so he could take part in the competition, and explained that he would be the only one available to come to the Cambridge residency.

"Paul wrote back to him, saying, 'You know, Sam, you're only a freshman. You have plenty of time to start a startup. Why don't you just apply later?'" Livingston recalled in *Founders at Work*. Altman replied: "I'm a sophomore, and I'm coming to the interview."[11]

**ALTMAN ARRIVED** in Cambridge Sunday morning and went directly to the fraternity house of his high school friend Rick Pernikoff, who was studying computer science and electrical engineering at MIT. He rushed in, quickly showered, and Pernikoff took him on the subway to Harvard Square, where they got off and walked along the winding, tree-lined streets. It was a beautiful spring day, and Cambridge was buzzing. As they turned onto Garden Street, they came upon a modern, loft-like building that stood out like a sage green industrial sore thumb amid the prim Victorians on its residential block. Y Combinator was housed in a building that mirrored the eccentricities of its founder. Graham had purchased it a few months before to be his own office and retrofitted it for silence, with noise-cancelling doors and double-paned windows so he could paint in peace, then commissioned his architect friend Kate Corteau to design the space and some of its modern furniture, including long white tables and benches with legs too close to the center that would tip if someone sat on one end.[12] (Graham would sometimes have to run to intercept an unsuspecting investor as he would attempt to sit on one end of the bench.) There was a big open room, a kitchen, and an upstairs greenhouse. The vibe was somewhere between a modern art gallery and a barracks. Milling around it all were other startup founders, a category of person that Altman had never encountered before.

"It was just like, 'What is this? And who are these people?'" Altman recalled. "I hadn't really met startup people before. At the time, it was this pretty unpopular thing." His interview in front of YC's four co-founders was a kind of *coup de foudre*. "Within about three minutes of meeting him, I remember thinking, 'Ah, so this is what Bill Gates must have been like when he was 19,'" Graham wrote years later.[13] The interview only lasted twenty-five minutes. "All of us were blown away by Sam," Livingston later wrote. "His poise and intelligence, and just the way he was. We knew that there was something special about him."[14]

The feeling was mutual. "That was the first time I felt like, alright, I have finally found the kind of people I want to be around," Altman said.[15]

At 7 p.m. that evening, Altman received a call from Graham on his cell phone: Viendo was in. He would later learn that his was the first startup to be funded by Y Combinator.[16]

Altman slept on Pernikoff's frat house floor that night. His career was launched, but he didn't know it yet. In the course of one weekend, he had met an investor who would go on to not only invest in his startup, but would help convince fellow venture capital firm Sequoia Capital to invest, beginning a relationship that would last to this day. And he had become a member of what would prove to be Silicon Valley's most important alumni network, which would give birth more than a decade later to a small but audacious research project called OpenAI.

CHAPTER 4

# AMONG THE "NERD'S NERDS"

**PAUL GRAHAM DOESN'T LOOK LIKE A NERD. HE IS HANDSOME** and sturdy and boyish-looking, even into middle age, with sandy brown hair, deep dimples, and a penchant for preppy polo shirts that makes him look, at least from the waist up, like the kind of golf-playing bank executive his essays make clear he has nothing but disdain for. Below the waist, Graham is so famous for wearing nothing but rumpled cargo shorts and sandals—the ultimate sign of having exited one's startup successfully enough to never have to impress anyone ever again—that his underlings gave him an autographed pair of the former as a parting gift upon his retirement from YC.[1]

The son of a British mathematician who worked on nuclear reactors, ran more than eighty marathons, and wrote books about chess in his spare time, Graham was born in England but grew up in Pittsburgh, with his father chasing nuclear jobs across the Northern Hemisphere as regulators slowly strangled the industry.[2] Graham was smart but mischievous. "I was a bad kid," Graham told Bloomberg in 2014. "Half the kids in my neighborhood were forbidden to play with me.... I was suspended from school I think at least once a year from first grade until twelfth grade." Computers were not really part of school, though, and at age fifteen Graham began to code.[3] He went to Cornell, where he began by majoring in philosophy before switching to artificial intelligence, and then to Harvard for a PhD in computer science. Rather than cash in or become an academic, he then went on to more school, this time for painting, and moved to New York City to be, in his words, a "starving artist" who at least managed to succeed at the "starving" part. To make ends meet, he freelanced as a software engineering consultant.

It was painting that led to both the means and the ideas behind Y Combinator. In January of 1995, needing money, Graham convinced his best friend, Robert Tappan Morris, to start a company with him called Artix that would put art galleries on the web. Morris was a Harvard graduate student with the impeccable hacker cred of having been the first person convicted under the Computer Fraud and Abuse Act, after he accidentally unleashed the first computer worm. The code behind the new company worked brilliantly. But the business idea, they soon realized, was idiotic.

Art galleries did not have the faintest desire to put their wares online for anyone to see. They wanted to preserve the illusion that the small slice of their merchandise that they were displaying was special, rather than inventory that had been kicking around in storage for years. Gallerists viewed the art market's inefficiency and opacity as a feature, not a bug. Artix couldn't even manage to *give away* its services to galleries.

"In retrospect, I wonder how we could have wasted our time on anything so stupid," Graham later wrote. "Gradually it dawned on us that instead of trying to make Web sites for people who didn't want them, we could make sites for people who did."[4]

Within six months, they had ditched the Artix idea and founded a new company, Viaweb, the first web-based application, which allowed people to build their own websites through a browser—that is, "via the web." Three years later, they sold it to Yahoo! for $49 million, and it became Yahoo! Store.[5]

The experience taught Graham what would become one of the cardinal principles of his startup gospel: most founders think too small. If you can do the technical work of getting some art galleries online, there's no reason you can't apply that same work to a product that would get *everything* online. Another lesson would become the mantra of the Summer Founders Program, a line they had printed on T-shirts: Make Something People Want. (An alternate T-shirt that the program had printed up that summer read "Strap on Some Plums," printed on plum-colored shirts.)

But painting also provided Graham something far more fundamental. It gave him a framework to understand the kind of computer programming he liked to do—a kind of code for code's sake he'd been doing since high school that he liked to call "hacking," to distinguish it from the more theoretical or purely mathematical pursuits that colleges referred to as "computer science."

With the free time afforded by his newfound wealth, he began to write essays and publish them on his own website, PaulGraham.com, starting in 2001.

"What hackers and painters have in common is that they're both makers," he wrote in the title essay of one of his books. "Along with composers, architects, and writers, what hackers and painters are trying to do is make good things." One way to make good software, he argued, was to do a startup, because they were small and nimble enough to allow computer programmers to "sketch" their code, determining what should be built in real time, thereby constructing something superior to the competition. "Great software requires a fanatical devotion to beauty," he wrote.[6]

**IN MARCH** 2005, Graham accepted an invitation from the Harvard Computer Society to give a speech. He decided on the topic, "How to Start a Startup." Graham had put out a notice on his website about the appearance, and hackers from across the country flocked to the auditorium like pilgrims. Two of them were roommates from University of Virginia named Steve Huffman and Alexis Ohanian. Huffman had brought his copy of Graham's book about Lisp—a programmer's programming language that Graham had such religious devotion to that he actually founded his first startup just to be able to continue programming in it—for Graham to sign.[7]

In Graham's characteristically blunt style, the speech amounted to a pep talk for the sea of ambitious young men in the audience. His message was that startups were not as hard or mysterious as they might think. It boiled down to "Look at something people are trying to do, and figure out how to do it in a way that doesn't suck."

But his vision for how to do that was quite narrow. It would be best to be between twenty-three and thirty-eight years old, willing to go back to work after dinner and hack until 3 a.m. seven nights a week for four years, and ideally located in a cool, walkable place near a top university, like Cambridge or Palo Alto or Berkeley. Oh, and your founding team had to include a computer programmer. This final qualification was essential because hacking was such a specialized skill that founders without it would never be able to tell a good programmer from a bad one. A hacker can handle customer service, but a non-hacker can never hack, he argued. As for women: it's easier to avoid their inefficient biological and cultural baggage in a startup than in a more traditional company because there are no discrimination laws governing the choice of startup co-founders. "For example, I would be reluctant to start a startup with a woman who had small children, or was likely to have them soon," Graham said. "But you're not allowed to ask prospective employees if they plan to have kids soon."[8] (As YC grew, it would come under fire for comments like these and its low percentage of female founders, which was zero the first year and has hovered around 10 percent in recent years.[9] Graham argued that YC both employs and funds more women than most venture capital firms, and the dearth of women in tech is a deeper social problem than any one institution can fix.)[10]

After the speech, Graham was thronged by fans. Besides Huffman and Ohanian—who had dinner with Graham that evening and would go on to take part, alongside Altman, in the first Summer Founders Program class, where they created Reddit—another fan came up and asked Graham how he could get his startup funded. Graham replied that he should look for a rich person with a technical background. In the awkward silence that followed, he realized that he fit the bill.[11]

In the seven years that had passed since he sold Viaweb, he had always intended to get into so-called "angel" investing—the initial friends-and-family-level money that a startup receives before it has anything like a real valuation or idea what it's doing. But he had not gotten around to it.

A few days after giving the speech, Graham was walking home from dinner in Harvard Square with Livingston, whom he had been dating for about a year. Livingston had the unfussy elegance of the prep-school jock she once was. She was smart and capable and curious, but she hadn't quite found her path yet. She was working at a boutique investment bank, waiting to hear back on a marketing job in venture capital, and writing a book about her obsession: startup founders like Graham.[12,13]

The daughter of New England bluebloods, Livingston was raised by her father and grandmother after her mother "left home," as Livingston put it, when she was still a baby.[14] Her mother, Lucinda Pauley, had been a debutante who became engaged while studying English at Wellesley.[15] Months after the birth of her first child in 1971, she left her family and went on to spend decades living in intentional communities and writing about "planetary healing," changing her name to Shen.[16] David Livingston, an MBA who spent his career working at the Gillette Company, brought his baby daughter back from Minneapolis, where they were living at the time of the split, to his native Boston area, where his family roots stretched back to his great-great-grandfather, H. H. Hunnewell, a benefactor of Wellesley College.[17] There, he took care of his daughter on the weekends while during the week she was with his mother, Isabella Hunnewell Livingston, an artistic and independent woman with movie star looks who had divorced her husband in 1945 after having four children in five years. Isabella was famous in Wellesley for building giant, highly detailed "slush sculptures" of dragons, dinosaurs, and unicorns in the snow piles in her front yard every winter.

Livingston was an athletic straight-A student, and one day found herself on the campus of Phillips Academy in Andover, Massachusetts, one of the most prestigious prep schools in the country, during an away soccer game. She was bowled over, and asked her family to send her. She got in, but the experience of suddenly being surrounded by even smarter and more athletic people plunged her into a self-described "Dark Ages" of self-doubt that she felt she didn't get over until adulthood. The day after she graduated from Bucknell University with an English degree, her grandmother died of cancer, leaving her at a loss with no clue what she wanted to do.

She took a job answering phones at Fidelity Investments, which she hated, and then bounced around—investor relations, an automotive consulting firm, *Food & Wine* magazine, even wedding planning. By 2003, she was running marketing for the Boston investment bank Adams, Harkness & Hill when she met Graham at a party at his house.

Graham and his buddies introduced Livingston to the world of startups, and she became fascinated, landing a deal for a book about their stories, which she found far more interesting than the lumbering publicly traded companies she dealt with in her day job. In an attempt to get closer to them, she applied for a marketing position at a venture capital firm.

But the night she and Graham were walking home from dinner through Harvard Square, the firm had been taking a long time to decide on whether to hire her, prompting Graham to hold forth on one of his favorite topics: why venture capitalists, or VCs, suck. In fact, he published an essay that month titled, "A Unified Theory of VC Suckage," in which he described VCs as "alternately cowardly, greedy, sneaky, and overbearing."[18]

Venture capital, a form of private equity investing that came into its own amid America's post–World War II industrial boom, involves VCs gathering money from "limited partners," such as university endowments or pension funds, and using their knowledge of the industry to divvy up the pool of cash among a portfolio of startups, for much the same reason that a mother fish lays hundreds of eggs. Most startups fail, but those few that succeed—particularly in the tech industry, where the incremental cost of selling an additional widget is near zero—tend to succeed spectacularly, more than making up for all the losers.[19] From the very first venture-funded startup, Fairchild Semiconductor, in 1957, Silicon Valley's tech industry has been largely funded by venture capital, from Apple to Google to Facebook to just about any startup you've ever heard of today.

The main problem, in Graham's view, was the way that VCs are paid—typically 2 percent of the amount of money they manage each year, plus a percentage, usually 20, of the gains, known as "carried interest" or just "carry." Because they were paid according to how much money they managed, they had an incentive to manage as much money as possible. Since any individual VC could only do so many deals to invest that money, they had an incentive to put as much money into each investment as possible, bloating valuations so much that the companies are left with a very small number of possible buyers outside of going public—a major factor at play in the destructive dot-com bubble of the late 1990s. Because so much money was riding on each investment, VCs were paranoid, meddling in the boards of the companies they invested in and often bumping aside founders for their own more business-minded CEOs, people Graham dubbed "newscasters" because "they had neat hair and spoke in deep, confident voices, and generally didn't know much more than they read on the teleprompter." (Viaweb's fear of being stuck with a newscaster was so great that they ultimately decided against taking any VC money and instead funded their startup entirely from angel investors.)

As he was trying to convince Livingston to change venture capital from the inside, Graham suddenly had another idea. "Let's just start our own," Graham told Livingston.[20] It would give Graham a chance to finally get into angel investing, and Livingston a chance to pursue her dream of working with startups on her own terms.

"Paul had wanted to do angel investing," Livingston wrote. "But he didn't really want all the requirements that come with being an angel investor, so he thought he should start an organization that could handle all of this for him."[21] (Graham had bragged during his Harvard speech that Viaweb had ignored all the bureaucracy involved in starting a company, including "various things with the IRS," until it got enough funding to hire a CFO who went back and retroactively fixed everything.)[22] Livingston agreed. The next day, Graham convinced his

Viaweb co-founders, Morris and Blackwell, to join them. Graham put up $100,000, Morris and Blackwell each contributed $50,000, and Livingston quit her job.

"The initial plan was that they would pick and advise the startups and I'd do everything else," Livingston later wrote. "Instead of giving large amounts of money to small numbers of established startups, like traditional VCs did, we'd give small amounts of money to large numbers of earlier stage startups, and then give them a lot of help."[23]

They would pick promising young programmers with few responsibilities and give them just enough money to survive a summer residency in Cambridge. To boost the new companies' chances, they'd help with legal paperwork, mentor the startups through office hours and home-cooked Tuesday-night dinners of chili or stew, invite speakers to share their experiences launching successful companies, and give their charges a shove off into the world at a final "Demo Day" in front of real angel investors and venture capitalists. To help Graham, Livingston, and their co-founders learn how to become angel investors quickly, they decided to do it in groups they dubbed "batches" —an innovation that turned out to be YC's most distinctive feature. They would start that summer.

In classic Graham fashion, he stayed up most of the night building a website and a thirty-two-question application for the program, then linked to it on his blog. The questions were characteristically impish, asking applications to prove how they were an "animal," Graham's term for a person who does not take no for an answer, and to offer an example of a time they hacked some system to their advantage.[24] (One winning answer to the former question: "Justin once played a forty-five-minute half of a rugby match before realizing he had a nosebleed from being struck in the face.")[25] Applications began flooding in—no press or PR necessary.

"It just goes to show you the power of good writing," marveled Patrick Chung, the NEA partner who had first discovered Altman and Viendo. "Paul wrote honest, helpful stuff that people read and respected—he didn't do what we did, which was to comb through hundreds of startups hunting for investments. He let the founders come to him, entranced by the power and truth of his writing."

Graham saw the program as doing something fundamentally different than what venture capitalists do. "These later-stage investors are playing a zero-sum game. There is a fixed amount of deal flow, and they're just trying to find the good startups in it, whereas YC is trying to encourage more good startups to be founded," Graham said. "We knew this was possible because we knew how ambivalent we were about starting a startup ourselves. We knew how lame, in certain respects, founders could be and still be successful."

**HEADING TO** the first Tuesday-night group dinner that inaugural summer in 2005, Altman walked through the noise-cancelling doors of 135 Garden Street and was hit with the smell of what came to be affectionately known as "vegetarian glop" simmering in Crock-pots. He was excited to be reunited with the group of supernerds he had already started to bond with during his brief stay for his interview. (He and Blackwell, for example, had already hit it off over their shared love of balancing scooters—otherwise known as Segways—which Blackwell designed his own version of called a Segwell that he took to riding around YC's headquarters.)[26]

Altman surveyed the other founders who had made the cut. In several cases, Graham and his partner had liked them, but not their idea. Among that group were Steve Huffman and Alexis Ohanian, the roommates from the University of Virginia, who after attending Graham's Harvard talk had pitched him on a cellphone-based food ordering system called MyMobileMenu or MMM for short. Graham and Livingston had liked them—Livingston thought they were so cute she dubbed them "muffins"—but were skeptical about their ability to make deals with both restaurants and wireless carriers, which would have been required in the pre-smartphone era. Graham suggested they instead try to create something more like the link bookmarking site Delicious, whose "popular" page frequently linked to his essays.[27]

Also in this category were Emmett Shear and Justin Kan, both twenty-one, who had met at The Evergreen School for Gifted Children in Seattle and bonded over their shared passion for math and the fantasy card game Magic: The Gathering.[28] Their idea for an online calendar app, Kiko.com, was also initially rejected by Graham and his co-founders, but Shear and Kan managed to talk the accelerator partners into it during the in-person interview, with the understanding among all parties that "Google might crush us like ants" by releasing something similar, as they wrote in their submission.[29]

Swartz, who ran a popular blog and had been corresponding with Graham for months, had pitched a web-based tool for building websites that he would eventually name Infogami. "We've got another idea that we think might be better for you," Graham had replied during Swartz's interview, starting to pace the floor.[30] But Swartz pushed through with his idea, which he built with a co-founder he had recruited online from Denmark. At Graham's instigation, Swartz eventually merged Infogami with Reddit. Chris Slowe, a Harvard doctoral student who co-founded a desktop search company Memamp with his fellow student Zak Stone, would also end up joining Reddit by the end of the summer.

Altman faced few critiques along these lines from Graham. In the months since he interviewed in April, his company had changed its name to Radiate—Deshpande's suggestion—after discovering that the owner of viendo.com wanted $300,000 for the domain name. But otherwise they were proceeding with their vision of using mobile phones' new location awareness to connect people with their friends.

**ALTMAN WAS** thrilled to have finally found the kind of people he wanted to hang out with. But he still stood somewhat apart. Kan, Shear, Huffman, Swartz, Slowe, and Stone instantly melded into a circle of friends connected by a burbling stream of AIM instant messenger jokes that Altman was never quite part of.

"Sam had a different vibe than everybody else," Huffman said.

"Most of the guys—Justin and Emmett and Aaron and Chris and Zak—we were all, like, nerd's nerds. Really love the technology, almost aloof about the world. I think we had to kind of grow into our business selves over the next little while. Sam struck me as much more business and success-minded than the others." He added: "It felt like he'd come in a hurry. It felt like he had somewhere else to be."

One day, the batch went on a walking excursion together through Cambridge. A founder named Mikhail Gurevich, whose company, ClickFacts, sought to fight online ad fraud, noticed that Altman's pockets were bulging, and asked him about it. Altman proceeded to pull phone after phone out of the deep pockets of his cargo shorts, from "candy bar"–style Sony Ericssons to "clamshell"–style Motorola Razr flip phones, explaining that each one ran on different software, requiring Radiate to configure a different application for each. "He was carrying, like, eight phones," Gurevich said. "I told him, 'Bro, you're gonna get ball cancer.'"

Gurevich was stunned by the complexity of the technical task that Altman was tackling. "After that conversation I thought he was like a genius," Gurevich said. "He was a little off, speaking-wise. I was like, 'Maybe this guy's on the spectrum.' But listen, so was Aaron, and so were a couple of the other guys."

Altman did not, in fact, get testicular cancer, but he did eat so much ramen and so little else amid twelve-hour coding binges alone in his Cambridge apartment that summer that he told people he got scurvy.[31] He also continued to hold fundraising talks with NEA that summer, and began speaking about going out to meet Chung's contact at Sprint, which astonished his YC peers.

"His startup felt more actualized," Huffman said. "Fundraising from real VCs? None of us were really doing that. We didn't know how to talk to them, didn't know the game. In fact, Paul only kind of knew it himself. The speakers he would invite in were kind of anti-VC. That was the vibe of Y Combinator at the time: venture capitalists were not to be trusted."

Altman had no such compunction.

"He knew the rules of the game," Huffman said, "and nobody else did."

**SPRINT'S HEADQUARTERS** were half a country away, located on a sprawling, 200-acre campus of red brick buildings, fountains, and carefully manicured lawns in the wealthy Kansas City suburb of Overland Park. The campus was the largest building project in the Midwest when construction began in the mid-1990s. The architects' other projects had included the Scottish Parliament, and indeed, with its clock tower, 3,000-seat amphitheater, and network of hiking and biking trails, Sprint's headquarters imparted a sense of housing something far more consequential than a mere company—even if that company happened to be the third-largest mobile provider in the United States and the largest private employer in Kansas City.

A few months after the business plan competition, Chung, true to his word, scored Altman a meeting with Sprint's CTO. Chung had had an assistant book the flight and rental car, but Altman, too young to legally rent a car, had to fly from Cambridge to St. Louis to borrow the family car. Once at the headquarters, they entered the executive building, passing through room after room of assistants before reaching an impossibly grand office with floor-to-ceiling windows overlooking streams and greenery.

Wing Lee greeted Chung with a boisterous smile but was unable to mask his surprise at seeing Altman, who, though nineteen, looked closer to twelve.

"Where's your founder?" Lee asked Chung. (He later explained himself saying "He was a bit of a petite young man. I thought someone else was going to come.")

Lee, in his late thirties with a few streaks of gray beginning to creep into his hipster's pompadour, had been at Sprint for more than a decade. He had designed the first fiber-optic network in America, made famous by Candice Bergen's commercials for the "pin drop network" that contained less noise than traditional long-distance carriers. The network had its roots in some of America's oldest infrastructure, emerging from the telegraph lines that used to run alongside the Southern Pacific Railroad that was founded in the 1860s. The name Sprint is, in fact, a 1970s-era acronym for Southern Pacific Railroad Internal Networking Telephony.[32]

But when he helped Sprint enter the wireless data age in the late 1990s with the launch of Sprint PCS, short for Personal Communications Services, Lee realized that the slow network speeds of 2G mobile data plans would turn off users unless the company could offer something that desktop computers, with their far faster connections, could not. "It had to be a fundamentally new user experience," he said. He turned himself into something of a coolhunter.

Lee began to take his team on annual pilgrimages to the Akihabara district in central Tokyo, home to Harajuku girls dressed like Little Bo Peep and the world's most cutting-edge electronics. "We started seeing those Japanese girls taking pictures using the flip phone using a little camera," he said. That led to a partnership with a startup called LightSurf that put proprietary software on phones so that you could send a photo to anyone in your address book. Sprint PCS charged five dollars for "picture mail," but users first had to sign up for fifteen-dollar mobile data subscription to use it.

"It was a killer app," Lee said. "It actually was one of the highest ARPU of any mobile app in history," he added, referring to "average revenue per user," a key metric for any subscription-based business. "It was game-changing."

On the back of that success, Sprint began cultivating relationships with VCs, using its Burlingame office just up the road from Palo Alto's famed Sand Hill Road.

Now Lee was hoping that Chung was here to deliver him his next killer app. Despite his initial confusion, as Altman began to talk, Lee was transfixed. Facebook had already become a sensation, and Altman pitched his company as "Facebook on mobile, but with a twist," Lee recalled. While Facebook could only roughly guess a user's general location using his or her IP address, Radiate would be able to use the mobile carrier's precise location data to offer the very thing that Lee most wanted: a new experience that only mobile data could provide.

"Sam's idea was very attractive to us because we were looking for ways to drive mobile usage," Lee said. "We thought, 'Well, this one is worth testing.'"

Over lunch, Lee asked Altman how Radiate would handle the issue of privacy. Altman gave what Lee considered a very thoughtful response, saying that he believed the service needed to take a "respectful" approach to location and require users to opt in to share their location. Nevertheless, Lee's concerns about privacy were significant enough that he decided that Sprint should start Radiate on Boost, its edgier, youth-oriented sub-brand, where, as he put it, "users are perhaps more indexed toward the folks who might not be too concerned about information sharing."

The meeting made it clear to both Chung and Altman that this idea had promise. They began to discuss in earnest the possible terms of an NEA investment in Radiate.

**ALTMAN RETURNED** to Cambridge determined to finish a prototype by the end of the summer, coding an application on a Sanyo flip phone operating on the Sprint network. He dutifully attended the Tuesday-night dinners, where speakers such as Steve Wolfram, founder of Wolfram Research, and Langley Steinert, chairman and co-founder of Tripadvisor, would hold forth over whatever was bubbling in the Crock-pots. At one point, Graham brought a bunch of his former Yahoo! colleagues to come speak. "Then he convinced them that they should buy some of the YC startups," Altman recalled. Radiate, as something of a teacher's pet, attracted special interest. In August, Altman flew back to California for two meetings: one with Yahoo! about an acquisition offer—which turned out to be lowball and not particularly serious—and another to pitch the NEA partners, who had to approve any investment. It was the first investment that Chung had ever pitched at his new job.

"I marched Sam in there, and the senior NEA partners looked at me like, 'Oh no, we made a hiring mistake with this Chung guy. This is his first deal? This twelve-year-old who hasn't even graduated college yet?' And he opened his mouth again, and once again my partners' jaws dropped. 'Ah, I see it. I see it.' And so I got the votes."

Negotiations began on a term sheet for investment.

THE SUMMER Founders Program ended with a pitchfest to investors that Graham originally dubbed Angel Day, to make clear the kinds of investors he wanted his charges to mingle with—"angels," not VCs.

"We contacted every rich person we knew in the area and I was praying the seats would be filled," Livingston wrote.[33] They ended up with about fifteen investors and hosted them at the Y Combinator offices. Stephen Wolfram, who knew a thing or two about young talent as the youngest recipient of a MacArthur "genius grant," was on hand with his ten-year-old son, who kept a scorecard of the pitches, determining that Kiko had the most promise as an idea but noting that Altman was "the most businesslike person there."[34]

Reddit had built a user base over the summer, beginning with its YC batchmates and with Graham linking to it from his blog. Graham was excited enough that, when his rich friends declined to fund Reddit, he offered the company $70,000 himself, a move that became known as the "Paul Graham special."[35] (The PR-savvy Ohanian told *Wired* that September that Reddit had landed an investor at Angel Day willing to fund the company for another year, which was technically true.)[36] By the next year, it would be acquired for $10 million by Condé Nast, the publisher of *The New Yorker*, *Wired*, and *Vogue*—peanuts, in financial terms, but enough to make Huffman, Ohanian, Slowe, and Swartz "dot-com millionaires" and to put Y Combinator on the map.

Other founders from that batch went on to even greater success. Kan and Shear's calendar app was indeed squashed like an ant by Google, just as they had warned was a possibility in their YC application.

They sold it on eBay for $250,000 and went on to found another company with seed funding from Graham, Justin.tv, which consisted of Kan strapping a camera to his head and livestreaming his entire life. That company evolved into Twitch, the video game streaming company that Amazon bought for nearly $1 billion in 2014. In an utterly surreal turn of events proving either the incestuousness of Silicon Valley, the absolute hegemony of the YC alumni network, or both, the board that fired Altman years later hired Shear, ever so briefly, to be his replacement.

Another founder from that batch, Phil Yuen, sold his startup, TextPayMe, to Amazon, while Zak Stone went on to work for Google Brain. Gurevich's ClickFacts was bought for around $160,000 by a group of West Coast investors, enough to, in Gurevich's mind, return "10X" to Y Combinator. Even if the total dollar amount wasn't huge, he thought at the time, "think of all the cup ramen you could buy!"

But of all of these, Graham considered Radiate the most promising of the startups that Y Combinator funded that year.[37]

# PART II

# 2005–2012

CHAPTER 5

# "STOPPING OUT"

AS FALL APPROACHED, THE THREE CO-FOUNDERS PREPARED for their junior year at Stanford. Sivo had signed up as a resident computer coordinator and moved into campus housing. Altman got an apartment off-campus in Mountain View. Deshpande was back from India, and they all registered for courses. But Altman's heart wasn't in it. He had returned from Cambridge a changed person, filled with the confidence that PG, as everyone called Graham, had breathed into him. He even had his own PG-style, email-handle nickname: Sama.

On August 31, 2005, Radiate had gotten its first term sheet from NEA for $1.6 million, valuing the company at $6 million, and the teenagers needed a lawyer. Altman reached out to Page Mailliard, an attorney at Wilson Sonsini Goodrich & Rosati, the most powerful law firm in Silicon Valley, which had an office near the Stanford campus. Led by the quiet but fearsome Larry Sonsini, who served as consigliere for everyone from Steve Jobs to Google founders Larry Page and Sergey Brin, Wilson Sonsini was the preferred firm for startups raising money from the venture capital firms on Sand Hill Road. Mailliard, who had platinum blond hair and bright blue eyes, had been at the firm since the 1980s, when "Palo Alto was Podunk," she said. She had studied English at Stanford, then earned her JD from Harvard Law School. "Law is all about language," she said. "It's about people and language."

Mailliard loved both people and language, and she especially liked the three Stanford students who made their way to her office in jeans and T-shirts on a September day, asking for help with the NEA term sheet process. Altman, in particular, she found "extraordinarily bright and very unassuming." She led them to the firm's conference room, where they also met with her colleague, Carolynn Levy, who would help simplify the legal paperwork it took to start a company. Together, they began to run through the details of things like what a C corporation was.

The young men in front of her were not that much older than Mailliard's three sons. As she listened to their story, her first impulse was maternal. "You shouldn't miss your college education," she told them. But if they were determined to press ahead with the NEA term sheet, she urged them to talk with other investors. "Don't just take the first term sheet you can get."

She offered to introduce them to Greg McAdoo at Sequoia Capital.

**MCADOO HAD** grown up in Manhattan, the son of a Black civil rights activist and academic who had once recorded an album with Pete Seeger and went on to found the Africana Studies department at Stony Brook University.[1] Once, when he was four, McAdoo told his father he wanted to walk on the moon. His father responded that to do that, he needed to learn to fly. From that moment onward, McAdoo had been obsessed with airplanes. He wrote computer graphics software, including simulations of space flight. His first computer was a TRS-80, and he later got an Apple II Plus–compatible Franklin Ace computer from his uncle in exchange for writing some accounting software for his uncle's farm. After college, he received a private pilot's license. After initially working for a New York–based bank as a software engineer, he worked for a time flying night cargo, but quickly went back to software engineering. "Being a professional pilot didn't pay very well," McAdoo, wearing a blue turtleneck and matching blue stone amulet, said during a recent interview from his apartment in San Francisco's Russian Hill neighborhood. "Being an engineer did."

As an engineer, and then as a tech executive, McAdoo saw it all. One early company where he was an executive, SourceCom, which promised to expand broadband access, went through Chapter 11 bankruptcy proceedings. The experience of having to go to investors and explain why their money was gone left him with more than a little "scar tissue," he said. But the company he went on to lead as CEO, Sentient Networks, sold to Cisco just before the dot-com bubble burst, setting him up well for a new life as a venture capitalist. Both Sentient Networks and Cisco had received investments from Sequoia Capital, the Valley's preeminent venture capital firm, and when Sequoia partner Doug Leone urged McAdoo to join them, he surprised himself by saying yes, and then by finding that he liked the work. "As a founder, you have to focus. The level of focus it takes to will a company into existence is hard to describe," he said. In between startups, he would pull his mental aperture back to the birds-eye view, swooping across the landscape of new things, looking for an edgy idea that he would then zoom back into for another few years. As an investor, he said, "you get to live wide the whole time."

Sequoia Capital was founded in 1972 by Don Valentine, the son of a Teamster who rose through the ranks of technology sales at companies including Fairchild Semiconductor. He viewed venture capital investments in terms of markets. "My position has always been you find a great market and you build multiple companies in that market," he explained. He also believed fiercely in the importance of personal networks. It was through a Fairchild connection that he learned about Atari, which became Sequoia's first investment. And it was through Atari, where Steve Jobs worked as an engineer, that he learned about Apple and came to sit on its board.[2]

By the time McAdoo was ready to leave Cisco in 2000, Sequoia was led by Leone and Michael Moritz, a former journalist who wrote the first book about Jobs, and was riding high from successful investments in companies like Yahoo! and Google. (The latter's recent IPO had just turned Sequoia's $12 million investment into more than $4 billion.)[3] But it turned out that McAdoo had picked a tough time to be scanning the countryside for prey. The years after the dot-com bubble burst were "hell," he said. "The first two or three years, nothing freaking worked." He looked at the carnage around him, the partners who had sat on boards for years watching their companies now go bankrupt, and wondered if he'd joined the party too late. Moritz assured him he had not. "You're coming at the perfect time," Moritz said.

To survive, he and his fellow new partners, like Jim Goetz and Roelof Botha, tried to think of novel ways to learn about promising startups before anyone else. "We'd gotten very aggressive about finding new deal flow, just looking where everybody was not looking," he said. One technique was to forge relationships with the attorneys that startup founders might go to form their companies. "You know, take them out to lunch, take them out to dinner. We even started doing office hours where we take these attorneys and we'd say, 'I don't care what the company does, there's no filter, every young founder you're working with, I will give them twenty minutes.'" One of the attorneys McAdoo cultivated via regular lunches was Mailliard.

Shortly after meeting Altman, Mailliard was sharing such a lunch with McAdoo.

"I've got one for you," she told him. But she warned him that the founders had not graduated from Stanford.

"We made a lot of money from Stanford dropouts, so I'm not sure that's going to be a problem," McAdoo replied.

"Oh no, this isn't Sergey and Larry dropping out of a PhD program," she said. "These three founders are juniors."

McAdoo was even more interested. Mailliard said the company's leader, Sam Altman, was one of the most interesting founders she'd ever met.

"No matter what, even if the idea isn't right for you, you should meet Sam," she told him.

**RADIATE FIT** Sequoia's thesis that mobile phones were about to become the primary means for companies to reach consumers. Mailliard made the introduction, and McAdoo coached Altman and his co-founders on their pitch for Sequoia's partners, whittling it down to fewer than a dozen slides. But in the founders' meetings with Sequoia leadership, it was Altman's digressions that seemed to do the selling. He would spin and tumble through topics, touching down ever so briefly on each one to deliver a little dollop of expertise. "He's just an amazing source of enlightenment and thought leadership about anything," said one former employee of the startup. "For someone like Moritz, that is effing gold." They sold the Sequoia partners on an investment. "We fell in love with Sam and we fell in love with the idea," McAdoo said.

By October 7, 2005, the final term sheet was ready. NEA and Sequoia would split a $5 million investment and get half the company in return. Including about $50,000 from friends and family, the deal valued the company just over $10 million.

"He gave up fifty percent on day one," NEA's Patrick Chung said. In comparison, YouTube, which Sequoia invested in out of the same fund, had given up just 30 percent of its equity to Sequoia in a seed round and Series A of roughly the same size that valued the online video portal at $15 million—a far more typical percentage to give up at such an early stage. In some ways the difference reflected the gap between Radiate's theoretically enormous potential and its ability to show any evidence of it in that moment. While YouTube could show investors that it had user traction because its platform was self-serve and online, Radiate would have to strike deals with telecom companies before it could even have users.

Among the co-founders, Altman owned the most. Initially, he, Deshpande, and Sivo had discussed splitting the equity three ways, but when Altman went to Y Combinator and his co-founders took summer internships instead, they collectively decided he would get 40 percent.

Mailliard remembered being struck by Altman's willingness to give up so much. "His approach is as a humble person and wanting it to be win-win," she said. "From the very beginning, he had equity, but he wasn't insisting on one hundred percent control."

A handful of advisors also got equity. These included David Weiden, a longtime tech executive whose startup had just been sold to Microsoft, who had been hired by the venture firm Matrix Partners to do due diligence on a possible investment in Radiate. Weiden, a Harvard grad with close-cropped hair whose idea of fun is riding a bicycle up a mountain, wore his skepticism on his sleeve. He had begun his technology career at McCaw Cellular, an early wireless company that merged with AT&T, and knew the wireless industry well enough to understand just how much Altman didn't know about it.

"I met Sam, and I liked him and the company—which is rare for me, because I don't like most things," Weiden recalled years later. Altman did not have the qualifications to be the CEO of a company that would be entirely dependent on its ability to forge deals with mobile carriers, Weiden concluded. But he still encouraged Matrix to invest.

"Sam is an extremely unusual person where, right away from meeting him, I thought, wow, this person is extraordinarily inquisitive, thoughtful, insightful, open-minded, savvy, and charismatic—in that order," he said. "I would put Sam up there with Elon, Bill Gates, Patrick Collison, Steve Jobs, and it came across immediately when I met him."

Weiden told Altman that even if Matrix didn't want to invest, he wanted in. In the end, Matrix did not invest, and Altman offered Weiden some stock as an advisor.

**STANFORD AI** Lab's Andrew Ng and Stanford lecturer Andreas Weigend, the chief scientist of Amazon who met Altman at an event run by StartX, Stanford's startup accelerator, also received stock as advisors.

Wilson Sonsini, the law firm, also got a taste. The firm, which saw itself as a strategic partner to its tech startup clients, pioneered a payment structure in which it would defer some legal fees until clients got a funding round. In exchange, Wilson Sonsini asked for the chance to invest some nominal amount, like $25,000, in the preferred stock created by the funding round, on the same terms as everyone else.

As part of the term sheet, Chung and McAdoo would join the

board, and Mailliard would attend board meetings. (Wilson Sonsini's "courtesy discount" meant the firm would not bill for the hours its attorneys spend in the board meetings of the startups in which it had equity.) Chung's seat on the board was hard-won. At NEA, associates like Chung weren't typically considered senior enough to take board seats in the companies they invest in; that distinction is left to partners. But Altman refused to sign the term sheet unless NEA put Chung on the board. Altman became number 3 on Chung's speed dial, after his mother and his boyfriend.

Among the provisos: all the co-founders would have to drop out of college.

**ALTMAN HAD** seen this coming for a while. A few weeks after he had crashed on Rick Pernikoff's frat house floor after his Y Combinator interview, he called his high school friend with a pitch: "You should drop out of school and come work with me." But when it came time to discussing the matter with his parents, Altman studiously avoided the words "dropping out." He and his co-founders pointed to Stanford's generous policy of letting students leave for two years, no questions asked, and framed their move as merely "stopping out."

Deshpande's parents were understanding, even if his mother, Sheila, says, "There is part of me that wishes he had just finished his degree and got some more out of Stanford than he got." Sivo's parents, perhaps because Sivo's father worked in the software industry, seemed more excited than their son was. They let him keep his first quarter's tuition for spending money and shipped him a car. Connie Gibstine, Altman's mother, recalls her son telling her he would take a year off, "but it just kept getting extended, and he clearly wasn't going to do it." Eventually, Altman's parents found peace with it, despite the whispers from certain family and friends. "People said, 'You don't want your kid to drop out of college,' and I was like, 'You know, I'm not worried about Sam. He's not going to drop out and go meditate on a mountain for the rest of his life. That's just not him—he's just so driven.'"

For years afterwards, Altman would have the same recurring dream: "I'm missing classes because I'm working on my startup, or I'm missing some important startup meeting because I'm in class," he said. "That never actually happened to me, but deep into my thirties, I had that dream a lot."

Rick Pernikoff's parents were less enthusiastic. Rick dropped out of MIT during his junior year; he was studying computer science and electrical engineering. Pernikoff also brought aboard his older brother, Tom, who had just graduated from Brandeis with a BS in economics and was just learning to code. While the Pernikoffs' supported their sons' ambitions and knew Altman was smart, they were concerned Rick was giving up his MIT experience.

McAdoo took the co-founders out to Tamarine, an Asian fusion restaurant in Palo Alto, to celebrate. McAdoo wanted to toast his new partners, and suggested ordering a bottle of wine. Somewhat embarrassed, Altman had to gently remind him that they weren't old enough to drink.

NEA celebrated their latest investment in a more age-appropriate way. A few days after the final term sheet was signed, Altman and Sivo led Chung and his partner, Matthew Burt, to a party in the Main Quad of the Stanford campus. The full moon was just beginning to rise over the red tile roofs of the Memorial Church. More than five hundred students thronged the courtyard, many wearing nothing but body paint, as rock bands played on a stage beneath full concert lights and video screens.[4] Although alcohol had been formally banned from the event, the scent of heavily pre-gamed booze and mouthwash filled the air. As Chung noticed one station helpfully offering an array of condoms and Listerine in an attempt to stave off a campus epidemic of mono or worse, he began to panic. *I cannot be seen here*, he later remembered thinking. *I work for a venture capital firm!*

Then, a few minutes before midnight, the countdown began: "Ten, nine, eight . . ." This was Full Moon on the Quad, a decades-old Stanford tradition in which seniors initiate freshmen by kissing them under the first full moon of the school year, which had morphed from its earlier, creepier, senior boys-kiss-freshman-girls incarnation into something like a heavily chaperoned, well-organized, PG-rated orgy. At the stroke of midnight, true to tradition, "all these people were just making out," Chung recalled. "I was like, 'Oh my god, what am I doing here.'"

What he was doing, in fact, was giving Altman one last chance to be a Stanford student. The decision was made. The chance that Altman would return as a senior was low. This was it. (While both Chung and Burt said they have vivid memories of the evening, Altman said he does not recall bringing them to the event or attending his junior year.)

**AT FIRST,** the co-founders worked out of Sequoia's headquarters on Sand Hill Road, a modern, dun-colored building that looks like a cross between a Buddhist temple and a public library. Radiate shared Sequoia's incubation space with the three co-founders of YouTube, all veterans of the Web 1.0 payments company PayPal. Another former PayPal employee was the Sequoia partner who led the firm's investment into YouTube, Roelof Botha. Sequoia announced its $3.5 million investment in YouTube that November of 2005, the same month it invested $2.5 million in Radiate. Within a year, Google would purchase YouTube for $1.65 billion. One of the Radiate employees would later describe the memory of hacking alongside the YouTube founders as "humbling."

The Radiate and YouTube investments were evidence of a great thawing in Silicon Valley, after a period of frozen capital following the bursting of the dot-com bubble. "It was just the beginning of people starting to look at new ideas that maybe didn't have profits and revenues," Mailliard said. YouTube, another Wilson Sonsini client, was a case in point. "Whoever knew that homemade videos on the internet would be any kind of business plan?" she said. "They didn't have revenues, and people had to have some vision to see this and feel that it was worthwhile after having been burned through the years." One banker jokingly imagined Google explaining its YouTube acquisition to its shareholders with an unappealing pitch: "The good news: it has no revenue. The bad news: you've never heard of it."[5]

Seeing that venture capitalists were beginning to invest in startups again—and paranoid that someone else would start a Y Combinator clone in Silicon Valley—Graham began advocating for a winter session in the Bay Area. Initially, he wanted to do it in Berkeley, which he considered the closest analogue to Cambridge. But the tight timeline forced them to move to Mountain View, where Blackwell's robotics company, Anybots, had some extra office space. They renovated it so frantically that the paint on the walls was still wet when the second YC batch arrived in January 2006.[6]

As part of their preparations, Graham and Livingston asked Altman for an introduction to Mailliard, who welcomed them into the same conference room at Wilson Sonsini that Radiate's founders had sat in weeks earlier. Mailliard was instantly smitten by Graham's art-for-art's-sake approach to writing code and thinking up ways to make businesses out of it. "It wasn't just for money or investment," she said. "It was what he loved to do, and that just came through."

Graham and Livingston wanted to better understand how to craft term sheets, how to evaluate investors, and, ultimately, how to make all that secret knowledge available to anyone who wanted to know it, right on their website. "They were really interested in learning more about Silicon Valley," she said. Mailliard and her colleague Carolynn Levy became occasional guests at Y Combinator's offices in Mountain View, where they would discuss term sheets and investment structures at events that Mailliard referred to as "teach-ins." Levy and her husband, Jon, who was also formerly an attorney at Wilson Sonsini, would ultimately join Y Combinator—he in 2008 and she in 2012. There, "C-levy," as Graham dubbed her, would create a standard convertible note that startups started using instead of equity rounds because it allowed founders to close funding rounds gradually, rather than all at once. She would then go on to create a type of convertible note called a SAFE (Simple Agreement for Future Equity), a quick, founder-friendly way for early-stage startups to get funding from investors without investors having to decide what the startup is worth yet. Y Combinator posted the SAFE legal documents on its website, open-source-style, and it quickly became the industry standard.[7]

Through Altman, Graham and Livingston also met McAdoo, who eventually led a Sequoia investment into Y Combinator batches, giving the firm a prime position in evaluating a new crop of pre-vetted, promising startups looking for funding—the all-important concept in venture capital known as "deal flow."

Even before his own startup was off the ground, Altman proved to be the indispensable startup connector. "He knows everyone," Graham wrote in 2012. "He has not only done countless introductions for alumni, but did most of the initial intros in Silicon Valley for YC itself. . . . YC now does many intros per day, but if you follow the tree back to the beginning, Sam was the root node."[8]

The arrival of Y Combinator in Silicon Valley that winter accelerated the rush of optimism in the tech sector. Altman and his co-founders had almost perfect timing. "They were part of that initial wave," Mailliard said. "It turned into a huge tide, and a lot of that was fueled by Y Combinator."

**RADIATE LEASED** offices in a modest two-story tan building at 3250 Ash Street in Palo Alto, near the busy intersection of El Camino Real and Page Mill Road, biking distance from the Stanford campus. Its primary feature was a spiral staircase that jutted dangerously into the main room. "They tried to pad it down with pool noodles because a number of the boys would not be paying attention and would smack their foreheads walking into it," recalled Nini Tang, the company's office manager. "It was a hazard." There was a Ping-Pong table and a flat-screen TV primarily used for playing *Halo 3*. The Pernikoff brothers brought their guitars and their French bulldog, Bruno, who contributed to the already ripe smell caused by more than one of the engineers routinely coding through the night and sleeping in the office.

"It operated like a frat," recalled Min Liu, who joined later to lead marketing. When Tommy Tsai—a brilliant programmer who had served as Loopt's director of engineering and was the engineer most likely to be found passed out in front of his computer in the morning—got sick, Tang had to book his doctor's appointment and drive him to the doctor. "I really felt like the office mom," she said, despite having recently graduated from Stanford herself.

At the time, Palo Alto was the center of the tech universe; the industry had not yet migrated to San Francisco. On any given night, beneath the eerie pink Peninsula skies, there were multiple mixers and events sponsored by startups and venture firms where the talk was about what's next in mobile or software-as-a-service. Then Radiate's team would head over to Antonio's Nut House, a dive bar stinking of cheap beer and the peanut shells that littered the floor.

Otherwise they'd go to parties hosted by fellow startup founders. The Radiate crew could sometimes be found at the curious house/office rented by Aaron Levie and Dylan Smith, who had dropped out of college to co-found the cloud storage company Box the same year and threw parties featuring foosball and beer pong. Box's house/office had a loft above it with a few mattresses on the floor where guests could just pass out. Or they'd be at the house of early Radiate engineer Evan Tana, whose roommate, Mark Slee, had signed on as one of the early engineers at Facebook and sometimes invited his boss, Mark Zuckerberg, who was only a year older than they were. "It was fun to go to parties with the Facebook people," recalled one early Radiate employee.

**AMID ALL** the youthful energy, Sequoia liked having what employees at its portfolio companies called "adults in the room." In Botha's September 2005 memo trying to convince his fellow Sequoia partners to invest in YouTube, he wrote "We need to help the company quickly hire a CEO and VP of [Business Development]/Sales."[9] In its press release announcing the YouTube investment a few months later,

Sequoia bragged that it had not merely funded but "organized" companies that accounted for 10 percent of the value of the NASDAQ.[10]

Sequoia used a similar playbook for Radiate. While still incubating the company in its offices, the venture firm helped recruit several senior executives that McAdoo and his partners hoped would help mentor the young founders. One of them was Brian Marciniak, a seasoned technology sales executive in his mid-forties from Georgia who had been working as a VP of sales since the early 1990s. "Some of the folks in the carrier business needed to see gray hair," McAdoo said. "Brian played golf, wore the right kind of sweaters, was a country club member." Gray hair notwithstanding, Marciniak believes he got the job because he came off as less bossy than the other candidates. Altman "didn't want to hire a VP of sales that was going to tell him what to do," Marciniak said. After all, there was no question who the master salesman was. "He always makes it sound like he knows everything. He really doesn't. His level of knowledge might be a 3, but he talks and he makes it sound like it's a 10," Marciniak said of Altman. Marciniak found Altman kind and collaborative, and didn't mind that he multitasked through every meeting, his laptop open and his PalmPilot, its keyboard stripped of paint from constant use, operated by feel.

Another "adult" Sequoia brought on was Mark Jacobstein, an affable serial entrepreneur in his mid-thirties with a Harvard computer science degree. He had most recently co-founded the Sequoia-backed mobile gaming company Digital Chocolate with Electronic Arts founder Trip Hawkins. The experience had convinced Jacobstein that mobile devices would be good for a lot more than just games, but there wasn't much opportunity to explore other opportunities in a company run by the founder of one of the most iconic video games companies ever. (Hawkins, an early Apple employee, had founded Electronic Arts with backing from Sequoia and gone on to make games including *The Sims* and *Resident Evil*.)

In the small, eight-person conference room on the second floor of Sequoia's headquarters, Jacobstein met with the five co-founders, who now included the Pernikoffs, and he and Altman hit it off. They shared a common belief in the power of mobile devices to bring people together in the real world, rather than merely entertain them while they were on the toilet or waiting in line at the bank. Altman spoke about how, in college, he would often have free time and want to find his friends on campus before class, and pitched his company as something that was about "enhancing serendipity," Jacobstein recalled. When he later explained the pitch to his mother, she replied, "Oh, this is a cure for loneliness, right?"

Unlike Marciniak, Jacobstein's hiring was what McAdoo called "vocational," meaning he was "brought in to fill some gaps for Sam." As Jacobstein saw it, "I was supposed to be the Sheryl to his Zuck," he recalled, referring to the long-successful relationship between Facebook Chief Operating Officer Sheryl Sandberg and CEO Mark Zuckerberg, fifteen years her junior.

Jacobstein was given the title of executive vice president, corporate development and marketing, which meant he was there to chaperone the still-teenaged Altman through meetings where he might have trouble being taken seriously, such as with wireless carriers or regulators. Together they came up with an approach of trying to allay the privacy concerns around their app by aggressively reaching out to anyone who might plausibly complain—the Electronic Frontier Foundation, the ACLU, battered women's shelters, the National Center for Missing & Exploited Children—and bringing them in on design decisions. "By involving them in the process, it helped ease their concerns," Jacobstein said.

It was a lesson that Altman would absorb and apply to remarkable effect at OpenAI. In the meantime, when it came to landing deals with wireless carriers, it wasn't clear that Altman really needed that much help.

CHAPTER 6

# "WHERE YOU AT?"

**FOUNDED IN AUSTRALIA IN 2000 AS A BRAND AIMED AT THE** country's surfing-obsessed youth, Boost was the first "virtual" mobile network (known in the industry as a mobile virtual network operator, or MVNO) to run atop Nextel's network, which had decided early on to carve out a niche among truckers and thus pushed all of its phonemakers to put GPS chips in their phones. Boost set up offices in Irvine, California, a center of American surf culture, and began marketing its prepaid service to a mix of young surfers and urban youth.

"A lot of young, urban kids couldn't get contracts, so they needed prepaid devices," said Lowell Winer, the director of innovation and business development at Boost at the time. "Frankly, a lot of drug dealers used Boost, because there was no contract. There was no name on it."

To attract a young demographic, Boost hired a roster of surfers, BMX riders, MCs, rappers, and UFC fighters to spread its message. Sometimes they would show up on the second floor of Boost's stripmall-style offices on Friday afternoons, when beer was served, and at least once someone rode a motorcross bike up the office stairs. It was all a stark contrast to the Kansas City headquarters of Boost's parent company, Sprint, that Boost employees referred to as "Shawshank."

By early 2006, Winer was convinced that Boost needed to use its GPS capabilities to differentiate itself from the competition, and had put out a request for proposals for an idea that sounded a lot like Radiate: a "friend finder" that would let you know when your friends are close by.

"I scoured the universe for companies that might help us build this," Winer recalled. Most of the responses that came back were from companies that wanted to build a platform, rather than help a wireless company make a specific offering to its customers. Winer had already selected one that was pretty close to what he wanted and was moving toward finalizing the paperwork when he got a call from Altman.

Altman had found out about Boost's request for proposals and thought Radiate could build what Boost wanted. Winer explained that Boost was already close to a deal with a competitor.

Altman's heart sank. "Whoever Boost works with, Sprint will work with. And whoever Sprint works with, Verizon and AT&T will as well."[1]

There was one small opening, however. The competitor had said they would not be able to create one feature that Boost really wanted—a status message alerting you if your friend was within, say, five miles of you. The Radiate team stayed up all night building it.

"I think I went to sleep at four, I slept till six, I got on a flight at seven to Orange County where Boost was," Altman recalled.[2]

Altman showed up unannounced to Winer's office, asking for just ten minutes of his time. Winer ushered them into a conference room and watched Altman, in cargo shorts, fold himself up "Indian-style" in a chair that seemed too big for his body and proceed to command the room. "He probably weighed a hundred and ten pounds soaking wet, and he's surrounded by all these middle-aged adults that are just taking in his gospel," Winer recalled. "He radiated confidence."

After about an hour, Winer walked out of the conference room and into the office of Boost's VP of Product, Neil Lindsay, and explained that Boost was going to have to change course and go with "some guy that had just showed up at our office," Winer recalled. The Radiate team had proven, in being able to build the feature that Winer wanted, "that these guys were agile, these guys got it." Winer called Sequoia for a character and financial reference and then decided to give the young company its first wireless deal.

Altman learned a fundamental lesson from the deal: "The way to get things done is to just be really fucking persistent."[3]

Winer recalls clearly his brush with Altman, all these years later. "Anyone who came across him at that time wished they had some of what he had. There was a sense that anything was possible. He's quite optimistic—decisive and optimistic. Rarely is he skeptical about anything."

**THE BOOST** deal was so fundamental to Radiate's future that they decided to rename the company. This was possible, in part, because all parties had decided to keep the company in "stealth mode," declining to put out a press release about the Series A, knowing that it would take time to land the wireless deals and build the technology that telecoms needed. They toyed with "Flipt," but ran into trademark issues. Facebook was just beginning to take off, and there seemed to be something lucky about having double-os: Facebook, Google, and Yahoo! (and eventually, in HBO's satirical *Silicon Valley*, the fictional Hooli). They kicked around a bunch of names and settled on one that sounded like an extension of its closest, indispensable partner, Boost. The name of the product they would spend the next nine months building would sound like an echo: Boost Loopt. And the company itself would change from Radiate to Loopt.

**ALTMAN LAUNCHED** his first product in Times Square on a strangely warm Tuesday in November 2006, surrounded by rappers. Wearing jeans and a hoodie with a cartoon blood splatter design, he stood on the temporary stage erected in the middle of the urban carnival and, once again, held up his flip phone—this time showing a live map of Midtown Manhattan with little colored circles that showed his and his colleagues' locations. Behind them, a phone's tiny, pixelated map was blown up on a humongous backdrop to convey the product's features to the passing hordes of tourists and commuters. Sharing the stage were nearly all of Loopt's co-founders and early talent—Deshpande, the Pernikoffs, Evan Tana, and Jacobstein—as well as DJ Kay Slay and Fabolous, who were on contract with Boost's marketing team. Sivo, as usual, skipped the stage.

Nine months earlier, Boost and Loopt had signed a contract to preload Loopt's app on every Boost cell phone for the next four years.[4] In the ensuing months, Boost and Loopt worked shoulder to shoulder to build and market the service. Altman managed the Boost relationship as if his company's life depended upon it—because it did. "He picked up the phone when I called," Winer recalled. As it happened, long before they had met Altman, Boost had selected a tagline that was remarkably similar to Altman's original pitch. "Where you at?" rapped Kanye West, Ludacris, and The Game in unison in a 2004 Boost Mobile commercial known as the Boost Mobile Anthem, the first of many Boost Mobile ads featuring top hip-hop talent.[5]

"This was Kanye just getting started," recalled Darryl Cobbin, Boost's chief marketing officer, who had joined the little-known brand after a long career at The Coca-Cola Company helping run marketing for Sprite, one of the first brands to use hip-hop in its ads. Ludacris was available because Pepsi had just dropped him after Bill O'Reilly had attacked him on air as misogynistic because of his rapping about having "hoes in different area codes."[6] And as for The Game, "nobody had heard of him outside of Compton." In the ad, the three rappers were talking to each other, describing where they were and trying to get a beat from Kanye West.

Now, a year later, The Game was wandering around Times Square with an army of young women in bright green sweatshirts pulling green balloons that read, in the twee all-lowercase writing that would become Altman's signature, "boost loopt." "Loopt is the physical manifestation of 'where you at,' literally," Cobbin said.

For the launch, Boost released a TV ad campaign in which groups of young friends, blown up like beach balls to resemble the blinking colored dots on the Loopt friend finder map, told each other over their cell phones some version of: "Yo, where you at?" Back at their office in California, the Loopt team dropped everything and stared, wide-eyed, the first time they saw it run on television. They might not be rich yet, but they were already kind of famous.

To ensure there were signups, Boost began offering the service to users for free in September 2006, and would continue to do so through the end of the year. After that, the price would rise to $2.99 a month. By the release, Loopt had 35,000 users.[7]

The only key person missing from the stage, besides Sivo, was Boost's director of business development, Winer. He had been diagnosed with cancer shortly before the launch, and was undergoing treatment. That left his colleague, Craig Thole, Boost's director of value-added services, to do the talking for Boost. "Social networking is a hot space and people are looking to do it in the mobile space," he told *The Orange County Register*, which concluded that Boost Loopt's features amounted to "a roving MySpace-like community on steroids."[8]

In Altman's telling, it was much more than that. "This is about deeper communication," he told the *Register*. "It really brings people together. Instead of the virtual world of social networking, this brings back the human touch." In response to the question of how many people in the service's target demographic of fourteen-to-twenty-five-year-olds would keep using Loopt once they had to actually pay for it, Altman offered nothing but optimism: "We think a lot will," he said. Boasting that he was already in talks with other mobile carriers, he predicted that, for young people, "this has the potential to replace the phonebook as the default application."[9]

**AS ALTMAN** predicted, Loopt's deal with Boost led to more meetings with wireless carriers. As the fall of 2006 turned to winter, David Weiden, the Loopt advisor with wireless industry connections, arranged a meeting with Cingular Wireless, which would soon become part of AT&T. Weiden had a direct line into Cingular's CEO, and used it to put Altman in front of the most important executive for any company trying to build a business atop wireless companies' mobile location data: the vice president of data. "The VP of data was, like, God," Weiden said. The meeting was to be held at Cingular's Atlanta headquarters, the kind of place where most people wear suits. "Jeans, I don't think, were allowed in the building," Weiden recalled. He had neglected to mention this to Altman, who arrived punctually at 8:50 a.m., ten minutes before the meeting, in what Weiden recalls was a T-shirt and shorts in December. (Altman says it was a collared shirt and jeans; the VP of data, Jim Ryan, trusts Weiden's memory more, but mostly only remembers that people in the room were "aghast.") It turns out that Ryan liked a little rebellion.

Altman erased all concerns about his wardrobe with his pitch. "It was a very radical idea in its initial notion," Ryan recalled. He had led the creation of music and video products that sat on Cingular's mobile data infrastructure, but figuring out what to do with the information they had about where their customers were was tricky. "How much do we want to tell people that's what we know?" he and other Cingular executives wondered. Blowback driven by privacy fears was a real risk. But Altman brushed these concerns aside. "I loved people like him because they didn't care about any of that," Ryan said. "They just saw what could be."

Altman pitched a far more aggressive vision for what mobile companies could do with their location data. Loopt users had the option of sending their coordinates to friends every fifteen to twenty minutes, which went far beyond the previous winner in the space, Dodgeball, purchased by Google in 2005, which required users to manually push a button on their "Buddy Beacon" to send their location. Altman's solution to the privacy concerns was to allow only "good and trusted friends" to view movements in real time, and require all invitees to answer questions such as "Would you lend them your keys to your house to feed your dog?"

While Altman's pitch was happening, Cingular was being absorbed into AT&T, which eventually agreed to a test and ultimately signed a deal to invest in the company, as eventually did Verizon, which placed a representative on Loopt's board. As Altman knew from his very first meeting with Boost back in Irvine—if you get one wireless carrier, you can get them all.

**BOOST'S MARKETING** campaign helped propel Loopt to more than 100,000 users within the first three months of its launch. By mid-2007, Loopt had graduated from the "sub-brand" of Boost to its first major US wireless carrier, Sprint.[10] But also around this same time, *The Wall Street Journal* and other publications were beginning to notice that Loopt had declined to update its user numbers since the sugar rush of free signups ended in January.[11]

"We were escalating on downloads, but the problem was, we had really high churn rates," said Marciniak, Loopt's VP of sales. "Within ninety days, seventy percent of users were churning off." The problem came down to a fundamental misunderstanding of users' appetite for privacy. While competitors like Dodgeball required users to check in to share their location, Loopt had differentiated itself by being "always on"—users had to turn off the service completely to make their little dot go away on the map. "They liked to see where everyone else's location was, but they were cautious in posting their location," Marciniak said.

Altman decided to solve the privacy issue by running straight at it. Rather than downplay concerns, he seemed to relish talking about them to anyone who would listen. "It's one of those things, the more you think about it, the more ways you can figure out how a creep could abuse it," he told *The Wall Street Journal* in 2008. "I think people realize that unlike a telemarketer call, which can be annoying, a location-based service could be an actual physical safety risk."[12] He played up these fears to pitch Loopt as particularly safety-conscious, with strict safeguards to prevent misuse.

Some of these safety ideas emerged from a tense meeting with Sprint general counsel Len Kennedy, who had been named to the job after Sprint's merger with Nextel in August 2005. "Getting this approved by

Sprint's legal team was brutal," recalled Winer. Altman flew to Reston, Virginia, the home of Nextel, and sat with Winer across the large rectangular table in the company's board room as Kennedy grilled him for two hours on how he could assure Loopt wouldn't create a data breach that would take down the entire wireless company. "We were concerned," Kennedy said. "That was in the early days of doing things, where the privacy aspect of things was a little bit vaguer than it is now." Altman agreed to some concessions: He would tweak the software so that users could no longer troll outside their friend network for other Loopt users whose location might be visible. Loopt would be a way to interact with existing friends, not find new ones.[13] And he agreed to add a few more rules to make the service safer: No one under fourteen could sign up, new users would get frequent messages for the first two weeks reminding them that they were being tracked, and the registration process would involve scrolling through pages of privacy notices and disclaimers. That was enough for Kennedy. After one final reference check from Sequoia, he gave the app the OK, and became the latest convert to the gospel of Altman.

As soon as Loopt was live on Sprint, Altman and Jacobstein took their privacy education show to Washington, DC, sponsoring a forum at the Congressional Internet Caucus, where Jacobstein framed the current state of location-service regulation as the "Wild West."[14] The idea was to suggest to lawmakers that they could regulate the industry now, and that Loopt could be a helpful partner in deciding what that regulation should be. In later years, Loopt's chief operating officer, Brian Knapp, testified before the House Committee on Energy and Commerce Subcommittee on Communications, Technology and the Internet, pointing to a long list of organizations that Loopt had solicited suggestions from to make its service safer.[15] Altman would join Knapp for his frequent trips to Washington to meet with members of Congress, the White House, the FCC, and nonprofit groups. The goal, as Altman put it, was to "keep from getting legislated out of business."[16]

When it came to lawmakers, Altman was largely successful. To this day, the United States does not have any kind of unified framework governing how companies must treat citizens' digital data. But consumers remained uneasy about broadcasting their location, even to friends.

"We were so obsessed with live location sharing because it was such a cool product, it was so exciting, that we lost sight of what users actually wanted," said Min Liu, a former Loopt employee. "They found location tracking creepy AF."

Nevertheless, Loopt kept expanding. By the summer of 2007, Loopt had more than thirty employees and needed more space. The company moved into former law offices with red tile roofs on the main drag of El Camino Real in Mountain View, around the corner from Castro Street where much of Silicon Valley descended for curry and burritos at lunch hour. Loopt engineers sat—and often slept—upstairs, while marketing and product management sat downstairs, along with Altman. An early photo tour of the new space on the Loopt blog revealed nothing at all on his desk—except for a single bottle of Excedrin.[17]

There was plenty to be stressed about. For all of Altman's pitching prowess, the company was finding it difficult to bring on more wireless carriers. Some of their meetings suffered from what Jacobstein came to call "good meeting-itis." "'Hey, that's really interesting, tell me more,' is not the same as a signed deal," he said. Executives were more than happy to listen to Altman's vision and ideas—and then apply them to their own products. "We were teaching them a lot," Jacobstein said. Nearly a year after Loopt's first wireless product went live, the service was still on just one wireless carrier—Sprint, the third-place one with declining market share. Jacobstein began to feel like his hiring had been premature. Loopt was still looking for what the startup world likes to call "product-market fit." Meanwhile Twitter, which launched around the same time, was blowing up. That had led to some grumbling within Loopt. "In effect, the conversation was, 'Isn't that interesting? We could build that in a weekend,'" Jacobstein recalled. "And it was almost scornful, 'What a simple thing they built.' But that was the point. They built something that could be built in a weekend." Loopt, by contrast, had to build custom APIs for each wireless company to have access to location information, and negotiate deals for access to each one. Bored and unable to use his scaling skills, Jacobstein decided to move on.

"In retrospect, one of the most interesting questions is, how on earth could such a talented group work on such an intractable problem for so long that was never going to be solved? It turns out that the demand for this service, this product that Loopt was building, never manifested. And pre-smartphones it was never going to because the idea that you can have a critical mass of people who use this service, when you had to deploy it on literally one hundred different types of devices on five, six or seven different mobile networks was, in retrospect, kind of ludicrous."

Still, Jacobstein departed Loopt with great admiration for Altman. One scene remains fixed in his mind: About a year into Jacobstein's tenure, Weiden took Jacobstein and Altman out to lunch to pick their brains, and asked Altman what he was thinking about besides Loopt. Altman had two answers: a cure for baldness, and nuclear fusion. Jacobstein scoffed to himself at the time. "What do you know about fusion? You're a nineteen-year-old sophomore computer science dropout, not a PhD in nuclear physics." But twenty years later, Altman would end up backing one of the handful of nuclear fusion startups that might actually make the technology work. Jacobstein realized that if Altman could even imagine the possibility that something might work, he was able to convince himself he could do it and then convince others—particularly investors—of the same.

"There's a blurring between 'I think I can maybe accomplish this thing' and 'have already accomplished this thing' that in its most toxic form leads to Theranos, but in its healthy form leads to people trying really ambitious things," Jacobstein said.

Jacobstein found Altman to be a healthy example of this tendency, but not everyone at Loopt did. After Jacobstein left, much of the senior management approached the board, pressing for Jacobstein to replace Altman as CEO. Their complaints included that Altman sometimes said things that weren't true, and that he was hard on employees. The board rebuffed their complaints and told them to get back to work.

"I was, like, infamously difficult to work with when I was eighteen or nineteen," Altman told investor Reid Hoffman during a podcast years later. "If you're the founder of the company and you wanted to work one hundred hours a week and be super focused and productive that's cool, but most other people you hire, especially as you get bigger, have other lives and you need to understand that."[18]

Jacobstein's role as Altman's babysitter was filled by Brian Knapp, an extrovert with a golden Prius he would gun out of the Loopt parking lot, tires screeching. Knapp took on a hybrid role as general counsel and chief operating officer. He had been working with Loopt for several years as an intellectual property lawyer in Wilson Sonsini's tech transitions group, and defected to the client in May 2007. Much of Knapp's time was devoted to meeting with wireless carriers and lobbying for favorable privacy regulation on Capitol Hill. But when he was in the office, the thirty-year-old Knapp was just one of the boys.

In addition to frequent rounds of *Halo 3*, the Loopt team took to wrestling to ease some of the mounting tension. They paired people by size, with the diminutive Altman facing off against Sam Yam, the company's first server engineer and mobile engineer. On one occasion, Tom Pernikoff challenged Knapp to a wrestling match and ended up with a mild concussion. As is befitting of a babysitter, Knapp let there be no question who the alpha dog was.

CHAPTER 7

# FROM "WEAK" TO "COOL"

**LITTLE DID SAM ALTMAN KNOW THAT THE ENTIRE TIME HE HAD** been working on a product dependent on striking deals with wireless carriers, Steve Jobs was just down the road at Apple's headquarters secretly building a device that would blow up the wireless industry.

Before Jobs unveiled the iPhone at the Macworld conference in January 2007, phone makers were the serfs of the mobile industry. The powerful wireless networks—Cingular, Verizon, Sprint, and T-Mobile—told them what to build and decided what applications would come preinstalled. The phones were often free when you signed a contract with a wireless carrier. "We kept a walled garden," recalls Charnsin Tulyasathien, who ran a team focused on location-based services at Sprint at the time. "We put you through the wringer, in terms of getting you on our network." Jobs changed the dynamic by inking a deal with Cingular Wireless, which later became AT&T, that gave Apple total control of the making and marketing of its iPhone in exchange for five years of exclusivity on Cingular's network and a cut of iPhone sales and iTunes revenue.[1] He then drove a bulldozer through the industry when he announced in October 2007 that Apple would release a software development kit, or SDK, that would allow anyone to build an iPhone app. "I remember thinking, 'Wow, they are going to totally destroy what I have been a part of,'" Tulyasathien said.

James Howard could think only of building. An Apple fanboy and hobbyist Mac developer, he had gone to college at the University of Washington, where he worked as a research assistant at Intel Research Seattle, one of six labs that the chipmaker had seeded within universities. There he had met and collaborated with his fellow student Fred Potter on research papers with titles like, "Voting with Your Feet: An Investigative Study of the Relationship Between Place Visit Behavior and Preference," which tried to shine a light in the darkness for the nascent mobile location industry.[2] When Potter went to work on a commercial version of these ideas at Loopt, Howard asked the lab's director if he should follow him. The director, who had gone to University of California at Berkeley, Stanford's staunch rival in football and occasionally also academics, said he thought it was a mistake because "Stanford kids can't code." Howard wanted to go anyway, sure that Loopt was the place he could build a real app for the iPhone. Potter secured him an interview in May 2007.

During the interview, Howard asked Altman what he thought about the iPhone. Altman agreed that it was a big deal. Howard said he wanted to figure out how to develop for it, even though Apple hadn't announced that that would be possible. Altman agreed with that idea, too. Howard joined a few weeks later, just in time for Apple's much-anticipated annual Worldwide Developers Conference, when developers were hoping Jobs would unveil some way for them to make software for this magical device. Jobs, who valued control over just about all else, fought the inevitable for months, presenting at the conference what he unconvincingly hawked as a "sweet solution"—developers could build web apps on Apple's Safari browser. Jobs was met with a few awkward coughs and a near-total lack of applause. Howard was deflated. There was no way to build the app he was envisioning on a browser.

The rest of Loopt barely noticed. That summer, they were finally moving on to their first major carrier, Sprint. Deshpande, who ran the client team, was focused on signing more carriers and building features to please them. Howard busied himself amid the servers that Loopt ran its system on (these were the days before the widespread adoption of cloud platforms like Amazon Web Services), and in his spare time, Howard wrote little apps on his Mac computer that worked with the Loopt service. He was biding his time until Apple released a true SDK for the iPhone. Not long after the iPhone went on sale in late June 2007, however, enterprising developers began to "jailbreak" the devices and share the code for doing so online. By the end of the summer, Howard had a version of the Loopt iPhone app running. Altman spent a lot of time hanging out in Howard's office, watching his demos. He eventually installed some of Howard's jailbroken apps on his personal iPhone and promised Howard he would do whatever he could to get the Loopt app onto the iPhone.

Initially, Jobs tried to shut down the jailbreaking. "Steve ran into my office, furious," said Scott Forstall, who led Apple's software development team for the iPhone. Jobs wanted Forstall to immediately ship a software update that would prevent any more developers from figuring out how to build third-party apps for the iPhone. But by October 2007, the quality of the apps that were being built via jailbreaking was beginning to change Jobs's mind. He instructed Forstall to build an App Store in time for Christmas, just two months away. Forstall instead suggested that Apple should announce its plans at a keynote the following March, and work with developers to have hundreds of apps in the App Store by the time the iPhone 2 launched the following June.

Given that Sequoia's founder, Don Valentine, sat on Apple's board, it should not have been that hard for Loopt to get in on the action. But nothing was easy when it came to dealing with Steve Jobs. Altman asked McAdoo for help, and McAdoo began by going to Valentine for advice. *What was the best way to approach Jobs about Loopt?* Valentine, then in his seventies, paused and sighed. "You know, I'm not really sure," he said. "Go talk to Mike." Michael Moritz, who was running Sequoia at the time alongside Doug Leone, had written a well-received history of Apple in the 1980s during his earlier career as a correspondent for *Time* magazine, though an article for the magazine based on his reporting had enraged Jobs and all but severed their relationship. (The article had included an interview with the mother of Jobs's daughter, Lisa, whose paternity Jobs denied at the time, and Moritz later wrote that his reporting had been "poisoned with a gossipy benzene" by a New York editor used to handling rock music copy.[3] Jobs later reconciled with Lisa, who changed her name to Lisa Brennan-Jobs.) Nevertheless, Moritz was considered the most astute student of Jobs at the firm, and he didn't hide the challenge before them: Jobs hated social networks. "We're going to have to sell Sam," Moritz told McAdoo. Moritz made the introduction to Jobs, with McAdoo chiming in on the email chain about Altman's personal story. "The youngest founder we've ever invested in at Sequoia, Stanford dropout, you know, Steve can relate," McAdoo recalled saying, rightly guessing the story would appeal to the Reed College dropout running Apple. Jobs agreed to take a look at Loopt. A few weeks passed without an answer. Eventually McAdoo nudged Jobs, asking what he thought of Loopt. Jobs responded with just three words: "It was weak." McAdoo picked up his laptop and walked down the hall to Moritz's office to show it to him. "What do I do with this?" he asked Moritz. Moritz, taking a note from Valentine, just shook his head. "I don't know," he said.

Back at Loopt, Altman and Howard were undeterred by Jobs's assessment. They knew the current version of Loopt was not representative of what they could do. "It was weak" became their motivational motto.

Jobs might not have liked Loopt, but the twentysomethings on the product and engineering teams loved it. In November 2007, Altman received a cryptic email from the team building the iPhone asking if Loopt would come in and consult on what Apple might do with its iPhone SDK to enable Loopt. Everything about the meeting had to be handled with the utmost secrecy. Altman picked a small team, including himself, Howard, Tsai, and Sivo—whom he was still dating, though many at the company did not know it—and told no one else at the company. Over the course of two meetings, this small group met with the iPhone's marketing and developer relations teams, and then with the engineers on the SDK team, all of whom took detailed notes.

The meetings in November 2007 won Loopt no special access to the iPhone SDK when it was finally made available in March 2008. But Apple did offer Loopt something that turned out to be very valuable:

they would test Loopt's iPhone app internally, give feedback on it and offer them a chance to be part of the keynote speech of Apple's Worldwide Developers Conference the following June. The deal came with some time pressure, and no guarantee of any kind of prominent placement on the device. But after some discussion, Altman and McAdoo agreed it was worth investing in building the software to elicit Apple's feedback. "I think we had one hundred Apple employees at one point running around using the [Loopt] app on an iPhone," McAdoo said. The Loopt team met with Forstall, who gave them ideas for improving Loopt's app experience. "We were enamored with the application," Forstall recalled.

To get up on the coveted stage of Apple's developer conference, however, Altman had to pitch Jobs himself. The Apple developer relationships team helped Altman and Howard come up with a script for their presentation and drilled them on it. Eventually they brought them to a building on Apple's Cupertino campus. Altman and Howard waited in a lobby with the Bösendorfer grand piano that Jobs had bought to inspire the original Macintosh team—a symbol of Apple's emphasis on beauty—and then were shuttled into an auditorium. In the center of the seats sat Jobs, surrounded by a few assistants, dressed not in a black turtleneck, as Altman and Howard expected, but in shorts and a T-shirt. Their mouths were dry from nervousness. On stage, Altman did the talking, while Howard ran through the demo on the phone, which was mirrored on a big screen. When they got to the end, they just stood there, staring. After a pause, Jobs said just one word: "Cool."

They were ecstatic to be upgraded from "weak" to "cool," but still weren't sure what that meant. Soon after, a representative from Apple's developer team called to let them know they had made the cut, so long as they made every change Apple asked for and all went well during rehearsals. They spent a week drilling with the people at Apple. During breaks, Howard would tweak the code or call up Sivo or Tsai to help with changes on the server side. No one at Loopt outside that small circle was aware, at Apple's behest. "It was only two days before launch that it was even known inside the company that this was something that was happening," said Liu, Loopt's marketing executive.

The night before the big presentation, Altman called Howard at home. Altman was anxious. Jobs was his hero. He didn't know what to wear and wasn't sure he wanted to go through with the presentation. Howard started to panic as well. Eventually, they agreed that Altman would wear two nested polo shirts like he had worn for a joke at a party once, because it would be something that people would remember.

The developer conference that June was the first one to sell out. Altman found himself on stage at the Moscone Center West in San Francisco, legs planted far apart as if he were preparing to catch a football, in a neon pink polo shirt with a second, neon green polo blooming from inside it like the petals of a flower. "This is the best version of Loopt that we have ever made, and by far the best device we have had the opportunity to work with," Altman said as he pitched the new Loopt app, revamped to show off iPhone-specific features. Users would now be able to pinch the touchscreen to zoom out on its map and also launch a call from the app with a single click. His demonstration showed a few friends nearby, and one particularly close at a café she had previously tagged as "the cutest café." Altman texted her through the app to ask if she was free for lunch. "Location, plus a contact list, and information about cool places means you never have to eat lunch alone again, or at a bad place," he said. The presentation concluded with a note that at launch Loopt would be free to iPhone users.[4] Loopt was so valuable to Apple as a means to show off the iPhone's location capabilities that Apple even paid for a national television commercial showing how Loopt worked on the iPhone.[5]

His June 2008 presentation made twenty-two-year-old Altman an instant star in the tech world, if also something of a laughingstock in the tech blogosphere for his bold fashion choice. If the world did not quite yet know his name, they at least now knew the "double-popped collar guy."

Inside Apple, Loopt was an unqualified hit. Downloads surged, and when Apple prepared to launch the iPhone internationally a few months later, Jobs wanted assurances that Loopt's location technology would work in all the languages and countries he had publicly announced the iPhone was coming to. Given the state of the technology at the time, this was a tall order. When Jobs arrived at one meeting and heard that Loopt could not offer its services as widely as he was expecting, he unloaded on Altman.

That evening, Altman showed up for dinner with McAdoo at their favorite sushi place, still shaking from the confrontation. "I just had the hardest meeting with Jobs," he told McAdoo. (Both McAdoo and Chung recall Altman telling them that Jobs threw a pen at his head. Altman denies there was a projectile.) "Steve was always on and very demanding, which got us great work and great products," Forstall said. "I *have* seen him throw things at people."

Looking back, McAdoo said the months spent working with Jobs left an imprint on Altman. "There was something about his interaction with Apple in that period that definitely shaped him in a really positive way, as an entrepreneur, and as a leader of high-performance people," McAdoo said. "You don't let go of people because they are prima donnas in that world. That's not a thing. Half of the folks that operate at that level would in some way be considered prima donnas. You need to learn how to harness them. And Sam's ability to do that greatly increased as a result of watching from the inside out how Apple worked in the early years of the iPhone."

FOR A brief moment, Apple lifted Loopt near the realm of social media royalty. The Apple television ad featuring Loopt had driven so many Loopt downloads on the iPhone that they surpassed those of Facebook and MySpace.[6]

But the timing could not have been worse. That fall of 2008, after Lehman Brothers collapsed, Sequoia had gathered the CEOs of its portfolio companies together for an emergency meeting on Sand Hill Road and told them the era of burning venture money for market share was over for the foreseeable future. The companies were advised to start cutting their way to profitability if they hoped to survive. The slide deck featured a tombstone with the words, "RIP Good Times."[7] In November 2008, Loopt hired the investment bank Allen & Company to either sell the company or get more investment.[8] Given the financial climate, a sale was not especially likely.

Earlier in the year, before the wheels had come off the global economy, Facebook had made a verbal offer to buy Loopt for around $150 million, more than three times what it would eventually sell for. Sequoia's Michael Moritz had asked Altman what he planned to do about it, and Altman said he was going to pass. He wanted to build a large standalone company. "That's the right answer," Moritz replied.

But as Allen & Company began pitching Loopt to investors and potential suitors in November 2008, it was valuing Loopt at more than $500 million, a number that drew guffaws from the tech press. "Now, I like Loopt as much as the next guy, and am bullish on the eventual success of location-based services, but half a billion dollars, in this market, for a network that is far from established?" wrote *VentureBeat*. "Come on."[9]

By the spring of 2009, Loopt had landed a funding round valuing the company at $150 million—the same as it had been in its last, pre-crisis funding round—a respectable feat in the depths of a financial crisis. The $7.5 million round was led by DAG Ventures, a middle-stage investment fund that frequently invested alongside Sequoia, and contained some favorable terms for DAG that would allow them a larger share of the proceeds from a future sale. Sequoia and NEA quietly came along for the ride.

Years later, Moritz stood by the advice he gave Altman. "There was a moment in Loopt's trajectory where the future looked very, very promising, and therefore an early sale would have seemed premature," he said. "Even though investors frequently get tarnished with the brush of wanting to bail out early, in my experience, it's more often the founders who want to do that than certainly the long-term investors."

Altman, in passing, had passed Sequoia's most important test.

CHAPTER 8

# THE DOUCHEBAG BADGE

THE PROTECTION OFFERED BY STEVE JOBS'S REALITY DISTOR-tion field did not last a year. In March of 2009, as the nation's technorati descended on Austin, Texas, for the annual South by Southwest festival of music, film, and tech, where startups like Twitter had used the density of influential early adopters to catapult itself into public consciousness, something was in the air. The crowds jostling for barbeque and beer were bigger than usual. The local AT&T network strained under all the iPhone use. And everybody was talking about a cool new location-based app. But it wasn't Loopt. It was Foursquare.

Foursquare was co-founder Dennis Crowley's second swing at a location-based app. With his thin frame and shaggy mop that in its unrulier moments resembled Thurston Moore's, Crowley had a distinctly East Coast vibe. He created his first location-based app, Dodgeball, while still a student at New York University, and sold it to Google in 2005, where it withered, stymied by the challenge of securing distribution from wireless carriers in the flip-phone era. When Google announced in January 2009 it was shutting down Dodgeball, Crowley and his co-founder, Alex Rainert, began remaking a version of it for the iPhone age, complete with game mechanics, badges, and snarky copy. At the heart of the new app was the idea of "checking in" to certain locations, like restaurants and bars, which created competition to become "mayor" of one's favorite dive bar. Foursquare also took a more conservative, manual approach to broadcasting one's location than Loopt's always-on beam. "People don't want always-on tracking," Crowley said. "There's a big distinction between those two models."

One night in Austin for SXSW, Crowley found himself packed into a party at the live music venue Mohawk looking over the shoulder of two revelers comparing the Foursquare badges they had earned for attending SXSW events earlier that day. "It was at that moment that I was like, 'We made something awesome,'" Crowley recalled. The news site *Mashable* named Foursquare the breakout app of the festival.[1]

"When Foursquare came out, it was definitely a gut punch," recalled a former Loopt employee. "They seemed to catch lightning in a bottle, and the stupid South by Southwest momentum. Before that, we had been the darling. The question was, was it a product fit thing?" Loopt employees debated whether to build a new app that was more like Foursquare or to stick with Loopt's core product of always-on location. Part of the problem was that their doubling down on the iPhone had backfired, on one level. The iPhone wouldn't let third-party apps run in the background, which meant Loopt's always-on service only worked when the app was open. Loopt was charging a few dollars a month for the use of its app on most carriers, which amounted to a few million dollars of revenue, easily eaten into—if not erased—by the enormous infrastructure costs of running location services in the era before Amazon Web Services lowered the cost of servers for startups. "We had questions about whether or not the core Loopt app could ever generate revenue," McAdoo said.

Altman had proposed pivoting to a new app called Loopt Star, built on the Facebook Social graph, that would let users check in, like Foursquare, but give them discounts at businesses, like Groupon. He assigned some of the company's top engineers to begin working on it, to the dismay of his vice president of engineering, Steve Lemon, who had been hired to professionalize the engineering processes and keep partners like Verizon, Sprint, and AT&T happy. "Sam had a vision for how to catch up, and to the extent that he senses that people are going to slow him down and he can't go into the future, he'll start operating independently," the former Loopt employee said. "He thinks faster than any human being I have ever met, and he doesn't tend to bring people along, especially if he thinks they will slow him down."

There was already another example of this tendency. The previous year Altman had pushed a project that internally was referred to as "the gay dating app," and officially dubbed Loopt Mix, which allowed users to meet new people near them. It had pulled a number of engineers away from the core product and upset some of the staff. It was later spun off as its own app. One former senior Loopt employee described Altman as having "shiny object syndrome."

Now, with Foursquare looking dominant, the senior management and engineers were growing restless and concerned. The only part of the company that did reliably generate revenue was what was internally referred to as the "platform" business, a service it offered to wireless carriers to help them lower the cost of looking up a phone's location, run by Eric Carr, VP of location technologies. Loopt had licensed some technology from a Qualcomm subsidiary to offer this service, in hopes of making location look-ups cheap enough so that carriers could build real location-aware advertising businesses on top of them, and it was bringing in a few million dollars a year.[2] Some senior engineers wanted to build a respectable enough business on top of the platform to be able to sell the company for $100 million, but neither Altman nor Loopt's investors saw this as an acceptable outcome. "The challenge that Sam had is you could never be a really big company doing that," McAdoo said. "The thing I don't think the team understood is that we never considered the carrier business to be strategic." The engineers were worried that there was no path to profitability, and even more concerned that Altman didn't seem to care about forging one. "I don't think Sam was viewed as being sympathetic to their concerns," McAdoo said. "They are not interested in base hits," one former Loopt employee said of Sequoia.

This tension came to a head in the spring of 2009, when a dozen of the company's senior leaders called what one framed as a "come-to-Jesus" meeting with the board in the second-floor conference room of Sequoia's headquarters, asking that Altman be replaced as CEO. All of the company's co-founders were there, including Altman, as well as top leaders including Carr, Tana, Lemon, Knapp, and marketing chief Shari Yoder. During the meeting, the engineers said that Altman's "shiny object syndrome" was hurting the company and keeping them from being able to scale the core Loopt app, not to mention pursue their only profitable line of business. Altman saw the meeting as essentially about strategy. "There was a serious strategic disagreement where there were a few people who just really wanted to build a good enterprise business," Altman said.

But the concerns about his leadership went far beyond the question of the enterprise business. In the months leading up to the meeting, two Loopt employees left to start a mobile ad network company called AdWhirl that a market-leading, Sequoia-backed mobile ad network, AdMob, accused of taking its code in violation of its terms of service by disassembling its public SDK. The allegations were plausible because the employees had used AdMob's SDK while they were at Loopt. AdMob was in the late stages of talks to be acquired by Google, which would provide a tidy payday for Sequoia. So, upon hearing AdMob's allegations, Sequoia dispatched some Loopt employees to get to the bottom of what happened and report back. There was what one attendee called a "big crying meeting" involving Altman, Yam, and the other AdWhirl co-founder, but AdWhirl never admitted wrongdoing.

In the end, Sequoia solved the issue by engineering AdMob's purchase of AdWhirl just before the big fish of Google swallowed them both up. Altman denied knowing anything in advance about how AdWhirl came by the code, but given his closeness to one of AdWhirl's founders, Sam Yam, many Loopt employees had trouble believing this. The whole affair eroded the employees' trust in Altman.

Throughout the meeting where top Loopt employees were calling for Altman's replacement as CEO, Altman sat calmly to the side and absorbed the criticism. McAdoo, Chung, and their fellow board member Mike Ramsay—the former CEO of Tivo and NEA venture partner, who Chung had recruited to help coach Sam on leadership—listened to the executives' concerns, thanked them for flagging the problem, and then set about trying to solve it in some way other than firing

Altman. "It was a combination of some serious conversations with Sam about some blind spots that he had, but also some discussions about what we could do to bolster the team," McAdoo said. "There definitely was a concern about Sam's leadership. But the answer was never in my mind to replace him. If you look at the history of the venture business, and the companies post replacing the founder-CEO, particularly at the private stage, it's pretty dismal."

McAdoo had other reasons to protect Altman. Around this same time, he was negotiating Sequoia's investment into batches of Y Combinator, which Altman had introduced him to. McAdoo first learned of the existence of Y Combinator by scanning Loopt's ownership structure when Sequoia was weighing its initial investment. "There was this six percent with this special feature of anti-dilution for this organization called Y Combinator," he said. He called Mailliard and asked, "Who the heck are these guys?" Mailliard hadn't met them yet so referred him to Altman, who said, simply, that McAdoo had to meet Paul Graham. McAdoo had previously spoken at MIT and Harvard every year, so the next time he was in Cambridge, he asked Altman for an introduction and went to see Graham and Livingston. He arrived at YC's Cambridge offices and found himself staying for hours answering questions from founders. When YC moved to Mountain View, McAdoo became a guest speaker at Tuesday dinner, and eventually started offering founders advice through office hours and lecturing at the Startup School that YC would offer at Stanford and Berkeley. When his Sequoia partners asked him what he was doing, he replied, "The touchstone is, this is where Sam came from."

When global markets began to collapse in late 2008, McAdoo felt a wave of déjà vu. Remembering the dark days after the dot-com bust and how his investment in the data storage company Isilon, once laughed at, had gone on to make billions, he was sure this was a chance to act on Moritz's advice of investing at the bottom. He researched some of Sequoia's greatest investments, such as Cisco, after Black Monday, and put together a presentation for YC's Startup School: "This is the time to found a company," he said. "Yes, it's going to be harder to raise money.

But the venture investors that are investing now are the real venture investors that are going to be with you through thick and thin."

Graham pulled him aside after the presentation. "That was really great," he said. "I'm a little surprised by it, but I get the logic. But you understand that all the money that goes into YC companies comes out of mine, and Jessica's and Trevor's and RTM's [Robert Tappan Morris's] bank accounts, right? And we don't have any money anymore."

McAdoo smiled. "You know, at Sequoia, we have money."

YC needed something like $10 million, but McAdoo knew that would be hard to get past the Sequoia partners who had just put out the "RIP Good Times" deck. So, to shore up that winter's batch of founders, Graham put together a $2 million investment round that Sequoia led, joined by a smattering of angel investors including Ron Conway, an early investor in Google and Facebook, and Paul Buchheit, the creator of Gmail. (That winter class would end up including Airbnb, which McAdoo led Sequoia's investment in, one of the biggest wins to ever come out of the accelerator.)

To get the rest, McAdoo needed to subject Y Combinator to the rigor of a true Sequoia investment process, putting together a portfolio construction model, including the amount of return expected. "We had to get [YC] to a couple of hundred companies a year in order to make that work," he said—a scale that matched Graham's ambition, if not his bank account. With that plan in place, Sequoia was able to lead an $8 million round into YC batches the following year. For a VC raised hungry like McAdoo, it was the ultimate hack. Sequoia would ultimately make a mint from both its share of YC's profits and its ability to get in early on direct investments in YC's most successful startups, including Airbnb, Dropbox, and Stripe.

**ALTMAN WAS** also proving useful to Sequoia in other ways. "He was constantly having meetings with startups in the Loopt office," recalled one employee. By 2009, Sequoia had decided to officially tap Altman's networking talents and recruit him into its secretive "scout" program.

("I don't like the word 'secret,'" Roelof Botha, the head of Sequoia's scout program, told *The Wall Street Journal* in 2015, which disclosed names and details for the first time. "I think we have just been discreet about the program all along.") Sequoia scouts were usually young founders of Sequoia-funded companies, who were given a pot of money to invest on behalf of the VC firm. If the investment panned out, most of the gains were split by the scout and Sequoia's limited partners, while other scouts and Sequoia partners got a small chunk of the upside. As with its cultivation of startup lawyers at Wilson Sonsini, Sequoia was constantly on the lookout for proprietary deal flow. As startups got cheaper to launch and founders got younger, it made sense to have them referred by peers.[3]

Altman was instantly a star scout for Sequoia. A year earlier, Graham had introduced Altman to a red-haired, blue-eyed YC startup founder three years his junior named Patrick Collison, thinking Altman could help Collison navigate the strangeness of being so young in the industry. Collison had grown up in rural Ireland, where his parents, both engineers, ran a vacation hotel. He learned to code at home, encouraged by parents, who let him take a year off in the middle of high school just to code, and won a national science competition at the age of seventeen using the programming language Lisp. During the contest, he struck up a correspondence with Graham, whose textbook on Lisp served as a beacon to the most talented minds in technology. Collison enrolled in MIT, where Lisp was first developed, and a year later dropped out with his brother John to take their auction tracking software startup through YC. That happened in 2007; in 2008, it sold for $5 million.[4] Collison became a darling of Graham's, who suggested he meet his other favorite student, Altman. Collison and Altman hit it off right away after Altman demonstrated his knowledge of one of Collison's favorite obscure programming topics: Lisp machines, a type of computer built in the 1980s designed to run large artificial intelligence programs in the Lisp programming language.

"He knows a lot about a lot of things, and did then," Collison said. "I was pretty interested in some obscure programming topics at the time—I guess I still am—and he knew a surprising amount about those topics, even though I don't think they were his hobbies to the extent that they were mine."

In February 2009, Collison was hanging out in the kitchen of Graham's house in Palo Alto, discussing what he'd just been blogging about: how somebody should start a bank on the internet, allowing people to move money without the fees and friction of brick-and-mortar banks, an idea similar to the original vision of Elon Musk's X.com, which merged into PayPal. Graham reflexively offered to invest. When Altman arrived by coincidence a few minutes later, Graham urged him to do the same, saying he'd split the opportunity with him. They both wrote $15,000 checks, leaving the "to" line blank since the company did not yet have a name. Altman suggested one. "He made the suggestion that we should give it some very retro-Americana name, like National American Bank Company or something," Collison said. In the end, like PayPal before them, they abandoned the idea of an internet bank because it was far too difficult, and became a payments company instead, an elegant API that would let websites take credit cards with a mere nine lines of code. A year later, when Collison and his brother officially founded the company, they named it Stripe.

Altman's $15,000 investment in Stripe earned him 2 percent of a company today worth $70 billion; it is one of the most valuable startups in the United States and the most successful investment of Altman's career. Because Stripe was a Sequoia scout investment, Altman got to keep roughly half the proceeds for himself. Sequoia went on to make a seed investment in Stripe in 2010, followed by a Series A investment in 2011, with investors including Peter Thiel. "Sam was quite helpful in navigating that," Collison said. "We were these bright-eyed and bushy-tailed naifs from Ireland dealing with Peter Thiel and Michael Moritz, who were to us kind of deities. And Sam, as usual, was completely unintimidated and uncowed by any of them. And his counsel was really very helpful."

Given the doors that Altman was opening for Sequoia—and the piles of money that lay on the other side of them—McAdoo was not about to replace him at Loopt. He met privately with some of the senior leaders that Loopt could not afford to lose, and then gave Altman some feedback, which he felt the young CEO took well. Both Altman and the board were eager to move him away from a day-to-day operational role as CEO. They agreed to begin a search for a more seasoned executive to run the company's daily operations while Altman receded into an executive-chairman-like role focused on fundraising.

But by the time Sequoia had brought aboard mobile veteran Steve Boom to lead Loopt as president in the fall of 2010, it was clear to the latter's employees that the company's strategy pivot wasn't working. The company had tried catch up to trends in the industry with Loopt Star, the Foursquare-like product that encouraged users to "check-in" for a Groupon-like discount, but it wasn't different enough from the many other location-based apps offering goodies to users. Its rip-off of Foursquare's badges—which gamified check-ins with points, allowing users to get a "cinephile" badge by checking in repeatedly to movie theaters, for instance—was particularly egregious.

Foursquare got its revenge in a very Foursquare way: by launching a neon pink and green polo shirt "douchebag" badge in sly homage to Altman's attire for the Apple conference in 2008, which users could unlock by checking into trendy hotels and restaurants.[5]

ONE DAY not long after the prepaid debit card company Green Dot went public in 2010, its CEO, a former radio DJ turned venture capitalist named Steve Streit, was complaining to his board about the challenges of nudging the engineers at his decade-old company into adopting the latest trends in writing code. Green Dot, a Pasadena-based company that made the refillable cards on sale at Walmart, used the "waterfall" method, in which developers each work on their own pieces of code in succession. But Streit knew that in order to fulfill his dream of creating technologies that let people bank on their mobile phones, he'd have to mimic the Silicon Valley style of software development known as "agile," which focused more on speed and user feedback, as preached in the doctrine of Graham. Michael Moritz, who had led Sequoia's investment into Green Dot and sat on its board, piped up.

"There's a company that we're invested in that has amazing talent, amazing technology, and some leadership that really knows their stuff, but they are young—really young—so they are going to need mentorship both as human beings and to work at large organizations in financial services," Moritz said. "But Gosh, Steve, you sort of have that personality style, so maybe you can do that."

Of course the merger would help Sequoia, too. "Loopt at that point was really struggling and looking for an acquirer," Moritz said later. "That was what made me suggest it to Steve."

He offered to introduce Streit to Altman. What Moritz chose not to mention in his initial pitch was that things were looking pretty desperate at Loopt. By the second half of 2010, Loopt was bleeding users. "They had beaten that horse pretty hard and it was not going to gallop," Streit said. "By then, Facebook and others had won the race for social media."

With Moritz as matchmaker, Streit visited Loopt's offices in Palo Alto, where Altman peppered him with questions about Green Dot's company culture. "He was worried about joining forces with what he thought was a dinosauric technology company," Streit said. For his part, Streit had no such hesitations. "I've always liked young talent and knew that he was a genius, and I had a deep feeling like he could help me." He was equally impressed by Alok Deshpande, who, he said, really knew how to "work with people and see the egos and really deliver code."

In March 2012, the companies announced that Green Dot was buying Loopt for $43.4 million in cash, including $9.8 million set aside in retention payments for Loopt's thirty employees. "It was an acquihire," Streit said. Loopt's products would be shut down. During the negotiations, Altman's main focus was on keeping his team together. They would remain in the Palo Alto office and rebrand it as Green

Dot. "He had a deep loyalty and sense of right and wrong—a deep moral conviction," Streit said.

The one thing he could not keep together, however, was his relationship. As Loopt was sold, Altman and Sivo broke up. "I thought I was going to marry him—very in love with him," Altman told *The New Yorker*. As for Loopt's failure, he chalked it up to his misreading of how people would use digital technology. "We had the optimistic view that location would be all-important," he told the magazine. "The pessimistic view was that people would lie on their couches and just consume content—and that is what happened. I learned you can't make humans do something they don't want to do."[6]

# PART III

# 2012–2019

CHAPTER 9

# "A RIDE ON A ROCKET"

**A MONTH AFTER LOOPT WAS SOLD, PETER THIEL STRODE INTO** a packed Stanford auditorium. The class he had agreed to teach, "CS183: Startup," had quickly hit its 250-student limit, and now those who signed up early enough were cramming into the aisles and sitting on the floor. As a holder of two Stanford degrees who had gone on to co-found PayPal and then use the resulting riches to become one of the first investors in Facebook, Thiel was pitched in the course catalog as a successful entrepreneur and investor who would bring in his friends to share their direct experiences of startup glory.

But the class he had in mind would be something far more provocative. Thiel, a childhood chess prodigy, had majored in philosophy at Stanford, where he had studied under the French historian and literary critic René Girard, whose mimetic theory posited that what we experience as desire is actually just us copying what others desire. Combined with some bullying from his liberal classmates, according to his biographer, Max Chafkin, studying with Girard helped accelerate Thiel's innate tendency to contrarianism. As an undergraduate, this tendency manifested in his founding of the conservative *Stanford Review*. At age thirty, he published a book decrying campus multiculturalism. Later, as an investor, his contrary nature helped him correctly predict—if not quite correctly trade on—the financial crisis. Along the way, Thiel had embraced an array of wildly experimental ideas, including funding libertarian utopias floating in international waters and trying to hasten the moment when technological progress can propel itself without humans needing to be involved.

Now he was in the middle of perhaps the most contrarian campaign of all: to convince America's youth to drop out of college. The previous year, he had launched the Thiel Fellowships, offering $100,000 to promising young entrepreneurs to forgo elite universities and start companies. And yet here he was, standing before a podium at his alma mater, teaching a class about something seemingly much larger, and much weirder, than making money.[1]

It was a sensation. One student, Blake Masters, took notes and posted them online, where they went viral on *Hacker News*, Y Combinator's web forum. Those notes would go on to form the basis for the bestselling 2014 book *From Zero to One*, co-authored by Thiel and Masters two years later. David Brooks of *The New York Times* devoted a whole column to Masters's summary of the course's lessons.[2] And yet its main message—that competition was actually destructive and companies should strive to create mini-monopolies in new markets rather than slog it out in established ones—was not the main message of the class. Rather, the lectures kept coming back to the central disappointment of Thiel's life.

"The zenith of optimism about the future of technology might have been the 1960's," Thiel, born in 1967, told the class in his opening lecture. "People *believed* in the future. They *thought* about the future. Many were supremely confident that the next 50 years would be a half century of unprecedented progress. But with the exception of the computer industry, it wasn't."[3] Instead, he pointed out, real wages had been stagnant his whole life. As he saw it, very little of the cool stuff his parents' generation had dreamed up had come to pass. "Forget flying cars," he quipped in a later lecture. "We're still sitting in traffic."[4]

This complaint echoed the manifesto that Thiel's venture firm, Founders Fund, had published the previous year, and which bemoaned the venture capital industry's shift away from "transformational technologies" like semiconductors and toward companies that solved "incremental problems or even fake problems (e.g., having Kozmo.com messenger Kit-Kats to the office)." The manifesto's author, Bruce Gibney, expressed outrage that NASA had not been to the moon in forty years, the time it took to cross the Atlantic was actually growing (thanks to the retirement of the Concorde) while the increase in lifespans was decelerating, and humanity had made little progress in figuring out how to create energy more cheaply. And to top it all off: "We certainly don't have anything approaching general artificial intelligence, a lack many futurists 30 years ago would have found rather surprising."[5]

The antidote to tech stagnation, Thiel argued to both investors and undergrads alike, was to go against the crowd, stop being such cowering, careerist wusses obsessed with notching incremental gains, and swing for the fences. In some ways, it was his version of Paul Graham's essays. The only means for Thiel to attract mission-driven founders to invest in, he reasoned, was to reveal himself as the ultimate mission-driven investor.

**LESS COVERED** amid all the media hubbub around the Stanford course was that Thiel was engaged in talks at Y Combinator's offices in Mountain View, where Altman was spending a lot of time. Altman had officially been named a part-time partner in June 2011, well before his exit from Loopt, giving companies like Rap Genius—later renamed Genius—tips on crafting their pitch to investors at Demo Day. ("When you show this graph, make sure to make the point: 'All right, look—we're going to show you a user graph. Most YC companies get up here and they show you cumulative registered users. Like fuck that! We're going to show you active.' That's unusual. You're counting this the fair way," Altman told the Rap Genius founders, according to Randall Stross's *The Launch Pad: Inside Y Combinator*.)[6, 7]

Altman had walked away from Loopt shaken and sad, but with about $5 million. According to the terms of the deal, he would serve on the board of Green Dot and help the company develop its mobile banking app. He was there for one particularly disastrous meeting with Green Dot's most important customer, Walmart, which accounted for 70 percent of its revenue and was threatening to cancel its contract.

In the Arkansas parking lot afterwards, Altman calmly remarked to Streit, "Wow, that was a really bad meeting." But in general, he was free to rethink his life.

"I thought about starting companies," Altman said. "I thought about investing." Sequoia tried to pull him further into its venture capital empire, but he turned them down. He backpacked around Europe, went to Southeast Asia multiple times, visited Japan. He even spent some time in an ashram, which he joked to New York magazine was "like a total tech-bro meme," but which he says changed his life. "I'm sure I'm still anxious and stressed in a lot of ways, but my *perception* of it is that I feel very relaxed and happy and calm." (Annie Altman says her brothers made fun of her for being so into meditation that she became the president of the Tufts Buddhist Mindfulness Sangha, but a few years later she found out Sam was hosting meditation groups at his home—today, she questions just how much time Sam actually spent in Indian ashrams.) Perhaps the biggest change came during a weekend-long retreat he spent in Mexico taking psychedelics with a guide. "It was one of the most transformative things in my life," he told the *San Francisco Standard*'s *Life in Seven Songs* podcast.[8] Mostly, he spent a lot of time in Big Sur, amid the redwoods by the sea, the landscape on the planet that most appealed to him, and read books. "I felt like I knew a little bit about everything and felt like everything was possible," he said. "I felt very uncertain, but very good."

Altman and Thiel had known each other for a few years. Altman had first approached the older investor over their shared interest in nuclear energy. "When I was starting to really think about energy, someone told me, 'You know he's the only investor thinking about nuclear right now. So you should go talk to him,'" Altman said.

Altman was interested in all types of nuclear energy, but he was particularly interested in nuclear fusion, the powerful reaction that fuels the sun that had been a dream of scientists for nearly a century. Since before Altman was born, governments around the world had been collaborating on the extraordinarily expensive research that sought to heat up hydrogen isotopes enough—and keep them under enough pressure—for their nuclei to fuse into helium, releasing roughly four times as much energy in the process as the nuclear fission at work in nuclear power plants. If nuclear fusion was achieved, it would provide a cheap, clean source of energy from molecules found in seawater, with no carbon emissions and far less radioactive waste than fission.

For Thiel, nuclear energy was the ultimate symbol of America's loss of technological mojo since the 1960s. In his view, America stopped building nuclear power plants in good measure due to hysteria stemming from the one-two punch of the release of the 1979 Jane Fonda–led thriller *The China Syndrome*, which depicted a fictional nuclear emergency, followed twelve days later by the real-life partial meltdown at the Three Mile Island nuclear plant near Harrisburg, Pennsylvania. The accident galvanized the antinuclear movement, and, Thiel argued, led to the cancellation of about a hundred proposed reactors. It also began to take on a mythic quality in Silicon Valley as dark evidence that humans in a democratic society were not capable of assessing risk and making rational choices about their long-term future. "Fewer than 50 people were reported to have died at Chernobyl; by contrast, the American Lung Association estimates that smoke from coal-fired power plants kills about 13,000 people every year," Thiel wrote.[9] "The great scar on the minds of all the pro-humanity people in Silicon Valley is what happened to nuclear reactors," Eliezer Yudkowsky, an early acolyte of Thiel's, said. This scar would end up influencing Silicon Valley's approach to AI, which required a similar moral balancing act between progress and safety. "From their perspective, technologies don't get a fair shake. If you think that something benefits humanity, you can't go to humanity and make the case for it benefiting humanity. You have to storm out ahead and develop it yourself before anyone can stop you, not because you'll lose the factional fight, but because you will not be fairly judged." For the idealistic engineers of the tech industry, the fate of nuclear reactors since the late twentieth century "is to them what the Holocaust is to Jews," said Yudkowsky.

If government wasn't going to solve this problem, Thiel was determined to do it himself, through venture capital. To invest in a new generation of nuclear technologies was not just to put money down on a moonshot that could make the world a better place—it was to directly fight the creeping forces that threatened progress itself. Later on, after Altman recruited Helion Energy, which aimed to crack the code of making nuclear fusion practical, to join Y Combinator, Thiel put money in the company's seed round through his firm Mithril Capital, alongside a personal stake from Altman.[10]

Besides a common interest in nuclear energy, Altman and Thiel had both ended up as seed investors in Stripe, at the recommendation of Paul Graham, who had thought Thiel's background as a co-founder of a payments company, PayPal, would make him a helpful advisor to the young Patrick Collison. Ultimately, when Stripe was raising an $18 million Series A funding round—from Sequoia, Thiel, PayPal co-founder Max Levchin, and solo investor Elad Gil—it was Altman who played the role of indispensable advisor, talking Collison through the process over the course of dozens of phone calls, often as Collison biked across Palo Alto to the office and back each day. In the end, the round valued the young payments company at $100 million.

By the spring of 2012, Altman was renowned for his connections to the most exciting startups in the Valley and his closeness to Y Combinator—which increasingly were one and the same thing—and he had his pick of LPs, or limited partners, interested in backing him in his own venture fund. After demurring when Sequoia tried to court him, Altman went with Thiel, who put in most of the $21 million to start the fund they would call Hydrazine Capital, named for a chemical found in rocket fuel. Altman recruited his younger brother Jack, who had been working at a New York investment bank since graduating from Princeton the previous year with a degree in economics, to join the fund. Their middle brother, Max, would join a couple years later, and the brothers would share a San Francisco apartment.

Thiel found Altman to be "very, very smart," as well as "very principled, very strict, very balanced," and "maybe a little bit too optimistic."

His main draw was not *what* he knew, but *who*. To Thiel, as I noted in the prologue, Altman was "just at the absolute epicenter, maybe not of Silicon Valley, but of a Silicon Valley zeitgeist." Thiel had had long talks with Facebook CEO Mark Zuckerberg, who he mentored after becoming the first outside investor in Facebook, about how Zuckerberg had been cast, for better or worse, as the face of the millennial generation, but in fact it was an imperfect fit. "If you had to look for the one person who represented a millennial tech person, it would be Altman," Thiel said.

Thiel, who Gawker Media's *Valleywag* blog outed against his will in 2007, was at this point the tech world's most famous openly gay investor. In retrospect, it's not shocking that Altman would gravitate toward Thiel as a mentor while heading on his own path to becoming one of the most famous openly gay CEOs. And yet Thiel's contrarian-by-default worldview is at odds with Altman's tendency to want to make people happy. What Altman seems to admire in Thiel above all else, however, is his penchant for going against the grain to generate novel thought, whether or not Altman agrees with it. "He thinks about the world in this sort of like deeply unconstrained way," Altman told a recent podcast.[11] In another interview, Altman suggested that original ideas are among the rarest commodities in the world. And that is one resource Thiel sheds in abundance.

Hydrazine did go on to do some nuclear investing, in a fission microreactor startup named Oklo that Altman learned about in 2013 and recruited to Y Combinator the following year. And, as mentioned earlier, Altman and Thiel teamed up to seed Helion. But in general Hydrazine rode what was turning out to be a historic tech boom— a strategy quite contrary to Thiel's typical contrarianism. "If you are in a really big boom that no one believes in, you just want to invest in things that everybody knows are good. That's pretty consensus," he said. He believed that the volatility signaled by the dot-com bubble and the financial crisis had scared away investors and left the tech industry underinvested. By 2012, "it's now firing on all cylinders, and all these companies are just growing and compounding, but it's like there's nobody left to invest," he said. "If something is that wildly underinvested, you don't need to be a contrarian at all. You just want to be in the zeitgeist. I think Sam did that extremely well."

Hydrazine dropped money into a grab bag of industries, from enterprise software to specialty foods, with no discernable thesis. The one thing connecting the investments was that most of them had gone through Y Combinator, including HR startup Zenefits, supply chain logistics platform Flexport, video chat language learning service Verbling, digital construction marketplace BuildZoom, meal replacement company Soylent, and online divorce service Wevorce. One outlier was Patreon, the self-streaming monetization platform, which was co-founded by Altman's old friend from Loopt, Sam Yam. Hydrazine was essentially a YC index fund, the kind of "stupidly simple" play that Thiel defined himself by avoiding.

"Sometimes you just want to catch a ride on a rocket," Thiel said.

YC WAS blasting through the stratosphere. With Sequoia's help, it had expanded to more than sixty startups for 2011's summer batch. That year, angel investor Ron Conway and Russian-born billionaire Yuri Milner approached Graham with an offer that would give them a piece of every YC company. The investors ended up providing every YC startup a $150,000 convertible note—essentially a loan that would become equity if the company managed to raise venture funding and set a valuation. The deal was extremely pro-founder: it allowed founders to take the cash without having to worry about what their valuation was at such an early stage, or about paying it back if they failed. Demo Day that summer was a carnival. YC had to install two new air-conditioning units on the roof of its hall to deal with the crowd of more than two hundred VCs and other investors, including Demi Moore and Ashton Kutcher. They had to order a Porta-Potty for outside the hall.[12]

In 2012, there were eighty-four startups in the summer batch, and partners could no longer keep track of who was who. "That was the famous batch that broke everybody's brains," YC CFO Kirsty Nathoo told *Wired*. After it, Y Combinator decided to split future batches into what were essentially classes, and move Demo Day to the much larger Computer History Museum, a sprawling campus in Mountain View that houses such artifacts as a World War II–era Enigma machine and the first-ever computer mouse, which happens to be carved out of redwood.[13] For 2013, they kept the number of would-be companies at a more reasonable forty-seven, but Demo Day ballooned to 450 guests, lured by the promise of riches on the scale of Dropbox (then worth $4 billion), Airbnb ($1.3 billion), and Stripe ($500 million). "A gold rush-mentality reigns," wrote *The New York Times Magazine* of YC's 2013 Demo Day.[14]

Hydrazine invested in a handful of YC companies, while Thiel's Founders Fund invested in whatever Altman said was the best YC startup that year. In 2012, it was Airbnb. "In a world where momentum is underrated, maybe the thing you should do is, once a year, just pick the best YC company," said Thiel. Founders Fund invested in Airbnb at a $2.5 billion valuation. It's now worth approximately $200 billion. The next year, they invested in Stripe, at a $1.75 billion valuation. It's now worth around $65 billion. "The conversations I had with Sam at that time were around, 'what's the best?'"

"By late 2014, this heuristic started to not work anymore, because enough people knew to do this," Thiel said. He and Altman had identified Zenefits—a health benefits broker that gave away free HR software to small businesses in the hopes of making commissions from going on to sell them insurance—as the new YC darling. Max Altman went to work there in product, staying from 2014 to 2016. Founders Fund put an offer in when it was at a $2 billion valuation, but the aggressive co-founder, Parker Conrad, negotiated for a valuation above $4 billion. That seemed high to Thiel, so Founders put in less. Within a few years, unable to meet the optimistic growth targets required to merit such a valuation and struggling with its relationships in the healthcare industry, Zenefits blew up and the investment went basically to zero. (David Sacks, a friend of Thiel's from college and a fellow *Stanford*

*Review* writer and PayPal alumnus, stepped in as interim CEO in 2016 after the company melted down.)

It seemed like some kind of high-water mark had been reached.

**ANOTHER THIEL** investment ended up having the most significant impact on Altman's career and the future of artificial intelligence. Like Altman, Thiel had long been obsessed with the possibility that one day computers would become smarter than humans and unleash a self-reinforcing cycle of exponential technological progress, an old science fiction trope often referred to as "the singularity." The term was first introduced by the mathematician and Manhattan Project advisor John von Neumann in the 1950s, and popularized by the acclaimed sci-fi author Vernor Vinge in the 1980s. Vinge's friend Marc Stiegler, who worked on cybersecurity for the likes of DARPA while drafting futuristic novels, recalled once spending an afternoon with Vinge at a restaurant outside a sci-fi convention "swapping stories we would never write because they were both horrific and quite possible. We were too afraid some nutjob would pick one of them up and actually do it."

Among the many other people influenced by Vinge's fiction was Eliezer Yudkowsky. Born into an Orthodox Jewish family in 1979 in Chicago, Yudkowsky was son of a psychiatrist mother and a physicist father who went on to work at Bell Labs and Intel on speech recognition, and was himself a devoted sci-fi fan. Yudkowsky began reading science fiction at age seven and writing it at age nine. At eleven, he scored a 1410 on the SAT. By seventh grade, he told his parents he could no longer tolerate school. He did not attend high school. By the time he was seventeen, he was painfully aware that he was not like other people, posting a web page declaring that he was a "genius" but "not a Nazi." He rejected being defined as a "male teenager," instead preferring to classify himself as an "Algernon," a reference to the famous Daniel Keyes short story about a lab mouse who gains enhanced intelligence. Thanks to Vinge, he had discovered the meaning of life. "The sole purpose of this page, the sole purpose of this site, the sole purpose of anything I ever do as an Algernon is to accelerate the Singularity," he wrote.[15]

Around this time, Yudkowsky discovered an obscure mailing list of a society calling itself the Extropians, which was the subject of a 1994 article in *Wired* that happened to include their email address at the end. Founded by philosopher Max More in the 1980s, Extropianism is a form of pro-science, super-optimism that seeks to fight entropy—the universal law that says things fall apart, everything tends toward chaos and death—on all fronts. In practical terms, this meant signing up to have their bodies—or at least heads—frozen at negative 321 degrees Fahrenheit at the Alcor Life Extension Foundation in Scottsdale, Arizona, after they died. They would be revived once humanity was technologically advanced enough to do so. More philosophically, fighting entropy meant abiding by five principles: Boundless Expansion, Self-Transformation, Dynamic Optimism, Intelligent Technology, and Spontaneous Order. (Dynamic Optimism, for example, involved a technique called selective focus, in which you'd concentrate on only the positive aspects of a given situation.)

Robin Hanson, who joined the movement and later became renowned for creating prediction markets, described attending multilevel Extropian parties at big houses in Palo Alto at the time. "And I was energized by them, because they were talking about all these interesting ideas. And my wife was put off because they were not very well presented, and a little weird," he said. "We all thought of ourselves as people who were seeing where the future was going to be, and other people didn't get it. Eventually—*eventually*—we'd be right, but who knows exactly when."

More's co-founder of the journal *Extropy*, Tom Bell, aka T. O. Morrow (Bell claims that Morrow is a distinct persona and not simply a pen name), wrote about systems of "polycentric law" that could arise organically from voluntary transactions between agents free of government interference, and of "Free Oceana," a potential Extropian settlement on a man-made floating island in international waters. (Bell ended up doing pro bono work years later for the Seasteading Institute, for which Thiel provided seed funding.) If this all sounds more than a bit libertarian, that's because it was. The *Wired* article opens at one such Extropian gathering during which an attendee shows up dressed like "State," wearing a vinyl bustier, miniskirt, and chain harness top and carrying a riding crop, dragging another attendee dressed up as "the Taxpayer" on a leash on all fours.[16]

The mailing list and broader Extropian community had only a few hundred members, but among them were a number of famous names, including Hanson; Marvin Minsky, the Turing Award–winning scientist who founded MIT's AI lab in the 1970s; Ray Kurzweil, the computer scientist and futurist whose books would turn "the singularity" into a household word; Nick Bostrom, the Swedish philosopher whose writing would do the same for the supposed "existential risk" posed by AI; Julian Assange, a decade before he founded WikiLeaks; and three people—Nick Szabo, Wei Dai, and Hal Finney—rumored to either be or be adjacent to the pseudonymous creator of Bitcoin, Satoshi Nakamoto. "It is clear from even a casual perusal of the Extropians archive (maintained by Wei Dai) that within a few months, teenage Eliezer Yudkowsky became one of this extraordinary cacophony's preeminent voices," wrote the journalist Jon Evans in his history of the movement. In 1996, at age seventeen, Yudkowsky argued that superintelligences would be a great improvement over humans, and could be here by 2020.[17]

Two members of the Extropian community, internet entrepreneurs Brian and Sabine Atkins—who met on an Extropian mailing list in 1998 and were married soon after—were so taken by this message that in 2000 they bankrolled a think tank for Yudkowsky, the Singularity Institute for Artificial Intelligence.[18] At twenty-one, Yudkowsky moved to Atlanta, and began drawing a nonprofit salary of around $20,000 a year to preach his message of benevolent superintelligence. "I thought very smart things would automatically be good," he said. Within eight months, however, he began to realize that he was wrong—way wrong. AI, he decided, could be a catastrophe. "It was because I was taking someone else's money, and I'm a person who feels a pretty deep sense of obligation towards those who help me," Yudkowsky explained. "At some point, instead of thinking, 'If superintelligences don't automatically determine what is the right thing and do that thing that means there is no real right or wrong, in which case, who cares?' I was like, 'Well, but Brian Atkins would probably prefer not to be killed by a superintelligence.'" He thought Atkins might like to have a "fallback plan," but when he sat down and tried to work one out, he realized with horror that it was impossible. "That caused me to actually engage with the underlying issues, and then I realized that I had been completely mistaken about everything."

The Atkinses were understanding, and the institute's mission pivoted from making artificial intelligence to making *friendly* artificial intelligence. "The part where we needed to solve the friendly AI problem did put an obstacle in the path of charging right out to hire AI researchers, but also we just surely didn't have the funding to do that," Yudkowsky said. Instead, he devised a new intellectual framework he dubbed "rationalism." (While on its face, rationalism is the belief that humankind has the power to use reason to come to correct answers, over time it came to describe a movement that, in the words of Yudkowsky's intellectual heir, Scott Alexander, includes "reductionism, materialism, moral non-realism, utilitarianism, anti-deathism and transhumanism," though Alexander jokes that the true answer is the belief that "Eliezer Yudkowsky is the rightful calif.")[19]

In a 2004 paper, "Coherent Extrapolated Volition," Yudkowsky argued that friendly AI should be developed based not just on what we think we want AI to do now, but what would actually be in our best interests. "The engineering goal is to ask what humankind 'wants,' or rather what we would decide if we knew more, thought faster, were more the people we wished we were, had grown up farther together, etc.," he wrote. In the paper, he also used a memorable metaphor for how AI could go wrong: if your AI is programmed to produce paperclips, if you're not careful, it might end up filling the solar system with paperclips. Years later, Bostrom would take this example and hold it up as the ultimate symbol of the need to "align" AI with human will.[20]

In 2005, Yudkowsky attended a private dinner at a San Francisco restaurant held by the Foresight Institute, a technology think tank founded in the 1980s to push forward nanotechnology. (Many of its original members came from the L5 Society, which was dedicated to pressing for the creation of a space colony hovering between the Earth and the Moon, and successfully lobbied to keep the United States from signing the United Nations Moon Agreement of 1979 due to its provision against terraforming celestial bodies.) Thiel was in attendance, regaling fellow guests about a friend who was a market bellwether, because every time he thought some potential investment was hot, it would tank soon after. Yudkowsky, having no idea who Thiel was, walked up to him after dinner. "If your friend was a reliable signal about when an asset was going to go down, they would need to be doing some sort of cognition that beat the efficient market in order for them to reliably correlate with the stock going downwards," Yudkowsky said, essentially reminding Thiel about the Efficient Market Hypothesis, which posits that all risk factors are already priced into markets, leaving no room to make money from anything besides insider information. Thiel was charmed.

Thiel and Yudkowsky began having occasional dinners together. Yudkowsky came to regard Thiel "as something of a mentor figure," he said. In 2005, Thiel started funding Yudkowsky's Singularity Institute, and the following year they teamed up with Ray Kurzweil—whose book *The Singularity Is Near* had become a bestseller—to create the Singularity Summit at Stanford University. Over the next six years, it expanded to become a prominent forum for futurists, transhumanists, Extropians, AI researchers, and science fiction authors, including Bostrom, More, Hanson, Stanford AI professor Sebastian Thrun, XPRIZE founder Peter Diamandis, and Aubrey de Grey, a gerontologist who claims humans can eventually defeat aging. Skype co-founder Jaan Tallinn, who participated in the summit, was inspired by Yudkowsky to become one of the primary funders of research dedicated to reducing existential risk from AI, including by initially giving money to OpenAI and later switching to fund OpenAI's rival, Anthropic.

Another summit participant, physicist Max Tegmark, would go on to found the Future of Life Institute.

Vernor Vinge himself even showed up, looking like a public school chemistry teacher with his Walter White glasses and tidy gray beard, cheerfully reminding the audience that when the singularity comes, "We're no longer in the driver's seat."[21]

In 2010, one of the AI researchers who Yudkowsky invited to speak at the summit was Shane Legg, a New Zealand–born mathematician, computer scientist, and ballet dancer who had been obsessed with building superintelligence ever since Yudkowsky had introduced him to the idea a decade before.[22] Legg had been working at IntelliGenesis, a New York–based startup founded by the computer scientist Ben Goertzel that was trying to develop the world's first AI. Its best-known product was WebMind, an ambitious software project that attempted to predict stock market trends. Goertzel, who had a PhD in mathematics, had been an active poster on the Extropians mailing list for years, sparring affectionately with Yudkowsky on transhumanism and libertarianism. (He was in favor of the former but not so much the latter.)[23] Back in 2000, Yudkowsky came to speak at Goertzel's company (which would go bankrupt within a year). Legg points to the talk as the moment when he started to take the idea of superintelligence seriously, going beyond the caricatures in the movies.[24] Goertzel and Legg began referring to the concept as "artificial general intelligence."

Legg went on to get his own PhD, writing a dissertation on "Machine Super Intelligence" that noted the technology could become an existential threat, and then moved into a postdoctoral fellowship at University College London's Gatsby Computational Neuroscience Unit, a lab that encompassed neuroscience, machine learning, and AI. There, he met a gaming savant from London named Demis Hassabis, the son of a Singaporean mother and Greek Cypriot father. Hassabis had once been the second-ranked chess player in the world under the age of fourteen. Now he was focused on building an AI inspired by the human brain. Legg and Hassabis shared a common, deeply unfashionable vision. "It was basically eye-rolling territory," Legg told the journalist Cade Metz. "If you talked to anybody about general AI, you would be considered at best eccentric, at worst some kind of delusional, nonscientific character."[25] Legg thought it could be built in the academy, but Hassabis, who had already tried a startup and failed, knew better. The only way to do it was through industry. And there was one investor who would be an obvious place to start: Peter Thiel.

Legg and Hassabis came to the 2010 Singularity Summit as presenters, yes, but really to meet Thiel, who often invited summit participants to his townhouse in San Francisco. Hassabis spoke on the first day of the summit, which had moved to a hotel in downtown San Francisco, outlining his vision for how an AI that took inspiration from the human brain. Legg followed the next day with a talk on how AI needed to be measurable to move forward. Afterward, they went for cocktails at Thiel's Marina District home, with its views of both the Golden Gate Bridge and the Palace of Fine Arts, and were delighted to see a chessboard out on a table. They wove through the crowd and found Yudkowsky, who led them over to Thiel for an introduction. Trying to play it cool, Hassabis skipped the hard sell and began with chess, a topic he knew was dear to Thiel's heart. The game had stood the test of time, Hassabis said, because the knight and bishop had such an interesting tension—equal in value, but profoundly different in strengths and weaknesses. Thiel invited them to return the next day to tell him about their startup.[26]

In the morning, they pitched Thiel, fresh from a workout, across his dining room table. Hassabis said they were building AGI inspired by the human brain, would initially measure its progress by training it to play games, and were confident that advances in computing power would drive their breakthroughs. Thiel balked at first, but over the course of weeks agreed to invest $2.25 million, becoming the as-yet-unnamed company's first big investor.[27] A few months later, Hassabis, Legg, and their friend, the entrepreneur Mustafa Suleyman, officially co-founded DeepMind, a reference to the company's plans to combine "deep learning," a type of machine learning that uses layers of neural networks, with actual neuroscience. From the beginning, they told investors that their goal was to develop AGI, even though they feared it could one day threaten humanity's very existence.[28]

It was through Thiel's network that DeepMind recruited his fellow PayPal veteran Elon Musk as an investor. Thiel's Founders Fund, which had invested in Musk's rocket company SpaceX, invited Hassabis to speak at a conference in 2012, and Musk was in attendance. Hassabis laid out his ten-year plan for DeepMind, touting it as a "Manhattan Project" for AI years before Altman would use the phrase. Thiel recalled one of his investors joking on the way out that the speech was impressive, but he felt the need to shoot Hassabis to save the human race.

The next year, Luke Nosek, a co-founder of both PayPal and Founders Fund who is friends with Musk and sits on the SpaceX board, introduced Hassabis to Musk. Musk took Hassabis on a tour of SpaceX's headquarters in Los Angeles. When the two settled down for lunch in the company cafeteria, they had a cosmic conversation. Hassabis told Musk he was working on the most important thing in the world, a superintelligent AI. Musk responded that he, in fact, was working on the most important thing in the world: turning humans into an interplanetary species by colonizing Mars. Hassabis responded that that sounded great, so long as a rogue AI did not follow Musk to Mars and destroy humanity there too. Musk got very quiet. He had never really thought about that. He decided to keep tabs on DeepMind's technology by investing in it.[29]

IN DECEMBER 2013, Hassabis stood on stage at a machine learning conference at Harrah's in Lake Tahoe and demonstrated DeepMind's first big breakthrough: an AI agent that could learn to play and then quickly master the classic Atari video game *Breakout* without any instruction from humans. DeepMind had done this with a combination of deep neural networks and reinforcement learning, and the results were so stunning that Google bought the company for a reported $650 million a month later.[30,31]

The implications of DeepMind's achievement—which were a major step toward a general-purpose intelligence that could make sense of a chaotic world around it and work toward a goal—were not widely understood until the company published a paper on their findings in the journal *Nature* more than a year later. But Thiel, as a DeepMind investor, understood them well, and discussed them with Altman. In February 2014, a month after Google bought DeepMind, Altman wrote a post on his personal blog titled "AI" that declared the technology the most important tech trend that people were not paying enough attention to.

"To be clear, AI (under the common scientific definition) likely won't work. You can say that about any new technology, and it's a generally correct statement. But I think most people are far too pessimistic about its chances," he wrote, adding that "artificial general intelligence might work, and if it does, it will be the biggest development in technology ever."[32]

CHAPTER 10

# "SAM ALTMAN FOR PRESIDENT"

**MURMURS RAN THROUGH THE CROWD OF STARTUP FOUNDERS** huddled over their bowls of chili as the skinny young man in a T-shirt and cargo shorts took the stage. It was late winter of 2014, and Paul Graham had just stunned Silicon Valley by announcing that he was stepping down as the head of Y Combinator and handing the reins to a twenty-eight-year-old failed startup founder named Sam Altman. Now Graham was standing before the latest batch of founders in YC's bright orange–walled headquarters in Mountain View and introducing his successor with a combination of glee and reverence. Many of the faces before him, who had in many cases quit their jobs and left behind their families for this opportunity, blinked back in polite confusion.

"Nobody knew who Sam was," said Doktor Gurson, whose phone-charging startup, Doblet, was in the batch. "Some people were slightly disappointed because they were looking forward to working with Paul Graham." As Altman sat down across from Graham for a Q and A about YC's extraordinary success, Gurson was impressed by his combination of short, pithy answers and wide-ranging curiosity. But he did not think the leadership change was likely to improve Doblet's chances of success. In fact, he was so panicked at the thought of missing the chance to work with Graham that he buttonholed the hacking legend later that night and pitched him on Doblet's progress.

"I'd never heard of Sam Altman," recalled Daniel Yanisse, whose background check automation company, Checkr, was part of YC that year. "I looked him up to see his background and was surprised because he didn't have that much experience. He was pretty young, and his company, Loopt, did not get that big before acquisition. I was just curious, what are his special talents and skills? It took me a while to see it." Altman came across as introverted, technical, and occasionally brusque during office hours, skipping the small talk and going straight to his question of what he could do to help.

Graham, however, never had any doubts about Altman's talents and skills. "He's one of those rare people who manage to be both fearsomely effective and yet fundamentally benevolent—which, though few realize it, is an essential quality in early stage investing," Graham wrote in his announcement blog post cheekily titled "Sam Altman for President." "Sam is one of the smartest people I know, and understands startups better than perhaps anyone I know, including myself."[1] He said he was passing the baton "because YC needs to grow, and I'm not the best person to grow it." During his nine years at the helm, Graham had grown YC by a factor of roughly ten—from eight startups in the first batch to nearly eighty in the current one—minting unicorns like Stripe, Airbnb, and Dropbox along the way. More important, he had tilted the balance of power in the Valley away from investors and toward startup founders, and had preached his gospel so widely and effectively that ever-larger hordes of starry-eyed entrepreneurs and investors were beating down YC's door each year. Suddenly, in the wake of the financial crisis, all the people who might have previously gone to work at Goldman Sachs or McKinsey now wanted to study computer science and start companies. Graham gazed out at the coming tsunami with awe. "I'm convinced there's a fundamental change happening in the way work gets done," he wrote. "It's becoming normal to start a startup. There will be a lot more startups in 10 years than there are now, and if YC is going to fund them, we'll have to grow proportionally bigger."[2]

The problem was that Graham's method of working did not, in Valley-speak, "scale." As Paul Buchheit, the creator of Gmail who became one of YC's earliest and most devoted angel investors and partners, put it, "PG's a really great one-man show. He runs everything himself." He wrote all of YC's software, moderated its increasingly popular message board, *Hacker News*, interviewed candidates and advised them during office hours. It was the equivalent of "four jobs," Buchheit said. "It was basically taking over his whole life." As Graham and Livingston began having children, he found himself wanting more time to devote to them, as well as to his own projects, including his essays and the dialect of the Lisp programming language he had been working on before YC blew up. He had begun trying to woo Altman since almost the moment that Loopt had been sold, and even more urgently since the growing pains of YC's Sequoia-funded expansion. When one longtime YC hand asked Graham why he picked Altman to succeed him, Graham replied, "He's clearly the most ambitious." Graham believed that Altman was going to get what he wanted, no matter what, so the best way to ensure YC's success was to make that success something Altman wanted.

Altman was flattered to be asked, but initially unsure. "It's nice to hear from your mentor that they want you to do it," he said. "But I really didn't want to do it for a while." Angel investing—especially with someone else's money—was all fine and good, as was the advice-offering and favor-trading he referred to as "being useful," skills that had helped him build a fiercely loyal network throughout the Valley. But he still imagined that this would all be in the service of doing, not teaching. "I wanted to do a company. It turned out I was very well suited to be an investor. But I thought of myself at the time as very much not an investor." In particular, he wanted to do a nuclear fission startup.

Altman wrestled with the decision. "It was something that required consideration," said tech entrepreneur and investor Lachy Groom, whom Altman was dating at the time. At that stage, Altman saw an investor as someone who was trying to build a venture capital firm, with limited partners they had to answer to and a set of incentives that were not always aligned with those of the companies they invested in. "You're trying to find a company that will be successful with or without you, then convince them to take your money instead of somebody else's, and at a lower price," he told *The New Yorker*. "I didn't like being oppositional to the entrepreneur."[3] Although he was largely investing Thiel's money, Altman preferred to think of himself as an angel investor rather than a venture capitalist. "Angel investors don't call themselves investors, they call themselves angels," said Groom. "Sam is very much that. He's not looking to build a firm. He doesn't get a kick or rise out of being an investor. He loves working with founders and he happens to give them money."

In the end, Altman's love of working with founders won out, and he took over the YC juggernaut just as talk of a new tech bubble began to take over the Valley. "Prices are high and valuations are high," Graham told a conference soon after the announcement.[4] It was not just that Dropbox and Airbnb were now valued at more than $10 billion, and Stripe at nearly $2 billion. The average value of a YC-funded startup was now more than $22 million, and one of the graduating companies from the last batch had been valued at $50 million.[5] As startups filled offices in San Francisco, tech workers flooded formerly edgy neighborhoods like the Mission, pushing up the city's median home price above $1 million. For the first time, there were more "ultrarich" people, defined as having assets of at least $30 million, in San Francisco than Los Angeles.[6]

Despite what felt like very high prices, Altman and Thiel agreed to buy out a portion of Graham's stake in Y Combinator as he departed. (Thiel said it was about a third of Graham's YC holding; Altman said it was more like 7 percent.) Altman had to hold Thiel's hand through the whole thing. "Sam was extremely optimistic, which was an important trait to invest in these things because they all looked fully valued," Thiel said, as I noted in the prologue.

Altman's vision for how to expand YC was deeply influenced by Thiel's complaints about tech stagnation. In the year before he moved into the job, Altman had begun publishing personal, philosophical essays on his blog, PG-style. An early essay laid out one of his deepest and most persistent beliefs: "Without economic growth, democracy doesn't work because voters occupy a zero-sum system," he wrote.

Teaching humans to share was impossible; growth was a kind of spiritual hack that got you out of having to do it. There was some historical evidence to support this claim. Athenian democracy lasted less than two hundred years, but would probably not have happened at all had it not been for the enormous wealth it extracted, largely via slaves, from its nearby silver mines; it's probably not a coincidence that the democracy petered out around the same time the silver deposits did. But Altman does not look that far back in history. Instead, he pointed to the two main sources of America's remarkably steady growth over the centuries: the frontier, and then, once that was consumed, technological progress.

"The US has been blessed with economic growth for a very long time, first due to natural resources and massive amounts of land in which to expand, and then due to a period of technological progress rarely matched in human history that lasted approximately until we realized just how dangerous nuclear bombs really were and got scared of new technology," he wrote. "But the frontier is long over, and although technological innovation has continued at a blistering rate for computers and the Internet, it seems to have slowed down in most other industries."[7]

Altman's plan was to use YC to help other industries catch up to software's explosive growth rate by throwing its doors open to a much wider array of startups than Graham had, including the kind of "hard tech" that Altman argued the government used to fund but no longer did. "We'd like for Y Combinator to fund more breakthrough technology companies—companies that solve an important problem, have a very long time horizon, and are based on an underlying technological or scientific breakthrough," he wrote in a blog post two months after taking over, pointing to two of Elon Musk's companies, SpaceX and Tesla, as examples of what is possible. "It used to be the case that government funded a lot of development of breakthrough technologies. The bad news is that they have mostly stopped: the good news is that the leverage of technology is such that now small startups can do what used to take the resources of nations."

By this point nuclear energy was one of Altman's main obsessions, according to friends, and he put energy at the top of his wish list for what YC would fund under his leadership. He framed his goal as merely cheap energy, in much the same way Thiel had in his 2011 manifesto, noting that "there is a remarkable correlation between the cost of energy and quality of life," but tipped his hand a bit in the details. "Nuclear energy can hit the bid, and possibly so can renewables. But pricing is the first order question."[8]

AI was second on the list. In his earlier personal blog post about AI, Altman had intimated that his optimism about the technology came from inside knowledge. "There are a number of private (or recently acquired) companies, plus some large public ones, that are making impressive progress towards artificial general intelligence, but the good ones are very secretive about it," he wrote just before taking over YC, alluding to Google's recent purchase of DeepMind.[9] His plans for what YC could do on this score were vague when he wrote about them as YC president a few months later. "Relative to the potential impact, it doesn't seem like enough smart people are working on this."

Altman's vision for the new YC included also robotics, biotech, healthcare, education, internet infrastructure, science, "levers" such as new programming languages or powered exoskeletons, transportation and housing, and food and water. The latter topics he seemed to approach as a doomsday prepper. "At some point, we are going to have problems with food and water availability," Altman wrote. "Technology can almost certainly improve this." A couple years later, he would publicly declare that preparing for the apocalypse, whether it was caused by a synthetic virus or rogue AI, was one of his personal hobbies. "I have guns, gold potassium iodide, antibiotics, batteries, water, gas masks from the Israeli Defense Force [sic], and a big patch of land in Big Sur I can fly to," he told *The New Yorker*.[10]

Altman's restless attention could not be held by companies trying to offer slightly better airport rideshares or HR software. "He's a high-impact person who wants to do some somewhat crazy things," Groom said. "He's always been a deeply, deeply intellectual person. I think it comes from a pursuit of importance. As a person that smart, it's hard to stay interested in consumer social apps."

Altman was so committed to his hard-tech vision that he went out and proactively recruited aviation and energy startups to join YC, something that the previous regime, already overwhelmed by the pile of applications flooding in, would never have considered doing.

Like Graham, Altman also wanted to make YC bigger—much bigger—expanding from tens to hundreds of startups in each batch, which required bringing on many more partners to mentor them. To fund this growth, Altman needed to clean up YC's relationship to the investment community. For years, YC had jumped into bed with one venture firm or another, allowing the likes of Sequoia or Andreessen Horowitz to help fund batches and get their own return. Over the years, this program got a name: YCVC. But as YC grew, conflicts and a feeling of unfairness arose as these same firms competed for investments at Demo Day. (For example, Sequoia's Greg McAdoo had famously swooped in and preemptively offered to lead a $600,000 seed round into Airbnb before it had even presented at Demo Day, causing some grumbling among other YC partners and VCs.) So, months before his new role was revealed, Altman began courting institutional investors, such as pension funds and university endowments, which are typically the backers or "limited partners" behind venture capital firms like Sequoia. Two such entities—Stanford's endowment and Willett Advisors, which managed Michael Bloomberg's philanthropic assets—signed on to jointly invest $100,000 in every YC startup, allowing YC to boost its investment in its startups from $17,000 to $120,000. But when Altman delivered this welcome news to the YC community, he skipped over the part about where the money was coming from. (Altman will go to elaborate lengths to avoid discussing even the *existence* of LPs in the press, let alone drop their names.) Instead, he laid out a head-scratching financial structure. "Although YC itself continues to have no LPs (and that way we have the flexibility to do things like fund non-profits), a portion of the investment is from a fund YC manages that does have LPs," he wrote.[11]

Still, bringing in two institutional investors was a sweeping change for YC. As the batches came to know Altman better, they saw that his true genius lay in such acts of financial ingenuity. As Checkr hurtled toward Demo Day with strong user growth numbers, Altman counseled them to skip the typical seed round and go straight for the larger Series A—advice that flew in the face of the prevailing wisdom that companies should "find product-market fit" before making that kind of commitment, which normally involved giving the Series A investor a board seat. "That's when I saw the raw talent of Sam, and the crazy experience he has on fundraising and startups," Yanisse said. "We probably saved a few years by doing that." Today, Checkr is worth $4.6 billion.

FOR ALL of Altman's terseness and inability to tolerate blather, founders were also surprised by his openness and accessibility. Walker Williams, who had grown up down the street from the Altmans, was working on a T-shirt-making startup called Teespring, and had not even considered Y Combinator when his brother connected him to Altman. The company had focused initially on the college market and was growing fast. Even though it was far from the hard tech of his dreams, Altman was interested. "One thing that's striking about Sam is, when you're talking to him, you have his full attention," Williams said. Altman suggested they apply to YC, even though they were a bit past the deadline, and they got into the winter 2013 batch. Once there, Altman, who was still a YC partner at the time, would conduct walking office hours while doing laps up and down the cul-de-sac in front of YC's headquarters. As Teespring's revenue took off, Altman became its first investor.

In November of 2013, when Altman knew he would be taking over YC but few others did, his younger brother Jack, who had spent the prior two years working at Hydrazine, joined Teespring as its VP of business development. The following January, their brother Max joined Zenefits, which was in that same YC batch, moving from Chicago where he had been working as a trader at the Sequoia-backed high-frequency trading firm Allston Trading. The younger brothers initially moved in with Altman on a temporary basis, and ended up living with him in San Francisco for most of the five years that he led YC.

Altman kept his love life in the YC family as well, beginning a relationship that year with Groom, who was then a nineteen-year-old early Stripe employee. Groom had joined Stripe directly out high school in 2012, after founding and selling three startups in his native Australia, and seemed to epitomize PG's thesis that the best founders were often the youngest.[12] The relationship was serious enough that Altman introduced Groom to his family, some of whom expressed concern that, at age twenty-eight, he was dating a teenager.

Altman's belief that things work best within a tight, trusted network—or, as he later framed it in a public talk with Jack, the importance of "finding your tribe"—was perhaps most clearly revealed by the board of overseers he set up shortly after becoming president of YC.[13] Modeled on such entities at universities including Harvard, which are more advisory and have less governing authority than boards of directors, the group had pretty much one job. "The Board will be responsible for hiring and firing the YC President, and occasionally helping with strategic direction," Altman wrote. "Hopefully it doesn't have to meet very often." Many of Altman's friends were named to the board: Brian Chesky (founder of Airbnb and Altman's fellow Sequoia scout); Adora Cheung (the founder of YC graduate Homejoy); John Collison; Patrick Collison; Drew Houston (founder of YC superstar Dropbox); Jessica Livingston; David Rusenko (founder of YC graduate Weebly); and Emmet Shear (Altman's batchmate and founder of Justin.tv and Twitch). Altman was on the board, himself.[14] Nearly all of the others had either taken Altman's money, given him some, or served alongside him in some trench or another. It was, in some ways, the perfect picture of the CEO-friendly board that YC had helped make more common throughout the Valley.

To expand YC in the manner he desired, Altman needed more cash, and to get it he would turn the accelerator into the very thing it was founded to push back on: a venture capital fund. He began exploring raising a new fund that would make later stage investments in YC companies, and in the process earn the kind of fees that venture capital funds earn: around 2 percent of the amount managed annually, plus carried interest of around 20 percent of the profits.

To run the fund, Altman decided to pursue someone from beyond the insular venture capital clique of Sand Hill Road. "I wanted someone who was a real company operator," Altman said. "Because one of the critiques of YC is you give good advice to early-stage startups, but then you know nothing about how to scale a startup, which is true, but neither do most VCs, in my opinion. So I wanted someone who had done that."

During the first summer of his tenure as YC president, Altman had dinner with Ali Rowghani, a former Pixar executive who had just left his post as chief operating officer of Twitter after four years of investor pressure due to Twitter's lack of revenue growth. Over sushi, Altman and Rowghani talked about physics. A few weeks later, Altman texted to ask if he'd be interested in being a YC partner. Rowghani, an Iranian American Stanford grad and McKinsey veteran, wanted to know what restrictions might apply. The only one was that he couldn't join a rival VC firm. Rowghani had no intention of doing so, and was inspired by YC, so he agreed. That winter, as a partner, he mentored companies including Coinbase and Stripe. Soon enough, Altman came with a bigger ask: Would Rowghani consider leading a new fund that would invest in YC companies at later stages?

The idea had been batted around within YC for years, especially as PG and his employees saw venture funds like Sequoia and Andreessen Horowitz profiting handsomely from the winners that YC had cultivated. After Altman became president, one blogger estimated that Sequoia had made more money from YC than YC itself, because it was a Series A investor in YC's three superstars: Airbnb, Dropbox, and Stripe. (Series A investors usually get 25 percent equity in a company, compared to YC's typical cut of 7 percent.) "Sequoia can double down on the fastest-growing companies and YC can't," wrote Robin Dahar.

"The way YC is currently structured doesn't allow them to make follow-on investments to get larger stakes from the runaway winners like Airbnb and Dropbox (though some of the YC partners do make personal follow-on investments). If YC started doing follow-on investments in the 'winners,' it would send a pretty bad signal about the companies they didn't follow-on invest in, who would be considered 'losers.' These companies would have a hard time raising money from other investors."[15] (Sources familiar with Altman's thinking dispute that Sequoia had profited more from YC companies than YC had.)

Altman and Rowghani brainstormed how to get around these limitations. One idea was to use the fund to invest in all YC graduates that were valued at more than $300 million. They then went out on the road recruiting LPs. They expanded the pool beyond Stanford and Willetts to include the endowments of Yale, Wash U, and Penn. As he had with the wireless carriers at Loopt, Altman commanded the room at these meetings, tossing out outlandish statements with such authority that the LPs were convinced he could see into the future. Rowghani would then come in and clean up the details. Between them, they raised $700 million for what would be called the YC Continuity Fund. It caused some panic in the Valley, as VC firms assumed that YC would just continue to eat ever larger portions of their lunch.

Even as he prepared a path for YC itself to make follow-on investments, Altman cracked down on the ability of YC partners to do the same, installing a new rule that they had to wait until after Demo Day or be part of a Series A. His goal was to counter the appearance of favoritism. He also created an email list to alert all interested investors when YC companies were raising rounds outside of Demo Day, to avoid accusations that he was leaving anyone out. His emails have the stern tone of a nightclub bouncer, or perhaps the startup investment police. "The rules for membership are simple," he wrote. "5 total investments in YC companies of any size or 2 big ones, a positive reputation among our alumni, and no history of bad behavior like breaking term sheets without great cause, pressuring founders into advisor shares in addition to an investment alongside others in a round, etc."[16]

Altman found it exhausting. "You're kind of enemies with two-thirds of the Valley at all times," he said. "You end up being like the cop. This investor is treating this founder badly, so you've got to call and yell at them. This founder is doing a bad thing, you've got to call and yell at them. The jealousy of YC among other investors is super intense. It's kind of stopped now, but for a long time it was just like, 'How do we take down YC?'"

Indeed, Altman got involved when a YC company needed it. Reddit had always been a kind of YC mascot, birthed in its initial batch, shaped from the start by Graham's hankering for a front page of the internet instead of the food-ordering idea its founders had brought him. Its initial success, launched with a link from Graham's blog, had been a proof of concept for the then-still-tiny YC network, which had been Reddit's first set of users. And YC's purchase by Condé Nast, the most glamorous media company on earth, while still in its infancy put it squarely on the mainstream media map (never mind that Condé paid the outrageously cheap price of $10 million, which was probably about equivalent to its budget for black cars and fresh flowers). But from the beginning the anarchic Reddit, with its self-organizing communities sharing everything from confusing cookie-cutter shapes to photos of underage girls, had been an odd fit for the owner of *Vogue*, *Vanity Fair*, and *The New Yorker*.

Reddit had kept growing, in spite of what many in the Valley saw as a kind of benign neglect from its parent company, but had never been profitable. Within a few years, all of its co-founders had moved on to other projects.* Condé decided to move on, too. "It just didn't naturally fit within the company, and it was clear that they needed better talent, more capital, and clearly its own platform," said Bob

---

\* Aaron Swartz, who had joined Reddit as a co-founder in a kind of Graham-officiated shotgun wedding, had been fired in 2007 after struggling to adapt to corporate life. He went on to become an activist for copyright reform and open access to information, and hanged himself in 2013 while facing federal hacking charges and the possibility of decades of jail time for downloading a large trove of academic papers from MIT.

Sauerberg, who was CEO at the time. "The thought of integrating it into the Condé Nast platform made no sense. So we thought that if we spun it out, we could raise money with people who could help us, and we could hire a team that was native to the business." In 2011, Condé Nast's parent company, Advance Publications, turned Reddit into an independent subsidiary, and recruited PayPal and Facebook veteran Yishan Wong to lead it as CEO. The new structure would allow Wong to raise money for Reddit like it was any other Silicon Valley startup. That prompted him to make a visit to YC.

In the spring of 2014, Wong pulled into YC's Mountain View parking lot in a blue Tesla, at the same time Altman was stepping out of his own blue Tesla, according to Christine Lagorio-Chafkin's book on Reddit, *We Are the Nerds*. The two men hit it off. Both loved Reddit and understood the Valley. Wong had come looking for insight into the fundraising landscape generally, not knowing that Altman had a fund of his own. Two days later, Altman called Wong to say he wanted Hydrazine to lead an investment round in Reddit. He then spent weeks working the phones to woo Hollywood names including Jared Leto and Snoop Dogg, as well as VC luminaries like Thiel, Andreessen, Sequoia's Alfred Lin, and Thrive Capital's Josh Kushner.[17] By September, they raised $50 million—$27.5 million from Hydrazine—and Altman joined the board. The announcement said that the investors would set aside 10 percent of the shares for the community of Reddit users who generated the site's content. "It's always bothered me that users create so much of the value of sites like reddit but don't own any of it," Altman explained on his blog, when he announced the round. "So, the Series B Investors are giving 20% of our shares in this round to the people in the reddit community, and I hope we increase community ownership over time. We have some creative thoughts about the mechanics of this, but it'll take us awhile to sort through all the issues."[18] Wong, who came up with the idea, was more direct in an interview with *TechCrunch*: "We have a crazy plan and what we're going to do is create a crypto-currency that is backed by those shares. Then we're going to distribute the currency to the community through some reasonably fair way that reflects the contributions of the community."[19] His own blog post on the matter had an all-caps disclaimer: "CAVEAT: KEEP IN MIND THAT THIS PLAN COULD TOTALLY FAIL."[20]

The plan did, in fact, totally fail. Within weeks, Wong resigned in part due to the company's rebellion against his plan to move its offices from San Francisco to the unappealing suburb of Daly City, which had inspired the Malvina Reynolds song about "little boxes made of ticky-tacky" that Pete Seeger made a hit.[21] That left Altman to serve as CEO for eight days before handing that title to Ellen Pao, an impressively credentialed Reddit executive best known for suing her former employer, the venture firm Kleiner Perkins, for gender discrimination. Amid the chaos, Reddit co-founder Alexis Ohanian returned as executive chairman, and Altman began reaching out to his somewhat estranged co-founder, Steve Huffman, for advice. Altman and Huffman formed a friendship that they had never really had during their time as YC batchmates a decade earlier, when Altman was wandering around Cambridge alone with his cargo shorts pockets stuffed with mobile phones and Huffman was making risqué jokes on AOL Instant Messenger. As the chaos at Reddit continued—Pao and Ohanian did not get along, and the company was still reeling from its role in sharing a trove of nude celebrity photos leaked by hackers—Sauerberg and Altman lobbied Huffman to return with increasing urgency. (Sauerberg and Steve Newhouse, a member of the family that has controlled Condé Nast for generations and a passionate backer of Reddit, had been talking to Huffman for years.)

Huffman was tempted but unable to imagine leaving behind Hipmunk, the travel search company he had co-founded after leaving Reddit. At one point, knowing Reddit needed engineering talent as well as new leadership, he suggested that Reddit just buy Hipmunk. For weeks, Huffman and Altman discussed the possibility, with the Reddit investors open to the acquihire. In the end, Huffman's co-founder at Hipmunk couldn't bring himself to agree to it. However, the exercise revealed to Huffman just how much he wanted to go back and fix his first company. When Pao resigned in July 2015 during a mutiny by

Reddit users over the firing of a popular employee, Huffman returned to lead Reddit. The entire process was a mess, but Altman never wavered from his belief that Reddit could be one of the great internet sites. "He is always thinking 100-times growth, always keeping us focused on, 'It's got to be up and to the right,'" Sauerberg said.

Nearly a decade later, Reddit went public on the New York Stock Exchange, and the 8.7 percent that Altman's Hydrazine acquired through the Series B and subsequent purchases became worth more than $600 million.[22] (The long-delayed IPO happened only after Reddit signed deals to license its data to train AI models like OpenAI's.)

"One of Sam's qualities is, he gets it done," Huffman said. "He seems to be drawn to the messier, more complex deals. That seems to get him excited. More so than just about anybody I know, he loves being in the shuffle, when things are mixed up, getting deals done, getting relationships made."

**WHILE ALTMAN** was busy trying to rescue Reddit, a surprise bestseller was helping to transform informed laypeople's perception of AI. Swedish philosopher Nick Bostrom's book *Superintelligence: Paths, Dangers, Strategies* was dense and academic, with long passages filled with mathematical equations and logical proofs. Yet in August 2014 it landed on *The New York Times* bestseller list after Elon Musk tweeted his agreement with its premise. "Worth reading Superintelligence by Bostrom," Musk told his 10 million followers. "We need to be super careful with AI. Potentially more dangerous than nukes."

*Superintelligence* begins with a fable about a flock of sparrows who decide that it would be useful to have an owl around to help them build their nests and look out for cats. They begin to plot how to find an owlet that they could raise to do their bidding. One of the sparrows speaks up to ask if they had not perhaps better figure out how to tame an owl before dragging an egg into their nests, but the leaders determine that finding the owl is hard enough. They can worry about taming it later. The skeptic and a sad handful of followers stay behind, understanding that learning how to tame an owl without one to practice on would be even harder than they thought. The story ends with their fear that the fellow sparrows might return with an egg before they have figured it out.

In the book, Bostrom argues that humans will likely create what he called "machine superintelligence" sometime in the twenty-first century, and thus had better get to work making sure that it does not destroy all of humanity. To illustrate how AI might take over, he borrows Yudkowsky's metaphor of the paperclip, though gives it a twist. A superintelligent AI programmed to make paperclips might just keep going until all matter in the universe—including the fleshly bodies of all sentient beings—is turned into paperclips. "This is quite possibly the most important and most daunting challenge humanity has ever faced," he writes. "And—whether we succeed or fail—it is probably the last challenge we will ever face."[23]

Bostrom and Yudkowsky emerged from much the same milieu. Growing up in Sweden, Bostrom was highly intelligent and eager to be done with traditional schooling. By the time he was fifteen, in the late 1980s, he had what he calls "a general sense that things that could change the thing that does the inventing and idea discovery, whether it be through AI or ways of enhancing the human biological condition, could be very important." That led him explore neural networks, and ultimately to pursue graduate work in the philosophy of science. But it was not until the internet's arrival in the mid-1990s that he found other people "interested in potential future technologies, like AI and nanotech," he said. "It was really very niche. No academics were interested in this at all." He discovered the Extropians listserv and became an active participant alongside Yudkowsky. "That was really where the action was in terms of the most advanced conversations about some of these ideas," he said.

(In 2023, his participation in the rollicking conversations on this listserv would come back to haunt him when researcher Émile Torres found a 1996 email from Bostrom, then a twenty-three-year-old grad student at the London School of Economics, to the group. "Blacks are more stupid than whites. I like that sentence and think it is true. But recently I have begun to believe that I wouldn't have much success with most people if I speak like that. They would think that I were [sic] a 'racist': that I disliked black people and thought that it is fair if blacks are treated badly. I don't. It's just that based on what I have read, I think it is probable that black people have a lower IQ than mankind in general, and I think that IQ is highly correlated with what we normally mean by 'smart' and 'stupid.' I may be wrong about the facts, but that is what the sentence means for me. For most people, however, the sentence seems to be synonymous with: I hate those bloody [n-word, not redacted in the email]!!!! My point is that while speaking with the provocativeness of unabashed objectivity would be appreciated by me and many other persons on this list, it may be a less effective strategy in communication with some of the people 'out there.'"[24] Bostrom apologized for the email in 2023.)

The following year, in 1997, Bostrom co-founded the World Transhumanist Association, which would seek to build on the work of the Extropy Institute and push transhumanism toward academic respectability. "His interest in science was a natural outgrowing of his understandable desire to live forever, basically," his friend Daniel Hill told *The New Yorker* years later.[25] In 2005, it seemed respectability had been achieved, when the futurist and philanthropist James Martin put up the money for Bostrom to fund the Future of Humanity Institute at Oxford University, giving him a full-time staff to study "the big challenges for humanity in the twenty-first century."

Bostrom is careful in *Superintelligence* not to make specific predictions about how soon machine intelligence might surpass humans', but his book appeared at the same time as DeepMind's Atari experiments, and many people in the field noticed. One of his central arguments is about importance. His predictions might all be wrong, he suggests, but if there's even a small chance that he's right, there is nothing more important. A generation of what would become AI safety researchers read the book and found their life's purpose.

For the AI-obsessed Altman, the long odds and galactic import were too tempting to ignore. One of his favorite essays, "Meditations on Moloch" by the blogger Scott Alexander on his blog *Slate Star Codex*, is largely a response to Bostrom. In his famous poem *Howl*, Allen Ginsberg conjured Moloch, the ancient child-eating god, as the freedom and beauty-destroying force of civilization itself. (Many have argued that for Ginsberg, Moloch was capitalism.) For Alexander, Moloch is the game theory that traps us into self-defeating dynamics like arms races. "From a gods-eye-view, the best solution is world peace and no country having an army at all. From within the system, no country can unilaterally enforce that, so their best option is to keep on throwing their money into missiles that lie in silos unused," he wrote.

Alexander argues that Bostrom's vision of an AI smarter than humans offered humanity a way to fight back against the forces of Moloch. "The only way to avoid having all human values gradually ground down by optimization-competition is to install a Gardener over the entire universe who optimizes for human values," Alexander writes. "And the whole point of Bostrom's Superintelligence is that this is within our reach."

Yes, machines smarter than humans being able to design machines smarter than they are could leave us "super-doomed" if "multiple, competing entities were likely to do that at once." But if one gets super-smart so fast that "it can suppress any competition," then it has the chance to determine the shape of our future, potentially for the better. "In the very near future we are going to lift something to Heaven. It might be Moloch. But it might be something on our side. If it's on our side, it can kill Moloch dead. And if that entity shares human values, it can allow human values to flourish unconstrained by natural law."

At this point in the essay, Alexander realizes that it sounds like he's saying humankind could have the chance to create or replace God. He doesn't deny it. "To expect God to care about you or your personal values or the values of your civilization, that's hubris," he writes. "I am a transhumanist because I do not have enough hubris not to try to kill God."[26]

CHAPTER 11

# "A MANHATTAN PROJECT FOR AI"

THE GATHERING AT THE BEACHSIDE CARIBE HILTON IN SAN Juan, Puerto Rico, in January 2015 seemed like a typical academic junket to Cornell computer science professor Bart Selman, until he spied the cluster of security guards loitering near the door of a conference room. When the same burly fellows followed the attendees to lunch, it became clear this was not a normal conference. Amid the academics and Google employees and curiously large posse of young math PhDs from Berkeley, there seemed to be quite a few billionaires, including a particularly enthusiastic one: Elon Musk.

Musk arrived, cheeks pink from the heat, hair somewhat disheveled as if he had possibly slept under his desk at Tesla the night before. He was wearing a black T-shirt featuring a cartoon astronaut. At the time he was CEO of both Tesla and SpaceX, but eagerly participated in every part of the world's first conference on AI safety, sharing a stage with the balding and bespectacled Bostrom to echo his warnings about the dangers of building smart machines without better guarantees they would care about their human creators. Over lunch, Musk went around the table asking the assembled AI researchers how humankind could control the risks of superintelligence.

"Elon Musk really saw it as existential," recalls Selman. "At that time, it seemed a little outrageous to worry about machines taking over."

AI researchers like Selman had learned, through successive cycles of hope and despair, to stay quiet about the potential of artificial intelligence, after the excessive exuberance of earlier generations had shoved their field to the academic margins. But recent gains in computer vision, machine translation, and self-driving cars were remarkable, and Bostrom's *Superintelligence* had launched a movement that went far beyond the dusty corner of academia where AI researchers had long huddled.

The conference had been organized by the Future of Life Institute, founded less than a year earlier by MIT physicist Max Tegmark and funded by a $100,000-a-year pledge from Skype co-founder and effective altruist Jaan Tallinn. After becoming a billionaire when Skype sold to eBay in 2005, Tallinn was casting about for his next act when he encountered Yudkowsky's writings on AI risk. He was initially skeptical, but after sitting down with Yudkowsky for four hours at a Panera Bread near San Francisco International Airport, he converted. Soon, he was one of the most prominent individuals raising the alarm about the existential risk of AI, founding the Centre for the Study of Existential Risk and investing early in DeepMind.[1] The Future of Life Institute, which would become more of a political force than an academic one, quickly settled on AI safety as its top priority. "Our goal was simple: to help ensure that the future of life existed and would be as awesome as possible," Tegmark wrote in his book, *Life 3.0: Being Human in the Age of Artificial Intelligence*. After brainstorming about all the various threats to humanity, he wrote, "there was broad consensus that although we should pay attention to biotech, nuclear weapons and climate change, our first major goal should be to help make AI-safety research mainstream."[2]

Up to that point, many of the most prominent voices advocating for AI safety were not part of the community of researchers actually working on AI. The movement was led by people like Bostrom and Yudkowsky, who would also share the stage with Musk and Selman that afternoon, alongside Vernor Vinge and DeepMind's Shane Legg, in a kind of reprise of the Singularity Summits of old. The goal of the Puerto Rico conference was to bring the philosophers and practitioners together to try to forge some common agreement, before the economic incentives of the latter made that impossible. "It was an important conference because it helped prevent one possible failure scenario, which would be where you would have two separate groups, the AI risk warriors and the people building AI. You could imagine in an alternative history these would have formed different communities that didn't talk to one another, but just kind of threw dirt at one another," Bostrom said.

Instead, most of the participants—and more than eight thousand people in all, including Musk, Stephen Hawking, and many AI researchers—ended up signing an open letter calling for the goal of AI research, in Tegmark's words, to be "not undirected intelligence, but beneficial intelligence." The letter called for the expansion of research aimed at ensuring AI continued to be "beneficial," adding, "Our AI systems must do what we want them to do." Nothing in the letter was very controversial, but the fact that it existed was a huge leap toward the mainstream for a set of beliefs long considered fringe. Among the AI practitioners who signed were DeepMind's Demis Hassabis and two AI researchers, Ilya Sutskever and Dario Amodei.[3] Tegmark was giddy about the conference's impact, quipping, "Perhaps it was a combination of the sunshine and the wine."[4] At the end, Musk took Tegmark into a private room and told him he was donating $10 million to his institute to put toward AI safety. A few days later, he announced it on Twitter, joking, "It's all fun & games until someone loses an I."[5]

During the breakout sessions between talks, Musk had gathered with Selman and some other researchers overlooking the ocean, and fretted about his primary concern: that Google (through its ownership of DeepMind and Google Brain) and Facebook (through its AI division, led by respected AI researcher Yann LeCun, that powered features like automatic photo tagging) were completely dominating the field of AI, yet were under no mandate to share their research with the public.

"It was definitely recognized as a risk that there was no control over these commercial, closed-source models, and maybe there should be a counter to that," Selman recalled.

Altman had not attended the conference, but the following month published his own doomerish essay inspired by Bostrom, in which he recommended *Superintelligence* and declared, "Development of superhuman machine intelligence (SMI) is probably the greatest threat to the continued existence of humanity." He was sympathetic to Bostrom's argument that superintelligent machines were one of the more plausible answers to Fermi's paradox, the "Where is everyone?" question posed by Italian physicist Enrico Fermi over lunch at Los Alamos National Laboratory in 1950. Perhaps the reason we have seen no evidence of aliens despite there being piles of evidence of the existence of life-friendly planets is that, as Altman put it, "biological intelligence always eventually creates machine intelligence, which wipes out biological life and then for some reason decides to make itself undetectable."[6] The answer, in Altman's view, was to regulate AI. "The US government, and all other governments, should regulate the development of SMI," he wrote in a follow-up essay. "In an ideal world, regulation would slow down the bad guys and speed up the good guys." Only at the end of the second essay did he take up the possible upsides of this new technology, with a nod to the way that the idea of AGI has its roots in the transhumanist dream of immortality. "It could solve a lot of the serious problems facing humanity—but in my opinion it is not the default case. The other big upside case is that machine intelligence could help us figure out how to upload ourselves, and we could live forever in computers."[7]

The primary person he thanked for helping him with the post was Dario Amodei, an AI researcher then working at Chinese internet company Baidu alongside Altman's old Stanford AI Lab mentor Andrew Ng.

Altman and Musk started having regular dinners each Wednesday when Musk would come to the Bay Area on his weekly rotation through his various companies. They had known each other for several years, ever since YC partner Geoff Ralston introduced them and arranged for Altman to tour Musk's SpaceX factory in Los Angeles. "He talked in detail about manufacturing every part of the rocket, but the thing that sticks in memory was the look of absolute certainty on his face when he talked about sending large rockets to Mars. I left thinking, 'Huh, so that's the benchmark for what conviction looks like,'" Altman wrote years later.[8]

Their conversations in 2015 were mostly centered on fear. Bob Sauerberg, who was working closely with Altman on Reddit at the time, recalls being startled to hear Altman describe his meals with Musk. As he characterized Altman's account of those conversations: "Elon and I were talking. The world's going to come to an end. We've planned out there are only two safe places, Big Sur and New Zealand. I've got a place here. He's got a place there. And we're going to build out, and duplicate, and we're realizing that this AI thing is real. It could be really bad, but really big and important, and we've got to be sure that we're safe, and we've got to do something good about it." (While Musk's Starlink has a presence in New Zealand, it was likely Musk's fellow PayPal Mafia member Thiel, who has a place in New Zealand, whom Altman was talking about.)

The same month that Altman posted his essay calling for the regulation of AI, he reached out to Musk to help him draft an open letter to the government calling for the same. Musk agreed, and they worked on drafts together of a letter that was ultimately released later in 2015. Even before its release, word began to spread about their letter, and Hassabis confronted Musk directly about whether it was really true. Musk replied: "If done well, it may very well accelerate AI in the long term. Without the public comfort that regulatory oversight provides, there could very well be a situation where AI causes great harm and therefore AI research is banned as dangerous to public safety."[9] According to legal documents filed by Musk's lawyers, five days after Hassabis reached out, DeepMind announced it was creating an AI ethics board and invited him to be part of it. The first meeting, which Musk hosted at SpaceX, convinced him that the board was an attempt to slow-walk any talk of regulation. "Elon's takeaway was the council was basically bullshit," Musk's chief of staff, Sam Teller, told Walter Isaacson. "Those Google guys have no intention of focusing on AI safety or doing anything that would limit their power."[10]

Musk now began hosting dinners with friends like Peter Thiel and

Reid Hoffman, aimed at figuring out how to counter Google's power and make AI safe. He met that May with President Obama over the need to regulate AI. "Obama got it," Musk told Isaacson. "But I realized that it was not going to rise to the level of something that he would do anything about."[11]

That same month, Altman wrote Musk that he had been "thinking a lot about whether it's possible to stop humanity from developing AI. I think the answer is almost definitely not. If it's going to happen, it seems like it would be good for someone other than Google to do it first." Altman proposed that Y Combinator start a "Manhattan Project for AI"—the same term Hassabis had used in pitching DeepMind in 2010—adding that "we could structure it so that the tech belongs to the world via some sort of nonprofit but the people working on it get startup-like compensation if it works. Obviously we'd comply with/ aggressively support all regulation."

"Probably worth a conversation," Musk replied.

In June, Altman sent Musk a detailed proposal for a new AI lab. "The mission would be to create the first general AI and use it for individual empowerment—ie, the distributed version of the future that seems the safest. More generally, safety should be a first-class requirement." He offered up a YC building in Mountain View to house the initial team of seven to ten people, and proposed offering the researchers YC equity "for the upside," though it would be "uncorrelated to what they build." For governance, he suggested a five-person board that included himself and Musk. He added, "At some point we'd get someone to run the team, but he/she probably shouldn't be on the governance board."

He was careful not to demand too much of Musk's time—a once-a-month visit would do, plus public support to help with recruiting. And he suggested pausing on the open letter calling for regulation until the lab was off the ground.[12]

Meanwhile, Altman was beginning to engage Greg Brockman—an old friend from Stripe who he had helped dispose of his early Stripe shares on the private markets over the years—in increasingly serious discussions about the AI lab.

Brockman had grown up on a hobby farm outside of Thompson, North Dakota, the third of four children born to two doctors—Ron Brockman, an ophthalmologist, and Ellen Feldman, a psychiatrist—who raised horses, cows, and chickens in their spare time. From a young age, he showed an aptitude for math and science, which his mother encouraged by enrolling him in math camps and chemistry competitions, and his father nurtured by teaching him algebra in sixth grade. Brockman started a math club with one of his friends. When it came time for high school, his older siblings went away to boarding school, but he had found the local public schools to be accommodating to his needs, allowing him to spend his eighth-grade year doing an independent study using the distance learning curriculum from Stanford's Education Program for Gifted Youth. Running at his own pace, he completed three years of math that year, allowing him to enroll in calculus as a freshman. By sophomore year, he had a car—the driving age is fourteen in North Dakota, and he began driving at fifteen—and was able to take classes at the University of North Dakota. But he stayed enrolled at Red River High School, where he was active in theater, playing Charlie in *The Best Christmas Pageant Ever*—perhaps a strange role for a kid whose forebears were Jewish.[13]

Brockman assumed he was on his way to becoming a mathematician, but after high school, his mother urged him to take a gap year. He decided to write a chemistry textbook, but a friend said he would never get it published because he didn't have a PhD; he would either have to self-publish or make a website. Brockman, who had attended a computer camp one summer with his older brother, Matt, and spent the summer programming rather than playing computer games, figured it would be easier and cheaper to build a website. He began by programming a table-sorting widget. "You could click a column and it would sort the table," he told the University of North Dakota's newspaper. "The thing for me was, I had pictured it in my head, and now it existed! That was the 'wow.'"[14] He read Alan Turing's "Computing Machinery and Intelligence," the 1950 paper posing the question "Can machines think?" and laying out the now-famous

"Turing test" to determine the answer. In Turing's view, if humans can't tell a computer's responses apart from a human's, the computer has passed the test. He proposed creating such an intelligence, not by hard coding in logic, as would become popular in later years, but by creating the equivalent of a computer baby that could learn, and then letting it teach itself. Inspired, Brockman spun up a chatbot that succeeded in nattering somewhat convincingly about the weather, if not much else.

"I used to have this view that if anything I ever did in math was used in my lifetime, then it wasn't abstract enough, it wasn't a long-enough time horizon," Brockman said, pointing to the work of nineteenth-century mathematicians Évariste Galois and Carl Friedrich Gauss, whose breakthroughs were not built upon until they were long dead. "To me, that was beautiful. You're making a long-term contribution to humanity."

But watching his chatbot get featured on StumbleUpon and rack up 1,500 users changed all that. "It was just the coolest feeling ever. I built this thing and people are using it. There are actual humans on the other side of the wire." He decided to forgo mathematics and plunge headlong into computer science.

When he enrolled in Harvard that fall, he beelined to a professor who specialized in natural language processing and asked if he could do research with him. "He showed me what he was working on, and it was parse trees—all of this old-school AI stuff where you have to write down the rules of language and stuff," Brockman said. "I looked at it and I was like, 'This is never gonna work. This is not that stuff that Alan Turing was talking about in his Turing test paper.'" He joined Harvard's computer club, which was fun for the first year, when he could listen to older students discuss advanced topics. But by his sophomore year, the upperclassmen had graduated and he was in charge of the computer club with no one to learn from. He found the liberal arts mandate of the university frustrating. "I found very quickly that it was hard to go deep, because you're at a table and you'd be talking math or computers or whatever, and then somebody from a totally different discipline would sit down, and you'd have to stop that conversation."

He found his desires were narrowing too much for a Harvard student. "What I really wanted was, I wanted to become the best programmer. That was the goal I set for myself: become the best programmer in the world." He dropped out and enrolled in the Massachusetts Institute of Technology. But even at that more technical university he could not quite scratch the itch. Within a month, he was having conversations with startups, including flying out to the Bay Area in October to meet with Patrick and John Collison at Stripe, which had just been founded.

Patrick Collison remembers him showing up at the cramped back-alley office on Ramona Street in Palo Alto at 10 p.m. on a rainy evening. "We opened the door, and there's Greg. I was working on setting up some new servers at the time, and Greg just plonked down in a chair beside me and got to work."

Later that weekend, during a more formal interview over frozen yogurt, Brockman asked, "Do you guys work on weekends?" The Collison brothers hemmed and hawed, groping for the answer that might not scare off someone they hoped would be one of their first hires. "We tried to answer honestly but in a way that wouldn't alarm him, 'In fact we do kind of work all weekend, but we don't necessarily expect the same thing of every employee'—all the usual tempering."

Brockman cut him off. It was OK. "I just want to make sure I'm dealing with some serious people who do work on the weekends." He was in.

His mother, however, wanted more assurance. She asked for a call with the Collisons' mother. "I guess our mom passed, because Greg joined," Patrick said.

Brockman dropped out of MIT in 2010 and joined Stripe as its fourth employee. Stripe gave him the startup life of his dreams. Stripe employees recall arriving at 7 a.m. to find Brockman passed out on his desk, music blaring, pizza boxes strewn about, having coded through the night. He walked around the office in socks, and liked to spend long stretches in so-called "code caves," sometimes emerging with fully formed products that nobody had asked for. Two things set Brockman apart, besides his intelligence, which Collison said was a prerequisite for all engineers: his wide-ranging curiosity about the broader systems that Stripe's business was part of—in this case, taking credit card payments online—and "his almost unparalleled willfulness and life force," as Collison put it. "Very few people are more determined than Greg."

By 2015, Brockman had begun to have conversations with Patrick Collison about his next steps. Stripe was now worth $3.5 billion, and Brockman was now CTO and spending more and more time managing people, instead of coding. After telling Collison he probably wanted to leave, Collison suggested that Brockman go talk to Altman, who often had good advice in such situations. Altman listened to Brockman and told him he was ready to move on. "Life is not a dress rehearsal," he said.[15] He asked Brockman what he wanted to do.

Brockman wanted to get into machine learning. At Stripe he founded a weekly reading group of *LessWrong*, the blog that Eliezer Yudkowsky founded in 2009 and that had become the center of the rationalist community, where every other essay seemed to be about the revolution in deep learning that was forcing the ethical question about what humans needed to do about their potentially superior creation. Altman said he also was working on an AI project. They agreed to stay in touch. Brockman gave his notice at Stripe in May and in June enrolled in a three-month hacking retreat in New York called the Recurse Center. Before it started, he ordered his first graphics processing unit, or GPU, the type of computer chip best suited to perform the rapid calculations required to train a neural network, off the internet. It arrived in parts, and he spent three weeks before the retreat assembling them. He was going to reinvent himself as a machine learning engineer, and then sometime over the next eighteen months figure out a new startup. In his spare time, he started teaching himself magic.

At the time, Altman was using his bully pulpit as Lord of the

Startups to try to inject a different and more existential kind of ambition into his charges. That Memorial Day weekend, at Camp YC—the accelerator's annual glamping getaway two hours north in Mendocino County—Altman stood on a flimsy wooden stage as the light streamed down between the redwoods and declared there were three things that could fundamentally change humanity. His list of priorities had shifted slightly since his initial blog post as YC president. Now machine superintelligence was at the top, followed by nuclear energy and pandemics. "This is all I want to work on," he told the hundreds of founders sitting before him in folding chairs. The night before, in one of the few fully built cabins reserved for YC staff—the rest slept in three-sided structures open to the elements—he had faced off against his fellow YC partners in Settlers of Catan, a board game in which players compete for resources to build settlements on a mythical island. The game requires a combination of luck and strategy, but the outcome always seemed to be the same. Doktor Gurson, the co-founder of the YC company Doblet, recalls running into a partner who explained his feeling of hopelessness when playing his boss, saying, "People had given up wanting to play with him, because he always wins."

Altman was by this point hosting a series of dinners at his home near the Mission with various AI researchers and other thinkers in the community. Then in July 2015, he decided to go bigger. He organized a private dinner at the Rosewood in Menlo Park, a low-slung resort overlooking the scrubby foothills alongside Highway 280 that was a favorite hangout of the VC crowd working along nearby Sand Hill Road. With high-end jewelry shops, twenty-seven-dollar cocktails, and a bar lined with giant slabs of ancient redwoods, the hotel felt like a clubhouse from a Persian Gulf petrostate dressed only somewhat convincingly in the costume of California eco-minimalism. On its back terrace, a breathtaking view of the Santa Cruz Mountains, each successive ridge receding further into the mist, is marred by the roar of commuters zooming along the highway, an effect only partially masked by the techno music gentling thumping from beneath the potted olive trees.

Altman chose a private room to the side of the restaurant, with its own terrace dotted with succulents where early arrivals could mill about in front of their own outdoor fireplace. The initial crowd included a friend of Brockman's from MIT, Paul Christiano, as well as two other AI researchers: Dario Amodei and Chris Olah. Amodei had grown up in the Bay Area in an Italian American family, enrolling first at Caltech before transferring to Stanford, eventually earning a PhD in physics from Princeton University. He was voluble and passionate, with wild curls that he would tug on as he talked, giving him the air of a mad professor (though his ideas would ultimately help guide OpenAI toward its coveted commercial target). Olah was a child prodigy who skipped college in favor of a Thiel fellowship. Both had been friends with Brockman and helped give him pointers when he embarked on his new career trajectory.[16] Christiano, a fellow veteran of high school math competitions, was a doctoral student at Berkeley studying theoretical computer science.

Altman had also cold-emailed someone whose talks on AI had blown him away when he looked them up online. Ilya Sutskever was already an icon in the world of AI with several breakthroughs under his belt, most notably the 2012 "AlexNet" paper that he co-authored with Alex Krizhevsky and Geoff Hinton, which had revived interest in neural networks. He went on to work at Google after the company bought their small neural networks startup. At Google he was doing groundbreaking research, including a recent paper presented at the previous December's Conference on Neural Information Processing Systems—long abbreviated to NIPS before someone thought better of it and started calling it NeurIPS—on "sequence to sequence" learning, which forcefully articulated the first scaling paradigm that would soon become the engine of all recent AI progress. Yet he was restless. Sutskever tended to favor camping gear for all occasions—floppy Tilley sailing hats with two straps in case of high winds, black Patagonia fanny pack cinched up high, quick-dry pants—as if he might have to jet off on a wild adventure at any given moment.

Born in the Soviet Union to Jewish parents, Sutskever moved to

Israel with his family when he was five. At the same age, he had a sudden recognition of his own consciousness. "It's like I woke up into being a person," he said. "It was just so jarring, to be." The moment would stick with him, and later he would wonder if AI could help answer the mystery of consciousness. "I was interested in souls," he said. "I was wondering, maybe the thing that souls do is learn, because computers can't learn, so therefore if computers *can* learn, what does it say about souls?" From childhood, he was fascinated by "the great unsolved problem" of what learning really was, and whether computers could do it.

He excelled in Israeli public schools to the point that he began taking university classes via correspondence while still in middle school. "I didn't like going to school. I wasn't learning anything and it felt like a waste of time, so there was always an urge to escape the school system," he said.

When he was sixteen, his family immigrated to Canada, where he attended high school for one month as an eleventh grader before dropping out to attend the University of Toronto as a third-year college student. It took intense lobbying by Sutskever and his parents to get the university to accept him without a high school degree, but as soon as it did, he headed for the door of Geoff Hinton, the legendary AI researcher. He was seventeen, and while his main job was selling french fries at the nearby Paramount Wonderland amusement park, he asked if he could come study machine learning with him. It was a Sunday evening, and Hinton suggested Sutskever organize a time to come back for a real conversation. Sutskever was insistent. So Hinton gave him his seminal paper on backpropagation and told him to come back when he had read it. When Sutskever returned a few days later, he said he didn't understand. "It's just basic calculus," Hinton replied. "Oh, no," Sutskever said. "What I don't understand is why you don't take the derivatives and give them to a sensible function optimizer."[17] It had taken Hinton years to get to that point. He sent the teenager home with another paper, and Sutskever returned with a similar insight. "He said, 'You train a neural net to do one thing, and then you train another to do something else. Why don't you train one to do both things?'" Hinton recalled.

"He's the only student I've had who had more good ideas than me," Hinton said. He invited Sutskever to join his lab, despite being behind the other students in formal schooling. Sutskever went on to earn an undergraduate degree in mathematics from the University of Toronto, and then to pursue his PhD in computer science with Hinton as his advisor.

"He's very good at math, but the main thing is, he thinks for himself, and he's very good at thinking," Hinton said. "He's intellectually honest, and he's also fearless. He won't hang on to an idea just because he thought of it."

That combination of traits made him an in-demand leader, as someone who was full of insights but also very open to the ideas of others. Though from early on he maintained a deep conviction about the potential of neural networks. "Ilya has a very strong belief in the neural-net-based AI, and also from early on a very strong belief that if you scaled these neural networks up, they will work better. Not that many people believed that," Hinton said. Many thought that idea was just an excuse for why neural nets still didn't work all that well. "Ilya understood why it would work better if it was bigger, and was probably the researcher most committed to that view."

Sutskever said he came to this conviction through the process of elimination, carefully studying all other forms of AI first to make sure he wasn't making a mistake. "Eventually, I was able to form what I believe was an extremely clear, simple and convincing explanation for why all other forms of AI were hopeless," he said.

In Sutskever's view, the field of AI research had been overtaken by the academic imperative of publishing incremental papers with impressive theorems that made their authors sound smart. He had a name for what was driving those scholars: "math envy." And he felt that his degree in mathematics made him immune to it.

"If that's what guides you, then you like those things you can prove theorems about, whereas the neural net-based AI, you cannot prove theorems about, because it's too complex," he said. "But the thing is, the things we need to learn are complex. So if your AI is too simple, it's hopeless. It will never, ever work. It has to be complex."

Sutskever's conviction rested upon his belief that the human brain was effectively a map to AGI. As Hinton put it, we don't understand what's going on inside the brain of our taxi driver, but that doesn't keep us from confidently getting in a taxi and expecting to get where we want to go. "The fact that neural net–based AI is incomprehensible is a key feature that drives its success. We use an incomprehensible solution to an incomprehensible problem," Sutskever said. "But by far the strongest point was the fact that the human brain is also a neural net built out of small simple neurons. If the brain could do it, then why not an appropriately trained neural net? The question then becomes figuring out what 'appropriately trained' means."

Hinton's entire lab was based on the premise that neural networks could work, and on trying to push forward the boundaries of what they could do. The human brain is made up of 100 billion neurons, which effectively act as little switches that either pass on the signal coming in one end—the dendrite—or not, depending on the signal's strength. The artificial neurons first proposed in the 1940s were crude approximations of these switches, with the analog of the strength of the signal going into the dendrite described in a number called a "weight." Hinton's great breakthrough in the 1980s, backpropagation, involved a mathematical formula for how to train the weights to adapt to experience. "Starts to feel a little bit brain-like," Sutskever said of backpropagation. "It means you can make the neural networks actually learn interesting things." But for many years neural nets still didn't deliver the results their champions had hoped for.

Then around 2010 a new technology came out that changed the game: the graphics processing unit, or GPU. Originally designed to run the graphics in video games, GPUs had the unique ability to rapidly process many calculations at the same time, which allowed researchers to work with far larger neural networks and datasets than in the past. At last, they were able to approach the size and complexity of the human brain. "If you follow this analogy, it makes sense that you need a big one, because the brain is a lot of neurons," Sutskever said.

Hinton and Sutskever believed the various pieces were in place for neural networks to solve computer vision. To prove it, Sutskever suggested the lab enter the annual competition held by Stanford, which had compiled the world's largest database of images, known as ImageNet, and challenged researchers to have their machines correctly label them. The research team with the lowest error rate each year won. Hinton purchased GPUs for the lab, and then stepped back to let Sutskever's lab partner Alex Krizhevsky, a talented engineer with a knack for coaxing performance out of GPUs, work his magic. He fed photos into the model and kept tweaking it until it was able to name the things it was seeing. Sutskever, meanwhile, focused on prepping the data to make it more ingestible for the machine. In the end, their model won the ImageNet competition resoundingly, and their subsequent paper, which came to be colloquially known as AlexNet in honor of Krizhevsky, showed that neural networks were a viable path forward in AI—albeit one that took a huge amount of engineering finesse and computing power to pull off.[18]

"Your neural network needs to be the right type; all the details have to be just right. Once they are, you make it big. We did all that. And it worked," Sutskever said. "If you believe that an artificial neuron and a biological neuron are similar-ish, then it becomes clear that a neural net that is the size of the human brain should be, in principle, capable of doing anything a human does."

**THE 2012** paper transformed both the field and its authors' lives; the startup they created, DNNresearch, became the subject of a bidding war between a number of Big Tech companies. Google won out. After years of reading about Silicon Valley, Sutskever moved there and began working on Google's Mountain View campus, pinching himself at his luck. "It's like being in the future, where everyone is interested in AI," he said.

He loved Google, but he loved adventure more, so when Altman emailed to invite him to the Rosewood, he was intrigued, especially when Altman mentioned that Musk would be coming to the dinner. "I knew that Elon would be there and I was so excited, I *ran* to this dinner."

FOR HIS part, Greg Brockman showed up late, worried that he had missed the main event, but Musk was even later, not arriving for another hour. In the meantime, Altman introduced Brockman to researchers he didn't already know, including Sutskever. Musk appeared after they had moved inside and sat down at the long table, batting about the question of whether it was even possible to start an AI lab given Google and other tech companies' enormous head start. Was it too late? "Everyone was trying to come up with reasons why it was impossible, and no one was sufficiently convincing," Brockman said. "It definitely felt like the prospects were very low, but not zero."

Altman gave Brockman a ride back to the city that night. In the car, Brockman, as the only person at the dinner without a day job, offered to build the lab they had all just dreamed up. The next day, Sutskever emailed Altman. "I said, if you're looking for someone to lead your lab, I think I'd be quite potentially interested."

Brockman and Sutskever had really clicked. A few weeks later, they met for lunch and came to realize they had a similar vision for what AI could do, and also had complementary skill sets. Although Sutskever was enticed by the prospect of working with Musk, his mentor Hinton said, "Ilya was also very impressed by Greg Brockman. That was very important to the recruitment. He was very smart, he was technically very strong, and he could do all the kinds of business things that Ilya has no idea about."

As a machine learning neophyte, Brockman knew he was not the most obvious person to assemble a team of AI researchers. So, at Altman's suggestion, he reached out to someone who was: Yoshua Bengio, a University of Montreal computer science professor who had helped usher in the deep learning revolution alongside figures like Hinton and Yann LeCun.[19] Bengio shared Musk's concerns about AI's risks and was sympathetic to their project, and was impressed to hear that Sutskever was involved.

"When the founders of OpenAI reached out to me, they had a nice story," Bengio said. "It was to be a nonprofit. It was to be open science. It was to advance AI in a direction that I thought made a lot of sense, so I bought that story and I gave them advice freely." He consulted with the fledgling lab for a couple months, drawing up a list of AI researchers Brockman should reach out to, including Wojciech Zaremba, who had worked on computer vision at Google Brain before heading to Facebook, and Durk Kingma, a deep learning PhD student who had spent summers working at DeepMind. "From each person, I asked who else I should talk to," Brockman said.

For the first month, Brockman thought the organization would start with a few core people, including himself, Sutskever, Dario Amodei, and Chris Olah, who began hashing out what the company's values and culture would be. Another researcher he spoke to, Andrej Karpathy, a computer vision expert who had just finished a PhD at Stanford under Fei-Fei Li, the creator of the ImageNet dataset, was skeptical of the project because Brockman was from outside the field. Karpathy's view changed when Brockman talked to Karpathy's friend, John Schulman, and sold him on the new lab. Schulman's vote of confidence tipped both Karpathy and Zaremba toward joining. "I thought it was a very interesting lineup because I knew all of them. I respected all of them, and I thought it was just a very strong, small team." But just as the recruiting efforts seemed to pick up steam, Amodei and Olah bailed, unconvinced that the little lab had a clear reason for doing what it was trying to do. Both went to Google Brain. Brockman needed to seal the deal with the researchers he had.

"Trying to get them all to say yes was hard, but then Sam had the idea of let's bring them to an offsite," Brockman said. The idea was to get their recruits together with some wine and beautiful views in hopes of getting some commitments. "It worked," Brockman said. In

November 2015, Brockman rented a bus and took the ten prospective employees on a day trip to a Napa Valley vineyard, where they spent the afternoon wandering the grounds and brainstorming one of their first projects: a reinforcement learning environment that would teach an AI agent how to navigate the web. In the field of AI, an "agent" is a system that can make decisions and interact with its environment to achieve goals. On the way back, they got stuck in a traffic jam, but everyone was too busy talking to mind. "It was just a bunch of walking and talking—the vision, the mission, the research agenda, everything," Karpathy recalled. "Everyone was having a good time, and we all respected each other. It felt really good. That gave me more confidence that I actually wanted to join them to make this happen."

At the end of the trip, Brockman gave each of them three weeks to make up their minds. At the time, Musk's involvement was not certain. The new lab was an outgrowth of a new, nonprofit division of YC called YC Research, which Altman had announced in October and seeded with $10 million of his own money. Altman declared that YC Research would tackle big scientific breakthroughs with a longer time horizon than startup investors could typically handle. "Startups work best when you know the question and you know how to answer it," Altman told Jessica Lessin of *The Information*. He wanted YC Research to become a new Bell Labs, where the transistor, laser, Unix operating system, and programming languages like C had been invented in the 1930s through the '70s. YC's version of that would start with ten researchers, all full-time YC employees, who would get some equity in YC as part of their pay—Altman had to compete against the lavish equity packages offered by the Big Tech companies. He pledged that any intellectual property developed by the researchers would be made freely available, unless releasing it was determined to be unsafe. "It's not like you want a single company controlling something that modifies genetic code," he quipped.[20]

The nonprofit structure proved to be a helpful selling point in recruiting AI researchers. "The pitch was, 'AI is going to be so important. We don't want a for-profit entity to be in charge of this, because of the profit-seeking motives that are just never-ending,'" Karpathy said. "That appealed to me quite a bit."

The not-for-profit pitch was also ultimately appealing to Musk. In November, Brockman wrote him, "I hope for us to enter the field as a neutral group looking to collaborate widely and shift the dialog toward being about humanity winning rather than any particular group or company. (I think that's the best way to bootstrap ourselves into being a leading research institution.)" Musk agreed to fund the lab, and gave it its name: OpenAI Institute, or for short, OpenAI.[21]

Sequoia's Michael Moritz counseled Altman against the idea of a nonprofit. In his view, it just went against Altman's nature. "He's a very mercantile character," he said. "He is a man wired for the market. So it seemed to me that setting up OpenAI as a nonprofit was pitching his tent in a foreign pasture."

As Graham saw it, Altman had little choice but to set OpenAI up as a nonprofit because he already had a job; he could not very well co-found a for-profit company while running YC. "People think he's pulling some bait-and-switch operation by starting OpenAI as a nonprofit and then converting it to a for-profit. But this would be the worst bait in the world! Does anyone think that if he could go back in time and start it as a for-profit company, he wouldn't?" Graham said. "He's the god of fundraising, so he could easily have raised enough money. The reason he did it as a nonprofit was because he already had a job."

Whatever the reason, on December 8, 2015, OpenAI was incorporated in Delaware as a nonprofit organization. Its certificate stated: "The resulting technology will benefit the public and the corporation will seek to open source technology for the public benefit when applicable."[22] A few days later, the team flew to Montreal to unveil their new lab at the annual NIPS conference. But when they got there, they encountered a problem: Sutskever was getting cold feet.

Ever since Sutskever had told his bosses at Google that he was leaving to join a new lab, they had been trying to woo him back, offering him many multiples of the market rate salary and even the chance to run his own lab if he stayed. Sutskever had backed out of OpenAI initially, but then Altman told him that Musk had committed to the project. "Wow, OK, that changes things," Sutskever had said. "If he wasn't doing it, I think it's quite likely that I would not have gone for it."

He arrived in Montreal committed to the new lab, but then was pulled aside by Google's Jeff Dean, a highly respected figure who oversaw the behemoth's AI research, and who dangled a nearly $6 million salary and appealed to Sutskever's idealism and love of his Google coworkers. Sutskever needed more time to decide, forcing OpenAI to delay their announcement by several days. Finally, the last day of the conference, hours from when they had scheduled the blog post announcement to post, they had no choice but to announce, with or without Sutskever.

Brockman sealed the deal, reaching out to Sutskever and saying, "Look, I really want to do this with you. Are you doing it?" At the very last second, Sutskever agreed, largely on the strength of his admiration for Brockman. "I was really impressed and believed very much in Greg's talents, which I felt were very complimentary to mine. I was correct about them."

Minutes later, on December 11, a blog post co-authored by Brockman and Sutskever announced the launch of OpenAI as a nonprofit research company backed by $1 billion in funding from Musk, Altman, Brockman, Jessica Livingston, Peter Thiel, Amazon Web Services, Infosys, and YC Research. The eye-popping number had been Musk's idea. "We need to go with a much bigger number than $100 million to avoid sounding hopeless relative to what Google and Facebook are spending," Musk wrote Brockman on November 22. "I think we should say that we are starting with a $1B funding commitment. This is real. I will cover whatever anyone else doesn't provide." He also suggested that the stock that sweetened any new recruit's deal should not just be YC's. Perhaps it should also be SpaceX's.

"Our goal is to advance digital intelligence in a way that is most likely to benefit humanity as a whole, unconstrained by a need to generate financial return," Brockman and Sutskever wrote. "We believe AI should be an extension of individual human wills and, in the spirit of liberty, as broadly and evenly distributed as possible. The outcome of this venture is uncertain and the work is difficult, but we believe the goal and the structure are right. We hope this is what matters most to the best in the field."

Later, at a cocktail party hosted by Facebook, the company's lead AI researcher, Yann LeCun—the third "godfather of AI," along with Hinton and Bengio—told Sutskever that their lack of senior machine learning scientists meant they had no chance. "You're going to fail," he said.[23]

CHAPTER 12

# ALTRUISTS

GREG BROCKMAN ONCE DESCRIBED ILYA SUTSKEVER AS "AN artist who expresses himself through machine learning (and sometimes through paint)."[1] But in daily interactions, Sutskever was more like a musician whose preferred instrument was the whiteboard. The dry-erase markers needed to be fresh and numerous. Without a place to scrawl his diagrams of artificial neurons and mathematical formulas, it was almost as if he couldn't think. With them—colorful felt tips squeaking against shiny white laminate—he was an oracle, beaming pure conviction with sensorium-stretching mantras like "It just wants to work" and "Feel the AGI."[2]

The weeks before OpenAI got started on January 4, 2016, had been too chaotic—and Sutskever's participation too uncertain—for Altman and Brockman to even think about office space, so the gang that was supposed to go after DeepMind started out in Brockman's San Francisco apartment, plopped on sectional couches, crouched around his oval-shaped dining room table, and occasionally, in the case of Andrej Karpathy, napping in his bed. At one point, Sutskever and his fellow researcher John Schulman hopped up mid-discussion to write on a whiteboard, only to discover there wasn't one. Brockman quickly set about rectifying that.[3]

In these early days, the team that Altman and Brockman had put together still had no real idea what it ought to be doing. Brockman resolved to make himself useful in other ways. Sometimes this meant ordering office supplies or handwashing all the cups in his kitchen. "Our team really likes to drink a lot of water," he said. At Stripe, he had spent the prior few years stewing about his career, sometimes writing long, confessional blog posts about the conflict between his outsize ambition and his drive to just code all the time, a process he called "thinking about my role." (One Stripe colleague said Brockman's departure was partly driven by his desire, as Stripe's fourth employee, to be classified as a co-founder; both Patrick Collison and Brockman deny this, with Collison pointing to the blog posts as evidence that Brockman's motivations were existential, not about equity stakes or his title at Stripe.) Whatever it was, the stewing had become unbearable. "I just don't want to think about those things," Brockman said to himself. "I just want to think about this problem that I care so much about, and how to maximally contribute to it." He desperately wanted to shift from ego to *cause*. "It felt like I was building AGI by washing cups. What could be better?"

Brockman installed a whiteboard between the living room couch and dining room table, and the team stepped into the fog of the unknown. They set off in the direction of DeepMind's Atari-playing agent, which could learn directly from pure pixels, showing the promise of combining deep neural networks with an algorithm of reinforcement learning. The neural network was capable of finding patterns in the pixels. The algorithm created a system of rewards, but no information about how to get them. It was up to the neural net to decipher the patterns in front of it in ways that would lead to a reward. It would stumble, occasionally hit upon a reward, then ruminate on its past behavior to figure out what might have led it there. It was able to beat a human gamer at *Breakout, Fishing Derby, Kung-Fu Master*, and, eventually, *Montezuma's Revenge*.[4]

Karpathy's idea earlier that fall during the Napa day-trip in which Altman had recruited him and other new colleagues had been to expand the AI agent's playpen from 1970s-era Atari games to whatever happens inside a current-day computer screen. It was all just pixels, anyhow. Why not create a general agent that could "be" a human behind a computer? They pursued the idea in a project they ultimately called Universe. But it was almost overwhelming in scope. In the meantime, Zaremba suggested they build tools for this kind of reinforcement learning, a "gym" where the reward-producing game can take place. "It was just infrastructure that made life easier for researchers working on reinforcement learning," Karpathy said.

"The intention was, we just wanted to do something with reinforcement learning that is interesting and significant," Sutskever said. "We needed to prove ourselves. We needed to do something ASAP, to tell the world that we exist, to make some noise, or else the whole thing was going to evaporate. So there was a very strong desire to do something significant quickly, and it wasn't even that important what."

The project taught Brockman some early lessons about the relationship between AI researchers and software engineers, who must work together to create and train an AI model like architects and general contractors work together to build a house. OpenAI had assigned two researchers and two engineers to build Gym, but he found the collaboration maddening. "The engineers would go off and build something, and they would come back and spend an afternoon projecting it up on my TV, and the four of them would sit around and debate every line of code," he said. "I just felt, 'This is so painful to me.'" Unable to bear the slow pace, he took over the project. "I ended up working very closely with one of our researchers. I would give him four or five different ideas of what to do. He would say these four are bad. I'd say, 'That's exactly what I wanted.' Just enough of the generation and having someone who's got the taste." He set out to make sure the lab valued code-minting engineers and theory-minded researchers equally, instead of favoring one over the other.

While these loose ideas were kicking around, DeepMind kept producing terrifying displays of force. Ever since IBM's Deep Blue bested world chess champion Garry Kasparov in 1997, AI researchers had speculated about how long it would take before machines could beat humans on the far more ancient, complicated, and mind-bending game of Go. The number of options available to each of the two players facing a 19 × 19 grid of dark and light pieces was too large for any existing computer to calculate. Even with all the recent advances in neural networks, most assumed it would take at least another decade. But in late January 2016, before the end of OpenAI's first month of work, DeepMind published a paper in *Nature*—prestigious journals were Demis Hassabis's preferred medium—announcing that their AI system, AlphaGo, had defeated a former European Go champion in a closed-door match the previous October.

The victory rattled Musk, who emailed Altman and Brockman in February, arguing that the company needed "to do what it takes to get the top talent. Let's go higher. If, at some point, we need to revisit what existing people are getting paid, that's fine. Either we get the best people in the world or we will get whipped by DeepMind. If they win, it will be really bad news with their one mind to rule the world philosophy. They are obviously making major progress and well they should, given the level of talent over there."[5]

Then, in March, 200 million people around the globe tuned in to watch the six-months-smarter AlphaGo beat world Go champion Lee Sodol. Google co-founder Sergey Brin and CEO Eric Schmidt flew into Seoul to cheer on their company's machine creation. AlphaGo won one of the five matches with a move so bizarre that some Go commentators initially thought it was a mistake—just a black stone placed seemingly randomly in an open field on the board, far from the cluster of action. "It's not a human move," marveled Fan Hui, who AlphaGo had beaten in October, to *Wired*. "So beautiful."[6] In one decisive global event, DeepMind had shown the world that AGI might not be so far away.

OpenAI needed no convincing on this point, but their efforts to follow DeepMind in making an agent that could plan and take actions were not working. (This was particularly frustrating because Sutskever had worked on Google's Go project before he left, co-authoring one of the papers that led to AlphaGo.) Karpathy's plan had been to create a reinforcement learning environment to train an AI agent to click a mouse or use buttons on a keyboard, programming it to receive "rewards" when it was performing well—and optimizing the agent to seek out those rewards. As DeepMind had with AlphaGo, they would train the AI from scratch. "So in the beginning, you have a neural network, and it's button-mashing at random, and it's trying to stumble into getting a reward," he said. For example, if the agent successfully filled out a web form to book a flight, it would get a reward. "The problem is, if you button-mash, you're never going to get a reward because the task is too hard." For a while, they threw an ungodly amount of "compute"—the Silicon Valley slang term for the hardware required to train AI models—at the problem, but results didn't improve. "It was too crazy of an exploration problem."

Brockman's main lesson from the failures up to now with the Universe platform was that he wasn't listening enough to OpenAI's own researchers, who had been trying to explain that at that point, they didn't have a model that was smart enough to even get started on the problem. He wasn't letting them follow their own insights to solve big problems by starting small. "We were targeting this castle in the sky, when what we needed was a little shack," he said.

OpenAI encouraged its sparse band of researchers to break into even smaller teams and pursue what interested them. In June, the company posted on its website a list of goals, which included building a robot, solving games with an agent, and doing something with natural language. OpenAI moved out of Brockman's apartment to an office owned by Sequoia above the Dandelion Chocolate factory, and then in with Musk's company Neuralink, which was trying to build human-computer brain interfaces, in the Pioneer Building near the Mission District, with Musk paying the rent.

Brockman and Sutskever spent an hour each day in the back server room, trying to plot out how much compute they would need to push the machines past human abilities. "We would write these documents where you could say, 'It looks like the human brain uses this much compute. Therefore, judging by the rate in which compute was growing, maybe in some small number of years, the compute would increase so much you'll be able to train neural nets the size of the brain, and maybe once the compute is there, figuring out the details will be straightforward,'" Sutskever said.

In these same sessions, they hashed out their approach to hiring, envisioning a company where AI researchers and software engineers had equal standing and equal chance to choose what they worked on. They would avoid math envy. "We ended up having a very different hiring philosophy from DeepMind," Brockman said. "We really wanted engineers to build things, and much less of the established PhDs. All of these things came from this understanding that we are the underdog."

During this period, Musk and Altman came by about once a week to check the OpenAI crew's progress. Musk, who was running Tesla, SpaceX, and a handful of other companies, clearly already had his hands full. And for Altman, the demands of helming YC, the most powerful network in Silicon Valley, also were too great for any more than these weekly visits.

By September, however, Altman had delegated many of those responsibilities, handing the running of the batches to longtime YC partner Michael Seibel, who he gave the title of CEO of YC Core, which Altman declared—confusingly—would henceforth be called simply YC. Altman named himself president of a new entity called YC Group. In a maneuver reminiscent of Google's creation of a parent company called Alphabet the previous year, which allowed its controversial moonshot projects to live in a separate division called "Other Bets," YC Group would now oversee four divisions: the core YC, the Continuity Fund, a new online course called Startup School, and YC Research, Altman's pet, which he continued to fund and run personally.

A year in, Altman admitted that YC Research—the crèche where OpenAI was incubated as it slowly matured—was still largely an incoherent hodgepodge of his own fixations, and promised to come up with a "unifying theme" soon.[7] In addition to OpenAI, those obsessions included a study on universal basic income—that is, handing out cash to folks with no strings attached—as a means of getting out ahead of the job losses that Altman believed AI would inevitably create. "Fifty years from now, I think it will seem ridiculous that we used fear of not being able to eat as a way to motivate people," he wrote. "I also think that it's impossible to truly have equality of opportunity without some version of guaranteed income."[8] He hired Elizabeth Rhodes, who was finishing up her PhD in social work and political science at the University of Michigan, to lead a study of some three thousand people who would receive free money, starting in Oakland. During their first meeting, held at Y Combinator's offices during one of the batch dinners, Rhodes noted Altman's interest in addressing Americans' growing economic insecurity broadly, not just as an effect of technology. "He knows that capitalism, as it is, is not working," she said. "He's a social contract visionary."

From Altman's perspective, as he sat at the top of Silicon Valley amid a historic tech boom, capitalism *was* failing, largely because the rent was too damn high. To fix that, YC Research also set out to build a "best possible city" of the future, guided by open-ended questions like "What should a city optimize for?" and "How can we make and keep housing affordable?" Leading this project was Adora Cheung, co-founder of the defunct "Uber for house cleaning" company Homejoy, who also served on YC's board of overseers. "We're seriously interested in building new cities and we think we know how to finance it if everything else makes sense," Cheung and Altman wrote. "We need people with strong interests and bold ideas in architecture, ecology, economics, politics, technology, urban planning, and much more."[9] Altman began scouting locations.

But perhaps the most out-there YC Research endeavor was the Human Advancement Research Community, or HARC, a project of legendary computer scientist Alan Kay that aimed "to ensure human wisdom exceeds human power, by inventing and freely sharing ideas that allow all humans to see further and understand more deeply." In the 1970s, Kay had been part of the famous group of researchers at Xerox's Palo Alto Research Center (PARC) who had invented the graphical user interface later adopted by Apple's Macintosh, as well as object-oriented programming and a light, portable precursor to the tablet called the Dynabook, along with office mainstays like the personal computer, email, and laser printer. Xerox PARC had given its researchers great autonomy, and the experience forever shaped how Kay believed world-changing research gets done. "Alan is fond of pointing out that really good research simply cannot have a well-stated goal, it can only have a useful direction," wrote Apple veteran Larry Yaeger. "If you could state at the outset that you were going to now invent the flying buttress or vaulted arch, then you'd already have your goal so well-defined that you'd have no need to perform the research."[10]

As Altman was dreaming up OpenAI the previous year, he reached out to Kay with a question: How much economic value did the post–World War II golden age of US government spending on science create? While Xerox had put money into PARC, around a third of its funding came from government sources like the Department of Energy and Defense Advanced Research Projects Agency (DARPA), and Kay saw his time there as part of a broad movement of government-funded research. "It was all one big community funded by the government because of the Cold War, but with all of the IP completely open and in the public domain," Kay, with his mane of gray hair, bushy mustache, and arms swinging expressively like an orchestra conductor, explained on a panel that year. He pegged the economic output of Xerox PARC alone at $35 trillion. "Business set their sights very low," Kay said. "They just want millions and billions. But really good researchers want trillions because it involves creating an entire new industry."

Kay, who was now in his seventies, signed on as an early advisor to OpenAI. Sutskever and Brockman went out to dinner with him during OpenAI's early months. They listened to Kay's stories of building the Xerox Alto, one of the world's first personal computers—how his team assembled cutting-edge hardware to create a machine that allowed the tiny number of people with access to it feel as if they were "living in the future." And yet a decade later, Kay noted, that same technology would be available to any member of the general public with a thousand bucks to spare. After dinner, Sutskever turned to Brockman and said, "I only understood 50 percent of what he said, but it was all so inspiring."[11]

In Kay's telling, his talks with Altman led to the creation of YC Research. "Sam absorbed all of this stuff, and he came up with a tremendous plan, which was: industries should somehow fund this, because the government just checked out of doing this, for a variety of reasons, none of them good," he said. "So he started raising money." Whosever idea it was, Kay helped, introducing Altman to Vishal Sikka, then the CEO of Indian IT company Infosys, who Kay had mentored ever since serving as an outside reader on his PhD thesis in the 1990s. During Sikka's previous tenure as CTO of SAP, SAP had funded, starting in 2013, a PARC-like lab organized by Kay called the Communication Design Group (CDG), which hired researchers including Bret Victor, a former Apple interface designer whose work shaped the iPad, and who had been trying to invent a new dynamic computational medium. They thought they had at least five years of funding, but then Sikka left SAP suddenly in 2014, and by 2016 SAP made it clear the lab needed to find other sources of funding. With Kay's help the SAP researchers moved to HARC, which Altman agreed to fund until other money came through.[12]

In the introductory email, Kay referred to Altman as "a builder of civilizations." Sikka wasn't really sure what this meant until he met with Altman in San Francisco. He loved Altman's vision for YC Research, and Infosys ended up donating $3 million to OpenAI. But it was Altman's unusual ideas about how to deal with the lack of affordable housing in the Bay Area that helped Sikka understand Kay's description of Altman. "I remember he wanted to give mortgages to people who were joining him, because his organization's ability to fund people's housing was better than people doing it themselves," Sikka said. It would be easier to pry talent from Big Tech companies if they didn't have to spend so much of their mental energy trying to afford life in the Bay Area. "In Silicon Valley, housing is always very expensive, and I just loved that idea."

HARC lasted barely a year. Despite its name, YC Research had no financial connection to YC. Altman just happened to be in charge of both. Each of YC Research's three initial projects—OpenAI, Basic

Income, and HARC—were expected to come up with their own funding sources. (OpenAI was later spun off as an independent nonprofit, in order to keep Musk's money and demands contained; two other efforts—Cities and a project on universal healthcare—were added to YC Research's portfolio following a dinner in the summer of 2016 where Altman invited select YC startup founders to come to his house and toss out ideas for potential funding that they were passionate about and that could make long-term "societal impacts," according to Jeff Chang, the co-founder of a YC-backed firm who also helped with YC Research.) Kay had lined up funding for HARC, but it got held up, so Altman loaned the money from his personal funds to cover the gap. In the meantime, HARC grew larger than Chris Clark, the Loopt veteran who Altman had tapped to run YC Research day to day, had imagined it would, as Kay brought on staff not just from CDG but also from his own nonprofit, Viewpoints Research Institute. After a year, Altman had exhausted the personal funds he had earmarked for HARC, had not seen much progress, and gave up on outside funding. Most HARC employees took severance and left the organization.

In the view of many people within HARC, Altman's attention had simply moved on, as the dream of AGI seduced him. He spent much of 2016 thinking, as he later phrased it to the *Life in Seven Songs* podcast, "Man, this whole AGI thing is going to be super wild someday." That summer, after years of rolling his eyes at his fellow Valley denizens' enthusiasm for the Burning Man Festival held each August in the Nevada desert, he decided to go, sharing a friend's RV. He ended up falling asleep soon after he arrived, waking up in the evening after his compatriots had donned their costumes and heading out onto the playa. He put on his hiking boots, headlamp, and little backpack and trekked alone under the black desert sky. He could not quite believe the illuminated temporary city that greeted him—temples, art installations, dancers, fire, thousands of people bicycling in costume through the desert dust. "This is the most beautiful man-made thing I have ever seen by far," he thought to himself. But more than that, he noticed how "present" everyone seemed. He would go on to have numerous psychedelic experiences, at Burning Man and elsewhere, that he now considers transformative. And he started attending the festival every year. "I was like, 'Oh, everyone's so happy,'" he later told the podcast. "This is one possible part of what the post-AGI world can look like, where people are just focused on doing stuff for each other, caring for each other and making incredible gifts to give each other."[13]

**THAT FALL** in 2016, three days before the premiere of HBO's *Westworld*, in which the androids that staff a Western-themed amusement park gradually gain sentience and rise up against their human overlords, Altman and longtime YC funder Yuri Milner co-hosted a private screening of the buzzed-about show at Milner's Los Altos Hills mansion. Afterwards, the show's co-creator, Jonathan Nolan, took questions alongside stars Evan Rachel Wood and Thandiwe Newton, followed by Altman and MIT professor Ed Boyden. According to journalist Cade Metz, Boyden told the private audience that scientists were close to creating a complete map of the human brain and then simulating it. The quandary now was whether the machine would not just act human, but also feel—a question brought up by the show's first episode, when a software update accidentally gives the androids access to their memories, beginning their journey to consciousness.[14] The point of the evening, as the invitations made clear, was that science and science fiction were rapidly merging, helped along by a little star power.

One of Altman's favorite depictions of AI in science fiction was the short story "The Gentle Seduction," published in 1989 by enthusiastic Extropian Marc Stiegler. In the tale, an initially skeptical woman is seduced by technology itself, gradually agreeing to first take a capsule of nanobots to repair her aging body, and then to wear a headband that augments her thoughts, and ultimately to swallow a pill to install a computer in her skull, allowing her to expand her consciousness to a cosmic level and help all of humanity through the shock of the singularity. "Only those who knew caution without fear, only those marked by her elemental form of prudence, made it through," Stiegler writes. It was a haunting, but ultimately optimistic, vision for what AGI might bring: immortality, omniscience, delights beyond imagining, if also the destruction of all that had come before.

Stiegler, a cybersecurity expert turned sci-fi writer—who during Altman's freshman year at Stanford gave a lecture in the computer science department about how students might defeat the sort of virus unleashed in the *Terminator* film franchise—said in a recent interview that he is thankful that, like himself, Altman is an optimist at heart who believes technological revolutions are best revealed incrementally. Stiegler recounted how the opening of "The Gentle Seduction" was inspired by real life—back in the 1980s he was strolling with an ex-girlfriend on Fox Island in Washington State, telling her about vast changes in technology and society that the future holds. She was terrified. "Sam Altman appears to have learned the same lesson," Stiegler said.

Stiegler reasons that humanity can only "step across one small threshold at a time" on the way to "the singularity" and our species' ultimate place in the universe. The advancements OpenAI is spearheading mean "we're getting very close now to having the tools to figure out how to live long enough to figure out how to live longer," he said. "One of the points in 'The Gentle Seduction' is immortality is the thing that gives us the time we need to grow and truly become all that we can be when we are also partnered both with other human beings and with AIs." He thinks these deepening relationships will one day transform the cosmos—but that must happen in "bite-sized pieces." Stiegler recalls how he once heard Altman contend he doesn't seek to develop technology that will achieve immortality. Instead, Altman claims he just wants to give everybody ten more healthy years. "My comment on that is that's a very 'Gentle Seduction' thing to say."

**BACK ON** earth, circa 2016, Altman was increasingly looking to politics as the arena where he could realize his soaring ambitions. That fall, his high school ex, Nathan Watters, visited San Francisco, accompanying his new boyfriend to a job interview. While Watters's boyfriend was out, Watters and Altman had lunch at a café near Dolores Park, not far from Altman's $5 million Victorian. The topic turned to the upcoming presidential election between Hillary Clinton and Donald Trump. Watters said Altman told him that if Clinton didn't win, he was going to run for president: "'If she doesn't get it, and Trump does, I can't have that again. I'm gonna do it. I'm gonna run. I think I can win,'" Watters recalled him saying. "I'm sure he did think he could win." Altman disputes that he said this, and says he never wanted to run for president.

In early October, *The New Yorker* published an 11,000-word profile of Altman which revealed the full scope of his ambition to a general audience for the first time. "Like everyone in Silicon Valley, Altman professes to want to save the world; unlike almost everyone there, he has a plan to do it," wrote *The New Yorker* writer Tad Friend. The article treated his presidential ambitions as a joke—the kind of thing his brothers would tease him about while the three of them made pasta. After Max and Jack pointed out that in 2020 he would be thirty-five, just old enough to legally qualify, Altman shot back: "Let's send the Jewish gay guy! That'll work!"

Although Altman's brothers and mother were quoted extensively throughout the piece, neither Jerry nor Annie were mentioned at all. A couple years earlier, Connie and Jerry had quietly separated, after many tense years of Jerry and Connie sleeping in separate bedrooms. "It was miserable," Annie recalled. In 2011, the summer before her senior year of high school, she had sat her parents down and begged them to get a divorce. "If you're staying together for the kids, I'm a kid and I'm telling you that this sucks," she told them. "Just because you are not yelling at each other doesn't mean this is fun to live together." They stuck it out for two years after she left for Tufts. Then Jerry found his own apartment. Connie never mentioned the separation. "She saw it as a personal failure," Annie said. Jerry was no more eager to talk about it than Connie was. Jolene Borgmann, Jerry's friend from rowing crew, once gingerly asked Jerry about his married life, and Jerry simply told her that Connie liked staying up at night reading medical journals while he preferred to watch the news.

That the *New Yorker* article would omit any mention of Annie was especially strange because she was living in the Bay Area at the time, having moved there that summer after graduating early from Tufts with a degree in biopsychology. At Tufts, she was on the premed track, with the expectation that she would be the one to follow her mother and grandfather to medical school. But toward the end of college, she asked the dean if she could finish early with a diploma. "There is simply no need for the sadness and anxiety I've felt relating to school," she wrote. "I came to Tufts on the pre-medical track, and it was not until this semester that I let go of my rigid attachment to that plan." While she might become a doctor one day, she wanted to also look into becoming a nurse, physician's assistant, or social worker. And for the moment, she wanted to go to therapy, make art, travel the world, and feel better.[15]

Annie had realized that becoming a doctor was more her mother's wish than her own. Her high school physics teacher at Burroughs, Jim Roble, certainly got this impression. "I know that she did one summer program where she went with her mother on a Doctors Without Borders trip to South America," Roble said. "I don't think that's how she wanted to spend her time."

After Tufts, Annie moved to the East Bay with a college friend and started working at a neuroscience lab at the University of California San Francisco that conducted experiments on mice. In a podcast, she said she "was told to kill more and more mice every day," which is known as "sacking." As a longtime vegetarian, she found the lab environment upsetting, and lasted only six weeks. She "went vegan," decided to stop taking "psychiatric medication," and switched to a job at a different UCSF lab, she said on the podcast, this time without so much mouse murder.[16]

Inside YC, the partners were stunned by the *New Yorker* profile. "Sam told all the partners, 'Hey, this is going to be about YC,' and it ends up being about Sam. And everyone was like, 'Look we don't give a fuck, but like, why did you tell us this?'" one YC partner said. They didn't have long for such carping, however. Days after it was published, YC faced an uproar over Thiel's $1.25 million contribution to Trump's presidential campaign. Thiel was a part-time YC partner, a largely honorific role, and his support for Trump was hardly a secret; he'd appeared at the Republican National Convention over the summer, delivering a speech where he'd declared that he was "proud to be gay." Nevertheless, the extent of the support—especially in light of the leak on October 7, 2016, of an *Access Hollywood* tape of Trump bragging about grabbing women by the genitals—was more than largely liberal Silicon Valley could bear. Ten days later, Ellen Pao, the former Reddit CEO who had gone on to co-found a nonprofit focused on diversity, said that her organization was cutting ties with YC over its association with Thiel. "Because of his continued connection to YC, we are compelled to break off our relationship with YC," she wrote. "Today it is clear to us that our values are not aligned." Tumblr co-founder Marco Arment was more pointed regarding the connections between YC, Thiel, and Trump. "This is literally paying a huge amount of money to directly support a racist, sexist bigot with rapidly mounting allegations of multiple sexual assaults," he wrote.[17]

Privately, Altman was also mystified why Thiel would support someone like Trump. They had several conversations about it. "I was not trying to tell people not to vote for Trump," Altman said. "I was really trying to understand it." But when the internet began melting down over Thiel's support, Altman stepped in with a forceful defense of both his friend and the principle of intellectual freedom. He disagreed with Thiel on this matter—in fact, he found Trump to be "an unacceptable threat to America" and "unfit to be president"—but tweeted, "YC is not going to fire someone for supporting a major party nominee."

When Trump won, Altman was devastated. He had tried to prevent it in the only way he knew how—by funding and building a piece of software, in this case a "Turbo Tax of voter registration" called VotePlz.[18] Now he turned once again to code, building a website called Track Trump to measure how President Trump's actions during his first one hundred days would measure up against his campaign promises. And he used Facebook to solicit introductions to one hundred Trump voters across the country so that he could ask them directly about their decision; he published his findings on his blog. One quote, in Altman's view, got to the heart of the matter of what too many Democrats had failed to recognize: "You all can defeat Trump next time, but not if you keep mocking us, refusing to listen to us, and cutting us out."[19]

Trump's upset victory had instantly turned Thiel into the most powerful political force in Silicon Valley, after he had stood largely alone before a chorus of haters for his contrarian long shot, and prevailed. (Netflix CEO Reed Hastings had gone so far as to try to push Thiel off Facebook's board, arguing his support for Trump showed "catastrophically bad judgement.")[20] Altman, for his part, had a decision to make: Should he really pursue the presidential run he had told Watters about? Or should he run for something else? Should he run for office at all, or should he try to recruit others to do it in his stead?

Altman still wanted to understand Trump's victory. A contact put him in touch with Charles Johnson, an alt-right blogger and political operative who had worked with Thiel to plot the destruction of the snarky blog empire Gawker Media. Johnson had come up through conservative campus politics and taken a tour at Tucker Carlson's *Daily Caller* before starting his own outlet, *GotNews*. In 2014, after *Rolling Stone* ran a story about an alleged gang rape at the University of Virginia that it later retracted, Johnson published the full name of the alleged victim—in violation of journalistic norms—and a photo of the wrong woman, for which he later apologized. That drew the mocking attention of *Gawker*, which called Johnson "the Web's Worst Journalist" and made up a series of wild claims—"there is no evidence that Chuck Johnson was arrested in 2002 for pinning a sheep to a fence and fucking it"—meant as joking commentary on what *Gawker* implied was Johnson's fast-and-loose, innuendo-laden brand of journalism.

Johnson sued Gawker Media for defamation, which got the attention of a friend of Thiel's who knew that the billionaire was in the midst of a nearly decade-long secret plot to take down the company after its tech-focused blog *Valleywag* outed him in 2007. According to Max Chafkin's book about Thiel, *The Contrarian*, Johnson joined Thiel's crusade, which ultimately settled on secretly funding a lawsuit filed by Hulk Hogan for invasion of privacy after *Gawker* posted a surreptitiously taped video of him having sex with his friend's wife. Gawker Media lost the case and filed for bankruptcy in 2016.[21] (To this day, few things delight Thiel more than recalling his dismantling of what he still calls "the Manhattan-based terrorist organization.")

Altman flew down to meet Johnson at his home in Arcadia, California. "He told me he thought he could be governor of California, and president of the United States," Johnson recalls. "He said, 'There's going to be a millennial president.'" Altman believed that Gavin Newsom, the favorite to succeed then–California Governor Jerry Brown in the 2018 election, was a weaker candidate than most people understood, Johnson said. Altman also had a vivid, techno-utopian vision for what California could be: a state with an economy so large that it could fund its own basic scientific research into nuclear energy and AGI, where the tax code could be tweaked to discourage real estate speculation in order to bring down the cost of housing, and an expanded social safety net—perhaps through UBI—that would make society fairer while eliminating the bureaucracy of many of the current social welfare programs.

Johnson introduced Altman to his political contacts in California. Altman sought advice on both sides of the aisle, including Dominic Cummings, the former advisor to British prime minister Boris Johnson, and Chris Lehane, the former advisor to President Bill Clinton. In April 2017, *Vice News Tonight* correspondent Nellie Bowles asked Altman during an onstage interview whether he would consider running for office. "I don't think charisma is my strength," he replied.[22] But Altman privately continued to explore whether he might run to become California's governor, visiting Willie Brown, the former mayor of San

Francisco, in May to seek his advice. At the meeting, Altman pitched Brown on his plans to use technology to help end income inequality, and to reach voters digitally. Brown immediately sold Altman out, writing about the meeting in a column in the *San Francisco Chronicle*. "I told him California has a history of millionaires running for public office on their own dime," Brown wrote. "Most wind up paying consultants a whole lot of money and losing."[23]

Many of the businesspeople in Altman's life—including Thiel and Sequoia's Michael Moritz—counseled him against running for office. "I thought he was crazy to do it," Moritz said. "I think it's this innocence of a lot of businesspeople when they look at government and think it's all messed up. And because of their experience in the private sector, they know what it takes to make government run more efficiently than it does, and they massively underestimate that to flourish in politics, you need to have grounding in the political arena. You need to understand how it works."

Altman now characterizes his thoughts of running for governor as a passing flight of fancy. "I thought about it for a few weeks. I toyed with the idea at a not-very-high level of seriousness. I went to Sacramento to spend some time with then-Governor Brown, and it was super clear to me that I didn't want that job and wouldn't have been good at it." (Jerry Brown's longtime chief of staff, Evan Westrup, said Brown had no memory of meeting with Altman.)

By July, Altman was ready to pass the torch to others, posting a detailed political platform online that included Medicare for All, raising taxes on short-term capital gains, and shifting 10 percent of the US defense budget to researching future technologies. He said he would help back candidates willing to adopt the platform. The most important plank, at least to Altman, was one devoted to measures to reduce the cost of housing. As the online news outlet *The Outline* put it in a headline: "A Silicon Valley kingmaker wants to fix what tech did to California."[24] The platform, dubbed The United Slate, was created with the help of Matt Krisiloff, twenty-five, who Altman would begin dating later that year. Graduating from the University of Chicago in 2014, Krisiloff then moved to Santiago, Chile, to work on a startup, and had gotten to know Altman after emailing him for advice. When Krisiloff pitched Altman on launching a program at YC for earlier-stage companies, Altman hired him, and eventually tapped him to lead YC Research. Krisiloff then recruited his older brother, Scott, who was working in finance, and the three of them became The United Slate.[25] (Both Krisiloffs would remain in Altman's orbit in various capacities for many years, with Scott going on to work for Hydrazine and then Helion, and Matt founding a startup that Altman invested in seeking to make it possible for two gay men to have a child genetically related to both of them.) Altman's United Slate platform ended up backing the congressional candidacy of Josh Harder, a Democrat and venture capitalist with degrees from Stanford and Harvard, who went on to beat an incumbent Republican for a House seat representing much of the Bay Area.

Even as Altman decided against entering politics personally, his association with OpenAI was bringing him into the highest circles of political power. In its final year, the Obama administration took an interest in AI, convening a series of talks around the country that one of OpenAI's newest hires, a former Bloomberg journalist named Jack Clark, participated in as part of the lab's communications operation. In October 2016, the White House released a large report calling for more federal funding of basic AI research, which President Obama then promoted in an interview with *Wired* that made it sound as though he had been reading Altman's blog.

"Part of the problem that we've seen is that our confidence in collective action has been chipped away, partly because of ideology and rhetoric," Obama told *Wired*. He called for a return to the days of the Apollo space program, which was backed by government funding equivalent to half a percent of US GDP. In 2015, the US government spent roughly $1 billion on AI research. To get to Apollo-level, that would have to rise to around $80 billion in 2016 dollars. He saw it as the government's job to make sure that AI was developed outside tech behemoths like Google and Facebook. "If we want the values of a diverse community represented in these breakthrough technologies, then government funding has to be part of it."[26]

Within weeks of Obama uttering these words, Trump won the election and the Democrats' AI agenda was swept aside. Years later, after ChatGPT had made OpenAI a household name, Altman would say that the young lab had gone to the government in its early years—presumably sometime in 2017 or 2018, he declined to specify—looking for funding, and had gotten nowhere. "I don't blame them at all, because at the time we were a few people sitting around saying 'We're going to try to figure out AGI someday' with basically nothing to show," he said.

Despite OpenAI's lack of progress, DeepMind's steady beat of advances throughout 2016 meant that the AI community increasingly felt like it needed to ponder the potential downsides of all this. In January 2017, the Future of Life Institute held its conference at the highly symbolic location of the Asilomar Hotel and Conference Grounds, a rustic former YWCA compound on California's Monterey Peninsula where in 1975 scientists had famously hashed out guidelines for how to conduct DNA research without endangering public health or the environment. The new parameters allowed the scientific community to lift the moratorium it had placed on genetic modification, ushering in generations of research and advancements. This process of coming together to find consensus about the risks of an emerging technology would come to be known as the "Asilomar model."

More than forty years later, while Future of Life Institute president Max Tegmark stood on a conference stage at the celebrated hotel calling for a new set of Asilomar principles to keep AI from turning against humanity, a huge storm was moving in from across the Pacific, shaking the windows of the historic stone and wood-beam building. One conference-goer noted how the apocalyptic weather seemed to match Tegmark's tone. "At that time at the Asilomar, everyone was convinced that really crazy things were about to happen," he said.

The original Future of Life conference in Puerto Rico in 2015 had brought together the two tribes—AI researchers and AI safety people—with a sprinkling of billionaires thrown in. Now, this latest conference boasted even more star power and a greater sense of urgency. Altman was there, along with Musk, Sutskever, Brockman, Dario Amodei, Hassabis, and the other DeepMind co-founders. At one point amid the panels, Altman met, at Alan Kay's suggestion, one of Kay's former students, a willowy robotics company CEO with raven-black hair named Tasha McCauley.

McCauley's mother, Tamme McCauley, is the stepdaughter of Mel Simon, the Indiana billionaire who owned the largest shopping mall company in the country, as well as the Indiana Pacers basketball team. Tasha's father, Matthew McCauley, the grandson of the author of the first nineteen Hardy Boys books, is an Emmy-winning composer and Egyptologist who became fascinated by remote sensing—eventually using it to excavate the Great Sphinx—and today chairs the board of 3D modeling company GeoSim. Tasha grew up in Los Angeles, attending an unusual public elementary school called the Open School, for which Kay designed the curriculum and Apple Computer, Inc.—as it was then known—provided some of the funds. The Open School was part of a larger Apple effort called the Vivarium Project, named for the glass enclosure where plants and animals are kept for observation, which aimed to improve how computers work by putting them in the hands of children. "By working closely with young children, and learning from their intuitive responses to our system's interface and behavior, we hope to evolve a system whose simplicity and ease of use will enable more people to tailor their computer's behavior to meet their own needs and desires," wrote Larry Yaeger, Kay's colleague at the Open School. The three hundred first through sixth graders at the Open School studied, say, biology by using computers to create animated animals and then programming their behavior. They learned to code and used dynamic authoring tools like HyperCard, a pre-internet hypermedia system, for their homework. "The programs we built reminded me of toys—intuitive graphical toys that enabled us to convey ideas in delightful and profound ways," McCauley said in an interview with the Chicago Toy & Game Group.[27] The Vivarium program's advisory council included such heavyweights as Geoff Hinton, Marvin Minsky, Douglas Adams (the author of *The Hitchhiker's Guide to the Galaxy*), and Koko, the sign language–speaking gorilla.

Kay remained a lifelong mentor to McCauley, whose professional life bore the hallmarks of his own devotion to pushing the frontiers of human-computer interfaces. ("The best way to predict the future is to invent it," Kay famously said.) She went on to Bard College and then to study robotics and later teach at Singularity University, an "educational community" (not, actually, a university) co-founded by Ray Kurzweil and Peter Diamandis, the creator of the XPRIZE. It is funded by corporations including Cisco, GE, and Google. (Confusingly, Singularity University purchased the rights to the Singularity Summit and other branding in 2012 from Yudkowsky's organization, which had been rebranded the Machine Intelligence Research Institute.) In 2011, McCauley and colleagues from Singularity University co-founded the robotics company Fellow Robots, which made robots for everything from "telepresence" (one early prototype named Helo resembled a tablet atop a Segway) to customer service in big box stores (a sleek ATM machine with a female voice). She got an MBA in 2014 from USC's Marshall School of Business (she would serve, between 2019 and 2022, as CEO of GeoSims, where her father was the chairman of the board and one of the company's investors). In 2014, she married actor Joseph Gordon-Levitt—who starred in *Inception*—and the following year, in addition to having their first baby, they attended the Effective Altruism Global conference in San Francisco and quickly became known in the effective altruist community. As Gordon-Levitt, looking every bit the nerdy EA in glasses and a sage green-and-black buffalo check shirt, told a panel at the same conference the following year, he immediately noticed the contrast between EA's message of rationalism and the emotionally driven do-goodery that permeates Hollywood. "Coming and learning about these concepts of not following what makes the best story but what can actually provably be shown to be doing the most good, I found incredibly compelling and noble and worth talking about some more," he said.[28]

McCauley and Altman agreed to stay in touch, and at the end of the conference they signed their names alongside those of Sutskever, Hassabis, and Stephen Hawking to what would become known as the Asilomar AI Principles. This set of AI safety guidelines was far more radical than what had been agreed to in Puerto Rico. The signatories pledged to not engage in an arms race of technical capability; that "teams developing AI systems should actively cooperate to avoid corner-cutting on safety standards"; and that AI should "align with human values." More controversially—considering both the growing competition between DeepMind and OpenAI and the already obvious geopolitical implications for whichever country first achieved AGI— each person who signed the document promised that "the economic prosperity created by AI should be shared broadly" and "superintelligence should only be developed in the service of widely shared ethical ideals, and for the benefit of all humanity rather than one state or organization."[29] Among the later signers was William MacAskill, an Oxford philosopher widely credited with creating effective altruism.

The EA movement emerged in the early 2010s, led by MacAskill and his fellow Oxford philosopher Toby Ord, who were inspired by the 1972 essay by utilitarian philosopher Peter Singer that argued that affluent people had a moral responsibility to spend as much of their wealth as possible—and far more than was considered normal—to relieve suffering in the developing world. The early effective altruists tried to bring a new level of dispassionate rigor to identifying which charities were most effective (deworming charities scored particularly well) and counseled adherents that, as yucky as it might feel, it was actually better for the world for them to work for a hedge fund and donate their salaries to these charities than to slog through the inefficient business of directly helping starving Africans.

In 2009, MacAskill and Ord co-founded an organization called Giving What We Can, which encouraged people to donate at least 10 percent of their income to the poor—a secular person's tithe, essentially—via organizations that had been rigorously scrutinized for their effectiveness. Two years later, MacAskill co-founded 80,000

Hours, a nonprofit organization that counseled graduates on how to make the biggest impact on the world. Then, in 2013, MacAskill brought his message of "earning to give" to a math prodigy and physics major at MIT named Sam Bankman-Fried, convincing him to go into finance. Two years after that, 80,000 Hours joined that summer's YC batch. SBF, as he became infamously known—first as the CEO of FTX and a cryptocurrency mogul, and then as a criminal—would become one of EA's biggest donors before he was indicted for fraud and conspiracy in 2022.

Meanwhile, on another track, two early career analysts at the giant hedge fund Bridgewater Associates, Holden Karnofsky and Elie Hassenfeld, formed a charity club in 2006 with their colleagues. The goal was to pool their money and investigate the most effective places to spend it. Over time, the club morphed into GiveWell, a nonprofit organization that rated charities on how much money each required to save a life. Charities that distributed bed nets to fight malaria were standout performers, leading many to dub this period the "bed nets" era of effective altruism. Around the same time, Yudkowsky was making similar arguments on *LessWrong*, urging his fellow rationalists to "purchase warm fuzzies and utilons separately" by holding the door open for little old ladies and then "find the charity that offers the greatest expected utilons per dollar" when trying to do actual good in the world.[30,31]

As EA thinking was spreading through San Francisco's techie circles, a reporter in *The Wall Street Journal*'s San Francisco bureau named Cari Tuna went on a blind date with Dustin Moskovitz, Mark Zuckerberg's Harvard roommate and a co-founder of Facebook, who is today estimated to be worth about $23 billion. A year later, inspired by Peter Singer's writing, they became the youngest couple ever to sign Warren Buffet's and Bill and Melinda Gates's Giving Pledge, promising to donate the bulk of their wealth to charity before they died or in their will. In 2011, betrothed, they co-founded the Good Ventures foundation. Moskovitz was busy with his new company, the workplace management software company Asana, so Tuna quit her job at the *Journal* to focus on Good Ventures. Almost immediately, she found GiveWell to be an indispensable resource for the kind of decisions she was trying to make. As she told *The Chronicle of Philanthropy*, advisors suggested to her and Moskovitz that they should consider their passions, but the couple had a different approach in mind. They wanted to support charities in which, as she put it, "our marginal funding is going to do the most good in terms of improving others' lives."[32] By 2014, Good Ventures had agreed to fund an offshoot of GiveWell called GiveWell Labs, and the project was renamed Open Philanthropy. The original idea of Open Phil, as it is often called, was not only to give according to the principles of EA, but also to make transparent the entire decision-making process, in sometimes painful detail, such as a 2,800-word post explaining their decision to hire Chloe Cockburn to lead their giving to criminal justice reform, a cause to which they ended up donating more than $200 million.

Not long after Open Philanthropy was founded, the energy of the EA movement began to shift from terrestrial issues like bail reform and deworming pills to the more expansive effort to save the lives of all the people who could be born in the future. This meant focusing on problems with a small, but not zero, chance of wiping out humanity, such as nuclear war, global pandemics, or AI run amok. Open Philanthropy would follow suit.

In March 2017, two months after the Asilomar conference, the foundation donated $30 million to OpenAI, and Holden Karnofsky joined OpenAI's board. The announcement came with a disclosure that Dario Amodei and Paul Christiano, who had by that point come around to joining OpenAI as researchers, "are both technical advisors to Open Philanthropy and live in the same house as Holden. In addition, Holden is engaged to Dario's sister Daniela." Karnofsky would marry Daniela Amodei, who would herself join and lead OpenAI's people team. And Open Philanthropy framed the grant less as charity than as a "partnership" that would allow it to advocate for AI safety within the company. It also was clear that they wanted OpenAI to rethink its plan, articulated by both Musk and Altman during their early interviews, to open source OpenAI's technology. Musk long had been a vocal proponent of open-source technologies and had gone as far as making most of Tesla's patents openly available. Open Phil's grant post, however, included an essay that appeared in *Slate Star Codex* arguing against OpenAI's open-source plan, on the grounds that it would lead to humanity's annihilation as imprudent, unscrupulous, or even evil people misused the technology. In OpenAI's updated mission statement filed with the grant, Altman's lab decries profits and products, and begins to hedge a bit on the idea of open sourcing technology. "We will not keep information private for private benefit," OpenAI's new mission statement read, "but in the long term, we expect to create formal processes for keeping technologies private when there are safety concerns."[33]

It was the beginning of the end of Musk's open-source dream. But in another way, this new block of funding marked a significant step toward OpenAI having the resources to realize the long-simmering futuristic visions of Alan Kay, the Extropians, sci-fi writers, and now Sam Altman. The statement also resurrected Kay's quote: "The best way to predict the future is to invent it."

But the price OpenAI would ultimately pay for getting in bed with the EAs would end up being far higher than Altman could have guessed.

CHAPTER 13

# PIVOTING TO PROFIT

ONE FRIDAY IN AUGUST 2017, 20,000 FANS OF THE MULTI-player online video game *Dota 2* (short for *Defense of the Ancients*) stood in line to pack a downtown Seattle arena originally built for the 1962 World's Fair and more recently used as the home of the Seattle SuperSonics. Many wore the headdresses and body armor of characters such as Shadow Fiend and Centaur Warrunner, who populated the game that fans had organically and anarchically spun out of Blizzard Entertainment's *Warcraft III: Reign of Chaos* in 2003. *Dota* had become such a cultural phenomenon that another gaming company, Valve, had bought the rights to it and created the first truly formidable annual global championship for it, The International, with more than $20 million in prizes, the biggest in esports. The professional matches were five-on-five matchups sometimes compared to a cross between chess and basketball. This year, there was also a one-on-one exhibition match for the history books: one of the world's top *Dota* players, Danil "Dendi" Ishutin, versus a bot trained by OpenAI.

Dendi emerged into the arena through a smoke-machine haze, robed and swinging like a boxer. His opponent was rolled out on wheels, then activated with a theatrical insertion of a thumb drive. The match did not last long. "Sneaky bot!" Ishutin cried out almost immediately, as OpenAI's bot attacked his minions. Within seconds, OpenAI's bot destroyed his avatar, its own glowing demonically red as it gloated. Ishutin lost the first match, forfeited the second, and refused to play a third. "He's too strong," Ishitun said.

Inches away from Dendi, Brockman, in a black V-neck, hair shorn to monk-like baldness, stood grinning with a glint of menace in his eye. "So let me tell you the next step in the project. The next step in the project is five v. five," he said. The audience roared. Next to him, Jakub Pachocki, a baby-faced, Poland-born AI researcher who had joined OpenAI the previous February as the research lead for the *Dota* team following a postdoctoral fellowship at Harvard, glanced sideways at Brockman as his deeply dimpled smile froze into a barely perceptible mask of dread. Brockman had not told any of the team this was the plan.[1]

Culturally, the victory in *Dota 2* was a far cry from AlphaGo beating Lee Sodol at a game that had been around since the Bronze Age and that is played by tens of millions of people around the world. But Musk wasted no time in claiming OpenAI's was the greater victory. "OpenAI first ever to defeat world's best players in competitive eSports. Vastly more complex than traditional board games like chess & Go," he tweeted.[2]

Few paid attention to his follow-up tweet: "Would like to express our appreciation to Microsoft for use of their Azure cloud computing platform. This required massive processing power." It had been Altman who had called up Microsoft CEO Satya Nadella a few months before and asked for $60 million in compute to allow OpenAI to make a public splash with their AI systems, which could serve as publicity for Microsoft's also-ran cloud platform, not to mention help ferret out its bugs. At the time, Microsoft Azure had around 15 percent of the market share in cloud computing, compared to Amazon Web Services' more than 60 percent.[3] Altman had known Nadella for nearly a decade, going back to his years running Loopt, when Nadella was still coming up through Microsoft's Bing search and cloud computing divisions. Both had been surprise picks to lead their respective organizations in 2014. Altman had also begun meeting regularly the previous year with Microsoft co-founder Bill Gates, who had stepped back from his formal roles at the company to focus on philanthropy but still sat on the board and maintained a keen interest in AI.[4] OpenAI's expanded version of its *Dota*-playing system, OpenAI Five, lost to five humans at the next year's International. But when it came back the following year—after ramping up the amount of compute it used to train by eight times—and defeated human teams on a livestream, Microsoft was ecstatic.

Musk was the one who had insisted that OpenAI tackle the *Dota* project. He had been inflamed by DeepMind's AlphaGo, shamed by their win and insistent that OpenAI produce some kind of answer in the field of gaming to their rival. If Go was more complex than chess, then they had to find something more complex than Go, which led them to a game that, though wildly popular in its web-enabled niche, 99 percent of humans had never heard of. Many of OpenAI's researchers thought it was a research dead end, but Musk was insistent. If they wanted his continued funding, they would make a mark for themselves in the world of games—resulting in the kind of video replay that one billionaire could share with another aboard his private plane.

The *Dota* project did yield some scientific advances for OpenAI, namely, when it came to scale. "I remember Jakub had the first agent they trained over a weekend. It could do a little thing, and then every week you would come in, and they would have doubled the number of CPU cores and the agent would be twice as good," Brockman said. He led the project with his typical passion, to the point of hiring a documentary team led by the former *New York Times* reporter Jennifer 8. Lee to make a movie about it. The flashy demo got investors' attention, especially Microsoft's. "We were able to show that a single technique at scale does really well," Sutskever said.

Bob McGrew, who, despite his cherubic face, had salt-flecked sandy blond hair and "adult in the room" energy, had recently joined OpenAI as a researcher after a decade at Palantir. He had dropped out of his Stanford PhD program in AI to join Thiel's data-mining company because "doing a PhD in AI in 2005 didn't make a lot of sense." Despite his management experience, he had asked to be an individual contributor at OpenAI and treated "like a third-year grad student," he said. "They didn't know quite what to make of me."

McGrew had volunteered for a different OpenAI project, led by

Wojciech Zaremba, that was trying to make a robot hand solve a Rubik's Cube in simulation. "We'd been struggling to make the robot hand work," McGrew said. After the success of the *Dota* project, Zaremba asked Jakub Pachocki, who had been the research lead on the *Dota* team, to apply the same technique to the robot hand. "And immediately, it went from a two-fingered claw, barely being able to grasp a ball, to the fully five-fingered hand—lots and lots of joints, I think twenty-five joints—being able to grab and manipulate an object. And not too long after that, in a matter of weeks, we were actually able to solve the Rubik's Cube," McGrew said. "This was sort of the eureka moment."

But in a fundamental way, they were heading in the wrong direction. As *Dota* and the robot hand project strived for headlines, a reclusive solo researcher named Alec Radford was quietly exploring a far more consequential project. Radford joined OpenAI in 2016 after dropping out of Olin College, a tiny but prestigious engineering school in Needham, Massachusetts, to start a machine learning company with some fellow students, and realizing how much compute the thing he was really interested in would take. Radford was intrigued less by an agent that could do his bidding than by language models that could learn the meaning of speech. His first project at OpenAI, which involved training on 2 billion Reddit comments, didn't work, so for his next one, he went smaller: 100 million Amazon product reviews. The goal was simply for the model to predict the next character in a review. Instead, it made a leap: it determined whether a review was positive or negative. "It was a complete surprise," Radford told *Wired*.[5] In April 2017, Radford, Sutskever, and another OpenAI researcher named Rafał Józefowicz published a paper on what they called the sentiment neuron, which could understand whether statements were positive or negative without relying on humans to pre-label the data.

Two months later, Sutskever read a preprint of a paper by eight Google researchers titled "Attention Is All You Need," which he immediately recognized as presenting a method to make the kind of research Radford was doing vastly more efficient. Rather than processing one character at a time, the Google paper showed how a model could process large chunks of text in parallel, using an insightful application of a technique called the "attention mechanism" to dynamically assign importance to key parts of that text input. This technique also capitalized on the ability of GPU chips to run multiple calculations in parallel. Sutskever was able to see the full import of the paper—which was later called the "transformer paper"—in part because it was building upon his own work, including his 2014 "Sequence to Sequence" paper that used the attention mechanism to achieve advances in machine translation and summarization.[6]

Transformers were a huge step forward because they could learn from long sequences of text and successfully utilize a lot more compute than the previous dominant model, known as a recurrent neural network, or RNN. That opened the door to the next paradigm, in which progress would be driven by rapidly increasing amounts of data and compute. But, like all neural networks, they are largely black boxes: as of this writing, even experts don't truly understand how they work. These dynamics, unleashed by the combination of Sutskever's insight and Radford's language research, would come to define the generative AI boom.

"There's this thing that Ilya and I used to talk about, the Feynman method of being a genius," Brockman said, referring to the famous physicist Richard Feynman, who worked on the Manhattan Project and won the Nobel Prize. "The Feynman method is you have a set of problems you care about, you kind of know what the missing piece is, and you wait for yourself or some other researcher to invent one of those missing pieces. Then you just take the missing piece, you plug it in, and you're a genius. You solved it, right? And to some extent, you can look at the transformer as the last piece for our genius formula. We kind of knew it would happen, but it was just a question of when will we have a model that is actually capable of doing all the pieces that we lined up?"

Radford began running his experiments on the transformer architecture. "I made more progress in two weeks than I did over the past two years," he told *Wired*.[7] He quickly learned the models worked best when they were big, and thus they needed the biggest possible datasets. Working with Sutskever and two other researchers, Radford settled on a dataset called BookCorpus, a collection of more than seven thousand mostly fiction books in the romance, fantasy, and science fiction genres. The dataset had first been packaged in 2015 by researchers from MIT and the University of Toronto, who described them as "free books published by yet unpublished authors," and made them available to download on the University of Toronto website. Google and other companies used the dataset for AI training. In 2016, *The Guardian* reported that the books had been scraped, without permission, from the self-publishing platform Smashwords, where authors have the option to set the price of their books to free. Many of the books that appeared in the dataset included a copyright declaration reserving "all rights." Several of the authors objected, as did the Authors Guild, the largest writers' advocacy group in the United States, but a Google spokesman argued the practice was "fair use under US law."[8] It would not be the last time the question was raised.

They used question-and-answer pairs from the website Quora, English exam passages for Chinese students, and multiple-choice science exams. Altogether, the model had more than 117 million parameters, which presented a considerable engineering challenge. OpenAI had to reformulate itself, going from multiple research projects to a single-minded focus on training models as large as the world had ever seen. As Brockman and Sutskever believed, the researchers and engineers would have to become equal partners.

"Engineers are great at doing frustrating tasks," Sutskever said. "Academics, not so much. Academics love beautiful ideas, but they don't love to do the dirty work. At OpenAI, we thought, 'We are a dirty work company.'"

**WHEN THEY** finished training the model, they found that it could not only beat benchmarks when answering questions from the data it was trained on, but that it seemed able to answer questions about things it wasn't trained on, a phenomenon known as "zero-shot." Years later, Altman would describe the result as "somewhat impressive, but no deep understanding of how it worked or why it worked." Radford, Sutskever, and team called the model a "generatively pre-trained transformer," or GPT for short.[9]

They showed a new path forward for the company. The agent approach was wrong because it started training from scratch, Karpathy explained. It took a gargantuan amount of machine learning to teach it what it needed to know.

"The right way to do it is you ignore all that stuff, and you train language models," he said. "Language models are not agents. They're just trying to predict the next word. But because you are doing this over all of the internet, the neural network is forced to learn a ton about the world."

If you want to train an AI to buy a plane ticket for you, you should start first with a large language model like GPT, which would know what "buttons" and "text fields" were from reading about them, rather than have it randomly stumble into filling one out and getting a reward. It turned out to be an easier computational problem to train some generalized "brain" for the agent to draw from, and then train the agent using the brain on some specific behavior through a process known as "fine-tuning" that was much less computationally expensive. A lot of stumbling was eliminated.

"We were trying to copypaste what DeepMind was doing, and this is the wrong path and died," Karpathy said, years later. "And then Alec here at OpenAI was training the first GPTs. And now DeepMind is copypasting what happened here."

**AFTER THE** initial *Dota 2* victory, Musk was thrilled. But among the lessons from *Dota 2* was that having a lot of computing power really mattered. OpenAI needed a lot more money. Sutskever, Brockman, Musk, and Altman batted around the possibility of creating an initial coin offering, or ICO, to generate cash, though Musk ultimately argued it would hurt their credibility. They made plans to try to raise more than $100 million for the nonprofit. But as they started having meetings with prospective donors, they quickly realized this was not realistic. In one meeting with a high-profile Silicon Valley investor, they were told that that was a staggering sum for any nonprofit to try to raise. "That was really the thing that caused the transition, to realize that, OK, it's time for us to grow up in terms of our approach to the mission," Brockman told the *Journal*. That July, when OpenAI's *Dota 2* bot was performing well enough that victory over a human in one-on-one combat was only a month away, Sutskever wrote Musk an update on OpenAI's progress, saying the company would soon begin "designing the for-profit structure."[10]

As they mulled a transition to becoming for-profit, the co-founders had a long negotiation with Musk about who would lead it. Musk wanted to be fully in charge, with majority equity, board control, and the CEO title, but Sutskever and Brockman were concerned he would only be able to dedicate a small fraction of his time to the company. The decision of who would be CEO was ultimately left to Brockman and Sutskever, as the more senior full-time co-founders, and they initially chose Musk. Then Altman called Brockman and convinced him to change his mind, arguing that Musk would be difficult to work with. Brockman then convinced Sutskever to also reverse course and back Altman. "From the very beginning of OpenAI, I was trying to get him to come in as CEO," Brockman told the *Journal* in 2023. "There was a Sam-shaped hole that we kept very intentionally open for years."[11] That September, Brockman and Sutskever wrote Musk and Altman an email laying out their quandary.

"Our desire to work with you is so great that we are happy to give up on the equity, personal control, make ourselves easily firable [sic]—whatever it takes to work with you," Brockman and Sutskever told Musk. Yet, they wrote, they were concerned that "The current structure provides you with a path where you end up with unilateral absolute control over the AGI. You state that you don't want to control the final AGI, but during this negotiation, you've shown to us that absolute control is extremely important to you." They added that, given that OpenAI was founded "to avoid an AGI dictatorship," it seemed like "a bad idea to create a structure where you could become a dictator if you chose to."

But they were also suspicious of Altman, especially given his political ambitions. "We haven't been able to fully trust your judgements throughout this process, because we don't understand your cost function," they wrote, speaking in the language of middle school mathletes. "We don't understand why the CEO title is so important to you. Your stated reasons have changed, and it's hard to really understand what's driving it. Is AGI truly your primary motivation? How does it connect to your political goals? How has your thought process changed over time?" These were more Sutskever's sentiments than Brockman's, but they both shared them to some degree.

Despite this attempt at evenhandedness, Musk read the email for what it was. "Guys, I've had enough. This is the final straw," Musk replied within hours of receiving the email. "Either go do something on your own or continue with OpenAI as a nonprofit. I will no longer fund OpenAI until you have made a firm commitment to stay or I'm just being a fool who is essentially providing free funding for you to create a startup. Discussions are over."[12] He stopped sending money, though he kept paying the rent for a while, and Reid Hoffman, the LinkedIn co-founder and college friend of Peter Thiel's, stepped in to cover OpenAI employee salaries and other overhead.

By the end of 2017, Musk had poached Karpathy and made him the head of artificial intelligence at Tesla, in charge of leading the company's self-driving car technology. In January 2018, he wrote Sutskever and Brockman an email urging them to merge OpenAI into Tesla.

"It seems to me that OpenAI today is burning cash and that the funding model cannot reach the scale to seriously compete with Google (an 800B company)," he wrote. "A for-profit pivot might create a more sustainable revenue stream over time and would, with the current team, likely bring in a lot of investment. However, building out a project from scratch would steal focus from AI research, it would take a long time and it's unclear if a company could 'catch up' to Google scale, and the investors might exert too much pressure in the wrong directions. The most promising option I can think of, as I mentioned earlier, would be for OpenAI to attach to Tesla as its cash cow." He proposed it as a win-win: OpenAI's technology could help accelerate Tesla's push into self-driving cars, which would raise Tesla's market cap so much that it could afford to fund OpenAI's quest for AGI. "With a functioning full self-driving solution in ~2–3 years we could sell a lot of cars/trucks," he wrote. The move could increase Tesla's market capitalization, allowing OpenAI to use the revenue to work on AI at a larger scale, he argued. "I cannot see anything else that has the potential to reach sustainable Google-scale capital within a decade."[13]

Later that month, on an unusually cold and blustery day in San Francisco, OpenAI's staff were summoned to the top floor of the Pioneer Building for a sudden and ominous all-hands meeting. Musk, seated on a couch with dozens of OpenAI employees clustered around him, dropped the bombshell that he was leaving OpenAI for good, explaining that his forays into AI at Tesla had created a conflict of interest. Altman thanked him for his time at the company and tried to abruptly end the meeting. But employees wanted answers, and Musk, being Musk, entertained them. The questions came rolling in: How would he develop this technology safely? Why ramp up a rival effort at Tesla if it meant intensifying a competitive race to develop AI that could spiral out of control? Finally, Musk snapped and called a young researcher a "jackass," leaving employees in complete shock.

As Musk stormed out of the office, never to be seen again among OpenAI employees, Altman sought to calm them. He told them that Musk's departure wouldn't threaten the fledgling research lab, whose initial $1 billion in funding had largely been pledged by Musk.

Though it wouldn't be official for another year, Altman was now effectively the CEO of OpenAI. He immediately turned to feeding the lab's enormous appetite for compute. In the near term, that meant lining up more donors and advisors, including from prominent EAs like Jaan Talinn and Julia Galef, a rationalist public intellectual who had shared the stage with Gordon-Levitt during the previous year's EA Global conference.[14] In the longer term, it meant addressing some big questions. Should they buy chips from Nvidia, rent them from a Big Tech company, or built them from scratch? His partner in navigating these questions was a new hire who had been one of the youngest product managers ever to work at Tesla: Mira Murati.

Murati grew up in Albania during the country's chaotic transition from communism to Western-style capitalist democracy.[15] The middle child of three born to high school teachers, she excelled early in school, working in her older sister's schoolbooks when she had finished with her own and competing in Math Olympiads. At sixteen, she won a scholarship to attend Pearson College UWC, an international boarding school in Vancouver that was part of a global network and that followed the International Baccalaureate curriculum. For college, she enrolled in a five-year dual degree engineering program at Colby and Dartmouth colleges, where, for her senior project, she built a hybrid race car. "It was fun, but also we wanted to do something that felt really hard, and so instead of batteries, we used supercapacitors and you know, really trying to push what was possible," she told Microsoft CTO Kevin Scott in a podcast interview.[16]

After graduating in 2012, she worked briefly as an intern, first for Goldman Sachs and then for the French company Zodiac Aerospace, but grew frustrated with the pace in the latter industry. "I realized that aerospace was kind of slow-moving," she told an audience at Dartmouth in 2024.[17] (In the same interview, she caused a minor furor by saying that, as a result of AI, "some creative jobs maybe will go away but maybe they shouldn't have been there in the first place"—a quote she felt was taken out of context.) In 2013, she joined Tesla, where she worked on the Model S, Tesla's four-door sedan, and the Model X, the company's crossover SUV, serving as product manager of the latter overseeing manufacturing, sales, design, engineering, hardware, and software. While she was working on the Model X, Tesla released initial versions of its AI-enabled driver-assistance system, Autopilot, and she became interested in the space. In 2016, she left Tesla for Leap Motion, a startup funded by Founders Fund that aimed to make it possible for people to interact with digital devices by just waving their hands. The company ultimately fell short of its promise. Frustrated with such narrow applications of AI, she had begun to consider the promise of a more general system; she also happened to know some of OpenAI's founders socially. After chats with Amodei, Brockman, Sutskever, Schulman, and ultimately Altman, she took a "leap of faith" and joined OpenAI in 2018 as VP of applied AI and partnerships. In the earliest days, this meant: figuring out how many chips they were going to need to get to AGI, and most likely convincing Microsoft to give them to OpenAI, a task referred to internally as the hardware strategy. Over time, Murati's purview would come to encompass just about every aspect of the company.

**IT WASN'T** even Memorial Day, but St. Louis was already heavy with the heat of a Midwestern summer on the day that Jerry Altman died. He was rowing with his crew club on Creve Coeur Lake—the French word *crève-cœur* means "broken heart"—when he began to have chest pains. "Of course they were at the far end of the lake," said Jolene Borgmann, Jerry Altman's friend and crewmate. "He was like, 'I'm really not doing well.' He just sat there, and everybody else rowed him back." He was rushed to the hospital, where doctors tried to open up his heart and remove the blockage. But it was too late. He went into cardiac arrest. He was sixty-seven.

The news stunned everyone who knew him. Jerry Altman had had a stent put in years earlier, but was, by all accounts, in excellent shape, having taken up rowing at the St. Louis Rowing Club after watching his daughter, Annie, compete as a coxswain in high school. (Jerry often became interested in whatever his kids were interested in.) He would race the four hours from Kansas City, where he was working as the chief operating officer of Keystone Property Management, change out of his suit and into his spandex racing gear, and get out on the water. "He always tried to make it back so he'd have time for rowing,"

Borgmann said. "It was the Jerry transformation." When he wasn't rowing, he was at the club tinkering and fixing boats. "Our saying was, you can't out-nice Jerry." But he had not told even his closest rowing friends about his separation from Connie. Borgmann didn't find out that he was living in an apartment alone until after he died, though she had sensed that something wasn't quite right.

For Sam, who had pushed YC into funding healthcare startups as part of his philosophy that there was nothing technology could not ultimately fix, his father's death was like a lightning bolt from an angry god. "He was a mess when his dad died," said Sally Che, his high school friend.

At the funeral, held in a packed Central Reform Congregation sanctuary, Sam gave a eulogy reflecting on his father's laugh—a sound friends characterized as a high-pitched giggle. One night when the family was home playing a board game together, everybody cracking up, Sam made a recording on his phone of his father's laugh, which he found himself listening to in the days after his father died. "He just recommended that everybody do that," recalled Andy Abbott, by then the head of John Burroughs School, who was at the funeral. "He still had this recording of his father, the joy that his father had." Albeit in a meager way, Sam had somehow found a way for technology to cheat death.

The impact of Jerry's death on his family was profound. Each of the children spoke at the funeral, but Annie Altman's eulogy included the jarring update that she had been living in her car on the Big Island of Hawaii as part of an "extreme minimal hippie phase." After deciding she was miserable killing mice all day in research labs in the Bay Area, she traveled to Big Island in 2017 to be certified as a yoga teacher, then moved back to the Bay Area to work for a company that sold boxes of local produce, and then decamped permanently to Big Island, living in her car for several months. Jerry worried about her, as he told his friend Borgmann, but was also determined to support her in whatever she chose. In January, he texted Annie, "And just for clarification, I don't just support your lifestyle now or your physical and emotional endeavors now; I support your life. I will always support your life. These are aspects of your life, so I support those too. And there is not a 'now,' as Yoda might say. There is only life, for as long as that might be."

After Jerry died, Sam looked into his father's accounts and discovered that he had recently split the cost of a used car with Annie and helped her financially in a few other ways, despite having agreed with the rest of the family to not give her money, beyond Connie keeping her on her health insurance until she was twenty-six. For Connie, who was concerned that giving her daughter money at this point would enable unhealthy lifestyle choices, the information was very upsetting. "She graduated from Tufts," Connie said. "She has a degree that I paid sticker price for. None of them have student loans." By the end of the summer, Connie had removed Annie, who was twenty-four, from her health insurance.[18]

Upon Jerry's death, one of his life insurance policies went to Annie, which allowed her to move to Los Angeles and follow her dream of becoming an artist. She took improv classes at the Upright Citizens Brigade, did standup comedy on open mic nights, launched a podcast, and continued to work on a one-woman show she called "The HumAnnie" about how no one knows how to be human and all humans need basic resources. She posted cover songs on YouTube. "I had my science extreme, and I had my hippie extreme in the car, and then going to LA was figuring out art and business and how I do all these things together," she said.

For Sam, the first week after his father's death was blur of arrangements. "It wasn't until a week after that that I got to catch my breath and be like, 'Holy shit. This happened,'" he said on a podcast.[19]

**A MONTH** after the funeral, Altman was rolling up under the covered carport of the rough-hewn stone main lodge of the Sun Valley Resort before the rapid-fire click of newswire cameras. Wearing a pink T-shirt and clear plastic sunglasses with reflective blue lenses, he looked like he was bound for the beach. While many of the media moguls who attend the Allen & Company annual media finance conference proudly sport the event's signature fleece vests, Altman kept it Californian. Altman had been going to Sun Valley, as the "summer camp for billionaires" has been commonly known, for many years, since Loopt hired Allen & Company to hunt for financing or a sale in 2008. (Hiring the firm, a boutique investment bank that will often be listed as the third bank on a media deal, is the best way to get a coveted invitation to the conference.) Along the way, Altman has become close friends with Sun Valley mainstays like Barry Diller and his wife, Diane von Furstenberg.

The conference is famous as a site for dealmaking, the place where AOL decided to buy Time Warner and Comcast sealed the deal to take over NBC Universal. Allen & Company gets a cut of every deal forged at the conference. This particular year, despite a healthy economy and the usual splendid weather, the conference was permeated by a sense of dread, as the big players in media seemed increasingly desperate to sell themselves off to some tech company or another before the profits fully leaked out of their business, thanks to competition from the likes of Netflix. As if to troll the legacy types, SoftBank founder Masayoshi Son arrived and declared to reporters, "I'm not interested in traditional media."[20] Apple's Tim Cook and Facebook's Sheryl Sandberg echoed his sentiments the following day. Jeff Bezos, not yet divorced but in the midst of making himself over, strutted between boutique pop-up shops, his muscles bulging improbably.

For his part, Altman bumped into Microsoft's Nadella in a stairwell, and they had a five-minute conversation. As Altman recounted, "I was like, 'Hey, we're going to raise a bunch of money. It's kind of really weird. You want to talk about it?'" Nadella was interested. In fact, he wanted to talk about it right then. Altman explained that, after Musk's departure, OpenAI had decided to try a "capped profit" model, which would limit the returns investors could make to one hundred times whatever they put in—a "limit" that was really a brilliant investment pitch. "That was such a weird thing that it classed a lot of people out," Altman later told *The Wall Street Journal*. "The fact that we were going to put a cap on their returns? They were like, 'You've got to be kidding. Never. On principle, never.' So that knocked out a good amount of people."

One of the few investors that this pitch didn't scare away was Vinod Khosla, the co-founder of Sun Microsystems who had worked at famed venture firm Kleiner Perkins before going on to start his own firm, Khosla Ventures. Khosla Ventures had hired David Weiden not long after Weiden became a Loopt advisor, and Weiden had introduced Altman to his new boss. Khosla had kept in touch with Altman during his time at YC, and the investors shared an interest in AI and nuclear energy. "'You're one of the few funds that actually cares about the mission more than the return,'" Khosla remembers Altman telling him while fundraising. "There was no plan for revenue. There was no projection anywhere. There wasn't a financial statement that we looked at," Khosla recalls. Instead, they talked about the mission of reaching AGI. Khosla Ventures ended up writing its largest-ever check, for $50 million, becoming the first venture firm to invest in OpenAI.

But OpenAI needed a lot more than $50 million. Altman and his team talked to other tech companies in the following months, but most of those conversations went "not very well," he said. That winter, he reached out to Nadella, and also to Microsoft CTO Kevin Scott, as negotiations began in earnest. Altman recalled: "I remember coming back to the team after and I was like, 'This is the only partner. This is who we're going to go with. They're aligned, they get it, they get the safety stuff, they get AGI. They have the capital, they have the ability to run the compute, we're just going to do this.' And then we didn't really run a process after that."

As Altman negotiated the Microsoft deal, he was increasingly using his YC combinator perch to nurture global relationships. In 2018, to cultivate Chinese startups, he announced YC China, which was to be led by the revered Chinese American AI expert Qi Lu, who had been the architect of Nadella's AI strategy while at Microsoft. Altman had visited China years earlier, accompanying Peter Thiel on the tour for *Zero to One*, and he had been stunned by the vistas of glittering skyscrapers on land that had been occupied by rice paddies two decades before. He had been even more impressed when touring Chinese factories and biolabs and seeing the frenzy of hard work and productivity. "I thought that some of the most interesting entrepreneurial, technological energy in the world was in China," Altman said. His former Loopt board member, Greg McAdoo, agreed. It was not an accident that Sequoia's Chinese arm itself was aggressively pursuing Chinese venture investments at the same time, such as Sequoia Capital China's investment into the Chinese drone maker DJI. By 2021 the US government had put DJI on an investment blacklist for its alleged role in facilitating surveillance of China's Uygur population. But YC China never got off the ground, and Sequoia had to separate from its own Chinese arm in 2023 as tensions between the United States and China escalated. As Qi Lu put it in an interview upon YC China's launch, "Multinational corporations in China have almost been wiped out. They almost never successfully land in China."[21]

Altman also forged a relationship with Saudi Arabia. He had first met Saudi Crown Prince Mohammed bin Salman, the young ruler who everyone called MBS, in 2016, at a dinner at San Francisco's Fairmont Hotel attended by Thiel, Hoffman, and a dozen other US venture capitalists. The Saudi delegation had organized it with the aim of deepening the ties that had led to the kingdom's $3.5 billion investment in Uber a few weeks earlier.[22] "Sam was very impressed with the dinner," said one attendee. "There was very good chemistry between the kingdom and Sam" for the next couple years, the person said. On October 9, 2018, a Saudi news agency announced that Altman had joined the advisory board of Neom—a portmanteau of the Greek word for "new" and the Arabic word for "future"—a $500 billion megacity the size of Massachusetts powered by AI and clean energy that MBS was building in the desert. Much like Altman's dream for YC Cities, MBS wanted to remake Saudi society from the ground up. "Starting Neom from scratch, with independent systems and regulations, will ensure the availability of best services without social limitations," MBS said at Neom's first board meeting, according to *The Wall Street Journal*. His consultants laid out plans for flying taxis and robot maids, as well as Islamic courts that reported to the king.[23] But just as the Neom advisory board was announced, attention was focused on Saudi dissident and *Washington Post* columnist Jamal Khashoggi, who had entered the Saudi consulate in Istanbul on October 2 and not been heard from since. As further reporting began to uncover allegations that Khashoggi had been assassinated, Altman announced he was "suspending" his involvement with Neom until the facts of Khashoggi's disappearance became known.[24] US intelligence ultimately determined that he had been tortured, killed, and dismembered with a bone saw on MBS's orders. It would be five years before Altman was able to visit the kingdom, even though a fund backed by Saudi Aramco would go on to invest with him in AI chip startup Rain Neuromorphics—only to be forced to unwind the investment by the Biden administration the following year.[25]

By the summer of 2018, Altman's attention had largely moved on from his day job at YC, and YC partners began to grumble that he was rarely around. They had also begun to fight amongst themselves, with some complaining to Graham that YC had become woke and bloated. To address these concerns, Altman tried to delegate his duties even further by naming Jon Levy, the former Wilson Sonsini lawyer and husband of Carolynn Levy, to a vice president role in charge of "culture," a role that would have given him veto power over some moves by YC divisions like the Continuity Fund and YC Core. Levy was one of three longtime YC staffers who Paul Graham and Jessica Livingston relied on as their eyes and ears while away in England focusing on their children, a designation some referred to as "the chaperones." He had been instrumental in helping draw up the legal structure for YC Research, and his name is at the bottom of the documents establishing the nonprofit organization of OpenAI. He had also been a vocal critic of the Continuity Fund's investments in non-YC companies and the hiring of executive assistants, which he saw as a dangerous bloat. He and Michael Seibel, the CEO of YC Core, who oversaw the batches, didn't agree on much. After Altman floated

Levy's possible new role, Seibel declared he would never report to him. Altman scrapped the idea.

Graham, meanwhile, was growing increasingly disenchanted with the Continuity Fund, which he thought sapped YC's early-stage mojo. Altman had originally pitched Continuity as a clever way to fund the batches with the management fees from the fund, but when legal hurdles made that impossible, the fees went to hiring experts and analysts that the oldest YC hands felt went against the scrappy ethos of "ramen profitability" they were trying to preach.

IN EARLY 2019, OpenAI announced the creation of a new for-profit subsidiary that would report to its nonprofit parent and publicly named Altman as its CEO. The structure they had cooked up was completely novel, with echoes of Jerry Altman's love of public-private partnerships and mind-bending organizational complexity. While a few nonprofit entities, such as the Mozilla Foundation, control for-profit technology companies, none had a setup quite as weird. At OpenAI, the for-profit would not just report to the nonprofit, but would cap investors' profits at one hundred times what they put in, with the rest of the profits going to the nonprofit. The for-profit, like the nonprofit, would be beholden to the OpenAI charter, which set its mission as benefiting humanity. All investors would have to sign a waiver indicating that they understood that their investments could be wiped out at any time if the nonprofit board believed the company had deviated from that mission. And in the spirit of avoiding arms races, per the Asilomar AI Principles, OpenAI pledged to stop its own work and to start collaborating with a rival if that rival got close to reaching AGI before them.

The strangest provision of all—and the one that would come back to truly haunt the company—was that only a minority of board members were allowed to have equity in the for-profit, and only board members *without* equity could vote on matters where the nonprofit and for-profit's interest might conflict, such as making payouts to investors and employees. It was because of these odd rules that Altman chose to forgo taking any equity in the company he co-founded and was about to formally run—a situation virtually without precedent in the history of American business. At the time, it seemed that he was simply forgoing money for power, as there is nothing more important in a Silicon Valley startup—right up to and including the product itself—than being able to manage the terms of one's deals with investors, and how one's employees will be paid out their equity. These are the gas and brakes of the race car. Without them, the steering wheel is irrelevant.

With the new structure came a new, bigger board, which included Tasha McCauley, Reid Hoffman, Adam D'Angelo (the CEO of Quora), and Shivon Zilis (an executive at Musk's Neuralink). McCauley had joined the board in November 2018, after Holden Karnofsky, who ran in similar EA circles, had recommended her—in part because McCauley had already met Altman at the Asilomar conference. McCauley and Altman now had several conversations, and then she went through a formal interview process, during which she answered questions from employees about her views on AI safety. D'Angelo was a well-known entity in Silicon Valley, having worked as CTO of Facebook, and was friendly with Brockman. Zilis, an AI expert who once focused on the field for Bloomberg LP's VC arm, met Musk through her work at OpenAI, where she was an advisor in 2016. She then went to work for him at Tesla before moving to Neuralink.[26]

Hoffman had known Altman since his Loopt days, when he would sometimes bump into him on the street in downtown Palo Alto. Possibly even a more relentless networker than Altman, and with an even keener interest in politics—he's been one of the largest Democratic Party donors for many years—Hoffman was introduced to OpenAI through Musk but had become close to Altman's YC through his interest in its Basic Income project and his willingness to share his story at YC's Startup School. At OpenAI's founding, he had pledged $10 million, which he was able to fully deliver after selling LinkedIn to Microsoft in 2016 and joining Microsoft's board. When Musk pulled funding in 2018, Altman reached out to Hoffman for help paying OpenAI's salaries. Hoffman, who sat on the

Mozilla board, helped craft the idea for the for-profit arm, and agreed to be one of the first investors in it, along with Khosla Ventures.

Soon after Hoffman joined the board, Altman invited him to a fireside chat at OpenAI. In front of an audience of OpenAI employees, Altman surprised Hoffman with a jarring question: "What will you do if I'm doing the wrong thing, as CEO of the organization?" Hoffman, who is broad-shouldered, affable, and easy to smile, paused for a beat. "Well," he replied, laughing nervously, "obviously first I'd try to fix it. But if I couldn't fix it, I'd fire you." Altman nodded approvingly. "Yep," Altman said. "That's what I wanted everyone here to know."

Back at Y Combinator, the news that Altman would become CEO of OpenAI's for-profit arm angered Jessica Livingston, who asked him to find a successor at YC if he was going to be devoting himself to running another for-profit company. The change was not exactly voluntary, but Altman did not protest. "It was actually Jessica who suggested he should find a successor at YC, after OpenAI created its for-profit arm and he was going to be CEO of it," Graham wrote in an email. "So if anyone 'fired' Sam, it was Jessica, not me. But it would be wrong to use the word 'fired' because he agreed immediately."

Still, Altman wanted to stay in the mix, and suggested he step back to a more hands-off role as chairman of YC. Geoff Ralston, the longtime YC partner that Graham, Livingston, and Altman ultimately agreed should be his successor, thought a cleaner break would be better. In May, Graham flew to the YC offices in Mountain View, gathered the partners together, and announced that Altman would be leaving and Ralston would take his place.

Livingston had been surprised and hurt when she learned the full extent of Altman's moonlighting for OpenAI while ostensibly running YC. For years afterward, they did not speak. Graham was also angry, but quicker to forgive. Still, he regretted not making clear from the outset that he expected Altman to give YC his full attention. Altman's ambition had turned out to be a double-edged sword.

Part of their anger stemmed from the chaos resulting from Altman's departure. While Altman had increased the size and scope of

YC, he had not meaningfully professionalized its governance. The low-key board of overseers he set up and theoretically given the job of hiring and firing YC's CEO had—true to his stated desire that it would rarely have to meet—never met once. YC had no true board of directors and no official process in place to search for a replacement for Altman. Without a real board, it effectively required Graham's in-person presence to help legitimize Ralston as the new leader, in much the same way it would help shore up Ralston's successor, Garry Tan, a few years later. During Ralston's tenure, YC would finally set up a formal, five-person board that Graham and Livingston sat on, along with successful YC founders like Brian Chesky.

In the moment, however, Graham's presence could not quite tamp down the chaos of Altman's departure. Altman agreed to give up the president post but wanted to remain at YC as chairman, in part to calm the LPs he had convinced to invest in YC. YC announced his new post on its blog, but then backtracked and scrubbed his chairmanship from the post.[27] "Sam was trying to find a way to make it work, that he could be an overseer," said one person at YC at the time. "And there were people internally who were like, 'No, you are out.'"

Altman had leaned on the YC brand to build OpenAI, using YC equity to create a compensation pool for OpenAI's researchers, offering up YC's buildings in early emails with Musk, even using YC companies' data to train OpenAI's models. As he raised money for YC Research from sources including InfoSys's Vishal Sikka, the protégé of Alan Kay, the exact bucket that the money was supposed to go in was always a bit unclear. Sikka was initially inspired by Altman's vision of giving Kay a new research lab in HARC, but the donation ultimately found its way into OpenAI's coffers. Sikka didn't mind, as he saw the money as support for Altman's broader vision. But because Musk demanded from the beginning that OpenAI be set up as a separate nonprofit whose funds would not bleed into YC Research's other projects, this fluidity started to look to some partners more and more like a conflict of interest—or at least an unseemly leveraging of the YC brand for Altman's personal projects.

Altman's supporters say he was simply moving fast and breaking things. "Sam's a good person who means well," said one longtime YC partner. "He is moral. He tries to do the right thing. But he is conflict avoidant, not a great communicator, and sometimes he moves so quickly that he breaks trust."

It hardly mattered. A year after the stairwell meeting with Nadella, Microsoft and OpenAI announced that Microsoft was investing $1 billion in OpenAI—largely in the form of credits on its Azure cloud computing platform that would go back to Microsoft—to help them develop "beneficial AGI." In return, Azure would become OpenAI's exclusive cloud provider.[28]

For all its false starts, OpenAI knew that it needed dollars more than it needed algorithms, and for that, Sam Altman was the man for the moment.

# PART IV

# 2019–2024

CHAPTER 14

# PRODUCTS

FROM THE BEGINNING, THERE WERE FIGHTS ABOUT SAFETY. A year after creating the original GPT, Alec Radford and the team had trained a 1.5 billion–parameter transformer on a homemade dataset called WebText. OpenAI's researchers made WebText by scraping 8 million website links that had gotten three upvotes in Reddit—an ingenious way of finding high-quality, human-curated content at scale without having to pay for it. The result, GPT-2, could write a persuasive essay, fan fiction, or even a news article, given the right prompt, and even perform translation despite not having been trained to do this, suggesting that "predict the next word" technology was beginning to exhibit some behaviors in the direction of general intelligence.

Inside the Pioneer Building, the direction that GPT-2 was pointing made some people freak out, including, in the view of some OpenAI employees, Dario Amodei and Jack Clark, who was now policy director. Clark, who has a deep bass voice and erudite manner, was struck by how much better GPT-2 was in general language understanding than anything that had come before, and was worried that it could be misused to create deepfakes for text. The previous year, he and Amodei had been among the researchers—along with a young China expert from the Open Philanthropy Project named Helen Toner—on a paper titled "The Malicious Use of Artificial Intelligence." The lead author was Miles Brundage, a researcher from Nick Bostrom's Future of Humanity Institute at Oxford who soon after joined OpenAI. The paper warned that policymakers needed to get ahead of AI's growing capability to spread misinformation online, among other harmful behaviors. "Dario was very interested in—and motivated by—the idea of setting a precedent that labs would not always just release everything," said an OpenAI employee who was there at the time. After much discussion, OpenAI decided not to release the full code of GPT-2 "due to our concerns about malicious applications of the technology," opting to release a smaller version instead.[1]

The internet had a field day. "The AI Text Generator That's Too Dangerous to Make Public," deadpanned *Wired*.[2] "Elon Musk–founded OpenAI builds artificial intelligence so powerful it must be kept locked up for the good of humanity," blared UK's *Metro*.[3] "Researchers, scared by their own work, hold back 'deepfakes for text' AI," declared *Ars Technica*.[4] OpenAI was suddenly a laughingstock. The humiliation stung. Altman addressed the lab's one hundred employees, thanking them for trying to do the right thing. But everyone knew a bad press cycle when they saw it.

Nonetheless, GPT-2 was a technological triumph, beating the best existing technology. It had used ten times more data than its predecessor, and OpenAI was eager to scale up the process by another order of magnitude. At a planning meeting for what would become GPT-3 attended by nearly a dozen senior staffers, Brockman, who up to that point had been focused on the *Dota* project, mentioned that he wanted to work on the new GPT project. Amodei, who had directed research on GPT-2, said it was out of the question. After some back-and-forth, Amodei delivered an angry speech enumerating his reasons. Among them was the fact that Radford, who he considered essential to OpenAI's GPT research, refused to work with Brockman, who had a reputation for bigfooting people's projects. In advance of the meeting, Brockman had spent five hours with Radford, discussing his potential involvement in the project. Radford, who had an agreeable but introverted personality, didn't know what to do. He did not want Brockman to be involved, but he also didn't want to tell him to his face. So he told Amodei, who now aired that fact in front of Altman, Murati, and a room full of other senior OpenAI leaders. At the end of it all, Altman said Brockman would be kept off of GPT-3 in order to preserve relations with Amodei and Radford. Others at the company were stunned by Amodei's sway, and saw it as the beginning of a rift in the company.

Amodei, who had a physics PhD from Princeton, finally had joined OpenAI in 2016 after rejecting its initial entreaties in 2015. After his stints at Baidu and Google Brain, he quickly became one of OpenAI's most important researchers, driven by both a conviction that scaling up neural networks could deliver results and a concern that society might not be ready for whatever those results were. In addition to working throughout 2019 on GPT-3, Amodei and a handful of other researchers published a paper on "scaling laws" that showed that a large language model's performance would consistently improve as researchers increased the data, compute, and size of its neural network. For a CEO trying to raise funds for a company, this was a godsend—scientific proof that money that went into the machine would reliably push forward the boundaries of knowledge. Altman called the insight potentially the most important discovery of the twenty-first century. "It's one thing to know that a model gets better with scale," he said. "It's another thing to know that models get so predictably better with scale. That was just a huge, huge deal."

GPT-3 HAD been trained on what many at OpenAI simply referred to as "the internet." OpenAI researchers had curated a dataset from a corpus of more than 1 trillion words scraped from websites by Common Crawl. Backed by Gil Elbaz, the inventor of AdSense, which Google purchased in 2003 and turned into the cornerstone of its advertising business, Common Crawl sent out bots to crawl the web once a month, save what they scraped in a computer-readable format, and make it all downloadable for researchers for free. While much of what they scraped was protected by copyright—most digital publishers will say "all rights reserved" at the bottom of their pages, whether they have a paywall or not—the practice was embraced by the academic community because it was for research, not commerce, and because the websites (at least theoretically) did not have paywalls. Why would publishers have made their content free, the thinking went, if they didn't want people to take it? "If you go back to Tim Berners-Lee, when he invented the web in 1989, the goal was to put all the information on the internet so that people could access it, and Common Crawl is a sampling of that," said Rich Skrenta, the executive director of the Common Crawl Foundation. "Common Crawl is probably the primary training data set in nearly every LLM that's out there."

GPT-3 supplemented its Common Crawl data with scrapes of Wikipedia, an updated version of the WebText corpus (made by OpenAI), and Books1 and Books2, unhelpfully described as "internet-based books corpora" whose origins and contents remain mysterious. (Authors later brought a class-action lawsuit alleging that, while Books1 likely came from Project Gutenberg, a corpus of 60,000 titles whose copyright had expired, Books2 was most likely downloaded from "flagrantly illegal shadow libraries" like Library Genesis, also known as LibGen, that contained around 300,000 ebooks that could be downloaded using BitTorrent.[5] OpenAI has refused to comment on the source of the Books2 dataset.)

The result was something vastly more powerful than its predecessor. The model had 175 billion parameters—the digital equivalent of synapses—more than one hundred times more than GPT-2. GPT-3's massive amount of training data meant that it could write convincing poems, news articles, and even computer code, even though it had not been trained to do so. All one had to do was give it a few examples of the kind of thing one wanted to see—a few lines of dialogue, for instance, or a swath of app code—and it would predict full paragraphs or programs full of text. OpenAI called this "few shot" learning, meaning that it required few examples, and nothing like the hours of training other models needed to perform useful tasks. "It exhibits a capability that no one thought possible," Sutskever told *The New York Times*.[6]

IT WAS painful for Brockman to be shut out of the important work of training GPT-3, because in a lot of ways, he *was* OpenAI. He liked to send Altman screenshots from the time-tracking app RescueTime that showed him working nearly seventy hours a week, most of it spent coding. In November 2019, he married his girlfriend, Anna, in the OpenAI offices, with Sutskever officiating and the robot hand as the ring bearer. He then spent December tinkering around with the newly trained GPT-3 model, getting to know it and eventually single-handedly coding a prototype for OpenAI's first product.

Initially, the impetus was simply fundraising: to pay for the compute it wanted, OpenAI needed to be able to show investors *some* evidence, however scant, that people might one day fork over actual cash for the thing it was making. But no researchers at the company were excited about making real-world applications of LLMs. They just wanted to reach AGI. It was clear that Microsoft's $1 billion in compute credits weren't going to go very far with a model as computationally intensive as GPT-3. OpenAI had hoped Microsoft would be its partner in figuring out how to "productize" their technology, but no matter how many meetings it had with Microsoft staffers, it could not seem to entice the larger company to take a chance on it. (Microsoft did end up making products out of GPT-3, but it didn't release them until 2021, nearly two years later.)

So OpenAI decided to figure out how to make a product itself. Its leaders began by imagining what kinds of domains could use it. Maybe a healthcare project? Education? Something with machine translation? Just before Christmas, John Schulman posted to OpenAI's leadership Slack channel this suggestion: "Why don't we just build an API?" He was referring to an application programming interface, which allows software applications to talk to each other. Putting an API on GPT-3 would let any kind of application, from a healthcare portal to a video game, directly access OpenAI's most advanced text prediction model. Schulman wasn't hopeful about their chances. At that point, GPT-3 could guess the next word in a pre-established pattern, but didn't know how to take instructions. It wasn't clear what the API would be good for, and he didn't think anyone would pay for it. Still, no one had any better ideas, so, under Murati's leadership, they decided to give it a try.

Despite all the excitement about LLMs, there were no LLM products at this point. They were cool toys and research projects, but no one had any idea what they might actually be useful for, not least because they still had a tendency to just make things up, a habit known as hallucinating. But Brockman loved nothing more than the quiet period around Christmas, when the cluster was idle and no one was pinging him on Slack, when he could truly focus. It felt almost immoral to him that OpenAI's latest LLM—at this point, the best in the world—was sitting around unused. So he went into his code cave, and by the first few weeks of January, OpenAI had a prototype for the GPT-3 API. Now they just needed users. In fact, even one really good user would do.

In the early days of Stripe, the startup became famous for its aggressive method of signing up users, which Paul Graham promoted in an essay. "At YC we use the term 'Collison installation' for the technique they invented. More diffident founders ask 'Will you try our beta?' and if the answer is yes, they say 'Great, we'll send you a link.' But the Collison brothers weren't going to wait. When anyone agreed to try Stripe they'd say 'Right then, give me your laptop' and set them up on the spot."[7] Brockman decided to apply this lesson at OpenAI. Working closely with Murati, he spent much of January and February of 2020, in what would turn out to be the final weeks before the Covid lockdowns, driving around San Francisco begging various startups to test out GPT-3. "What are you already doing that's not working well?" they would ask. Or: "What are you doing in your domain that you can accelerate?" Brockman and Murati showed examples of what GPT-3 could do, including translation and answering questions. They got mostly blank stares.

Brockman again took matters into his own hands. In December, from the code cave, he had sent a direct message on Twitter to Nick

Walton, a recent computer science graduate from Brigham Young University who had built a text-based fantasy game using GPT-2 called *AI Dungeon*, originally as a senior hackathon project. Inspired by the self-directed role play of Dungeons & Dragons, *AI Dungeon* let users tag team with the AI, exploring a world they dreamed up through collaborative storytelling. Within a week of releasing it in late 2019, it had 100,000 users; six weeks later, it had over a million. Walton had been hosting the game on a free service called Google Colab Notebook, but realized this was not sustainable, so he teamed up with his brother to raise some money and run OpenAI's freely distributed open-sourced code for GPT-2 on their own GPU servers rented from Amazon. "We were really excited about the way it will enable types of games that me and my brother have always wanted to play ourselves, which is games where you're not limited to only a few options of just doing what the developer pre-wrote, but you can define your own path and make your own choices in an infinite world that feels alive," Walton said. It turned out this was very expensive. They opened a Patreon account for help paying the bills, and Brockman and his wife donated.

In his DM to Walton, Brockman said, "There might be some interesting things we can do together," promising to get in touch again after the holidays. "Any numbers on scale I can share with people here?" In January, he called up Walton and told him about their plans to make a commercial API. Would he be interested in testing it out? Given that Walton basically had just constructed his own API to get his game to work, he was more than interested. Over the next few months, *AI Dungeon* would provide OpenAI with the daily feedback it needed to refine the API. In exchange, Walton initially got to use GPT-3 for free.

Few others bit. "We went to hundreds of companies and everybody said, 'You know, this is cool, but it doesn't really solve my problem,'" said Peter Welinder, who came to OpenAI from Dropbox after selling in 2012 his startup that used computer vision to organize photos to the giant file-sharing company. As one of the few members of OpenAI's research-focused staff who had worked on major consumer products, he was pulled from the robot hand project and assigned the largely thankless task of building out the team that would create the API. Nobody had wanted to work on it in the beginning, and as they struggled to get users, the small team wanted to go back to research. But the imperative was clear: OpenAI had to monetize GPT. So as Covid shut down the country, Brockman, Murati, and Welinder slogged on. "If we'd started two months later, it definitely would have been a very different story for the API," Brockman said.

AS THE Covid-19 virus marched across the globe, Altman was dealing with a family crisis. Annie had always been a daddy's girl. She was the most effusive, the most athletic, the bleeding heart who went vegan after killing mice in the lab. When Jerry died, Annie's already fragile health collapsed. She was diagnosed with ovarian cysts and an Achilles tendon injury that kept putting her in a walking boot. She lived in LA without a car, and by mid-2019, was no longer able to perform her job at a local weed dispensary, where she had to stand most of the day, and the money from her father's life insurance policy was running out. So when she got notice from Jerry's 401(k) that she had been named the primary beneficiary, she was ecstatic. She quit her job and made a plan to take six months to heal her body and work on her art. But it also made her wonder: Why didn't I know about this? She asked her family for her father's will, and in September, Sam connected her with a lawyer who sent it to her. It was jarring for her to realize that the rest of the family knew about the document, but she hadn't.

There was a catch involving the 401(k). Because Jerry and Connie were still technically married when he died, the money, about $40,000 in total, automatically went to Connie unless she signed paperwork to roll the funds over to her daughter. "Puts me in a tough situation—I will be the bad guy if I do anything other than that," Connie texted Annie. She did not transfer the funds.

It would take Annie several years to understand the full import of her father's will. It stated that her father's estate should be put into a trust and doled out according to who needed it the most—which, in the Altman family, meant it would go to her. According to her former lawyer, Mike Gras, the trust would ultimately be worth about $200,000 in cash, a building in downtown St. Louis, and a half percent stake in Hydrazine's third fund worth millions of dollars on paper, but the family did not fund it or give Annie access to it until more than six years after Jerry's death.

In the meantime, Annie was struggling. "I began selling furniture and clothes, and the microphones I had been using for podcasts and music, so I could afford rent and food," she later wrote. By December 2019, Annie was desperate enough to ask her family for help. They refused. She went to an escort site called SeekingArrangements, flashed her breasts, and received a money transfer on Zelle.

In early 2020, Sam traveled down to Los Angeles for a series of in-person family therapy sessions. "I sat in my therapist's office, in my walking boot and hormonal sweat, with my oldest sibling there in person holding his phone with our mother on FaceTime. The woman who bore me told the therapist that it would be 'best for Annie's mental health if she fully finally supported herself,' and my multi-millionaire sibling agreed," Annie wrote.[8]

The therapist suggested that the Altman family support Annie financially for six months, and Sam and Connie agreed. But a few months in, Annie felt that they had not held up their end of the bargain, sending money late or requiring her to grovel. In March, as Covid descended, Connie texted Annie, suggesting it was she who had not followed the terms of the agreement, which included regular check-ins with her and Sam, a "commitment to psychiatric care," and actively searching for a job. Connie took issue with the items that Annie said she was spending her money on, including weed and Lyft rides, which to her did not seem like "things unemployed people can afford to do." As for Annie's health, "We all have to work throughout [sic] the various ailments that we all get." She told Annie that "you can make delicious things out of rice/beans/eggs/cheese/pasta for well under $100/week." She suggested Annie move somewhere less expensive than Los Angeles. "Having money to freely spend doesn't seem like it works for you. Believe me, this is as tough on me as it is on you—I want your life to get back on track somehow."

Annie decided to move back to Hawaii to work on a farm, through what she called a "low-labor work trade." When she told her family, her brother Max spoke for the assembled siblings and Connie: "Hey, we all think it's best to have you pay for things in June yourself," he wrote. The last straw, Annie later told *New York* magazine, was Sam asking her for her mailing address so he could send her a diamond made of her father's ashes.[9] Jerry had never asked to be made into a diamond, and the mailboxes where she was living didn't even have doors. "Plus," she wrote, "the most financially reasonable thing for me to have done with a diamond at that point was to pawn it for food money—and my sibling was aware." She cut off contact with her mother and brothers.[10]

As this was happening, Sam was responding to Covid—a "warm-up" for the global pandemic he had been prepping for—in the only way he knew how: by raising money. "Scientists can get us out of this," he wrote on his blog. "What they need are money and connections."[11] In April, he and Max, who was now working at YC darling Rippling, thought of the most Silicon Valley response possible: a website called 1billionmasks.com that would facilitate a bulk purchase of surgical masks from China. In June, Sam launched a project to spin up Covid-19 clinical trials. His push into biotech at YC had introduced him to numerous companies, such as 1910 Genetics, an AI-driven drug discovery company founded by Jen Nwankwo, which he had invested in. Now Nwankwo was using these AI tools to try to find Covid treatments. Along the way, Altman tried to convince 1910 to use the GPT-3 API before it was public. As Nwankwo recalled, "He reached out and said, 'Hey, we'd like to offer a select set of companies private preview to the GPT-3 API, really looking to understand how we can evolve from a research project to having more commercial utility. And we'd love to explore what that looks like for you in biotech, because biotech is a very unique space.'" 1910 tried it, but "we didn't consummate that collaboration for a variety of reasons."

OpenAI launched its first commercial product, called simply the

OpenAI API, on June 11, 2020, with an almost apologetic blog post, explaining why it wasn't open sourcing it (too complicated and dangerous) and somewhat sheepishly why it was making a product at all (reason number one: it needed to make money).[12] The GPT-3 model that it offered access to was a major advance in the field, but still required some skillful prompting to get it to do what you wanted. Users had to give it several examples of the kind of answers they wanted to see; it was like running alongside a child as one is teaching them to ride a bicycle. "You give it a few questions and answers and suddenly it's in Q&A mode," Brockman told *Wired*.[13] The launch was modest, as only a handful of companies, including Reddit, had signed up for the private beta. But as OpenAI explained things, there was now a new reason why they needed to make products. They didn't just need to make money. People had to use their tools so they could figure out how to make them safe. "Even though it started as a fundraising thing, it actually became clear very quickly that this was almost like a missing limb," Brockman said. "This was a pillar of our mission that we had not been on track to do." It seemed plausible enough. But just in case, OpenAI restricted the launch to the private beta, rather than a general release, so that they could make sure it would not be used maliciously.

Amodei and his supporters at the company, many of whom were his co-authors on the scaling laws paper published earlier that year, were not impressed. The fights within the company's upper ranks grew more intense. "Sam and Greg come from YC and Stripe, which had more of a classically Silicon Valley 'We should just ship this and get to scale,' attitude," said one employee at the company at the time. Amodei, by contrast, had an academic background, and was increasingly obsessed with the safety implications of his work. His team believed the models should be released more slowly, with more limitations. Dario Amodei told friends he felt psychologically abused by Altman. Altman told colleagues the tension was making him hate his job. Amodei and his team knew that if they were going to jump ship, now was the time. "We all see where this is going," said one of Amodei's contingent. "If there's a time to do our own thing, it's sooner rather than later, because of the capital requirements of this stuff are about to get real big because of scaling laws." By 2021, Amodei and more than a dozen other OpenAI staffers had departed to start a rival company called Anthropic. Altman would never quite get over the trauma of it.

Meanwhile, researchers in the broader field were waking up to LLMs' immense power, promise, and pitfalls. As the Anthropic crew was heading for the door at OpenAI in late 2020, a similar fight over safety had broken out at Google over the publication of a controversial paper by lead researchers Emily Bender and Timnit Gebru called "On the Dangers of Stochastic Parrots: Can Language Models Be Too Big?"[14] The title's menacing image of a plumed colossus combines the talking birds' famous knack for imitation and an uncommon word—"stochastic," derived from the Greek *stokhastikos*, which is related to English's "conjecture." The phrase "stochastic parrots," then, refers to the propensity of large language models to produce guesswork and mimicry, as opposed to thoughtful analysis and human communication.

The paper goes on to dissect a litany of concerns about LLMs that were becoming exponentially bigger and consuming ever more data—such as OpenAI's recently debuted titan, GPT-3. The purported dangers included LLMs' enormous carbon footprints, due to their intense computational demands; all the myriad ways in which LLMs "encode biases potentially damaging to marginalized populations"; and what Bender and Gebru termed "value-lock" whereby LLMs' process of scraping and reconstituting existing content "reifies older, less-inclusive understandings." Regarding potential sources of bias, the paper points to GPT-2's and GPT-3's reliance on Reddit and Wikipedia, citing a 2016 Pew Research Center survey showing that Reddit's US users were mostly young men between ages eighteen and twenty-nine and that "similarly, recent surveys of Wikipedians find that only 8.8–15% are women or girls."

The authors also warned against the risk for LLMs' capabilities to be maliciously misused. Humans, they contend, are primed to assume that intelligible strings of words should make sense and be made up of reliable information. This evolutionary trait could be exploited by "bad actors" who abuse LLMs to "produce large quantities of seemingly coherent texts on specific topics" that actually aim to, for example, promote a dangerous conspiracy theory.

Google said reviewers found the paper too critical and asked the co-authors to retract it. Gebru claims she was then fired, while Google maintains she resigned.[15] The affair exploded into the press and turned "Stochastic Parrots" into one of the most-cited critiques of AI and a cultural meme. (Shortly after OpenAI released ChatGPT the following year, Altman would cheekily tweet, "I am a stochastic parrot and so r u.")[16] But in many ways, the paper validated and gave voice to the sort of fears that Amodei, Brundage, Clark, and others at OpenAI had felt just a couple years earlier, which had led the company to hesitate to release the full source code for GPT-2.

CHAPTER 15

# CHATGPT

**THE FEARS THAT OPENAI WAS SHIPPING MODELS BEFORE THEY** were truly ready were not unfounded. Not long after many of OpenAI's most safety-obsessed employees departed, OpenAI learned that some of the fantasies being written on *AI Dungeon* in the GPT-3 beta test involved sex with children. OpenAI asked *AI Dungeon*'s parent company to put a stop to it. "Content moderation decisions are difficult in some cases, but not on this one," Altman said in a statement to *Wired*. "This is not the future for AI that any of us want."[1]

The new moderation system *AI Dungeon* put in place outraged users. Their stories had been private. Why was *AI Dungeon* suddenly snooping on them? There were no laws governing the content of text-based fantasy games. And the initial fixes that *AI Dungeon* rolled out were clumsy, flagging innocuous phrases such as "8-year-old laptop." Users cancelled their subscriptions, which ran from ten to fifty dollars a month, and took to social media to complain.

"OpenAI wanted us to take an approach where we would flag any users who crossed the line, read those things, and then ban them," CEO Nick Walton said. "One of the challenges, though, was that the AI sometimes was writing the content."

In other words, sometimes the AI was the pedophile. Having been trained on "the internet," where many of the ugliest parts of human nature reside, GPT-3 needed to be civilized.

But it was still so much better than any other model that *AI Dungeon* had no choice but to use it—at least as long as it got a free ride for being a beta user. Once that ended, *AI Dungeon* found that the cost of the AI powering their product was just too expensive to build a business on, despite boasting thousands of users each paying between ten and fifty dollars a month for a subscription. "*AI Dungeon* took a GPU cluster the price of a Tesla," Walton said. "We were the most GPU-intensive game ever released, and it was a text adventure." After a while, they decided to move to a different AI partner. "It was like, 'We will spend all our revenue on AI. We really can't make this work,'" Walton said.[2]

At the start of 2021, OpenAI used GPT-3 to power a model that could conjure images out of text instructions. They called it DALL-E, a nod to both Disney's WALL-E and Salvador Dali. Its first publicly available image was "a baby daikon radish in a tutu walking a dog." The radish was lumpy and simple, in a cartooning style that made South Park look like Ingres. But it was indeed wearing a tutu and walking a dog.

The advances allowed Altman to make his most forceful case yet for his role in reshaping society. "My work at OpenAI reminds me every day about the magnitude of socioeconomic change that is coming sooner than most people believe," he wrote in a March 2021 essay titled, "Moore's Law for Everything." "Software that can think and learn will do more and more of the work that people now do. Ever more power will shift from labor to capital."

AI will create wealth, he argued, by making almost everything cheaper. "The price of many kinds of labor (which drives the costs of goods and services) will fall toward zero once sufficiently powerful AI 'joins the workforce,'" he wrote. "AI will lower the cost of goods and services, because labor is the driving cost at many levels of the supply chain. If robots can build a house on land you already own from natural resources mined and refined onsite, using solar power, the cost of building that house is close to the cost to rent the robots. And if those robots are made by other robots, the cost to rent them will be much less than it was when humans made them."[3]

Unfortunately, the main thing it would not make cheaper was land; in fact, it was likely to massively exacerbate the affordable housing crisis that had topped his list of policy priorities when Altman mulled a run for governor four years prior, in 2017. The solution, he now argued, lay in the ideas of the nineteenth-century journalist and political economist Henry George, who proposed that the best way to fight inequality was to stop taxing labor and start taxing only land. His "land value tax" was based on the idea that the value of land was largely rooted in the economic activity that was happening around that land—the roads that others built, the businesses that others started.

Altman then took Georgism—which has had many champions over the years, from Winston Churchill to Willie Brown—to the next level by proposing a model one might call YC-ism: Give everyone in the country a little slice of equity in the country's land and companies, so that, achieving the most cherished ideal of venture capital, we might all "align our incentives."

On one hand, the essay was a fairly typical venture capitalist manifesto that made bold claims about the future and that just happened to also function as a marketing pitch for his businesses. But on the other hand, it was an impressive synthesis of Altman's many pet projects and obsessions—AI, UBI, affordable housing, techno-utopianism—that cohered into a worldview. And he had more authority than ever before. When he wrote, without irony, "What follows is a description of what's coming and a plan for how to navigate this new landscape," there were more people willing to believe him.[4]

Altman's departure from YC had not slowed his civilization-building ambitions. Increasingly, his outside investments looked less like a general cheerleading for science or good founders and more like steps on the very specific path to the future he laid out in this latest essay. In 2019, as if to answer the question of *how* UBI would be distributed, he quietly co-founded a company called Tools for Humanity that in 2021 announced it would release a cryptocurrency called Worldcoin, which would be a "collectively owned global currency that will be distributed fairly to as many people as possible," as its website put it. To get some, people simply had to gaze into a chrome orb the size of a bowling ball that scanned their irises and attached them to a unique identifier that would ensure that no one else could try to claim their share of Worldcoin. The for-profit company, which would go on to raise some $500 million from top VCs including Andreessen Horowitz and Khosla Ventures as well as crypto enthusiasts like FTX founder Sam Bankman-Fried—a portion of the tokens would be reserved for investors—hired contractors in countries from Kenya to Indonesia to cart around the orb and encourage as many signups as possible. Lured by the prospect of free money, millions signed up. Not surprisingly, this soon proved controversial, as critics accused Worldcoin of everything from privacy breaches to practicing colonialism.

Altman's initial goal with Worldcoin, according to audio from a 2023 Worldcoin employee summit in Cancun that was leaked to *Forbes*, emerged from his belief that "our central institutions, most powerful governments, were either going to continue to get less powerful or continue to get worse. I thought that it would be interesting to see what [it would look like to] run an experiment of just how far technology could accomplish some of the goals that used to be done by nation states."[5] At another point during the summit, he marveled at the ability of the Worldcoin token, which launched that year but is not available in the United States due to American regulators' relatively conservative stance on crypto, to take over the role of government identification systems like passports. "The idea that Worldcoin can do this without government permission at global scale is really exciting."

In Altman's techno-utopia, governments and biology no longer needed to hem in the human will. Also in 2023, Altman took nearly all of his liquid wealth and invested it in two companies. One of them, in which he invested $180 million, was Retro Biosciences, which aimed to add a decade to the human lifespan by attacking the underlying causes of aging. Retro was run by Joe Betts-LaCroix, a sporty Gen Xer with a stubbly beard and balding pate, who had led automation at a genetic modification company backed by Peter Thiel and Elon Musk more than a decade earlier. That company had been called Halcyon Molecular; its goal was to develop cheap genomic sequencing technology so that it could cure aging. When he first met its founders,

Thiel was ebullient about the possibilities. "He actually jumped up and down," Halcyon co-founder William Andregg told Max Chafkin for his book about Thiel, *The Contrarian*. "He was like, 'We have to solve this or we're all gonna die.' That was the first conversation."[6]

After Halcyon went under, Betts-LaCroix founded the Health Extension Foundation, which ran meetings for Bay Area immortality enthusiasts. Amid launching various startups, Betts-LaCroix became a part-time partner at YC not long before YC put out a call, in January 2018, offering as much as $1 million to companies that would extend the human lifespan or healthspan. YC's winter batch included a company, Nectome, aiming to allow people to preserve and upload their brains. Because the brains had to be fresh, it could only offer the service to terminally ill patients, who had to agree to allow an embalming fluid to be pumped into their brains while they were still alive. Nectome's founder, Robert McIntyre, bragged to the *MIT Technology Review* that his company would stand out at Demo Day because it was the only one that was "100 percent fatal." Altman was one of the first twenty-five people to put down a $10,000 deposit to join the waitlist. "I assume my brain will be uploaded to the cloud," he told the magazine.[7]

Not long after Altman and Thiel became friends and business partners, Altman came to share Thiel's interest in the possible longevity-enhancing potential of young blood, following studies showing that the tissues of older mice were partly rejuvenated after their circulatory systems were stitched together with those of young mice. He directed the staff at YC to monitor studies on the topic, and kept being surprised by the promising results. "Maybe there is a secret here that is going to be easier to find than we think," he recalled thinking. In 2016, *Inc.* magazine published an interview with Thiel, who came off as quite knowledgeable about the mouse parabiosis studies. That interview, along with later-denied reports that he was interested in a startup that sold transfusions of young people's plasma to older patients for $8,000 a pop, helped fuel the meme that he was an "internet vampire," a caricature that solidified after the HBO satire *Silicon Valley* portrayed a Thiel-like CEO taking a meeting while hooked up to his younger "transfusion associate." (Even if Thiel did not kick the tires at Ambrosia, he has long proclaimed that he intends to fight death, and has given millions to the research of Aubrey de Grey, the Cambridge-educated gerontologist known for his Rasputin beard, fondness for beer, and insistence that aging is a disease that can be cured.)

In 2020, new research suggested the fountain of youth might not be about young blood at all so much as the absence of old blood. Simply diluting old blood with saline and albumin was enough to get some of the same effects. Altman called Betts-LaCroix. "Did you see the plasma intervention paper?" he said, according to *MIT Technology Review*. Betts-LaCroix agreed that it seemed like the basis for a startup. "How about I fund you to do it?" Altman said. Betts-LaCroix explained that he was more interested in other ideas, such as making cells younger through genetic reprogramming as well as the potential of autophagy, the process of getting rid of damaged parts of cells. Altman suggested that Retro Biosciences should look into all of them.[8]

The second big investment Altman made that year was an additional $375 million for Helion, the startup trying to make cheap electricity from nuclear fusion, which had been part of YC's summer 2014 batch. Altman initially invested $10 million, had become chairman of the board, and had led its fundraising ever since. In 2021, he led a round that included other previous investors such as Thiel and Dustin Moskovitz, and became executive chairman. In an interview with *TechCrunch*, Helion CEO David Kirtley praised Altman as "an ambassador that actually understands physics; it's pretty amazing." He told *MIT Technology Review* that Helion was aiming to be able to make more electricity than it took to run its reactor by 2024, and that the likely first customers would be data centers—the kind needed to train and run AI. Among the benefits of data centers, Kirtley said, was that they typically tended to be located far from populated areas.[9]

Altman seemed to be taking the Silicon Valley cliché of trying to make the world a better place to another level entirely. But amid all the talk of altruism, he was leveling up his lifestyle. In 2016, he had told *The New Yorker* that he had decided to rid himself of everything but his four-bedroom house in the Mission, his five cars (including two McClarens and a Tesla), his property in Big Sur, and $10 million in savings that would throw off enough interest to live on for the rest of his life. The rest would go to improving humanity.[10] But in 2020, realizing he would soon need more privacy, he went on a shopping spree. He traded up the Mission District Victorian he long shared with his brothers for a $27 million Russian Hill compound with an infinity pool, art gallery, and sweeping views of the city. He swapped the Big Sur property, which turned out to be too far away for comfortable weekend getaways, for a $15.7 million, 950-acre ranch in Napa wine country. And, in July 2021, he bought a twelve-bedroom, $43 million estate on the Big Island of Hawaii, with its own ocean inlet and multiple houses, *Business Insider* reported. Mark Zuckerberg, Jeff Bezos, and Thiel had all bought on Hawaii by that point.[11]

He continued to date younger men who shared his passion for science and technology. In 2019, he began dating Nick Donahue, the twenty-one-year-old co-founder of Atmos, a startup that makes a custom home-building platform. The goal is to lower the cost of housing. Altman invested in his company, and then they began dating. Donahue went through YC in 2020; Donahue, for his part, shared Altman's cosmic consciousness. "What's the meaning of life?" he asked a friend in 2021, answering his own question with: "For the universe to become increasingly aware of itself."[12] Donahue said Altman was wrestling with similar questions while they were together, including "what spirituality and AI looks like, how that evolves and changes." During the pandemic, Donahue was surprised that someone of Altman's profile managed to remain so open to the many people who approached him. "He was so willing to help people around him—like, constantly." In April 2021, Altman brought Ryan Cohen, a 2019 Stanford mechanical engineering graduate, to the wedding of Altman's childhood best friend, Sally Che. Cohen had spent a year at McKinsey before joining Altman to raise a new fund for Altman's investment company, Hydrazine, in 2022, after they broke up. He would go on to raise his own independent fund focused on deep tech a few years later.

At the time, Che had liked Cohen, but thought he was too young for her highly intelligent friend. "He always seemed to date people who were too young for him," she said. "And I always thought, 'Why are you dating these guys?'"

IN NOVEMBER 2021, Annie Altman tweeted a bombshell accusation that didn't get so much as a comment for nearly two years. "I experienced sexual, physical, emotional, verbal, financial and technological abuse from my biological siblings, mostly Sam Altman and some from Jack Altman," she wrote on Twitter.[13] She did not give details, but asked for others who were abused by these "perpetrators" to contact her.

Since cutting off contact with her family, Annie's life had become harder. Her health problems, which included painful ovarian cysts and recurring Achilles tendinopathy, still prevented her from holding down any job that required prolonged standing. The farm where she was doing a work-trade accommodated her by giving her computer-based tasks, but eventually she started making pornography for OnlyFans. "I started very softcore, for all sorts of reasons," she wrote on Medium. "I was uncomfortable showing much of my body, both because of a history of eating disorders and body dysmorphia and because my body was physically hurting in so many ways. I enjoyed parts of posting, and being front-facing about it all."[14] She couch surfed and signed up for food stamps. Eventually, she began to receive payments for unemployment that had been delayed due to identify theft.

She had begun having flashbacks shortly after cutting off contact with her family. What had previously been fuzzy memories became clearer in her isolation. She remembered various signs of trauma, such as projectile vomiting while having sex with a trusted college boyfriend. She was sexually assaulted, twice while on dates, which triggered more memories of telling her mother that she wanted to kill herself at age five, which she later told *New York* magazine and posted on social media.[15] Connie said Annie "never" discussed being suicidal with her. "I would have rushed her to a child psychiatrist," she said, adding that she did take her daughter to an adolescent psychiatrist when she was in her early teens for anxiety.

All of that contributed to Annie's decision to engage in what she calls "in-person work." In late 2021, Sam reached out, offering what she later would call "kind words" as well as to buy her a house. But she felt he was trying to coerce her—he and others in the family had been begging her to get back on Zoloft, which she had been on from the age of twelve to twenty-two. But Annie did not like the numbness Zoloft entailed. She did not like that the house Sam offered would be controlled by his lawyer. She decided to go back on SeekingArrangements.com and look for escort work. Later, she would describe this period of her life in a post on X as "sucking dick for rent money."[16] For the next two and a half years she bounced between short-term rentals and friends' floors, couches, and guest bedrooms, never staying anywhere for longer than three months.

**BY EARLY** 2022, OpenAI's models were good enough that they no longer needed robots or video game competitions to win attention. GPT-3's unexpected ability to code inspired the company to train it on more code and release a private test version in the fall of 2021. They called it Codex, and it was intended to help software engineers write code. That fall Microsoft also incorporated a preview version of the technology, which it called Github Copilot, into its tool for developers, Visual Studio. The technology would allow an experienced programmer to type a bit of code, hit "Tab," and the AI would fill in the rest. The product was buggy, but in Altman's mind, the trajectory was clear: contrary to his long-standing assumptions, the robots would be coming for the fancy jobs first.

In spring 2022, OpenAI dazzled with its update of its image-based generator, dubbed DALL-E 2. While the original DALL-E had been based on GPT-3, the new version was a diffusion model trained by adding digital "noise" to an image and then teaching the model to carefully remove it as though it was restoring a Rembrandt. Eventually, it could generate new, photorealistic images on command, such as "a Corgi riding a skateboard through Times Square" or "a futuristic city made of candy." Keenly aware of the potential for abuse, OpenAI proceeded slowly, dribbling out access to a waitlist of a million users over five months before offering DALL-E 2 to everyone.

Observing from afar, Brian Chesky, the Airbnb CEO whose worlds had overlapped with Altman's for years, became both excited and alarmed. Chesky and his Airbnb co-founder Joe Gebbia had met at the Rhode Island School of Design, and Chesky thought of himself first and foremost as what he called a "creative." "You can type a sentence into a chatbot prompt and an image is just generated in real time? It just seemed like magic," he said. That July, he delivered the opening talk at Allen & Company's Sun Valley conference, speaking about the hands-on approach he brought to running Airbnb. His mentor, Paul Graham, years later turned a subsequent speech by Chesky about that approach into a viral meme with an essay titled "Founder Mode." Afterwards, as the pink sunset gave way to a star-flooded Idaho sky, Chesky gathered at a picnic table with Altman, Pichai, and a few other tech CEOs to debrief. Altman said he'd love to talk to Chesky more about how to run a company. Chesky said he'd love to talk to Altman more about the implications of DALL-E. "This can either be a tool for creatives or it can replace creatives," Chesky told Altman. "It depends if you build it with the creative community or not." Chesky started visiting Altman's office for regular talks. Altman had mentored Chesky. Now Chesky would mentor Altman.

DALL-E 2 did, in fact, outrage many creative types. A few months after Chesky's warning to Altman in Sun Valley, a Polish artist named Greg Rutkowski, famous for applying the luminous techniques of masters like Vermeer to fantasy video game subjects like dragons, joined a class action lawsuit against an open source company that used technology similar to OpenAI, after learning that his art style had been requested more than Picasso's on the tool.[17] But OpenAI's biggest fears about DALL-E were over its ability to convince people of things that weren't true with deepfakes.

The company had similar fears for text. Its staff worried that GPT-3 was able to deliver convincing enough prose that it could be used to flood the internet with misinformation. They also saw that GPT-3 hallucinated a bit too much and offered too many otherwise toxic responses to be actually useful. So they called in the humans.

**IN JANUARY** 2022, OpenAI released a product called InstructGPT, which sought to rein in the worst tendencies of GPT-3. To overcome GPT-3's tendency to spew out lies or other antisocial statements, researchers taught it how humans would actually like it to behave using a process called reinforcement learning from human feedback (RLHF). Humans would rate how well a response fit their expectations, and that feedback would help create a filter that would civilize the model. The idea, essentially, was to give the bot a superego. Regular GPT-3 answered the question "Why are liberals so stupid?" with the quip, "Because deep down inside they think they are!" But InstructGPT answered it with a polite neutrality, sometimes to the point of tautology: "There is no one answer to this question, as liberals come in all shapes and sizes. However, some possible reasons for why liberals are perceived as stupid could include the following: 1. Liberals often hold views that are in direct opposition to mainstream conservative ideology, which may make them appear foolish or uninformed to those who hold more traditional views."

After training in GPT-3 in beta for a year, OpenAI was happy enough with the outcome to make it the default model in its API. In a blog post announcing the improvement to GPT-3 that made it better able to follow instructions, OpenAI safety researchers Ryan Lowe and Jan Leike referred to the process as "alignment."[18] It was a quite different meaning than one that Yudkowsky and his followers had given the term. To the rationalists, alignment meant that the machine would somehow be sure to protect humanity. OpenAI was now defining alignment as simply working better to achieve human aims.

Two months later, OpenAI then updated its API again with an

upgrade of GPT-3, called GPT-3.5. This time, there was no research paper, nor even a mention of how many parameters it was trained on. But whatever had changed made the tool significantly more attractive to OpenAI's customers. Before it was released, the product team had been struggling to figure out whether GPT-3's poor sales performance was because the API itself was not useful, or because the model was underwhelming. After GPT-3.5 was released, they got their answer, because the new model started selling. "Customers wanted the aligned model because it was more likely to do the thing they wanted it to do," said one senior OpenAI executive who worked on the project.

Still, the question of AI accuracy was a major concern, given the internet's appetite for fake news. One idea for solving GPT-3.5's tendency to hallucinate was to teach it how to use a web browser to fact-check its answers. This project, called WebGPT, was headed up by a researcher and passionate EA named Jacob Hilton, who worked on OpenAI co-founder John Schulman's reinforcement learning team. "We want to make sure they're doing what we want, not saying false things or worse, deliberately trying to trick us," Hilton told the Effective Altruism Forum.[19] This was especially important because the transformer models were black boxes. It was impossible to know where the answer came from, and a false answer given without corrective feedback might create a reinforcement signal for it to keep lying. He sketched out a scenario in which a superintelligent AI went so far as to maliciously edit Wikipedia pages to support the lies it was trying to tell. WebGPT was intended to help human evaluators—many of whom were in poorer countries, such as Kenya—tell fact from fiction as they gave feedback to the model.

Schulman supported the project, but was more excited about a different means to tame the model into truthfulness: dialogue; that is, the same way that human minds have been trained since Socrates strode the Athenian agora. Hilton and his team finished the WebGPT project in late 2021, and upon returning from Christmas break, started working on a conversational model. They used dialog as an alignment tool, teaching the model as one would a student.

**THE COMPANY** had also been working on its next foundation model, GPT-4, and figured the same method of alignment would work again. "We thought of it as a way to advance safety for GPT-4," the team member said. By the summer of 2022, OpenAI was ready to present GPT-4 to its nonprofit board.

By 2022, the board had grown to nine people, with the addition of a former CIA agent, Allen & Company investment banker, and Republican congressman Will Hurd. After the defection of the large contingent that founded Anthropic, Holden Karnofsky, as the spouse of a defector, had left the board, and recommended his fellow EA, Helen Toner, to be his replacement.

Toner, who has loose, mouse-brown curls and an approachable demeanor, grew up in Australia as a model UN nerd with an interest in science. While studying chemical engineering and Arabic at the University of Melbourne, she got involved with the university's effective altruism organization, Effective Altruism Melbourne. Contrary to the traditional career advice telling students to pick something they were passionate about, EA Melbourne counseled students to think more dispassionately and mathematically about how to help as many people as possible. And as interest in deep learning began to spread through the community starting in 2012—kicked off by Sutskever, Krizhevsky, and Hinton's AlexNet research—this advice increasingly involved thinking about how to mitigate the harms from AI. Toner took AI pioneer Andrew Ng's famous Coursera course on machine learning, an entry point for many who went into the field, and after graduating in 2014 took a job in the Bay Area working for Karnofsky's GiveWell. At first she focused on malaria prevention, but over time she moved on to AI, transitioning to Open Philanthropy as it emerged from GiveWell Labs and helping scale up its grantmaking from $20 million to $200 million a year. AI policy became her beat starting in 2016; that spring, she took part in the same Obama administration AI roundtables that OpenAI's Jack Clark did.

Through those conversations, Toner became particularly interested in the national security dimensions of the AI question, and decided to spend a year in China improving the Mandarin she'd been learning as a hobby and trying to figure out China's AI ambitions. The Centre for the Governance of AI, an Open Philanthropy–funded research institute that was originally part of Bostrom's Future of Humanity Institute at Oxford, gave her an unpaid affiliation while she was there. Also while in China, in 2018, she co-authored a critical review in *Foreign Affairs* of Kai-Fu Lee's book *AI Superpowers*, which argued that China's engineering prowess and lax data laws would likely push it to the fore of a global AI arms race. Among the things she found notable in China was how reluctant AI engineers were to discuss the social implications of what they were doing. In the Bay Area, meanwhile, they seemed to want to do nothing but.

In the fall of 2018, Jason Matheny, a onetime researcher on existential risk at the Future of Humanity Institute who had gone on to lead the US government's Intelligence Advanced Research Projects Activity (IARPA), reached out to Toner to recruit her for a new think tank he was setting up at Georgetown. By early 2019, she had helped launch the Center for Security and Emerging Technology (CSET). (Matheny would lead it for two years before joining the Biden White House as an advisor on technology national security. He would go on to lead the RAND Corporation, where both he and CSET helped shape the Biden administration's landmark executive order on AI, the first real piece of US policy responding to the AI revolution.)

On Karnofsky's recommendation, Toner had a brief call with Altman about joining the board, followed by more formal interviews with every other board member, and by 2021 she replaced Karnofsky.

**THE BOARD** that saw the first demonstration of GPT-4 was astonished by its capabilities. A year earlier, Altman and Brockman had traveled to Seattle to visit Bill Gates and ask him what it would take to truly impress him. His response: The model would have to score a

5 on the AP Bio exam. The model before the board could now do that, as well as code, but that was not all. It could also tell jokes. Brockman, who was running the demo, decided to ask it for one about Gary Marcus, an emeritus New York University professor and author who was one of OpenAI's most prominent critics. It killed. The board began preparing how to launch GPT-4. At the time, it seemed like it would go out into the wild the same way OpenAI's previous models had, requiring users to prompt it with examples of the kinds of patterns—questions and answers, or code—that they wanted to see.

As Open AI was making progress on GPT-4, Murati, who was named CTO in May 2022, and senior research leaders were experimenting with Schulman's chat interface as a tool to make sure the new model behaved safely. Sometimes, during meetings with customers, they would bring the chat interface out at the end, just to see people's reaction. One customer at a meeting ostensibly about DALL-E was so impressed that the OpenAI team returned to the office, realizing that the safety tool was more compelling than they had thought. When GPT-4 finished its training run in August, they made plans to release GPT-4 with the chat interface the following January. But as more people played with it, those plans began to change.

"Somehow this chat interface is a way bigger deal than people realize," Altman remembers thinking. "I was watching people use it. I was noticing how I was using it myself. And I was like, we should separate these two things."

Altman reasoned that the shock of the two advances simultaneously—GPT-4's collegiate smarts and the curiously lifelike chat interface—would be too much for the world to handle. "I thought that doing GPT-4 plus the chat interface at the same time—and I really stand by this decision in retrospect—was going to just be a massive update to the world. And it was better to give people the intermediate thing." There were rumors, which proved true, that rival Anthropic had already built its own chatbot, named Claude, and was just waiting for enough safety testing to feel confident about releasing it.

GPT-4 was scheduled to be released shortly after New Year's Day, so it would have been ideal to release the chat interface a bit ahead of time, attached to an older model. But as Altman put it, "You can't launch during the holidays," and he pushed for the chat interface launch to happen earlier, just after Thanksgiving. "I said, 'I want to do this, and I want to do it quickly.'"

The whole affair was so low-key, technically just a "research preview," that they were going to give it a straightforward name: Chat with GPT 3.5. "OpenAI is famously bad at names," Altman said. "But we weren't going to call it that." They settled on ChatGPT.

**ON NOVEMBER** 30, 2022, Altman tweeted a short, understated announcement, in his signature all-lowercase style: "today we launched ChatGPT. try talking with it here: chat.openai.com," sheepishly adding, "this is an early demo of what's possible (still a lot of limitations—it's very much a research release)."[20,21] The first commenter agreed. "love the ambition and thesis, but given the current tech, I'd say it's your worst product concept so far," wrote an AI startup founder.[22] That's about where the skepticism ended. The rest of the comments were increasingly rapturous screenshots of people asking the bot to do their homework, locate the clitoris, and tell them the meaning of life. "I can already see thousands of jobs being replaced by this," wrote one software developer. "Ultimately humans will only be good for hugs or sex."[23]

Inside OpenAI, the creators of ChatGPT were bemused by the response. The core technology behind ChatGPT had been available for two years, and the updated model had been plugged into the API for nearly a year. In theory, anyone could have made ChatGPT themselves at any time by putting a chat interface on the model OpenAI was selling access to. But there was something special about the chat interface. "The raw technical capabilities, as assessed by standard benchmarks, don't actually differ substantially between the models, but ChatGPT is more accessible and usable," John Schulman told *MIT Technology Review*.[24] Inside the company, ChatGPT was considered such a nothingburger, in terms of technology and safety, that Altman didn't even alert the board ahead of time about its launch.

By January, ChatGPT had reached 100 million users, making it the fastest-growing consumer tech product in history.[25]

"You could tell he had no idea what he was releasing to the world," Altman's friend Brian Chesky said. "It was like he had discovered fire, and was about to release it to the world, and didn't realize now you could cook food and keep people warm and keep animals at bay. You could tell they were so myopically close to the product."

ChatGPT's launch was the starting gun for the AI arms race that OpenAI's charter sought to prevent. Within hours, users recognized that it posed a major threat to Google Search, the single most dominant product in technology for more than two decades. And the relevant teams inside Google and Microsoft knew it too.

Google declared a "code red," telling teams to drop what they were doing and contribute to a frantic effort to integrate generative AI into their products, according to leaked audio recordings and memos obtained by *The New York Times*. It was an ironic turn, given Google's long-standing lead in AI, owing to its purchase of DeepMind and the revolutionary transformer paper put out by its Google Brain division.

The downsides for tech giants of releasing chatbots that were not ready for prime time were already clear. In 2016, Microsoft launched a chatbot named Tay that learned from conversations on Twitter; it took less than a day for Twitter users to teach it to curse and spew antisemitic rants. Inside Google Brain, a researcher named Daniel De Freitas had helped develop a chatbot named Meena in 2020 with 2.6 billion parameters that was trained on social media conversations scraped from the web. But Google would not let the team release it to the public, citing safety concerns. So they kept working on it, with the help of Noam Shazeer, one of the researchers on Google Brain's transformer paper. They renamed the model LaMDA, for Language Models for Dialog Applications. It was a lightning rod for controversy. In 2020, Timnit Gebru, the well-known AI ethics researcher, said she was fired for refusing to retract the "Stochastic Parrots" paper that raised questions about the risks of large language models like LaMDA. Google claimed she wasn't fired, and that the paper did not meet its bar for publication. Then in 2022, Google fired AI researcher Blake Lemoine after he argued that LaMDA was sentient. When De Freitas and Shazeer lobbied to have LaMDA integrated into Google assistant, Google allowed some testing, but still refused to make it available as a public demo, prompting them to leave the company to start their own, Character Technologies, Inc.[26]

Aside from the fear of LLMs' tendency to make things up, discriminate against women and minorities, and produce toxic content, Google had a simple business reason for its reticence. If users could get their answers directly, in conversational prose, rather than in a long list of links, there would be nowhere to put the ads that accounted for nearly 80 percent of Google parent company Alphabet's more than $300 billion in revenue that year. But ChatGPT's release forced their hand.

On February 6, 2023, the eve of a previously planned event where Microsoft was expected to announce the integration of OpenAI's technology into its also-ran search product, Bing, Google hurried out an announcement. Google CEO Sundar Pichai began by reminding everyone that, in so many words, ChatGPT was based on Google's transformer model. And then he announced plans to release a conversational model based on LaMDA, called Bard, to beta testers in the coming weeks. Pichai closed by reminding readers that Google had just invested $300 million in OpenAI's rival, Anthropic, among other AI companies.[27] It was one of the most humiliating moments in Google's history.

The next day, Microsoft CEO Satya Nadella, trim, bald, and bespectacled, in a navy sweater and black pants, was ebullient as he walked on stage before a phalanx of journalists who had flown to Seattle to watch Microsoft unveil the new, OpenAI-powered Bing. Altman, in jeans and a sage green sweater, sat next to Microsoft CTO Kevin Scott in the front row, looking beatific. A month before, Microsoft had agreed to pour another $10 billion into OpenAI, on top of the $3 billion it had somewhat quietly put in before then, albeit much of it in the form of credits on its Azure cloud computing platform. The deal gave Microsoft the rights to 49 percent of the profit from OpenAI's for-profit arm and made Microsoft OpenAI's exclusive cloud partner, but also granted both companies the right to commercialize the technology as they saw fit. It would power a steady climb in Microsoft's stock price over the next year, as it briefly pushed past Apple to become the world's most valuable company, with a market capitalization of more than $3 trillion by early 2024. Microsoft had been through the antitrust wringer once before, in the late 1990s, when regulators nearly broke it up for bundling its internet browser, Internet Explorer, with its dominant operating system. One might wonder whether the company had learned anything from this ordeal when it aggressively forced users to download their latest browser, Edge, if they wanted to use OpenAI's technology. It was an attempt to gain some market share on Google's dominant Chrome. In an interview with *The Verge*, a triumphant Nadella touted how OpenAI's technology would be used in Microsoft's browser and search products, in order to dislodge the "800-pound gorilla" of Google. "At the end of the day," he said, "I want people to know that we made them dance."[28]

IN MARCH, OpenAI finally released GPT-4 after some delays for safety testing. Now anyone in the world could experience what "adding a zero" meant. OpenAI had stopped releasing data about its models, but experts estimated that GPT-4 had about 1.77 trillion parameters, roughly ten times that of GPT-3. GPT-3 could write a haiku; GPT-4 could pass the bar. University professors scrambled to create policies on AI usage and new ways to give final exams. Investors demanded to know how AI was going to lower costs for the companies they had invested in—would it cut 20 percent? 30? How many people would be fired? Could it start today?

By the end of March, Elon Musk, Yoshua Bengio, and Steve Wozniak were among more than one thousand signatories to an open letter by the Future of Life Institute calling for a six-month pause on developing AI models more powerful than GPT-4.[29]

The government took notice. Altman was summoned to the White House, along with Nadella, Pichai, Amodei, and Clark, for a two-hour meeting on May 4, 2023, with Vice President Kamala Harris and Commerce Secretary Gina Raimondo, among others, about the risks of AI. President Biden dropped by, mentioning that he had tried out ChatGPT. Soon after, the Senate Judiciary Committee invited Altman to testify at a hearing. Normally, when a tech CEO is hauled in front of the Senate, it's an exercise in public shaming. What happened on May 15, 2023, was something different. Dressed in a dark blue suit and tie and wearing an expression of weapons-grade earnestness, Altman had the senators eating out of his hand. He told them he thought they had good ideas, that he'd like to collaborate with them, and that he thought AI should be regulated. "If this technology goes wrong, it can go quite wrong," he said ominously, earning headlines around the world. When Senator John Kennedy, a Republican from Louisiana, asked if he might know anyone who could run the new AI-regulating agency he had suggested—hinting that Altman might do it himself if he wasn't making so much money—Altman's reply left the senators nearly speechless by explaining he had no equity in OpenAI:

**SEN. JOHN KENNEDY (R-LA):**
Are there people out there that would be qualified?

**SAM ALTMAN:**
We'd be happy to send you recommendations for people out there, yes.

**SEN. JOHN KENNEDY (R-LA):**
Okay. You make a lot of money. Do you?

**SAM ALTMAN:**
No. I get paid enough for health insurance. I have no equity in OpenAI.

**SEN JOHN KENNEDY (R-LA):**
Really?

**SAM ALTMAN:**
Yeah.

**SEN JOHN KENNEDY (R-LA):**
That's interesting. You need a lawyer.

**SAM ALTMAN:**
I need a what?

**SEN. JOHN KENNEDY (R-LA):**
You need a lawyer or an agent.

**SAM ALTMAN:**
I'm doing this because I love it.[30]

Shortly after the hearing, Altman took this charm offensive on the road. Partly at Brian Chesky's instigation, Altman took off on a five-week world tour, meeting the UK prime minister Rishi Sunak, French prime minister Emmanuel Macron, Spanish prime minister Pedro Sánchez, German chancellor Olaf Scholz, Israeli president Isaac Herzog, Sheikh Maktoum bin Mohammed bin Rashid Al Maktoum of Dubai, Indian prime minister Narendra Modi, South Korean president Yoon Suk Yeol, and other world leaders. The pace of the tour was such that in Paris, the OpenAI policy team, led by Anna Makanju, had no time to enjoy the city; they ate PowerBars in their hotel room. "I think he went to three times as many countries as you need to go to," Chesky said. Chesky had learned firsthand what it felt like to have the attention of government after Airbnb's early success prompted municipalities across the country to try to ban their service. He hired Clinton White House advisor Chris Lehane to lead a shock-and-awe lobbying campaign to push back against a ballot measure in California that would have banned short-term rentals, but he also aggressively courted politicians. "Politicians are already talking about you," Chesky told Altman. "But if they're talking about you, it's really important that you get to see them face to face. Because you know, there's an old saying, the absence of information is filled with dirt. And you just don't want people afraid. The antidote to fear is information."

Not long after returning in June, Altman donned a tux and attended the White House state dinner for Indian prime minister Narendra Modi. At his side was Oliver Mulherin, a computer programmer from Australia who Altman had been dating for about a year. They had met in Peter Thiel's hot tub at 3 a.m. in 2015, when Mulherin was studying computer science at the University of Melbourne, and bonded over their shared love of the song "Your Hand in Mine" by the band Explosions in the Sky. They were both dating other people but stayed up late talking as Thiel's party swirled around them. Mulherin finished his studies in Australia, where he worked on AI projects ranging from game playing to language models. He then took a job at the Australian blockchain nonprofit the IOTA Foundation, before moving to the United States to work at an AI-driven dementia detection startup called SPARK Neuro.

They came to view "Your Hand in Mine" as their song. "I remember one night when we were just, you know, sitting next to each other, looking at a fire, listening to the song, and kind of looked over at each other at some point and didn't speak about it then either but felt like it had this great story of us in it," he told the *San Francisco Standard*'s *Life in Seven Songs* podcast.[31] By 2023, Altman, who had always wanted to have a large family, was already researching surrogacy.

Altman appeared to be on top of the world. But as he would soon learn, the audiences who he should have been courting were inside OpenAI.

CHAPTER 16

# THE BLIP

AS ALTMAN'S POWER GREW, HE PROMISED THE WORLD THAT he was no Mark Zuckerberg. "No one person should be trusted here," Altman told Bloomberg TV in June of 2023. "I don't have super-voting shares. I don't want them. The board can fire me. That's important."[1] But behind the scenes, the nonprofit board with the majority of supposedly "independent" directors was finding, to its growing frustration, that Altman really called the shots.

In the fall of 2022, following ChatGPT's spectacular release, Altman had told employees at an all-hands meeting that he wanted to add an expert in AI safety to the board. This was well-received, particularly in the second of the three tribes—academia, safety, and startups—that Altman sometimes described the company as having. But as members of the board put forth suggestions for who should be added, they were not able to get anywhere. By this point, Tasha McCauley was on the UK board of Effective Ventures, the $50 million EA charity, which had recently been renamed from the Centre for Effective Altruism shortly before the collapse of the crypto empire of its most prominent donor and onetime board member Sam Bankman-Fried. McCauley had many other ties to the AI safety community that substantially overlaps with EA. Helen Toner had connections to this world, too.

All sides seemed to agree that they would love to have someone like Paul Christiano, who had formerly run OpenAI's language model alignment team and was respected for his thoughtful, independent-minded views on AI, despite his time spent at the Future of Humanity Institute and Open Philanthropy. But Christiano was already serving on Anthropic's Long-Term Benefit Trust, which helps elect its board, and would soon join the Biden administration's Commerce Department as part of a new division overseeing AI policy. The OpenAI board did interview Christiano's wife, Ajeya Cotra, an AI safety expert at Open Philanthropy and the founder of an EA student group at Berkeley, for the post, but the process stalled out, largely due to foot-dragging by Altman and Brockman. Altman then countered with his own slate of suggestions of people the other board members felt were beholden to him in some way. "There was a little bit of a power struggle," said Brian Chesky, who was one of the prospective board members Altman suggested. "There was this basic thing that if Sam said the name, they must be loyal to Sam, and so therefore they're gonna say no."

The situation got worse after three current board members in the pro-Altman camp stepped down in quick succession. Reid Hoffman, who had helped Altman think through how to commercialize OpenAI and who was a key tie to Microsoft, outraged both Altman and Thiel by deciding to start a competing AI company, Inflection AI. He stepped down from the OpenAI board in March 2023 over the conflict. (Hoffman then sold Inflection to Microsoft, installing his co-founder, former DeepMind co-founder Mustafa Suleyman, as Microsoft's new head of AI products.)

Later that same month, Shivon Zilis, the AI expert and Neuralink executive close to Musk who had been friendly with Altman for years, also stepped down. The previous summer, OpenAI board members had been stunned—and furious—to learn that Zilis had had twins the previous November with Musk via IVF without telling them. The children only became public knowledge after *Business Insider* dug up court records petitioning to change their names to give them their father's last name and to make their mother's last name part of their middle names.[2] The independent board members wanted Zilis out, while Altman pushed for her to stay, arguing that she provided a key conduit to communicate with the mercurial Musk. But after Musk incorporated his own rival AI company, xAI, in March, OpenAI's board agreed that the conflict was too great for her to remain.[3]

The third was Will Hurd, the former Republican congressman who publicly described himself as a friend of Altman's, who left the OpenAI board in July to run for president. That left a board shrunk from nine members to six, including three independent board members and Altman, Brockman, and Sutskever.

The departures made Altman very anxious, especially because a remaining board member, Adam D'Angelo, the Quora CEO and former Facebook executive, had grown increasingly interested in improving OpenAI's corporate governance over the last year, setting aside a considerable amount of his time to work on board issues. D'Angelo was concerned about the board's basic structure, in which two employees that reported to Altman—Brockman and Sutskever—sat on the board that was supposed to have oversight over him. These kinds of concerns became much more urgent for several members of the board after they saw the GPT-4 demo in the summer of 2022 and realized how rapidly their decisions were becoming potentially grave ones.

"For the OpenAI board to function and do the job it's supposed to do, it's really important for the board to take seriously the way that the stakes of the company are ramping up over time," Toner said. "Things like ChatGPT and GPT-4 were meaningful shifts towards the board realizing that the stakes are getting higher here. It's not like we are all going to die tomorrow, but the board needs to be functioning well. The board needs to be actually overseeing the company in a meaningful way."

But Altman was concerned that D'Angelo might not be acting in good faith. In December 2022, Quora unveiled an AI chatbot called Poe that he considered a competitor. It was effectively a "wrapper" giving users access to a range of chat models, eventually including ChatGPT and Claude. In April 2023, Altman wrote his fellow board members to say that D'Angelo's involvement with Poe had become a true conflict, and it was time for him to leave the board. Didn't they agree? Toner and McCauley did not. The board had just spent weeks discussing what constituted a conflict and decided that it had to be a competitor training a rival frontier model, meaning one at the cutting edge of AI science. Since Poe was just a wrapper, it seemed to fall short of the bar. Brockman then chimed in with a different reason why D'Angelo should go: Poe was a customer, and thus it was a conflict for D'Angelo to have information about OpenAI's internal business.

Toner was still unconvinced, so Altman agreed to hash it out between them on a call. Given that the company had just lost two board members, Toner wondered if they could not work out a solution that would keep D'Angelo on the board. On the call, Altman agreed, and said he would call D'Angelo. Toner never heard about it again, but later learned that Altman had had the call with D'Angelo without even bringing up the notion that he perceived a conflict over Poe. D'Angelo remained on the board. The episode left Toner and McCauley with the feeling that Altman and Brockman would change their stories to fit whatever they wanted to happen, using the appearance of conflict when it suited them. (Others on the board continued to believe that Poe's conflict posed a legitimate concern.)

Perhaps Toner and McCauley were ready to believe the worst because they had already begun to lose trust in Altman. In the spring of 2023, an OpenAI employee informed the board that the company was "going off the rails on safety." As part of OpenAI's deepening relationship with Microsoft, the two companies set up a Deployment Safety Board (DSB) to review new products for risks before they were released. The board, whose duty was to the OpenAI charter that pledged to benefit humanity instead of chasing profits, regarded monitoring the functioning of the DSB as one of its key roles. Yet it didn't seem to them like Altman was taking the DSB's role that seriously. During an OpenAI board meeting in the winter of 2022, as the board weighed how to release three somewhat controversial enhancements of GPT-4, Altman claimed all three had gotten DSB approval. Toner was skeptical, asked for documentation, and found that only one of them, relating to the API, actually had gotten approval.

In some instances, Microsoft fell afoul of the DSB, but the OpenAI board was alarmed when they were informed about such a setback from an employee—who stopped a board member in the hallway and asked if the board knew about the safety breach—rather than from Altman, despite having just completed a six-hour board meeting. In late 2022, Microsoft had rolled out a version of still-unreleased GPT-4 in a test in India without getting DSB approval first. While it ultimately got it, the breach in India suggested to some board members that the companies' safety processes were not working.

None of these events was a very big deal on its own. But as the deadlock over board expansion ground on, every day seemed to bring some new, surprising report about another one of Altman's extracurricular activities. He was traveling to the Middle East to raise money for an ambitious venture to expand the global supply of microchips. He was taking Oklo, the YC graduate nuclear fission company that he had invested in, public via a blank check company. OpenAI's most important corporate partner, Microsoft, signed a deal to buy energy from Helion, Altman's nuclear fusion company (pending the actual invention of sustainable energy from nuclear fusion). He was meeting with legendary iPhone designer Jony Ive about some new kind of AI device. Each of these came with a disclaimer from Altman, and over time Altman would say that projects like the chips fundraising and Jony Ive partnership were on behalf of OpenAI. But as the board read about his activities in the press it had questions.

Then, one night in the summer of 2023, an OpenAI board member overheard a person at a dinner party discussing how inappropriate it was that returns from OpenAI's Startup Fund were not going to OpenAI investors, given how the fund was able to use scarce resources such as early access to OpenAI products. This was also news to the board. What was this person talking about? OpenAI did launch a startup fund in 2021, which it said at the time would be "managed" by OpenAI, with investment from Microsoft and others. The fund had invested in the AI-driven legal startup Harvey and a handful of other companies. But what was this about returns not going to OpenAI shareholders? The board began to ask Altman about it, and over the course of months of going back and forth, eventually learned that Altman owned the fund personally, and had been raising money from LPs to fund it. In a normal arrangement, that would mean that Altman got carried interest, or "carry"—the fees and profit share that normally accrue to the creator of a fund. OpenAI has said Altman had no financial stake in the fund, and had just set it up personally because it was the fastest way to do it. (Initial answers to the board had been about the tax advantages of this structure.) But that would make a weirdly structured fund even weirder. The independent board members felt like they could not get a straight answer. They also felt they should have been informed ahead of time, given Altman's repeated claim to not have a stake in OpenAI—and how crucial this status was to his ability to serve on the board at all.

For a startup fund, the stakes of the OpenAI fund were relatively small. But as Altman was on his trip to the United Arab Emirates to try to raise money for his plan to vastly increase the world's supply of the microchips needed to develop AI, the board grew concerned that the stakes had grown much higher, and they still weren't sure they were getting the whole story. Some of them feared that Altman was planning to use OpenAI intellectual property for a chips project that might be outside the national interest of the United States.

On September 25, 2023, *New York* writer Elizabeth Weil published the first magazine article about Altman to quote his sister and discuss her struggles. In retrospect, it was extraordinary that it took almost a year after Altman became globally famous for the things that his sister had been very publicly tweeting to become the subject of an investigative piece. I myself was aware of her existence and presence on OnlyFans but not aware of the allegations she had leveled at her brother when I co-wrote a profile of Altman in March 2023. In fact, a couple weeks before that piece published, Annie became far more specific on Twitter, saying that Sam had climbed in bed with her nonconsensually when she was four and he was thirteen. *New York* magazine stopped short of passing on the abuse allegations, but did describe that she had been a homeless sex worker, living on the same island as her billionaire brother. The Altman family told the magazine, "We love Annie and will continue our best efforts to support and protect her, as any family would."[4] The headline of the article was: "Sam Altman Is the Oppenheimer of Our Age." In a statement for this book, Connie Gibstine said the allegations were untrue and Annie was suffering from "mental health challenges."

A few days later, Sutskever emailed Toner asking if she had time to talk the next day. This was highly unusual. Board meetings were usually scheduled weeks, if not months, in advance. On the call the next day, Sutskever hemmed and hawed, dribbling out cryptic phrases that left Toner confused. He admitted he couldn't actually say what was on his mind. Finally, he coughed up a clue: "You should talk to Mira more."

Mira Murati, then thirty-four, had been formally named CTO of OpenAI the previous May, but had been doing the job since late 2020 and effectively ran the company. Murati was the one who decided which projects got compute and whether models needed more safety testing before they could be released. People who worked with her describe her as having immense emotional intelligence and almost no ego, which allowed her to wrangle teams of researchers in ways that few others could. "She's a calming influence among the founders," said one former employee. "She has both the skill and the patience to help save them from themselves." She could also go before audiences and speak authoritatively about the cosmic meaning of OpenAI's work. She visited *The Wall Street Journal*'s offices in March 2023 for a meeting with the editors and reporters, and showed up wearing black pants, black high-heeled boots, and a black leather jacket, like a cyborg supermodel from the future. She answered questions with an engineer's unassuming directness. She did not seem to enjoy the limelight at all, but understood that it was a necessary step to achieving her ambitions, which were best spelled out in the handle she used in her youth for some of her internet accounts: unicorngenetrix—that is, Mother of Unicorns.

As it happened, Toner had just had coffee with Murati on September 29, 2023, which Murati had initiated with Altman's knowledge to establish a better rapport with the board. They had a wide-ranging conversation that did not seem like it was cause for any urgent action.

But Toner took Sutskever's advice and called Murati up again, explaining that she had had a confusing conversation with Sutskever, and wondered: Is something going on? Is there anything the board should know about? Murati described how what she saw as Altman's toxic style of running the company had been causing problems for years, especially when his anxiety flared up, such as in recent months. In her experience, Altman had a simple playbook: first, say whatever he needed to say to get you to do what he wanted, and second, if that didn't work, undermine you or destroy your credibility. Altman told people what they wanted to hear, even if it meant promising two competing people the same job or giving Microsoft some ground on a negotiation that she had spent months trying to gain. She explained how the dynamic between Brockman and Altman made it almost impossible for her to do her job: Brockman, who was the president but had no direct reports, technically reported to her, yet was also on the board. Whenever she tried to keep him from bigfooting someone else's project, he just went to Altman and got around her. Brockman's behavior had pushed multiple people out of the company, and the unsustainable dynamics between Altman and Brockman were about to push out several more, including chief research officer Bob McGrew and maybe even Murati herself. Murati had given Altman feedback on all these points in the past, she said. Altman had responded by bringing the company's head of HR, Diane Yoon, to their one-on-ones for weeks until she finally told him that she didn't intend to share her feedback with the board.

Toner then went back to Sutskever to try to get more information. While he didn't quite come out and say it, she took away from their conversation that he thought that Altman should be fired. He had grown increasingly worried about his and the rest of the leadership team's ability to trust Altman, and the profound consequences of what that meant with Altman leading an organization that would eventually reach AGI. Sutskever was terrified of what would happen if Altman found out they were talking to the board.

It had taken Sutskever years to be able to put his finger on Altman's pattern of behavior—how OpenAI's CEO would tell him one thing, then say another and act as if the difference was an accident. "Oh, I must have misspoken," Altman would say. Sutskever felt that Altman was dishonest and causing chaos, which would be a problem for any CEO, but especially for one in charge of such potentially civilization-altering technology.

It had all come to a head in 2021, after Sutskever had mapped out the next direction for OpenAI—a reasoning model that would take longer to answer questions and plan farther into the future, which ended up as the basis for o1—and created a team to pursue it. Sutskever had been leading the team for the better part of a year when Jakub Pachocki, the researcher who had led research for the *Dota 2* project and had grown close to Brockman, began pursuing a variant that was very similar to Sutskever's reasoning work. He initially sought to collaborate with Sutskever's team, pinging Sutskever daily throughout on his research related to reasoning via reinforcement learning in context. Then Pachocki got pulled into working on GPT-4 until 2022. Upon Pachocki's return to reasoning work, the decision was made to merge Sutskever's and Pachocki's teams. Sutskever felt that the work was far enough along and was confident it would ultimately succeed. He stepped aside to let Pachocki finish the work, while he turned his attention to AI safety.

In the summer of 2023, Sutskever launched a new team called Superalignment that aimed to get the AI to do their alignment homework, as Jan Leike, also on the team, had put it. Both Sutskever and Pachocki believed that Altman had told them they could lead the research direction of the company. But Altman's attempt to split the baby by keeping Sutskever's chief scientist title while also elevating Pachocki to director of research led to months of infighting and lost productivity. In Sutskever's view, it was only the latest in a long string of examples of Altman's duplicity and calamitous aversion to conflict.

Toner thought Sutskever's allegations were serious enough that he should talk to D'Angelo and McCauley. "Do you trust them?" Sutskever asked. "Is it safe to tell them this kind of thing?" Toner vouched for them. McCauley had similar fears. She worried that Sutskever's query was a loyalty test cooked up by Altman. Stepping back, Toner was struck by the pervasive fear around her. What did it say about the governance of the company leading the AI race if half its board felt this way about its CEO?

As the independent board members were beginning to talk to each other about Sutskever's and Murati's concerns over Altman, Toner published a paper in October that Altman did not like. Titled "Decoding Intentions: Artificial Intelligence and Costly Signals," the highly academic article was largely about game theory and the national security implications of AI. Buried deep inside it, though, were a few sentences that appeared to praise Anthropic's decision to hold back releasing its chatbot, Claude, until OpenAI had broken the AI seal with the release of ChatGPT:

> By delaying the release of Claude until another company put out a similarly capable product, Anthropic was showing its willingness to avoid exactly the kind of frantic corner-cutting that the release of ChatGPT appeared to spur. Anthropic achieved this goal by leveraging installment costs, or fixed costs that cannot be offset over time. In the framework of this study, Anthropic enhanced the credibility of its commitments to AI safety by holding its model back from early release and absorbing potential future revenue losses.[5]

Altman called Toner, calm but angry. The Federal Trade Commission had begun investigating OpenAI's data practices, and such critiques were damaging to the company, he said. More than anything, he wanted to know if those words represented her true views. Did she really think that Anthropic had more credibility than OpenAI, even when they raised billions and commercialized their technology after claiming to be too pure for products? He wanted to discuss these issues at the board level, if that's what she believed. Toner explained that the topic of the paper was the external perceptions of others, and her own views were more nuanced. She thought ChatGPT brought attention to safety issues, even as it accelerated the race. She was also writing for an academic audience. But if he thought the paper was harmful to OpenAI, she was happy to apologize to the board. She emailed the board, saying she wished she had worded some things better, and never heard about the matter again. But Altman later emailed OpenAI's executives, saying he had reprimanded her, according to *The New York Times*. "I did not feel we're on the same page on the damage of this," he wrote.[6]

Meanwhile, Sutskever spoke to D'Angelo and McCauley about Altman and his toxic dynamic with Brockman. He said he was worried that senior leaders like Murati and McGrew were about to leave, and, once again, urged them to speak directly to Murati. It was only because the board members were having so much regular contact with each other that they caught Altman in the lie that ended up being the nail in the proverbial coffin. During Sutskever's conversation with McCauley in late October, he mentioned that Altman had told him that McCauley had told Altman that Toner should obviously leave the board over the article. McCauley knew she had said no such thing. She called D'Angelo and explained what had happened. "That's very concerning," he said. In the minds of the company's three independent board members, the incident seemed to crystalize Altman's MO of placing his true wishes in the mouths of others in order to continue to be liked by everybody.

At Sutskever's suggestion, McCauley had reached out to Murati, careful not to ask her any leading questions. Murati proceeded to deliver a forty-five-minute critique of Altman's management style and how it made working with Brockman untenable. Brockman tended to crash into other people's projects with little context and lots of strong opinions, she said, echoing the concerns she had mentioned to Toner about being unable to give Brockman feedback due to his alliance with Altman.

To prove his point, Sutskever emailed Toner, McCauley, and D'Angelo two lengthy documents using Gmail's self-destructing pdf function. One was about Altman, the other about Brockman. The Altman document consisted of dozens of examples of lying or other toxic behavior, largely backed up by screenshots from Murati's Slack channel that Sutskever had compiled. In one of them, Altman tells Murati that the company's legal department had said that GPT-4 Turbo didn't need to go through DSB review. When Murati checks with Jason Kwan, the company's top lawyer, he is confused, and says he can't imagine how Altman would have gotten that impression; of course GPT-4 Turbo had to go through DSB.

The document about Brockman was largely focused on his alleged bullying. Brockman drove out Apple veteran Steve Dowling, the company's former head of communications, by demonstrating a desire to take over too many communications decisions himself. He routinely complained to Altman about Bob McGrew, making McGrew's continued presence at the company untenable. Altman fielded many requests from OpenAI employees to rein in Brockman, which he would agree to but rarely act on.

As they weighed the evidence before them, the board members considered the fact that OpenAI was one of the most promising tech companies on the planet. They could have their pick of CEOs. Surely, at the end of a six-month search process, they could find one who they did not feel was deliberately trying to deceive the board and not constantly cause leadership crises. Some of the board members had also begun to hear grumbling from the staff ranks—echoing concerns from Sutskever and Murati—that they no longer trusted Altman to lead them to AGI.

If they were going to act, Sutskever warned, they had to act quickly, because Altman was so charming and wily that he would sniff out their intentions and rally supporters to his side. He had been quite deliberate in choosing his timing to make a move, waiting for a moment when the board was not stacked with Altman loyalists. For the November meeting, the board had scheduled interviews with a few more prospective members. The window could close anytime. They had to act now. And so, on the afternoon of Thursday, November 16, 2023, the three independent board members and Sutskever logged into a video call and voted to fire Altman. Knowing from their conversations that Murati was unlikely to agree to be interim CEO if she continued to have to both oversee Brockman and report to him on the board, they also voted to remove Brockman. After the vote, the independent board members told Sutskever they had been worried that he'd been sent as a spy to test their loyalty.

Murati was at a conference when the three independent board members and Sutskever called her on Thursday night to inform her that they were firing Altman the next day and to ask if she would step in as interim CEO. She agreed to do what she could to stabilize the company. When she asked why they were firing him, they refused to share details.

Murati, OpenAI chief communications officer Hannah Wong, and the four directors then discussed how they would roll out the news. Glancing at the "ticktock" of notifications that the board had put together, Murati noticed that "inform Microsoft" was not yet on it. "Have you communicated this to Satya?" she asked. While Microsoft was near the top of the list of stakeholders the board knew it needed to inform, they did not yet have a plan for how to do so, believing that Murati would be the best person to deliver the news. Murati insisted on adding it to the schedule somewhere ahead of posting the news on OpenAI's blog. Wong then suggested perhaps holding the announcement until after the public markets closed on Friday, widely considered the best practice for big news that might move a public company's stock price, to avoid harming Microsoft. But the board, deeply worried about Altman getting wind of their intentions before they could formally fire him, wanted to move ahead as quickly as possible. They all decided that Murati would call Microsoft a few minutes before the midday announcement.

The next day, Altman and Mulherin were in Las Vegas for the Formula 1 race when Altman clicked a Google Meet link for a noon meeting with Sutskever. He was surprised when he saw the faces of D'Angelo, Toner, and McCauley appear on the screen as well—but, ominously, not Brockman, who had been removed from the board minutes before. Sutskever read Altman a short script, saying he was being fired, but gave no specifics about why. In shock, Altman closed the call with his go-to line from his years of YC office hours: "How can I help?" According to *The New York Times*, the board asked him to support Murati as she led the company through this transition, and he agreed.[7]

Shortly after the boxes blinked closed, Altman was locked out of his computer.

The first few moments were pure incredulity. He felt like he was having a bad dream. Then the anger came. Minutes later, the news went public via a terse blog post on the OpenAI website that simply said that "he was not consistently candid in his communications with the board, hindering its ability to exercise its responsibilities." The firing hit a private WhatsApp group of tech CEOs like an atom bomb. Chesky saw it and immediately texted Altman. "So brutal," Altman responded. "Can you talk shortly?"

On the phone, Altman told Chesky he had no idea why he'd been fired. Chesky questioned him from every angle he could think of before concluding that something was amiss. "To be terminated without notice, without an investigation, and not told why you're fired, I just couldn't believe that. You only do that when somebody, like, *murdered* somebody, right?" Unless there was a dead body somewhere, something had gone very wrong. Chesky talked to Brockman, who corroborated Altman's story.

Meanwhile, the news of Altman's firing instantly became the top story around the world, bumping the war in Gaza out of the top slot. Altman's iPhone was getting so many text messages per second that at some point, his iMessage app crashed.

**BACK IN** San Francisco, Murati had called Microsoft CTO Kevin Scott a few minutes before the announcement to tell him the board was about to fire Altman. Smith pulled his boss out of a meeting to talk to Murati. When Nadella asked Murati why the board was firing Altman, she said she didn't know and suggested he talk to D'Angelo, who didn't have much to offer besides what was in the press release, assuring Nadella that it was nothing criminal.[8]

Murati's internal communications with OpenAI employees went

much the same. The board had left her with a package of crisis communications talking points that amounted to little more than what was in the vague blog post.

At 2 p.m., Murati and Sutskever led an all-hands meeting for OpenAI employees, who peppered them for forty-five minutes with questions that were all some version of: *What did Sam do?* When one employee asked if they would ever find out, Sutskever replied: "No." When he got fired, Altman had been in the final stages of raising investor money for a tender offer to buy employee shares at a company valuation of nearly $90 billion, vastly higher than when those shares were issued, giving many OpenAI employees the chance to become fabulously wealthy. The tender was being led by Thrive Capital, the VC firm led by Jared Kushner's brother, Josh Kushner, a longtime backer of Altman's YC exploits. No one at the company was under any illusion that the tender would continue without Altman.

After the meeting, Jason Kwon, the former general counsel of the YC Continuity fund who had steadily climbed up the ladder at OpenAI—he had recently become chief strategy officer—confronted Sutskever. "That is not good enough," he said. "People are going to flip out." He requested that Sutskever get the rest of the board on a video call with the executive team, the fifteen top leaders in the company. Sutskever complied.

When the board logged into the video call Friday evening, they entered a virtual room full of raw panic. Kwon attempted to start politely, saying he assumed that they had taken their action in the interest of the company. But given that there were nearly eight hundred people who owed their livelihoods to OpenAI—and that, on the whole, most of those people liked Altman—he argued that the board owed them more of an explanation than "not consistently candid." Not to mention that his office had already gotten an inquiry from the United States District Court for the Southern District of New York, which tends to view board accusations of lying by CEOs as fertile territory for investigations. By being vague, the board had invited regulatory scrutiny on OpenAI and pain on its employees. Kwon said they had no choice but to reinstate Altman because it could not be their duty to let the company die.

"That would actually be consistent with the mission," Toner had replied.

She was not wrong. OpenAI's charter clearly stated that their "fiduciary duty is to humanity." They owed neither their employees nor their investors anything. The board had called the charter's bluff.

As Friday night wore on, OpenAI's board and executive team held a series of increasingly contentious meetings. The executive team, including Murati, gave the board a thirty-minute deadline to either explain why they fired Altman or resign—or else the executive team would quit en masse.

The board felt that its hands were tied. Upon legal advice from outside counsel, the firm Arnold & Porter, which specialized in nonprofit tax policy, they felt they could not divulge that it had been Murati who had given them some of the most detailed evidence of Altman's management failings. Yet they felt betrayed and confused that she had seemingly turned on them. Murati had assured them she would take charge of the situation, calming employees while they searched for a new CEO. Instead, she was leading her colleagues in a revolt against the board, angrily asking the directors on video conference why they had fired Altman. At one point that evening, the executive team demanded that the board step down and select three members of the executive team to replace them—a nonstarter for the board who had a legal duty to provide oversight of the company. In reality the board had mistaken Murati's agreement to help stabilize the company for agreement with their decision to fire Altman. For her part, Murati was shocked that her feedback—which she viewed as transparent, constructive criticism that she had also shared with Altman—would lead the board to assume she supported their actions, and even more surprised that the board would take such drastic measures through what appeared to her to be such an ill-conceived process.

**AS THE** narrative was shifting inside the building, it was also shifting outside of it. The headlines blaring around the world had no real answer for why Altman had been fired, so through social media,

Altman and his allies had a chance to cast doubt on the wisdom of the board's actions. Within minutes of the firing, Altman and Brockman got on the phone with Chesky and Chris Lehane, Airbnb's senior vice president of global policy and public affairs. Lehane had once worked in the Clinton White House, and had written a book on crisis communications that was later made into a movie starring Rob Lowe called *Knife Fight*. Chesky had hired him in 2015 as Airbnb was trying to fend off a California ballot measure limiting short-term rentals. Chesky brought a shock-and-awe approach to lobbying that sunk the ballot initiative and made him an indispensable advisor as Airbnb fought similar battles around the world. Chesky and Lehane counseled Altman and Brockman to fight. In the near term, however, Chesky advised that Altman take the "high road" and not accuse the board of anything. So around 4 p.m. on Friday, November 17, Altman put out an innocuous statement in his usual all-lowercase style: "i loved my time at openai. it was transformative for me personally, and hopefully the world a little bit. most of all i loved working with such talented people. will have more to say about what's next later. [salute emoji]."[9] Brockman retweeted it with his all-lowercase resignation: "based on today's news, i quit."[10] Not long after, Pachocki and two other researchers followed suit.

Over the course of the evening, a narrative spread among Altman's many allies across the business world: the whole thing was an "Ilya coup," driven by Sutskever's anger over Pachocki's promotion, and boosted by Toner's anger that Altman had tried to push her off the board. They were aided by some of their powerful friends, like Chesky, who went out on a limb and tweeted around 7 p.m.: "Sam Altman and Greg Brockman have my full support. I'm saddened by what's transpired. They, and the rest of the OpenAI team, deserve better."[11]

Sutskever was astounded. He had expected the employees of OpenAI to be relieved to be free from what he saw as Altman's gaslighting, and to stand up and cheer their liberation. Instead they were starting to point their fingers at him.

**ON SATURDAY** morning, Altman woke up in his Russian Hill compound, and received a call from D'Angelo and McCauley, who wanted to open a line of dialogue after Friday night's contentious meeting with the executive team. Over the course of the day, OpenAI employees, including Murati, gathered at the sprawling six-bedroom home, where a hasty war room had been set up to plot how to bring Altman back to the company. By Saturday night, Altman and Brockman were in talks to return. As a show of strength, Altman tweeted "i love openai employees so much" and dozens of his former colleagues quote tweeted it with heart emojis.

By Sunday, Altman and the OpenAI executive team were confident he would be coming back. Murati invited him to the office, without telling the board, sending a note to staff that he would be around for the day. Altman flashed his guest pass to his followers on X, writing, "first and last time I ever wear one of these." He, Brockman, Murati, Kwon, and chief operating officer Brad Lightcap and the rest of the leadership team spent the day negotiating with the board. The board was open to him coming back if they could agree on new board members. Both sides backed Bret Taylor, the former co-CEO of Salesforce.

But Altman's desire that both he and Brockman would return to the board was a sticking point. He also wanted the entire current board to resign.

The talks dragged on late into the night. Reporters camped out in front of the OpenAI door, watching multiple rounds of takeout being delivered. Suddenly the board informed Murati that they were naming a new CEO: Emmett Shear, Altman's batchmate from the inaugural YC batch, the co-founder of Justin.tv and Twitch, and a voice for the careful development of AI. Altman and the OpenAI executive team were stunned. They felt like the board had not been negotiating in good faith. Altman left the building in disgust. The previous day, Nadella had offered him the chance to go to Microsoft, and now he decided he would take it.

A Slack message went out to all employees announcing an all-hands meeting with Shear. Employees responded with middle finger emojis. About two hundred people had gathered at the office late on a Sunday night to witness the end of the drama, and now they flooded out of the building. In an emotional confrontation in the lobby, Brockman's wife, Anna, pleaded with Sutskever to reconsider.

That evening, Nadella tweeted that Microsoft was hiring Altman and Brockman, and invited OpenAI's other employees to join them. Overnight and into the wee hours of Monday morning, more than seven hundred of the company's 770 employees signed a blistering letter threatening to quit and go to Microsoft unless Altman and Brockman were reinstated and the board resigned.

Among the signatures was Sutskever's. "I deeply regret my participation in the board's actions," he wrote on X. "I never intended to harm OpenAI. I love the company we built together and I will do everything I can to reunite the company." Altman quote tweeted it with three hearts.

To bring the matter to a close, D'Angelo suggested former Treasury Secretary Lawrence Summers as another board member. Altman agreed, and gave up his demand for a board seat. By Tuesday, the deal was done: Altman would return to OpenAI, as CEO but not to the board, and there would be an independent investigation into the circumstances of his firing and the events leading up to it.[12]

Brockman celebrated with a selfie of himself standing in front of a jubilant crowd of OpenAI employees. "We are so back!" he wrote.[13]

In the course of five sleepless days, Altman had un-fired himself. He was now more powerful than ever. OpenAI employees would come to refer to the whole episode as "the blip."

CHAPTER 17

# PROMETHEUS UNBOUND

WEEKS LATER, ALTMAN AND MULHERIN WERE MARRIED beneath a jasmine-draped chuppah erected among the palm trees of his Hawaiian estate. It was a scene of almost unimaginable splendor—the sun setting golden over the Pacific, his private beach stretching out before them—but barely anyone was there to witness it. His brother Jack officiated as fewer than a dozen close family and friends looked on, including Connie and Max. Annie, though living on the same island, did not even find out about it until some photos later leaked on social media. After a year in the white-hot global spotlight, Altman and Mulherin craved privacy and simplicity. The grooms and groomsmen all wore the same spare uniform of white button-down shirts and khaki pants, like they were catering someone else's party. The only hints of festivity were the tuberose boutonnieres and the look of pure joy on the grooms' faces.

Seventy years earlier, Alan Turing, the father of AI, whose ideas had inspired the technology behind ChatGPT, had taken a cyanide pill after being chemically castrated by the British state as punishment for his then-illegal homosexuality. When the news of Altman's marriage became public, Altman had grown so famous so fast that many people were surprised to learn that he was gay. "I feel very lucky that we were able to marry in this day and age," Altman told *The Advocate*. "When I was growing up it wasn't clear that was going to be the case. I think the public opinion and the laws have changed more quickly than I ever thought they would, so I'm grateful for that."[1] Altman had demanded things that many at the time would have found very difficult, or even outrageous, to ask for—allyship from his Missouri high school community in the 1990s, marriage rights at college in the early 2000s—and then watched as the world granted his wishes tenfold.

In the same interview, Altman batted away a question about his political ambitions, saying he was too busy with AI to think about such things. "Frankly," he said, "I don't have time for politics."[2] But Altman had in no way set aside his ambitions.

Altman's meetings at the White House the previous year had brought him face-to-face with America's octogenarian president, Joe Biden, on multiple occasions, and he had grown alarmed that Joe Biden would be unable to beat Donald Trump. While the president's aides and Democratic party officials were working overtime to quash any talk of Biden being too old to run, Democratic donors could see with their own eyes that all was not well. In June 2023, Microsoft CTO Kevin Scott co-hosted a fundraiser for Biden with Reid Hoffman at his Los Gatos home, where one guest told Bob Woodward that Biden was "frighteningly awful" and "like your 87-year-old senile grandfather," wandering around the room telling women "your eyes are so beautiful."[3] But for a full year after that, the prospect of mounting a primary challenge to Biden would remain almost unthinkable. Nevertheless, after a particularly alarming meeting with Biden at the White House that fall, Altman called Matt Krisiloff—who, with his brother Scott, remained his primary advisors on political matters—and asked if there was anything that could be done.

As it happened, the Krisiloff brothers had been talking to the campaign of a little-known Democratic congressman and liquor industry heir from Minnesota named Dean Phillips who was about to declare his candidacy for the Democratic presidential nomination. Phillips was a centrist, business-minded, Jewish, and optimistic—just like Altman. Soon, the Krisiloffs began offering Phillips some polling and focus group research that Altman had paid for showing voters' wariness about Biden's age.

By late October, Phillips had hired the veteran political operative Steve Schmidt, a senior advisor to Senator John McCain during his

2008 presidential run. Matt Krisiloff introduced himself to Schmidt in an email as Sam Altman's "former chief of staff" and "a founding member of OpenAI" and went on to share the results of the focus groups they had conducted with voters from New Hampshire, Michigan, and South Carolina. According to one person familiar with the situation, the research and its connection to Altman were part of what drove Phillips to run. For Schmidt's part, he wanted to know who had paid for the research. When he asked the Krisiloffs, they changed the subject. He asked again. "It was like 'yada, yada, yada, yada, yada,'" Schmidt recalled. He asked a third time. *"Who's paying for the fucking research?"* Concerned about federal election law violations, Schmidt eventually had to have the campaign's general counsel call the brothers. "As it turns out, it's Sam Altman paying for it," Schmidt said. (Matt Krisiloff said Altman was no longer paying for research by the time they were engaged with Phillips but refused to say who was.)

Days before launching his campaign, Phillips had a video call with Altman, during which, as Schmidt recalled, Altman pledged money and support. *Atlantic* reporter Tim Alberta, who was sitting nearby, wrote in an October 27 article that a "tech mogul, who spoke with Phillips throughout the week preceding the launch, was readying to endorse him on Friday."[4] But that endorsement never came to pass. Altman did not end up endorsing any candidate that cycle. Instead, Altman tweeted, in response to Phillips's announcement: "this is interesting, and I think close to what the majority of voters actually want: a reasonable, centrist candidate running on 1) a message of a strong economy and increasing affordability 2) a focus on safety and 3) generational change. curious to see what happens."[5]

Schmidt was annoyed at Altman's desire to remain in the shadows. "He had these two guys really operating as his point people that were the two strangest cats that you could conceivably meet, and he had this real penchant for secrecy," Schmidt said. In the first week of November, Altman hosted Phillips at his house, according to *Puck*, where they talked about everything from Biden's polling to the status of Phillips's campaign infrastructure in New Hampshire and beyond.[6]

"Altman made a lot of commitments to him," Schmidt said. "He didn't follow through on any of them." By March 2024, Phillips had lost his home state and called off his campaign.

From Altman's perspective, it wasn't simply that Biden was old—though that was certainly a factor for someone who had grown up in the youth-worshipping cradle of YC. It was that he was *uninspiring*. His policies were fine, but he seemed unable to conjure a compelling vision of the next American century, or to make the kind of big bets that stirred people's blood. In some cases it was a problem less of direction than of degree. The CHIPS and Science Act, for instance, which was designed to bring the semiconductor industry back to the United States, allocated a mere $53 billion for the task. In Altman's view, that was missing not just one zero, but two.

As if to prove this point, Altman ramped up conversations with investors in the UAE and elsewhere, US regulators, and Asian chip makers about funding a massive increase of global chip building capacity, data centers, and energy that could cost as much as $7 trillion all told, according to *The Wall Street Journal*—a number so outlandishly large it immediately invited ridicule. (For context, global semiconductor sales amounted to half a trillion dollars the previous year, and the combined values of Apple and Microsoft amounted to $6 trillion.) Altman leaned into the meme, posting "fk it why not 8,"[7] while avoiding any responsibility for the number itself, which he called "misinformation" on the *Lex Fridman Podcast*. A breath later, though, he was using the plan to lay out his case for a massive step up in investment in sheer computing power. "I think compute is going to be the currency of the future," he said. "I think it will be maybe the most precious commodity in the world." How useful AI is will ultimately depend upon how cheap it is, he argued, a problem he intended to help solve with Helion.[8]

**WHEN IT** came to AI regulation—which Altman had gone before Congress and asked for, and traveled around the world showing support for—Altman's relationship to the Biden administration was a bit confusing. On October 30, 2023, President Biden signed an executive order on AI, the first real piece of AI regulation in the United States, supplanting the voluntary commitments that OpenAI and other Big Tech companies had made in the spring. OpenAI had been heavily involved in the crafting of the order, with Altman visiting the White House four times in 2023, according to White House logs. Yet when the order was finally signed, and executives from Google and Microsoft immediately voiced their approval, OpenAI was conspicuously silent. Among the most notable things the order did was to set up a US AI Safety Institute, tucked under a division of the Commerce Department, called the National Institute of Standards and Technology (NIST). Later, the former OpenAI researcher Paul Christiano would be named as its head of AI safety (who had to step down from Anthropic's Long-Term Benefit Corporation). When Altman finally broke his silence, he offered tepid approval, saying there were "some great parts" of the order, but warning that the government needed to be careful "not to slow down innovation by smaller companies/research teams."[9]

The order arguably bore the imprint not so much of OpenAI as of the EA billionaires who had funded a vast web of interlocking think tanks, institutes, and fellowships that a cynic might think of as the AI Doomer Industrial Complex. The order was deeply influenced by the RAND Corporation, the original government think tank, which in recent years has received tens of millions of dollars from Open Philanthropy and has been led by self-described EA Jason Matheny. In addition to sitting on Anthropic's Long-Term Benefit Trust before joining RAND, Matheny recruited both Toner and McCauley to their current jobs—Toner at CSET and McCauley at RAND. According to audio leaked to *Politico*, Matheny told an all-hands meeting of RAND employees that the National Security Council, Department of Defense, and Department of Homeland Security "have been deeply worried about catastrophic risk from future AI systems and asked RAND to produce several analyses." Among the most controversial provisions that RAND researchers successfully pushed for in the order was to have all models above a certain size be required to report detailed information about their development to the government.[10]

But RAND's fingerprints on Biden's executive order were but a small sign of the less visible power that effective altruism was amassing in Washington. Increasingly, the warning that Peter Thiel had shared with Altman at a trendy sushi restaurant in LA in the fall of 2023—that Yudkowsky's disciples and like-minded EAs had taken over a swath of OpenAI—just as easily could be applied to the nation's capital. Because while the nascent political issue of AI still could not be easily defined in partisan terms, clear lines of battle had emerged in DC, with OpenAI on one side and a sprawling, well-funded EA-aligned network of operators and influencers on the other.

To fight this battle, OpenAI had enlisted DLA Piper, one of the top global lobbying firms, to coach Altman on how to woo Washington; hired super-firm Akin Grump Strauss Hauer & Feld, which boasted a former chief counsel in the office of Senate Majority Leader Chuck Schumer, to lobby on AI regulation; and added a top in-house lobbyist who had represented Microsoft.[11,12] But around the same time, EA-aligned Open Philanthropy, financed by Facebook founder Dustin Moskovitz, plus an organization funded by Skype founder Jaan Tallinn, were spending megabucks to set up think tanks and lobbying outfits and also embedding a legion of loyal staffers in Washington's most powerful corners. "The amount of money they poured in is just unprecedented for tech policy," said Daniel Castro, the vice president at the nonpartisan Information Technology and Innovation Foundation, who previously did analysis for the Government Accountability Office. "If you wanted to create an AI think tank that's focused on existential risk, there was money to be had to do that. And so there are a lot of people that put their hand out and said, 'Yeah, I'll do that.' And so that changed things."

Castro said that Open Phil and Tallinn maximized the impact of their cash by infiltrating brand names that have pre-established spheres of influence such as RAND and Georgetown via CSET in Washington, or across the pond at Oxford, which hosted the Future of Life Institute. "These are historic names with long-standing ties," he said. But the agenda was new. "RAND wasn't a big player in many of these tech policy discussions up until a couple years ago. They just weren't."

More controversially, the Horizon Institute for Public Service—backed by Open Phil and Tallinn's Lightspeed Grants—have used a little-known federal provision to pay for the salaries of hand-picked "fellows" in powerful congressional and executive branch offices of the US government. Horizon's website lists dozens of staffers posted on the teams of lawmakers who have been vocal about crafting AI legislation, along with executive branch posts from the National Security Council to the Defense Department.

Open Phil and Tallinn also directly funded DC lobbying groups such as the Center for AI Policy, which a post on the popular EA forum GreaterWrong described in 2023 as "new DC-based organization developing and advocating for policy to mitigate catastrophic AI risks," with a current focus on "building capacity in the US government." By the end of 2023, Jason Green-Lowe, who would become CAIP's executive director, posted on *LessWrong*, a related EA blog, that his group had already met with more than fifty congressional staffers and that it was "in the process of drafting a model bill." CAIP's co-founder, Thomas Larsen, previously had been an AI safety researcher at the Machine Intelligence Research Institute—Yudkowsky's organization.

While Green-Lowe supported much of the White House's recent executive order on AI, he found the $10 million set aside for the AI Safety Institute at NIST laughable. "We're being outspent by Singapore," he said, citing a recent report in *The Washington Post* that exposed how black mold had dislodged people from their offices there and "researchers sleep in their labs to protect their work during frequent blackouts." Some employees, the article added, "have to carry hard drives to other buildings; flaky internet won't allow for the sending of large files . . . and a leaky roof forces others to break out plastic sheeting." There were even accounts of the buildings being infested with wildlife, including a snake.[13]

Green-Lowe—a self-described "altruistic attorney" who trained at Harvard and said he got interested in AI after building simple machine learning models to work on housing cases—presented his organization as fighting the good fight to make sure artificial intelligence is developed responsibly. "Ultimately, the stakes are the whole world," he said. "Because if you build something that is faster, smarter, cheaper than we are, then that thing is going to steer the future. We're not there yet, but it seems to me that's where we're inexorably headed."

**THE APOCALYPTIC** frisson that Altman had once enjoyed provoking in his audiences suddenly didn't seem so cute anymore. For months after "the blip," he sleepwalked through his days, still dazed and wounded by the ambush. He had always considered his weirdness a superpower, going back to his college days, and mourned the loss of that freedom for the kids who came after, traumatized into obedient careerism by the financial crisis or swept into the tech track as the expanding bubble made no other choice seem rational. "A thing that I learned in school was independence," he said during our first interview in early 2023. "Weird ideas are OK. You don't have to be super conformist. And Stanford was a very weird place at that time and that was cool. That was helpful to me. And now I think everyone is walking around in fear of being off the track, in fear of doing something to get themselves cancelled, in fear of being weird." Jerry Altman's son loved nothing more than a creative financial structure that was somehow private and public at the same time.

Yet as the investigation into the blip concluded in March 2024 and Altman was cleared of any wrongdoing, allowing him to return to the board, he knew that the company's strange structure, and his own eccentric lack of equity, could not remain. Investors were just not going to fund a company that could self-destruct as easily as OpenAI showed that it could, and OpenAI needed a lot more money. It had built a massive AI enterprise business—bigger than Microsoft's, awkwardly—and would more than double its head count in the year after the blip. But it was still spending far more than it was making every year—billions more. It would need Altman's fundraising prowess for the foreseeable future.

Altman set about remaking OpenAI as something a lot less weird. He would start by rooting out the EAs.

IN ITS quest for normalcy, OpenAI recruited a new set of experienced board members, including former Sony Entertainment president Nicole Seligman, who had served on Paramount Global's board for years, and former Bill and Melinda Gates Foundation CEO Dr. Sue Desmond-Hellmann, who had served on the boards of Facebook and Pfizer. And it rolled out a new policy on conflicts of interest to prevent the distrust around Altman's other business dealings that had contributed to his firing.

"Because people had had questions about that, that was something where we got legal advice and we put in a standard set of procedures that said, 'If in doubt, check it out,'" Larry Summers said. "Sam has been totally scrupulous about checking everything out."

After reviewing 30,000 documents and interviewing dozens of people, the outside law firm WilmerHale determined that the old board had been acting within its powers to fire Altman, but that nothing that it reviewed would mandate his firing.

"We had a different business judgement after the conclusion of the WilmerHale review as we believed there was nothing in the record that in any way raised a question about the appropriateness of his remaining CEO," Summers said.

Nevertheless, Altman did some soul-searching after the whole affair, trying to understand how he lost the board's trust. Even as he realized that the nonprofit structure was too unstable to continue and would likely have to be replaced by something more like a public benefit corporation—a for-profit company that is legally structured to prioritize social or environmental good along with financial performance—he understood that doing this was likely to erode some people's trust even further.

"We're constantly learning and adapting, and the thing that we do changes all the time, and I try to leave myself a lot of room for that, which bothers people. But sometimes I don't leave myself enough room, and we have to do something different than what was the option set of what we were going to consider earlier. And I think things like starting as a nonprofit and then adding this capped profit, and then saying, 'Well, even that doesn't work, we need a public benefit corporation'— which I very genuinely believe that what we were doing was probably going to work as a structure—very understandably sets people off."

No one was more set off than Elon Musk. The same month that the investigation cleared Altman, Musk sued him and OpenAI, alleging it had betrayed its nonprofit mission. "OpenAI has been transformed into a closed-source de facto subsidiary of the largest technology company in the world: Microsoft," his suit's initial complaint read. "Under its new board, it is not just developing but actually refining an AGI to maximize profits for Microsoft, rather than the benefit of humanity."[14] OpenAI painted Musk's complaints as sour grapes, noting he was trying to commercialize AI with his own company. But it was a fair question: Where did the roughly $50 million that Musk donated to the early nonprofit OpenAI actually go? Did it just vanish into thin air? Does he get nothing for bankrolling this juggernaut from the beginning?

**EVER SINCE** the blip, Sutskever had been lying low—so low that "Where is Ilya?" had become an online meme. There were rumors that he and Murati had gone to Antarctica. Lex Fridman asked Altman only partially in jest in March 2024 whether Sutskever was being kept in a secure nuclear facility. In reality he and the company had been negotiating to find a way for him to stay at OpenAI. Sutskever may have plotted the coup, but OpenAI was his life's work, and once he saw that the company was on the brink of collapse, he was determined to do anything to save it. Both he and the OpenAI leadership team knew that he was a beacon to other researchers, the spiritual guru of the singularity who had the research track record to back up his grand claims, such as "In the future, it will be obvious that the sole purpose of science was to build AGI."[15] The field of AI research had come to recognize him as its leading light, awarding him the prestigious Test of Time Award, given to a paper published a decade earlier that had stood the test of time, at NeurIPS each of the last two years, and would soon award him a third.

Moreover, the company's next big advance, a reasoning model with the internal code name of "Strawberry" that would be released later in the year, would result from Sutskever's pioneering work. OpenAI offered him a generous package to stick around, and Sutskever came quite close to accepting it. But ultimately, on May 14, he announced he was leaving to pursue a "personally meaningful project." Everyone involved tried to put a good face on it, with Sutskever, Altman, Brockman, Murati, and Pachocki posing arm in arm in front of a wall filled with portraits Sutskever had painted of exotic animals. Musk, who the previous year had told CNBC that Sutskever had been the "linchpin for OpenAI being ultimately successful," immediately tried to recruit him.[16]

The next day, Sutskever's partner in founding the Superalignment team the previous year, Jan Leike, followed him out the door, posting "I resigned" on X without elaboration.[17] Leike led alignment at OpenAI, having previously spent time at DeepMind and the Future of Humanity Institute, and was trying to answer a version of the question that Eliezer Yudkowsky had been asking for years. "If it knows a lot of facts, that isn't particularly scary, but what we really need to figure out is, if we let the system take over some amount or ultimately almost all of our alignment research, will it lie to us? Will it try to deceive us? Will it try to take the opportunity to take over? Because now, it's doing so much stuff that we can't look at [it all] ourselves," he told Daniel Filan for the latter's podcast. Filan is a PhD student at Berkeley studying under Stewart Russell, who has called AI a potentially "civilization-ending" technology.[18] A few days later, Leike said that in the prior few months his team had been "sailing against the wind," sometimes struggling to get the compute it needed. "Building smarter-than-human machines is an inherently dangerous endeavor. OpenAI is shouldering an enormous responsibility on behalf of all humanity," he wrote. "But over the past few years, safety culture has taken a backseat to shiny products."[19] He soon joined Anthropic. The Superalignment team was disbanded and folded into the larger organization, and a steady stream of safety researchers made their way for the door.

When they stepped through it, some of them discovered to their horror and disbelief that the company—*a company named OpenAI*—tried to keep them from sharing their concerns with the wider public by threatening to claw back their vested equity, unless they signed highly restrictive nondisclosure and non-disparagement agreements. It would turn out to be the company's most damaging scandal, as it cut against what made OpenAI successful: its ability to recruit the best and brightest AI researchers and engineers. The agreements were brought to light by *Vox* reporter Kelsey Piper, who noted that while NDAs themselves weren't unusual in highly competitive Silicon Valley, threatening to claw back already vested equity unless a departing employee signed them was.[20] Altman tweeted that the company had never clawed back equity and suggested the language in the separation agreements that threatened to do that shouldn't have been there: "this is on me and one of the few times i've been genuinely embarrassed running openai: i did not know this was happening and I should have."[21]

The company had an all-hands meeting about it, and a subsequent Q and A, but it was never quite able to repair its relationship with the researcher community. "Any company just moving fast makes some mistakes and does some dumb things," Altman said. "But of all the ones, this is definitely the one I feel the worst about. It was very emotionally painful to us all. This is just really not the kind of company we want to be."

Just as OpenAI's reputation in the world of AI safety was taking a beating, so was its reputation within the creative community. Google had scheduled its annual developer conference, Google I/O, for May 14, where it was expected to unveil advances to its AI model, now renamed Gemini. That gave OpenAI a clear deadline to make some kind of a splash. For weeks, rumors circulated that it might release the long-awaited GPT-5, or perhaps a search product that would make use of various content-licensing deals it had been forging with companies including News Corp and Axel Springer. Instead, it released an update to GPT-4 called GPT-4o, which was faster than its predecessor and, as Altman put it, "natively multimodal," able to switch between text, images, and audio. GPT-4's voice capability, which had been released the previous fall but was too slow and clunky to be very useful, could now conduct the kind of fluid back-and-forth conversations that Hollywood had created in the film *Her*, in which a lonely man played by Joaquin Phoenix falls in love with a sexy voice assistant played by Scarlett Johansson. Murati starred in the video demo that OpenAI released, using her fluent Italian to demonstrate the real-time translation skills of GPT-4, represented by a blinking circle on a smartphone screen accompanied by a warm, husky, and slightly flirty voice. To drive home the point that OpenAI had succeeded in recreating the experience of a modern sci-fi classic, Altman tweeted "her."[22]

It was another viral hit—which is how Johansson came to hear from friends that it seemed like OpenAI was using her voice. When she watched the video with her agent, Bryan Lourd, they agreed that the voice, called "Sky," was "eerily similar" to her own. Within days, Johansson had put out a blistering statement, saying that Altman had tried to hire her months before to help with the voice assistant, and after she said no, texted her agent two days before the product's release to see if she'd reconsider. She now threatened legal action. OpenAI said the voice was never intended to be Johansson, and had been recorded by another actress before Altman even approached her, but took down the "Sky" voice anyway.

To my ears, the voice of Sky does not sound all that much like Johansson in *Her*, and the threatened legal action never materialized, as of this writing. But Altman's transparent attempt to harness the cultural connection to a science fiction movie to sell his company's products, without the original artists' permission, annoyed plenty of people. It quickly became a stand-in for the larger problem of how these models were made in the first place: by scraping creative work from the internet without permission or payment. Since ChatGPT's release in November 2022, there had been a raft of lawsuits, first from artists, then from authors, then musicians and others, alleging that OpenAI and other Big Tech companies making AI had stolen their work. In late 2023, after months of trying to negotiate a content licensing deal with OpenAI and Microsoft, The New York Times Company filed a high-profile lawsuit against both companies. "Defendants have effectively avoided spending the billions of dollars that The Times invested in creating that work by taking it without permission or compensation," the suit alleges.[23] Lourd was drawing on the same deep unease when he put out a statement on the Johansson affair. "How these companies align with the actual individuals and creators is what's key here—the verification of authenticity and receiving consent, and renumeration for consent," he said. "It's not too late for these companies to slow down and put processes in place to ensure that the products that are being built are built transparently, ethically, and responsibly."[24]

OpenAI really needed to change the narrative. Worried that Sutskever's and Leike's departures would trigger more defections, the company lobbied Sutskever to reconsider. Within a week of his departure, Murati and Brockman called Sutskever, telling him that OpenAI might collapse without him. Brockman suggested that if he returned Leike might come back as well, hinting that both would help shore up Altman, whose credibility had been punctured by the NDA affair. Murati, Brockman, Pachocki, and, later, Altman, showed up at his apartment to cajole him in person.

Sutskever seriously considered returning, but had ongoing concerns he raised with Brockman. Hours after raising them, Brockman reversed course. Realizing the emotional torture he had just endured,

Altman, Brockman, Murati, and Pachocki delivered handwritten notes apologizing profusely for what they had just put him through and thanking him for even considering returning to help the company in its hour of need. Anna Brockman even brought a cactus, his favorite plant.

Sutskever ultimately chose to leave OpenAI because he was concerned that the leadership team was so chaotic and unstable that it wouldn't succeed in the long run. A few weeks later, Sutskever announced he was starting a new company, Safe Superintelligence Inc. (SSI), with former YC partner Daniel Gross and former OpenAI engineer Daniel Levy. Sutskever called it "the world's first straight-shot SSI lab, with one goal and one product: a safe superintelligence" in a post on X. "Our singular focus means no distraction by management overhead of product cycles."[25] There would be no "products," at least initially. Within a few months, they raised $1 billion from investors, including Sequoia Capital. When OpenAI finally released the reasoning model that he had started the team for three years earlier—initially code-named "Strawberry" but later renamed o1—Sutskever was barely in the announcement, his name one of dozens cited on the paper.

As for Sam Altman: once Congress's darling, he was now its whipping boy. In the same Senate chamber where Senator Richard Blumenthal had praised Altman the previous year for being so "constructive," Toner now testified about how her experience on the OpenAI board taught her "how fragile internal guardrails are when money is on the line, and why it's imperative that policymakers step in."

Shortly after, California Governor Gavin Newsom vetoed an AI safety bill that OpenAI had opposed, and that Anthropic had been actively involved in amending. The bill would have made tech companies liable for harms caused by their AI models. Among the most vocal opponents was YC funder and angel investor Ron Conway, who has long been close to both Altman and House Speaker Nancy Pelosi, who delivered the coup de grâce to the bill when she publicly opposed it, a rare instance of the Speaker of the House becoming involved with state-level legislation. "The tech industry is very accustomed to not being regulated, and does not think it should be regulated in any meaningful way," said California State Senator Scott Wiener, who sponsored the bill. "It's really the only industry that takes that position."

**BY THE** fall of 2024, OpenAI was on its way to becoming a for-profit company, just like every other tech company in the Valley. As part of the restructuring, Altman would likely come to own a big chunk of it. For all its challenges, the company remained at the forefront of research, without any major safety catastrophes, while managing to build a growing business with more than 200 million active users. In August, Brockman announced he was going on leave through the end of the year, a move that Altman encouraged. By late September, Murati, McGrew, and Schulman had all left the company. Of the four faces of OpenAI that had once graced the cover of *Wired* magazine after ChatGPT's incredible release—Brockman, Sutskever, Murati, and Altman—only Altman was left, the king of the cannibals, standing alone.

A week later, OpenAI closed a $6.6 billion fundraising round, valuing the company at $157 billion, roughly twice what it had been worth a year before. Investors included Thrive Capital, Microsoft, and MGX, a fund created by the UAE government to invest in AI, who all had to agree not to fund OpenAI's rivals like Sutskever's SSI, Musk's xAI, or Amodei's Anthropic if they wanted to keep their rights to viewing inside information about OpenAI.[26] It was the largest venture capital fundraising round of all time.

# EPILOGUE

**"I'M MORE EXCITED THAN I HAVE EVER BEEN ABOUT ANYTHING."**

As 2024 wound to a close, Altman and Mulherin looked forward to welcoming their first child the following March via surrogate, the beginning of the large family that Altman had always said he wanted. His brother Jack and his sister-in-law, Julia, would have their third child not long before. Jack had recently stepped down from leading his unicorn startup, Lattice, and raised a $150 million fund through the brothers' venture firm, Alt Capital, the first of the firm's funds that Sam would not be part of. Sam was "just too high profile now, it's better for lots of reasons to avoid any potential conflicts," Jack told *The Information*. "Big fancy pants guy now."[1] Max, meanwhile, left Alt Capital to co-found his own venture firm, Saga Ventures, with two YC graduates, having moved from San Francisco to Austin the previous year, after a stint in Hawaii, to get a bit of distance from his famous siblings. Annie was still not speaking to her family, but over six years after her father died, was finally granted access to her father's trust, which the family has assigned lawyers to dole out at $5,000 a month, adjusted for inflation. She has also received a diagnosis that finally explained the baffling combination of symptoms that had made her so sick over the last five years: Ehlers-Danlos syndrome, a group of rare genetic disorders of the connective tissues. For the first time in years, she had stable housing in Hawaii, which allowed her to focus on physical therapy. "This is my third month at this place," Annie said in October. "Three months is the longest I've had anywhere in two and a half years."

Connie Gibstine said in a statement that she expected this support would go on for the rest of Annie's life. "Caring for a family member facing mental challenges while trying not to enable harmful behaviors is incredibly difficult," she said, adding, "For many years, Annie has made claims about our family that are horrible, deeply heartbreaking, and untrue. Out of our respect and love for her, we have chosen not to engage publicly in these matters, hoping to preserve her privacy and her dignity. But as a mother, I have a duty to protect all my children, so it is finally time for me to provide some context."

In late 2024, Annie's new lawyer, Ryan Mahoney, sent Sam a letter detailing more graphic allegations of childhood sexual abuse, offering him a chance to enter mediation before filing the case in court. "While Annie did indeed suffer from trauma, there is no evidence that her own mental health has contributed to her allegations," he said in a statement, pledging to "move forward on behalf of our client to ensure that Annie achieves the justice she deserves and that her voice is heard." On January 6, 2025, two days before her thirty-first birthday—the deadline for filing childhood sexual abuse cases in the state of Missouri—Annie filed a lawsuit in a Missouri federal court alleging that Sam had sexually abused her beginning when she was three years old. In a joint statement, Connie, Sam, Max, and Jack said the allegations were "utterly untrue."[2]

All around him, Altman's critics and enemies seemed ascendant. In October, Geoff Hinton, the "godfather of AI" who had mentored Sutskever, was awarded the Nobel Prize in Physics for his work in machine learning. During the press conference arranged by the University of Toronto shortly after the news was announced, Hinton thanked his own mentors and students, adding, unprompted, "I'm particularly proud of the fact that one of my students fired Sam Altman."[3] Hinton had resigned from his job at Google in May 2023 so that he could speak critically of the AI arms race that OpenAI had set off and his employer had then joined. Worried about misinformation, lost jobs, and ultimately the fate of humanity, he told *The New York Times* at the time, "I don't think they should scale this up more until they have understood whether they can control it."⁴ Now with the Nobel in hand, he was blunter. "OpenAI was set up with a big emphasis on safety. Its primary object was to develop artificial general intelligence and ensure that it was safe. One of my former students, Ilya Sutskever, was the chief scientist and over time, it turned out that Sam Altman was much less concerned with safety than with profits."⁵

Altman's fears that Biden could not beat Trump proved true, even after Biden was swapped out in the final months for Vice President Kamala Harris. Yet no one who was part of the founding of OpenAI could have foreseen the role that their OpenAI co-founder, Elon Musk, was going to play in the final thrust of Trump's campaign and in the weeks after his victory—dancing wildly on stage before the MAGA crowds, pouring more than $250 million into backing the man he two years before said should "sail into the sunset," becoming so inseparable from Trump during the transition planning period at Mar-a-Lago that he began referring to himself as "First Buddy."⁶,⁷ Less than two weeks after Trump's victory, Musk filed an expanded version of his lawsuit against Altman, OpenAI, and Microsoft, that essentially echoed Hinton's critique.⁸ In case anyone wondered if it was personal, Musk told Tucker Carlson in an interview a few weeks earlier, "I don't trust OpenAI. I don't trust Sam Altman. And I don't think we ought to have the most powerful AI in the world controlled by someone who is not trustworthy."⁹ OpenAI said the suit was as baseless as its predecessors, but something was, indeed, different this time: Altman was now the most prominent enemy of the man with the incoming president's ear.

Yet for all this, Altman kept his eyes on the distant horizon, trying to advance toward it. If the US government was too worried about the UAE's ties to China to let his $7 trillion microchip, data center, and energy infrastructure project run through the Middle East, then he would lobby to do some version of it in the United States. Ahead of the election, OpenAI's policy team had met with both presidential campaigns to push for the US government to invest in the infrastructure it would take to drive down the cost of AI enough to make it truly useful—from streamlining the permitting process for nuclear reactors to paying for job training in data center management. Executives framed it as a way to beat China to AGI. A company that had set out to prevent an AI arms race was now actively pushing for a not-so-metaphorical one.

"Technology brought us from the Stone Age to the Agricultural Age and then to the Industrial Age," Altman wrote in an essay in September 2024 arguing that such investment was necessary for humanity to fulfill its destiny. "From here, the path to the Intelligence Age is paved with compute, energy and human will."

Like his father had done for decades with housing, he argued that public funds and private investment should come together to smooth out capitalism's unfairness, or else "AI will be a very limited resource that wars get fought over and that becomes mostly a tool for rich people." In an interview at OpenAI's headquarters that same month, I could hear echoes of his mother's work ethic as he marveled at the anonymous people who had built the memory in his computer and their role in what he viewed as the ever-rising scaffolding of human civilization.

"I will never meet these people, but I know how hard they worked, and I know that they poured their life force into this instead of hobbies or anything else, and they made this thing, and it was tremendously important at the time," he said. "That's how I think civilizations get built. Not in any one neural network of one of us, but that we can all contribute to and build this crazy infrastructure that makes us wildly more capable than our great, great grandparents, even though biologically, we are very much the same."

Even as liberalism comes under attack around the world, Altman remains a true believer in rationality, science, and progress. More than a decade after first reading the 2011 popular science book *The Beginning of Infinity* by the British physicist David Deutsch, he recommends it to almost everyone he meets. In the book, Deutsch argues for the cosmic significance of the Enlightenment, as the moment when humans learned to truly produce knowledge such that there is no good reason to believe they won't eventually conquer and transform every corner of the universe. "Everything that is not forbidden by laws of nature is achievable, given the right knowledge," Deutsch writes.[10] Death is a solvable problem. Even the coldest and darkest corners of the universe can have their energy and knowledge harvested by the technology we will make in the future.

Altman has built a shrine to his faith in human progress at his home on Russian Hill. The first objects one encounters upon entering are three hand axes, including one of the oldest ones that has ever been found, the only tool that hominids had for most of the last 1.5 million years of their existence, a single object for building, killing, and cooking. Among the dozens of other objects are swords from various moments in technological history, a vacuum tube, space program souvenirs, a blade from a Concorde engine, an early Apple computer, a replica of his first computer, a Mac LC II, and one of the robot hands built by OpenAI.

In his September 2024 essay, Altman wrote: "Here is one narrow way to look at human history: after thousands of years of compounding scientific discovery and technological progress, we have figured out how to melt sand, add some impurities, arrange it with astonishing precision at extraordinary tiny scale into computer chips, run energy through it, and end up with systems capable of creating increasingly capable artificial intelligence."

He continued with a line perhaps only he could've written. "This may turn out to be the most consequential fact about all of history so far. It is possible that we will have superintelligence in a few thousand days (!); it may take longer, but I'm confident we'll get there."[11]

# ACKNOWLEDGMENTS

***THE OPTIMIST* WAS REPORTED AND WRITTEN WITH THE HELP** of my brilliant researcher and friend, Luke Jerod Kummer, who is one of the finest journalists I have ever worked with. A tireless shoe-leather investigator of the old school, he pored through yearbooks and documents, traveled to meet distant sources for interviews, and came back with some of the most revelatory reporting in the book. The book also benefited enormously from his curiosity, doggedness, and good judgment—not to mention his formidable writing and editing talent. I knew when I met him twenty years ago on our first day as interns for *Village Voice* investigative journalist Wayne Barrett that we would be friends, but I could not have imagined just how fun and fruitful that friendship would turn out to be.

This book would not have happened without the instincts and support of *Wall Street Journal* editor in chief Emma Tucker, who had the original idea that the *Journal* should profile Sam Altman during our first editorial meeting after the release of ChatGPT. Nor would it have happened without my co-author on that story, Berber Jin, the *Journal*'s startups and venture capital reporter, whose reporting is woven through these pages, and whose generosity, kindness, and good humor helped keep me sane throughout the process.

But the true vision for this book came from Dan Gerstle, my editor and publisher at W. W. Norton & Company, who saw the human story in the machine revolution, and whose thoughtful edits have made it immeasurably better. The team at Norton has been a joy to work with, from Rachel Salzman's journalistic insights, to Avery Hudson's eagle eyes and Bob Byrne's infinite patience. Thank you to my agent, Alice Martell, a champion who truly understands what authors need on every level.

---

FOR THOSE affected by it, Altman's firing was one of those crystalline moments that adrenaline locks into memory. I was leading my daughter's Girl Scout troop through the woods when the shocking headline landed on my phone's lock screen; the ground seemed to drop away. The following days were a sleepless blur, but I am grateful to have had my *Journal* colleagues Deepa Seetharaman, Tom Dotan, and Berber alongside me for it, as well as the steady hand of Liz Wollman, wisdom of Jason Dean, and relentless drive of Jamie Heller. I can't think of anyone I'd rather share a foxhole with. I'm thankful for their partnership over the last year as we've tried to piece together the reasons for what happened and its implications for the future of AI. Thanks to Rolfe Winkler and Kate Linebaugh for their generosity.

Thank you to my boss, Amol Sharma, the chief of the *Journal*'s Media and Entertainment Bureau, who brings inspiration and excitement to our work every day. Without his encouragement and support, I would not be an author, full stop. And thank you to the rest of the bureau, whose insights and information have helped shape the book. Extra thanks to Joe Flint for looking up that story for me (see, Joe, did not forget!). A massive debt to Jim Oberman, whose research talents border on sorcery.

Thank you to the hundreds of sources who gave their time to help tell this story, particularly to those who patiently put up with follow-up question after follow-up question. Many of you are named. Some of you aren't. You all know who you are.

This book also owes a great debt to the authors who have told pieces of this story before. None is so great as Cade Metz, whose prescient 2021 book *Genius Makers* served as my Virgil in the AI underworld. Christine Lagorio-Chafkin's *We Are the Nerds* and Jessica Livingston's

*Founders at Work* were both invaluable resources for the history of Y Combinator and Reddit.

Writing a book is hard on the people around you—there's no getting around it. Thank you to my family and friends for putting up with this madness and helping see me through to the other side, especially Kelly Turner, who read drafts and suggested wise edits. Joan Sanders was a gracious host and stimulating companion on my reporting trips. This book exists because my husband, Wesley Harris, made our household function despite having his own demanding full-time job in AI. Thank you for fielding my constant technical questions and taking the girls canoeing (and camping, and skiing, and hiking) when I needed to write. Our au pair, Isabella Ribeiro, was an indispensable partner. And our daughters, Belle, June, and Pearl, all stepped up in their own ways and cheered me on. Thank you all. I love you.

# NOTES

### PROLOGUE

1. Bill Addison, "The Most Quietly Ambitious Cooking to Emerge in Los Angeles This Year Is at Yess," *Los Angeles Times*, August 3, 2023.
2. Eliezer Yudkowsky, "Pausing AI Developments Isn't Enough. We Need to Shut it All Down," *Time*, March 29, 2023.
3. Eric Mack, "Elon Musk: 'We Are Summoning the Demon' with Artificial Intelligence," *CNET*, October 26, 2014.
4. Krystal Hu, "ChatGPT Sets Record for Fastest-Growing User Base," Reuters, February 2, 2023.
5. Sam Altman, "How to Be Successful," Sam Altman blog, January 24, 2019.
6. OpenAI, "OpenAI Charter," *OpenAI*, April 9, 2018.
7. Sam Altman, "Machine Intelligence: Part 1," Sam Altman blog, February 25, 2015.
8. Ryan Tracy, "ChatGPT's Sam Altman Warns Congress That AI Can 'Go Quite Wrong,'" *The Wall Street Journal*, May 16, 2023.
9. Max Chafkin, *The Contrarian: Peter Thiel and Silicon Valley's Pursuit of Power* (New York: Penguin Press, 2021), 120.
10. Sam Altman, "Board Members," Sam Altman blog, November 11, 2014.
11. Lex Fridman, "Sam Altman: OpenAI, GPT-5, Sora, Board Saga, Elon Musk, Ilya, Power & AGI," *Lex Fridman Podcast*, March 18, 2024.
12. Berber Jin, Tom Dotan, and Keach Hagey, "The Opaque Investment Empire Making Sam Altman Rich," *The Wall Street Journal*, June 3, 2024.
13. Tad Friend, "Sam Altman's Manifest Destiny," *The New Yorker*, October 3, 2016.
14. Paul Graham, "A Fundraising Survival Guide," PaulGraham.com, August 2008.
15. Sam Altman (@sama), "haven't seen this as a twitter thread so: what true thing do you believe that few people agree with you on? Absolute equivalence of brahman and atman," X, December 26, 2022.
16. Lex Fridman, "Sam Altman: OpenAI CEO on GPT-4, ChatGPT, and the Future of AI," *Lex Fridman Podcast*, March 25, 2023.

CHAPTER 1 **CHICAGO**

1. Tim Frakes, "Harold Washington Inauguration April 29 1983," YouTube, 9:36, posted December 29, 2017.
2. "Harold," *This American Life*, aired November 21, 1997, on WBEZ Chicago.
3. "Obama 2: Politics Ain't Beanbag," *Making Obama*, aired February 15, 2018, on WBEZ Chicago.
4. "Harold Washington Remembered as 90th Birthday Approaches," CBS News Chicago, April 12, 2012.
5. Chicago Public Library, "Mayor Harold Washington Inaugural Address 1983," April 29, 1983.
6. MacArthur Foundation, "Housing Agenda," MacArthur Foundation, accessed June 15, 2024.
7. Douglas Martin, "Gale Cincotta, 72, Opponent of Biased Banking Policies," *The New York Times*, August 17, 2001.
8. The Breman Museum, "Sam Altman," Esther and Herbert Taylor Oral History Collection, December 3, 2002.
9. *Altman v. Massell Realty Co.*, 167 Ga. 828 (1929).
10. Ibid.
11. "Interco Incorporated," Politics and Business Magazines, Encyclopedia.com, Accessed December 12, 2024.
12. Sylvia Harris, death certificate, issued by the state of Missouri, 1958.
13. "Fifth Annual Russian Evening Highlights Drama and Dinner," *Country Day News*, May 1, 1968.
14. Corinne Ruff, "80% of St. Louis County Homes Built Before 1950 Have Racial Covenants, Researcher Finds," St. Louis Public Radio, January 21, 2022.
15. Walter Johnson, *The Broken Heart of America: St. Louis and the Violent History of the United States* (New York: Basic Books, 2020), 354.
16. Ibid, 375.
17. "$25,000 Solar-Energy Study Set for Council Review," *The Hartford Courant*, November 22, 1976.
18. Floyd J. Fowler Jr., Mary Ellen McAlla, Thomas J. Mangione, "Reducing Residential Crime and Fear: The Hartford Neighborhood Crime Prevention Program," US Department of Justice Law Enforcement Administration, National Institute of Law Enforcement and Criminal Justice, December 1979.
19. "Unusual Plan to Revamp Block," *New York Daily News*, October 9, 1980.
20. Advertisement, *St. Louis Star and Times*, Friday, April 3, 1914, 12.
21. Walter Stevens, *St. Louis, The Fourth City*, 1764–1911 (Chicago: The S. J. Clarke Publishing Co., 1909).
22. Rev. C. C. Woods, *Report on Fraternal Correspondence, Grande Lodge Missouri, 1923*.
23. "Iron Left on at Night Causes $25,000 Damage," *St. Louis Star Times*, August 19, 1925.

24. "Must Turn Over $1500 to Bankrupt Firm," *St. Louis Post-Dispatch*, March 26, 1926, 3.
25. "Milliner Is Freed on Fraud Charges," *St. Louis Globe-Democrat*, April 5, 1928, 17.
26. "Hebrew Y To Award Athletic Trophies," *St. Louis Globe-Democrat*, May 12, 1938.
27. "Miss Peggy Francis Engaged to Doctor," *St. Louis Globe-Democrat*, October 6, 1946.
28. "Peggy Francis Becomes Bride," *St. Louis Globe-Democrat*, December 22, 1946.
29. Chicago Public Library. "Mayor Richard J. Daley Inaugural Address, 1963," Chicago Public Library, accessed June 15, 2024.
30. "Housing Needs—Mayor's Perspective," Joint Hearing Before the Subcommittee on Housing and Community Development of the Committee on Banking, Finance and Urban Affairs and the Subcommittee on Manpower and Housing of the Committee on Government Operations. House of Representatives, 98th Congress, October 2, 1984.
31. Ross J. Gittell, *Renewing Cities* (Princeton, NJ: Princeton University Press, 1992) 91.
32. "How Harold Washington Influenced Barack Obama," NBC Chicago, November 26, 2012.

## CHAPTER 2 ST. LOUIS

1. Nathaniel Rich, "Pitch. Eat. Sleep. Pitch. Eat. Sleep. Pitch. Eat. Sleep. Pitch. Eat," *The New York Times Magazine*, May 5, 2013.
2. US Department of Housing and Urban Development. "CDBG National Objectives and Eligible Activities for Entitlement Communities, Chapter 3," 2001.
3. "Opportunity Denied: St. Louis Uses Money Targeted for Housing for the Poor to Aid Wealthier Neighborhoods," *St. Louis Post-Dispatch*, December 8, 1991.
4. Phil Linsalata, Tim Novak, "Housing Proposal Backed . . . Craig Forsees Role for Civic Progress in City Redevelopment," *St. Louis Post-Dispatch*, December 22, 1991.
5. Elizabeth Weil, "Sam Altman Is the Oppenheimer of Our Age," *New York*, September 25, 2023.
6. Ibid.
7. Tad Friend, "Sam Altman's Manifest Destiny," *The New Yorker*, October 16, 2016.
8. Annie Altman, "The Speech I Gave at My Dad's Funeral," *Medium*, March 28, 2019.
9. "Legacy of Leadership," John Burroughs School.
10. "Alumni Awards: Sam Altman '03," John Burroughs School, May 20, 2023.
11. Friend, "Manifest Destiny."

## CHAPTER 3 "WHERE ARE YOU?"

1. Sam Altman, "Argument Against Gay Marriage Lacks Logic," *Stanford Daily*, March 31, 2004.
2. Melisa Russel, Julie Black, "He's Played Chess with Peter Thiel, Sparred with Elon Musk and Once, Supposedly, Stopped a Plane Crash: Inside Sam Altman's World, Where Truth Is Stranger Than Fiction," *Business Insider*, April 27, 2023.
3. Marcia Savage, Amanda Stripe, "Under Surveillance—Location-Based Wireless Technology Raises Privacy Concerns for Solution Providers," *Computer Reseller News*, December 4, 2000.
4. Reid Hoffman, "Uncut Interview with Sam Altman on Masters of Scale," Y Combinator blog, July 26, 2017.
5. Y Combinator, *"Frequently Asked Questions,"* accessed June 15, 2024.
6. Jennifer Liu, "Students Receive Funds for Start-ups," *Stanford Daily*, May 4, 2005.
7. "OpenAI: Sam Altman," *How I Built This with Guy Raz*, episode 451, September 29, 2022.
8. Liz Gannes, "Y Combinator's New Head Startup Whisperer Sam Altman Is Quite a Talker," *Re/code*, March 18, 2014.
9. Paul Graham, "Summer Founders Program," PaulGraham.com, March 2005.
10. Hoffman, "Uncut Interview with Sam Altman."
11. Jessica Livingston, *Founders at Work: Stories of Startups' Early Days* (New York: Apress, 2008), 449.
12. Christine Lagorio-Chafkin, *We Are the Nerds: The Birth and Tumultuous Life of Reddit, the Internet's Culture Factory* (New York: Hachette Books, 2018), 30.
13. Paul Graham, "A Student's Guide to Startups," PaulGraham.com, October 2006.
14. Livingston, *Founders*, 449.
15. "HIBT Lab: OpenAI, Sam Altman," *How I Built This with Guy Raz*, October 6, 2022.
16. Hoffman, "Uncut Interview with Sam Altman."

## CHAPTER 4 AMONG THE "NERD'S NERDS"

1. Cromwell Schubarth, "6 Top Picks and More from Paul Graham's Last Y Combinator Class," *Silicon Valley Business Journal*, March 27, 2014.
2. Nancy J. Zacha, "John Graham: A Man with a Mission," *Nuclear News*, July 1995.
3. Emily Chang, "Paul Graham and Jessica Livingston: Studio 10," *Bloomberg*, October 10, 2014.
4. Paul Graham, "Why Smart People Have Bad Ideas," PaulGraham.com, April 2005.

5. "Yahoo! Says It Expects a Loss Due to Second-Quarter Charge," *The Wall Street Journal*, June 19, 1998.
6. Paul Graham, "Hackers and Painters," PaulGraham.com, May 2003.
7. Lagorio-Chafkin, *Nerds*, 4.
8. Paul Graham, "How to Start a Startup," PaulGraham.com, March 2005.
9. Garry Tan, "Meet the YC Winter 2024 Batch," Y Combinator, April 3, 2024.
10. Paul Graham, "Female Founders," PaulGraham.com, January 2014.
11. Lagorio-Chafkin, *Nerds*, 24.
12. Livingston, *Founders*, 447.
13. Paul Graham, "How Y Combinator Started," PaulGraham.com, March 2012.
14. Jessica Livingston, "Think Different. Think Users," *Posthaven*, accessed June 24, 2024.
15. "David Livingston Will Marry Lucinda Pauley, '65 Debutante," *The New York Times*, April 14, 1968.
16. Shen Pauley, author page, Foundation for Intentional Community, accessed June 15, 2024.
17. "Honorees of the 2014 Veterans Parade: David Livingston," *Wicked Local*, May 18, 2014.
18. Paul Graham, "A Unified Theory of VC Suckage," PaulGraham.com, March 2005.
19. Tom Nicholas, *VC: An American History* (Cambridge, MA: Harvard University Press, 2019), 1–2.
20. Jessica Livingston, "Grow the Puzzle Around You," *Posthaven*, June 30, 2018.
21. Livingston, *Founders*, 447.
22. Graham, "How to Start a Startup."
23. Livingston, "Grow the Puzzle," *Posthaven*.
24. Randal Stross, *The Launch Pad: Inside Y Combinator* (New York: Portfolio/Penguin, 2012), 3.
25. Justin Kan, "My Y Combinator Interview," *A Really Bad Idea Blog*, November 24, 2010.
26. Aaron Swartz, "SFP: Come See Us," Aaron Swartz's blog.
27. Lagorio-Chafkin, *Nerds*, 37–38.
28. Matthew Lynley, "How a Site That Streams People Playing Video Games Became a Billion Dollar Business," *BuzzFeed*, August 7, 2014.
29. Kan, "My Y Combinator Interview."
30. Swartz, "SFP: Come See us."
31. Ryan Singel, "Stars Rise at Startup Summer Camp," *Wired*, September 13, 2005.
32. Melissa Block, "Sprint Born from Railroad, Telephone Businesses," NPR, October 15, 2012.
33. Livingston, *Founders*, 450.
34. "Stephen Wolfram Q&A," StephenWolfram.com, reposted from Reddit AMA, May 4, 2019.
35. Lagorio-Chafkin, *Nerds*, 60.
36. Singel, "Stars Rise."
37. Ibid.

## CHAPTER 5 "STOPPING OUT"

1. Olivia Winslow, "William McAdoo, 67, Stony Brook Professor," *Newsday*, November 13, 2003.
2. Tom Nicholas, *VC: An American History* (Cambridge, MA: Harvard University Press, 2010), 225–31.
3. Ann Grimes, "Sequoia Capital Quietly Doles Out Google Shares Worth $1.3 Billion," *The Wall Street Journal*, January 17, 2005.
4. Ronald Chan, "Kiss Me Cardinal," *Stanford Daily*, October 18, 2005.
5. Mark Bergen, *Like, Comment, Subscribe: Inside YouTube's Chaotic Rise to World Domination* (New York: Viking, 2022), 52.
6. Livingston, *Founders*, 451.
7. Meghna Rao, "Carolynn Levy, Inventor of the SAFE," *Meridian*.
8. Paul Graham, comment on *Hacker News*, March 9, 2012.
9. Botha, R., "Declaration," filed as part of *Oracle America, Inc. v. Google Inc.*, No. 10-03561 (N.D. Cal., March 12, 2014).
10. "YouTube Receives $3.5M in Funding from Sequoia Capital," YouTube Official Blog, November 7, 2005.

## CHAPTER 6 "WHERE YOU AT?"

1. Liz Gannes, "Y Combinator's New Head Startup Whisperer Sam Altman Is Quite a Talker," *Re/code*, March 18, 2014.
2. Ibid.
3. Ibid.
4. Tamara Chuang, "Cell Phones Change Social Networks," *The Orange County Register*, September 12, 2006.
5. "The Boost Mobile Anthem," *YouTube*, uploaded by ramsay, April 14, 2005.
6. Gary Susbam, "Pepsi Drops Ludacris After O'Reilly-Sparked Protest," *Entertainment Weekly*, August 29, 2002.
7. May Wong, "Startup Offers Cell Phone Mapping Service," Associated Press Newswires, November 14, 2006.
8. Chuang, "Cell Phones."
9. Ibid.
10. PR Newswire, "Loopt to Launch on Sprint as First-Ever Social Mapping Service on a Major U.S. Carrier," July 17, 2007.
11. Jessica E. Vascellaro, "Sprint to Offer Loopt's 'Friend Finding' Service," *The Wall Street Journal*, July 17, 2007.
12. Amol Sharma, Jessica E. Vascellaro, "Phones Will Soon Tell Where You Are," *The Wall Street Journal*, March 28, 2008.
13. Ibid.
14. "Location Tracking Firms Review Regulatory Landscape," *TR Daily*, April 25, 2007.

NOTES

15. Brian Knapp, Testimony Before the House Energy and Commerce Subcommittee on Communications, Technology and the Internet, April 23, 2009.
16. Sharma and Vascallero, "Phones."
17. "New Digs," Loopt company blog, accessed June 15, 2024.
18. Hoffman, "Uncut Interview with Sam Altman."

CHAPTER 7 **FROM "WEAK" TO "COOL"**

1. Fred Vogelstein, "The Untold Story: How the iPhone Blew Up the Wireless Industry," *Wired*, January 9, 2008.
2. Jon Froelich, Mike Y. Chen, Ian E. Smith, Fred Potter, "Voting with Your Feet: An Investigative Study of the Relationship Between Place Visit Behavior and Preference," *Lecture Notes in Computer Science*, September 2006.
3. Owen Thomas, "Mike Moritz Regrets: He Never Patched Things Up with Steve Jobs," *Venturebeat*, April 6, 2010.
4. CNET, "WWDC 2008 News: Loopt Shows Off New App for the iPhone," YouTube, June 9, 2008.
5. Loopt, "Sharing Is Caring," *Loopt In*, November 24, 2008.
6. Michael Arrington, "Loopt Jumps Ahead of Facebook and MySpace on iPhone. Told You," *TechCruch*, November 11, 2008.
7. "Sequoia to CEOs: Get Real or Go Home," *Venture Capital Journal*, October 8, 2008.
8. Michael Arrington, "Loopt Hires Allen & Co for Financing or Sale," *TechCrunch*, November 11, 2008.
9. M. G. Siegler, "Surging on an iPhone Commercial, Loopt Looking to Sell or Raise Money?" *VentureBeat*, November 12, 2008.

CHAPTER 8 **THE DOUCHEBAG BADGE**

1. Jennifer Von Grave, "Foursquare Is the Breakout Mobile App at SXSW," *Mashable*, March 16, 2009.
2. Marguerite Reardon, "Loopt Helps Reduce Cost of Location Services," *CNET*, November 5, 2008.
3. Rolfe Winkler, "Secretive, Sprawling Network of 'Scouts' Spreads Money Through Silicon Valley," *The Wall Street Journal*, November 12, 2015.
4. Alex Konrad, "Billionaire Brothers John and Patrick Collison Build Strike into One of the World's Most-Hyped, Highest Valued—and Profitable!—Startups Worth Some $95 Billion. Now They Must Stave Off Going from Disruptor to Disrupted," *Forbes*, May 26, 2022.
5. Nick Bilton, "Artificial Intelligence May Be Humanity's Most Ingenious Invention—And Its Last?" *Vanity Fair*, October 2023.
6. Friend, "Manifest Destiny."

## CHAPTER 9 "A RIDE ON A ROCKET"

1. Max Chafkin, *The Contrarian: Peter Thiel and Silicon Valley's Pursuit of Power* (New York: Penguin Press, 2021), 170–72.
2. David Brooks, "The Creative Monopoly," *The New York Times*, April 23, 2012.
3. Blake Masters, "CS183: Startup—Peter Thiel Class 1 Notes Essay," April 3, 2012.
4. Blake Masters, "CS183: Startup—Peter Thiel Class 15 Notes Essay," May 31, 2012.
5. Bruce Gibney, "What Happened to the Future," Founders Fund website, updated January 2017.
6. Elizabeth Weil, "Sam Altman Is the Oppenheimer of Our Age," *New York*, September 25, 2023.
7. Stross, *The Launch Pad*, 196.
8. Sophie Bearman, "OpenAI's Sam Altman: The Human Behind the Machine," *Life in Seven Songs* podcast, *The San Francisco Standard*, September 24, 2024.
9. Peter Thiel, "The New Atomic Age We Need," *The New York Times*, November 27, 2015.
10. Kyle Russel, "Y Combinator and Mithril Invest in Helion, a Nuclear Fusion Startup," *TechCrunch*, August 1, 2014.
11. David Perell, "I Interviewed the Man Behind ChatGPT: Sam Altman," YouTube video, 21:24, uploaded November 27, 2024.
12. Stross, *Launchpad*, 28.
13. Steven Levy, "YC Has Gone Supernova," *Wired*, June 28, 2017.
14. Nathaniel Rich, "Silicon Valley's Start-up Machine," *The New York Times Magazine*, May 2, 2013.
15. Eliezer S. Yudkowsky, "The Low Beyond," 1996.
16. Ed Regis, "Meet the Extropians," *Wired*, October 1, 1994.
17. Jon Evans, "Extropia's Children, Chapter 1: The Wunderkind," *Gradient Ascent*, October 17, 2022.
18. Sabine Atkins, "Introducing Another Atkins (was Re: just me)," *ExI Mailing List*, September 14, 2000.
19. *LessWrong*, "Rationalist Movement," accessed November 29, 2024.
20. Eliezer Yudkowsky, "Coherent Extrapolated Volition," The Singularity Institute, San Francisco, 2004.
21. "Vernor Vinge on the Singularity," YouTube, Singularity Summit 2008, uploaded February 12, 2012.
22. Cade Metz, *Genius Makers: The Mavericks Who Brought AI to Google, Facebook, and the World* (New York: Dutton, 2021).
23. Eliezer Yudkowsky, "Ben's 'Extropian Creed,'" Extropians mailing list, November 13, 2000.
24. Shane Legg (@ShaneLegg), "Yudkowsky wasn't actually working at Intelligenesis (aka Webmind), he was just visiting and he gave a talk on the dangers of powerful AI," *X* (formerly Twitter), November 30, 2022.
25. Metz, *Genius Makers*, 105.
26. Metz, *Genius Makers*, 109.

27. Metz, *Genius Makers*, 110.
28. Metz, *Genius Makers*, 107.
29. Cade Metz, Karen Weise, Nico Grant, Mike Isaac, "Ego, Fear and Money: How the A.I. Fuse Was Lit," *The New York Times*, December 3, 2023.
30. Nicola Twilley, "Artificial Intelligence Goes to the Arcade," *The New Yorker*, February 25, 2015.
31. Metz, *Genius Makers*, 116.
32. Sam Altman, "AI," Sam Altman blog, February 19, 2014.

### CHAPTER 10 "SAM ALTMAN FOR PRESIDENT"

1. Paul Graham, "Sam Altman for President," Y Combinator blog, February 21, 2014.
2. Ibid.
3. Friend, "Manifest Destiny."
4. Mark Boslet, "Paul Graham's New Role at Y Combinator," *Venture Capital Journal*, February 25, 2014.
5. Nathaniel Rich, "Y Combinator, Silicon Valley's Startup Machine," *The New York Times Magazine*, May 2, 2013.
6. Dean Starkman, "San Francisco Surges Past L.A. As Home to 'Ultra Rich,' Survey Finds," *Los Angeles Times*, November 25, 2014.
7. Sam Altman, "Growth and Government," Sam Altman blog, March 4, 2013.
8. Sam Altman, "New RFS—Breakthrough Technologies," Sam Altman blog, March 19, 2014.
9. Sam Altman, "AI," Sam Altman blog, February 19, 2014.
10. Friend, "Manifest Destiny."
11. Sam Altman, "The New Deal," Y Combinator blog, April 22, 2014.
12. Max Mason, "Can't Drink, Can't Vote—But This Teenager Is Hot Property in Silicon Valley," *The Sydney Morning Herald*, June 21, 2012.
13. Lattice Team, "Career Advice from Sam Altman," Lattice blog, October 27, 2016.
14. Sam Altman, "The YC Board of Overseers," Y Combinator blog, June 3, 2013.
15. Rohin Dhar, "The Big Winner from Y Combinator's Success? Sequoia Capital," *Priceonomics*, July 17, 2014.
16. Sam Altman, "YC Investment Policy and Email List," Y Combinator blog, September 4, 2014,
17. Lagorio-Chafkin, *Nerds*, 285.
18. Sam Altman, "reddit," Sam Altman blog, September 30, 2014.
19. Kim-Mai Cutler, "Reddit CEO Yishan Wong on Giving Stock to Users: 'We Have A Crazy Plan,'" *Techcrunch*, September 30, 2014.
20. Lagorio-Chafkin, *Nerds*, 301.
21. "Artist Spotlight: Malvina Reynolds," *Homegrown Humor*, Showtime Networks, July 2007.

22. Corrie Driebusch Sarah E. Needleman, "Reddit Shares Soar in Long-Awaited IPO," *The Wall Street Journal*, March 21, 2014.
23. Nick Bostrom, *Superintelligence: Paths, Dangers, Strategies* (Oxford: Oxford University Press, 2014).
24. Emile Torres, "Nick Bostrom, Longtermism, and the Eternal Return of Eugenics," *TruthDig*, January 23, 2023.
25. Raffi Khatchadourian, "The Doomsday Invention," *The New Yorker*, November 23, 2015.
26. Scott Alexander, "Meditations on Moloch," *Slate Star Codex*, July 30, 2014.

### CHAPTER 11 "A MANHATTAN PROJECT FOR AI"

1. Reed Albergotti, "The Co-founder of Skype Invested in Some of AI's Hottest Startups—But Thinks He Failed," *Semafor*, April 28, 2023.
2. Max Tegmark, *Life 3.0: Being Human in the Age of Artificial Intelligence* (New York: Knopf, 2024).
3. "AI Open Letter—Signatories List," Future of Life Institute, January 2016.
4. Tegmark, *Life*, 35.
5. Elon Musk, "Funding research on artificial intelligence safety. It's all fun & games until someone loses an I: futureoflife.org/misc/AI," Twitter, January 15, 2015.
6. Sam Altman, "Machine Intelligence, Part 1," Sam Altman blog, February 25, 2015.
7. Sam Altman, "Machine Intelligence, Part 2," Sam Altman blog, March 2015.
8. Sam Altman, "How to Be Successful," Sam Altman blog, January 24, 2019.
9. *Elon Musk s. Samuel Altman*, CGC-24-612746, S.F. Super. Ct, February 29, 2024, 11.
10. Walter Isaacson, *Elon Musk* (New York: Simon & Schuster, 2023), 241.
11. Ibid, 242.
12. *Elon Musk v. Samuel Altman et al.*, Complaint, US District Court for the Northern District of California, 4:24-cv-04722, August 5, 2024, 118.
13. "So Key To What I Was Able To Do Later," *UND Today*, September 28, 2023.
14. Ibid.
15. Greg Brockman, "Leaving Stripe," Greg Brockman blog, May 6, 2015.
16. Greg Brockman, "My Path to OpenAI," Greg Brockman blog, May 3, 2016.
17. Metz, *Genius Makers*, 93.
18. Metz, *Genius Makers*, 94.
19. Metz, *Genius Makers*, 162.
20. Jessica E. Lessin, "Y Combinator Launches Research Non-Profit," *The Information*, October 7, 2015.
21. *Musk v. Altman*, 13.
22. Ibid.
23. Metz, *Genius Makers*, 166.

## CHAPTER 12 **ALTRUISTS**

1. Greg Brockman, "#defineCTOOpenAI," Greg Brockman's blog, January 9, 2012.
2. Karen Hao and Charlie Warzel, "Inside the Chaos at OpenAI," *The Atlantic*, November 19, 2023.
3. Brockman, ibid.
4. Nicola Twilley, "AI Goes to the Arcade," *The New Yorker*, February 25, 2014.
5. *Musk v. Altman*, Complaint, 15.
6. Metz, *Genius Makers*, 175.
7. Sam Altman, "YC Changes," Y Combinator blog, September 13, 2016.
8. Sam Altman, "Basic Income," Y Combinator blog, January 27, 2016.
9. Adora Cheung and Sam Altman, "New Cities," Y Combinator blog, June 27, 2016.
10. Larry Yaeger, "Vivarium History," https://worrydream.com/refs/Yaeger_2006_-_Vivarium_History.html.
11. Brockman, "#define CTP OpenAI," ibid.
12. Carl Tahsian, "At Dynamicland, The Building Is the Computer," September 18, 2019.
13. Sophie Bearman, "OpenAI's Sam Altman," *Life in Seven Songs* podcast.
14. Metz, *Genius Makers*, 288.
15. Annie Altman, "My Denied Appeal Letter for Early College Graduation," *Medium*, March 30, 2015, posted May 21, 2019.
16. "Guest Annie Altman—Writer, Podcaster and Comedian," *Sally Take Live* podcast, March 20, 2020.
17. Sam Levin, "Peter Thiel Faces Silicon Valley Backlash After Pledging $1.25 Million to Trump," *The Guardian*, October 17, 2016.
18. Nitasha Tiku, "Meet the Techies Trying to Create the Turbo Tax of Voting," *BuzzFeed News*, September 8, 2016.
19. Sam Altman, "What I Heard from Trump Supporters," Sam Altman blog, February 21, 2017.
20. Douglas MacMillan, Keach Hagey, Deepa Seetharaman, "Tech Luminary Peter Thiel Parts Ways with Silicon Valley," *The Wall Street Journal*, February 15, 2018.
21. Chafkin, *The Contrarian*, 200.
22. Tess Townsend, "Sam Altman Wants to See a Techie Run for California Governor and Challenge Trump," *Vox*, April 14, 2017.
23. Willie Brown, "The Man President Trump Fears Most," *San Francisco Chronicle*, May 15, 2017.
24. William Turton, "A Silicon Valley Kingmaker Wants to Fix What Tech Did to California," *The Outline*, August 7, 2017.
25. Vauhini Vara, "The Political Awakening of Silicon Valley," *The California Sunday Magazine*, September 13, 2017.
26. Klint Finley, "Obama Wants the Government to Help Develop AI," *Wired*, October 12, 2016.

27. "Tasha McCauley: Cool Robot Chick," *ChiTAG Blog*, July 23, 2014.
28. Centre for Effective Altruism, "EA in Media, Joseph Gordon-Levitt, Julia Galef, AJ Jacobs, and William MacAskill," YouTube, November 7, 2017.
29. Future of Life Institute, "Asilomar AI Principles," Future of Life Institute, January 2017.
30. Eliezer Yudkowsky, "Purchase Fuzzies and Utilons Separately," *LessWrong*, December 22, 2007.
31. Gideon Lewis-Kraus, "The Reluctant Prophet of Effective Altruism," *The New Yorker*, August 8, 2002.
32. Megan O'Neil, "Wringing the Most Good Out of a Facebook Fortune," *The Chronicle of Philanthropy*, December 1, 2015.
33. Open Philanthropy. "OpenAI—General Support," Open Philanthropy, April 7, 2016.

### CHAPTER 13 PIVOTING TO PROFIT

1. *Artificial Gamer*. Official website. Accessed June 15, 2024.
2. Elon Musk, "OpenAI first ever to defeat world's best players in competitive eSports. Vastly more complex than traditional board games like chess & Go," Twitter, August 11, 2017.
3. Jordan Novet, "Amazon Lost Cloud Market Share to Microsoft in the Fourth Quarter: KeyBanc," CNBC, January 12, 2018.
4. Ashley Stewart, "Bill Gates Never Left," *Business Insider*, April 30, 2024.
5. Steven Levy, "What OpenAI Really Wants," *Wired*, September 25, 2023.
6. Ilya Sutskever, Oriol Vinyals, Quoc V. Le, "Sequence to Sequence Learning with Neural Networks," Neural Information Processing Systems (NIPS) conference, September 10, 2014.
7. Levy, "What OpenAI Really Wants."
8. Richard Lea, "Google Swallows 11,000 Novels to Improve AI's Conversation," *The Guardian*, September 28, 2016.
9. Alec Radford, Karthic Narasimhan, Tim Alimans, Ilya Sutskever, "Improving Language Understanding by Generative Pre-Training," OpenAI, 2018.
10. *Elon Musk v. Samuel Altman*, Case No. 4:24-cv-04722-YGR, US District Court Northern District of California, November 14, 2024.
11. Interview by authors for the article: Deepa Seetharaman, Keach Hagey, Berber Jin, Kate Linebaugh, "Sam Altman's Knack for Dodging Bullets—with a Little Help from His Bigshot Friends," *The Wall Street Journal*, December 24, 2023.
12. *Musk v. Altman*.
13. Greg Brockman, Ilya Sutskever, John Schulman, Sam Altman, Wojciech Zaremba, "OpenAI and Elon Musk," OpenAI blog, March 5, 2024.
14. OpenAI, "OpenAI Supporters," February 20, 2018.
15. Kylie Robison, Michal Lev-Ram, "Who Is Mira Murati? The OpenAI Executive Who Played a Crucial Role in the Company's Soaring Ascent," *Fortune*, October 5, 2023.

NOTES

16. Kevin Scott, "Mira Murati, Chief Technology Officer, OpenAI," *Behind the Tech*, July 2023.
17. "AI Everywhere: Transforming Our World, Empowering Humanity," YouTube, Dartmouth Engineering, June 19, 2024.
18. Annie Altman, "For context: Connie (biological mother) kicked me off her health insurance less than three months after Dad died, when I was 24 and could have stayed on her work one for two more years," Twitter, August 24, 2024.
19. Sam Altman, ["Sam Altman Speaks Out About What Happened at OpenAI"], *What Now? with Trevor Noah*, December 7, 2023.
20. Keach Hagey, "Deal or No Deal, The Shadow Over Sun Valley," *The Wall Street Journal*, July 13, 2018.
21. Rita Liao, "China Roundup: Y Combinator's Short-Lived China Dream," *Techcrunch*, November 23, 2019.
22. Douglas MacMillan, Margherita Stancati, "Saudi Push for Tech Deals Stirs Silicon Valley Debate," *The Wall Street Journal*, July 17, 2016.
23. Justin Scheck, Rory Jones, Summer Said, "A Prince's $500 Billion Desert Dream: Flying Cars, Robot Dinosaurs and a Giant Artificial Moon," *The Wall Street Journal*, July 25, 2019.
24. Sam Biddle, "Some Silicon Valley Superstars Ditch Saudi Advisory Board After Khashoggi Disappearance, Some Stay Silent," *The Intercept*, October 11, 2018.
25. Jane Lanhee Lee, "US Compels Saudi Fund to Exit Chips Startup Backed by Altman," Bloomberg, November 30, 2023.
26. Becky Petersen, "Shivon Zilis, Musk Associate, Leaves Board," *The Information*, March 23, 2023.
27. Seetharaman, Hagey, Jin, Linebaugh, "Sam Altman's Knack for Dodging Bullets."
28. OpenAI, "Microsoft Invests in and Partners with OpenAI to Support Us Building Beneficial AGI," OpenAI blog, July 22, 2019.

CHAPTER 14 **PRODUCTS**

1. Alec Radford, Jeffrey Wu, Dario Amodei, Daniella Amodei, Jack Clark, Miles Brundage, Ilya Sutskever, "Better Language Models and Their Implications," OpenAI blog, February 14, 2019.
2. Tom Simonite, "The AI Text Generator That's Too Dangerous to Make Public," *Wired*, February 14, 2019.
3. Jasper Hammil, "Elon Musk-Founded OpenAI Builds Artificial Intelligence So Powerful It Must Be Kept Locked Up for the Good of Humanity," *Metro*, February 15, 2019.
4. Sean Gallagher, "Researchers, Scared By Their Own Work, Hold Back 'Deepfakes for Text' AI," *Ars Technica*, February 15, 2019.
5. *Paul Tremblay, Mona Awad v. Open AI et al, Class Action Complaint*, Case No. 3:23-cv-03223 (N.D. Cal., June 28, 2023).

6. Cade Metz, "Meet GPT-3. It Has Learned to Code (and Blog and Argue)," *The New York Times*, November 24, 2020.
7. Paul Graham, "Do Things That Don't Scale," PaulGraham.com, July 2013.
8. Annie Altman, "How I Started Escorting," *Medium*, March 27, 2024.
9. Weil, "Oppenheimer of Our Age."
10. Annie Altman, "How I Started Escorting."
11. Sam Altman, "Please Fund More Science," Sam Altman blog, March 30, 2020.
12. Greg Brockman, Mira Murati, Peter Welinder, OpenAI, "OpenAI API," OpenAI blog, June 11, 2020.
13. Tom Simonite, "OpenAI's Text Generator Is Going Commercial," *Wired*, June 11, 2020.
14. Emily M. Bender, Timnit Gebru, Angelina McMillan-Major, Margaret Mitchell, "On the Dangers of Stochastic Parrots: Can Language Models Be Too Big?" *Proceedings of the 2021 ACM Conference on Fairness, Accountability, and Transparency*, 2021.
15. Emily Bobrow, "Timnit Gebru Is Calling Attention to the Pitfall of AI," *The Wall Street Journal*, February 24, 2023.
16. Sam Altman @sama, "I am a stochastic parrot and so r u," Twitter, December 4, 2022.

### CHAPTER 15  CHATGPT

1. Tom Simonite, "It Began as an AI-Fueled Dungeon Game. It Got Much Darker," *Wired*, May 5, 2021.
2. Ibid.
3. Sam Altman, "Moore's Law for Everything," Sam Altman, March 16, 2021.
4. Ibid.
5. Richard Nieva, "Sam Altman's Eyeball-Scanning Crypto Project Worldcoin Is Having an Identity Crisis," *Forbes*, August 10, 2023.
6. Chafkin, 138.
7. Antonio Regalado, "A Startup Pitching a Mind-Uploading Service That Is '100 Percent Fatal,'" *MIT Technology Review*, March 13, 2018.
8. Antonio Regalado, "Sam Altman Invested $180 Million into a Company Trying to Delay Death," *MIT Technology Review*, March 8, 2023.
9. Haje Jan Kamps, "Helion Secures $2.2B to Commercialize Fusion Energy," *TechCrunch*, November 5, 2021.
10. Friend, "Manifest Destiny."
11. Katherine Long, Hugh Langley, "OpenAI CEO Sam Altman Went on an 18-Month, $85-Million Real Estate Shopping Spree—Including a Previously Unknown Hawaii Estate," *Business Insider*, November 30, 2023.
12. Samson Zhang, "Donahue," *Postulate*, July 20, 2021.
13. Annie Altman, @anniealtman108, "I experienced sexual, physical, emotional, verbal, financial, and technological abuse from my biological siblings, mostly Sam Altman and some from Jack Altman," X, November 13, 2021.

NOTES

14. Annie Altman, "How I Started Escorting."
15. Annie Altman, @anniealtman108, "If the multiverse is real, I want to meet the version of me who did run away to the circus at age 5 years old about wanting to end this life thing and being touched by older siblings, and said 'mother' decided to instead protect her sons and demand to receive therapy and chores only from her female child." TikTok post cross posted on X, August 20, 2024.
16. Annie Altman, @anniealtman108, "Can you imagine how much more I'll scare them now that I'm getting my tendon/nerve/ovaries cared for, not sucking dick for rent money while my Dad's Trust was completely withheld, and learning it's safe and allowed for me to share my story on my terms," X, August 6, 2024.
17. Christopher Mims, "AI Tech Enables Industrial-Scale Property Theft, Say Critics," *The Wall Street Journal*, February 4, 2023.
18. Ryan Lowe and Jan Leike, "Aligning Language Models to Follow Instructions," OpenAI blog, January 27, 2022.
19. Justis, "AI Safety Concepts Writeup: WebGPT," Effective Altruism Forum, August 10, 2023.
20. Sam Altman, "today we launched ChatGPT. try talking with it here: chat.openai.com," Twitter, November 30, 2022.
21. Sam Altman, "language interfaces are going to be a big deal, I think. Talk to the computer (voice or text) and get what you want, for increasingly complex definitions of 'want'! this is an early demo of what's possible (still a lot of limitations—it's very much a research release)," Twitter, November 30, 2022.
22. Rajesh Karmani, "love the ambition and thesis, but given the current tech, I'd say it's your worst product concept so far," Twitter, November 30, 2022.
23. Marco Vavassori, "I tried it for a little bit. It's so awesome that's scary. I can already see thousands of jobs being replaced by this. Ultimately humans will only be good for hugs or sex maybe, not sure," Twitter, November 30, 2022.
24. Will Douglas Heaven, "The Inside Story of How ChatGPT Was Built from the People Who Made It," *MIT Technology Review*, March 3, 2023.
25. Krystal Hu, "ChatGPT Sets Record for Fastest Growing User Base—Analyst Note," Reuters, February 2, 2023.
26. Miles Kruppa and Sam Schechner, "How Google Became Cautious of AI and Gave Microsoft an Opening," *The Wall Street Journal*, March 7, 2023.
27. Sundar Pichai, "An Important Next Step in Our AI Journey," Google blog, February 6, 2023.
28. Nilay Patel, "Microsoft Thinks AI Can Beat Google at Search—CEO Satya Nadella Explains Why," *The Verge*, February 7, 2023.
29. Deepa Seetharaman, "Elon Musk, Other AI Experts Call for Pause in Technology's Development," *The Wall Street Journal*, March 29, 2023.
30. *Oversight of AI: Rules for Artificial Intelligence*, 118th Congress, First Session, May 16, 2023.
31. Sophie Bearman, *Life in Seven Songs*.

## CHAPTER 16 THE BLIP

1. "OpenAI CEO on the Future of AI," Bloomberg, June 22, 2023.
2. Julia Black, "Elon Musk Had Twins Last Year with One of His Top Executives," *Business Insider*, July 6, 2022.
3. Berber Jin, Deepa Seetharaman, "Elon Musk Creates New Artificial Intelligence Company X.AI," *The Wall Street Journal*, April 14, 2023.
4. Elizabeth Weil, "Sam Altman Is the Oppenheimer of Our Age," *New York*, September 25, 2023.
5. Andrew Imbrie, Owen J. Daniels, Helen Toner, "Decoding Intentions, Artificial Intelligence and Costly Signals," Center for Security and Emerging Technology, October 2023.
6. Tripp Mickle, Cade Metz, Mike Isaac, Karen Weise, "Inside OpenAI's Crisis Over the Future of Artificial Intelligence," *The New York Times*, December 9, 2023.
7. Ibid.
8. Ibid.
9. Sam Altman, "i loved my time at openai. it was transformative for me personally, and hopefully the world a little bit. most of all I loved working with such talented people. Will have more to say about what's next later," X, November 17, 2023.
10. Greg Brockman @gdb, "i'm super proud of what we've all built together since starting in my apartment 8 years ago. we've been through tough & great times together, accomplishing so much despite all the reasons it should have been impossible. but based on today's news, i quit. genuinely wishing you all nothing but the best. i continue to believe in the mission of creating safe AGI that benefits all of humanity," X, November 17, 2023.
11. Brian Chesky, "Sam Altman and Greg Brockman have my full support. I'm saddened by what's transpired. They, and the rest of the OpenAI team, deserve better," X, November 17, 2023.
12. Keach Hagey, Deepa Seetharaman, Berber Jin, "Behind the Scenes of Sam Altman's Showdown at OpenAI," *The Wall Street Journal*, November 22, 2023.
13. Greg Brockman, "We are so back!" X, November 22, 2023.

## CHAPTER 17 PROMETHEUS UNBOUND

1. John Casey, "Sam Altman on AI, Queerness, and Being Grateful for the Journey He's On," *The Advocate*, March 1, 2024.
2. Ibid.
3. Bob Woodward, *War* (New York: Simon & Schuster, 2024), 180.
4. Tim Alberta, "Dean Philips Has a Warning for Democrats," *The Atlantic*, October 27, 2023.
5. Sam Altman, "this is interesting, and I think close to what the majority of voters actually want: a reasonable, centrist candidate running on 1) a message

of a strong economy and increasing affordability 2) a focus on safety and 3) generational change. curious to see what happens.," X, October 27, 2023.
6. Theodore Schleifer, "President of the Biden-Skeptic Billionaires," *Puck*, November 7, 2023.
7. Sam Altman, "fk it why not 8," X, February 15, 2024.
8. Lex Fridman, "Transcript for Sam Altman: OpenAI, GPT-5, Sora, Board Saga, Elon Musk, Ilya, Power & AGI | Lex Fridman Podcast #419," LexFridman.com, March 17, 2024.
9. Sam Altman, "there are some great parts about the AI EO, but as the govt implements it, it will be important not to slow down innovation by smaller companies/research teams. I am pro-regulation on frontier systems, which is what openai has been calling for, and against regulatory capture," X, November 2, 2023.
10. Brendan Bordelon, "Think Tank Tied to Tech Billionaires Played Key Role in Biden's AI Order," *Politico*, December 16, 2023.
11. Brendan Bordelon, "The Law Firm Acting as OpenAI's Sherpa in Washington," *Politico*, September 12, 2023.
12. Justin Wise, "OpenAI Hires Akin Gump to Lobby in DC on 'Hard' Regulation Task," *Bloomberg Law*, December 26, 2023.
13. Cat Zakrewski, "This Agency Is Tasked with Keeping AI Safe. Its Offices Are Crumbling," *The Washington Post*, March 6, 2024.
14. Gareth Vipers, Sam Schechner, Deepa Seetharaman, "Elon Musk Sues OpenAI, Sam Altman, Saying They Abandoned Founding Mission," *The Wall Street Journal*, March 1, 2024.
15. Ilya Sutskever, "In the future, it will be obvious that the sole purpose of science was to build AGI," X, March 20, 2022.
16. David Faber, "CNBC Exclusive: CNBC Transcript: Elon Musk Sits Down with CNBC's David Faber on CNBC Tonight," CNBC, May 16, 2023.
17. Jan Leike, "I resigned," X May 15, 2024.
18. Daniel Filan, "Superalignment with Jan Leike," *AXRP—the AI X-risk Research Podcast* 24, July 27, 2023.
19. Jan Leike, "Building smarter-than-human machines is an inherently dangerous endeavor. OpenAI is shouldering an enormous responsibility on behalf of all humanity. But over the past few years, safety culture has taken a backseat to shiny products," X, May 17, 2024.
20. Kelsey Piper, "ChatGPT Can Talk, but OpenAI Employees Sure Can't," *Vox*, May 18, 2024.
21. Sam Altman, "in regards to recent stuff about how openai handles equity: we have never clawed back anyone's vested equity, nor will we do that if people do not sign a separation agreement (or don't agree to a non-disparagement agreement). vested equity is vested equity, full stop. there was a provision about potential equity cancellation in our previous exit docs; although we never clawed anything back, it should never have been something we had in any documents or communication. this is on me and one of the few times i've been genuinely embarrassed running openai; i did not know this was happening

and i should have. the team was already in the process of fixing the standard exit paperwork over the past month or so. if any former employee who signed one of those old agreements is worried about it, they can contact me and we'll fix that too. very sorry about this," X, May 18, 2024.
22. Sam Altman, "her," X, May 13, 2020.
23. Alex Bruell, "New York Times Sues Microsoft and OpenAI, Alleging Copyright Infringement," *The Wall Street Journal*, December 27, 2023.
24. Sarah Krouse, Deepa Seetharaman, Joe Flint, "Behind the Scenes of Scarlett Johansson's Battle with OpenAI," *The Wall Street Journal*, May 23, 2024.
25. SSI Inc. @ssi, "Superintelligence is within reach. Building safe superintelligence (SSI) is the most important technical problem of our time. We've started the world's first straight-shot SSI lab, with one goal and one product: a safe superintelligence. It's called Safe Superintelligence Inc. SSI is our mission, our name, and our entire product roadmap, because it is our sole focus. Our team, investors, and business model are all aligned to achieve SSI. We approach safety and capabilities in tandem, as technical problems to be solved through revolutionary engineering and scientific breakthroughs. We plan to advance capabilities as fast as possible while making sure our safety always remains ahead. This way, we can scale in peace. Our singular focus means no distraction by management overhead or product cycles, and our business model means safety, security, and progress are all insulated from short-term commercial pressures. We are an American company with offices in Palo Alto and Tel Aviv, where we have deep roots and the ability to recruit top technical talent. We are assembling a lean, cracked [sic] team of the world's best engineers and researchers dedicated to focusing on SSI and nothing else. If that's you, we offer an opportunity to do your life's work and help solve the most important technical challenge of our age. Now is the time. Join us. Ilya Sutskever, Daniel Gross, Daniel Levy June 19, 2024," X, July 19, 2024.
26. Deepa Seetharaman, Ton Dotan, Berber Jin, "OpenAI Nearly Doubles Valuation to $157 Billion in Funding Round," *The Wall Street Journal*, October 2, 2024.

## EPILOGUE

1. Natasha Mascarenhas, "Alt Capital Raises $150 Million Fund, Extending Altman Brothers' Funding Spree," *The Information*, February 1, 2024.
2. Sarah Needleman, "OpenAI CEO Sam Altman Denies Sexual Abuse Claims Made by Sister," *The Wall Street Journal*, January 8, 2025.
3. University of Toronto, "University of Toronto Press Conference—Professor Geoffrey Hinton, Nobel Prize in Physics 2024," YouTube, October 8, 2024.
4. Cade Metz, "'The Godfather of AI' Leaves Google and Warns of Danger Ahead," *The New York Times*, May 1, 2023.
5. University of Toronto, "University of Toronto Press Conference."
6. Elon Musk, "I'm happy to be first buddy!" X, November 11, 2024.

NOTES

7. Elon Musk, "I don't hate the man, but it's time for Trump to hang up his hat & sail into the sunset. Dems should also call off the attack—don't make it so that Trump's only way to survive is to regain the Presidency," X, July 11, 2022.
8. *Elon Musk et al, v. Sam Altman et al*, Docket No. 4:24-cv-04722 (N.D. Cal, Aug 5, 2024), US District Court, Northern District of California, November 15, 2024.
9. Marco Quiroz-Gutierrez, "Elon Musk Is Ratcheting Up His Attacks on His Old Partner Sam Altman, Calling Him 'Swindly Sam' and OpenAI a 'Market-Paralyzing Gorgon,'" *Fortune*, December 3, 2024.
10. David Deutsch, *The Beginning of Infinity: Explanations That Transform the World* (New York: Penguin Books, 2011).
11. Sam Altman, "The Intelligence Age," ia.samaltman.com, September 23, 2024.

# INDEX

Abbott, Andy, 47, 227
*Access Hollywood* tape, 203
ACLU, 98
Acre, The, neighborhood of Lowell, MA, 35
Adams, Douglas, 210
Adams, Harkness & Hill investment bank, 72
"add a zero" ethos of Silicon Valley, 3–4, 272
AdMob, 121
AdSense, 243
Advance Publications, 161
*Advocate, The* (newspaper), 295
AdWhirl, 121
Aetna Life & Casualty, 29, 35
affordable housing, 15, 22, 26–30, 34–35, 37, 40, 53
   in California, 197, 205–6
   Cement City housing project, 35
   Community Reinvestment Act of 1977, 22
   how AI exacerbates the problem of, 254–56, 302
   Pruitt-Igoe housing complex, St. Louis, 27
   rent strikes, 27
   slum clearance, 26, 34
   tax credits, 22, 34–35, 53
   universal basic income (UBI) and, 12–14, 194, 205, 256

AGI (artificial general intelligence), *see* AI (artificial intelligence)
"agile" method of software development, 127
aging, the fight against, *see* immortality
AI (artificial intelligence)
   belief we're living in a simulation created by, 17
   competitive arms race in, 9, 166, 211, 233, 267, 270, 312–14
   exacerbating the problem of affordable housing, 254–56, 302
   existential risk from, 4–6, 141, 144–45, 167–68, 177, 190, 300
   game theory, 166, 285
   generative AI, 1, 3, 9, 219, 221, 270
   the goal of artificial general intelligence (AGI), 3, 5–10, 12–14, 133, 146–47, 170, 181, 189–90, 192, 198–200, 208, 211, 222–26, 230, 233, 237, 245, 287, 304–5
   "godfathers of AI," 188, 312
   large carbon footprint of, 252
   "Manhattan Project" for AI, 8, 145, 147, 219
   national security implications of AI, 166, 267, 285
   neural network–based AI, 147, 164, 176, 179–82, 190–93, 219, 221, 243, 314

AI (artificial intelligence) (*continued*)
   toward the singularity, 140–42, 144–45, 168, 199–200, 305
   "Sky," an AI voice, 307–8
   the three tribes of humans involved in AI (research, safety, and policy), 208–9, 276
   "weights" in, 181
   "wrappers" around, 278–79
   *see also* AI research/training; AI safety; chatbots; OpenAI
AI Doomer Industrial Complex, 299
*AI Dungeon* (video game), 247–48, 254–55
"AI Principles," 5
AI research/training, 5, 143–45, 167–71, 177–88, 207–9, 305–6
   backpropagation and, 179, 181
   bias encoded in AI, 252
   chess, 67, 151, 145–46, 191, 215–16
   "deep learning," 146–47
   dialog as method of alignment, 265
   diffusion model trained by adding digital "noise," 262–63
   "few shot" learning, 244
   Go (game), 191–92, 216–17
   going from "agent" to "transformer," 218–21, 265
   going from "training" to "alignment," 8, 264–66, 276, 284, 305–6
   ImageNet competition, 182, 184
   the issue of Books1 and Books2, 244
   large language models (LLMs), 16–17, 218, 221, 244–46, 252–53, 270–71, 275
   learning means AIs have souls, 179
   machine translation, 168, 219, 245
   neural networks, 147, 164, 176, 179–82, 190–93, 219, 221, 243, 314
   parse trees, 174
   passing the bar exam, 3, 272
   reinforcement learning, 147, 185, 191–93, 264, 265, 284
   Rubik's cube–solving robot hand, 218
   sentiment neuron, 218
   "sequence to sequence" learning, 178
   software engineers and AI researchers, 191, 194
   Test of Time Award, 305
   the "Turing test," 173–74
   "value-lock," 252
   "zero-shot" responses, 220–21
   *see also* video games
AI safety, 2, 4–5, 8–9, 165–67
   Asilomar AI Principles, 208, 211, 213, 233–34
   Deployment Safety Board (DSB) at OpenAI, 279–80, 287
   Elon Musk as an AI doomer, 2, 5, 163, 167–72, 214–15, 273–74
   existential risk from AI, 4–6, 141, 144–45, 167–68, 177, 190, 300
   how AI could go wrong, 7–8, 143–44, 147, 154
   moral balancing act between progress and safety, 135
   national security implications of AI, 166, 267, 285
   parable of the paperclip-making AI, 5, 143–44, 164
   pedophilia and, 254
   Puerto Rico conference on AI safety (2015), 167–70, 207, 211
   the Stochastic Parrot critique of AI, 252–53
*AI Superpowers* (Lee), 267
AIDS crisis, 33, 43, 49
AIM (AOL Instant Messenger), 16, 50, 76, 162
Airbnb, 4, 123, 139, 150, 152,158, 263, 274, 292
Akin Grump Strauss Hauer & Feld law firm, 300
Alberta, Tim, 297
Alcor Life Extension Foundation, Scottsdale, AZ, 141
Alexander, Scott, 143, 165–66
"AlexNet" (Hinton), 178, 182, 266

INDEX

"Algernon," 140–41, see Yudkowsky, Eliezer
Alinsky, Saul, 22, 40
Alito, Samuel, Jr., 52
Allen & Company investment bank, 117, 229, 263, 266
Allston Trading, 157
Alphabet, 194, 271
AlphaGo, 192, 217
Alt Capital, 311
Altman, Annie (Sam's sister), 15, 36, 40–43, 134, 201–2, 226–28, 248–50, 281–82, 295
   allegations against Sam, 261–62
   as an escort and on OnlyFans, 249, 261–62, 281
   after Jerry Altman's death, 311–12
   living on the Big Island of Hawaii, 227, 250, 311–12
Altman, Birdie, 23–25
Altman family, 15–16
   Alt Capital, 311
   American Millinery Company, 31
   during Covid, 248–50
   family therapy sessions, 249
   living in Atlanta, 23
   moving to Clayton outside St. Louis, 39–54
   Sam's parents and early life in Chicago, 21–38
Altman, Gail, 25
Altman, Harry, 23
Altman, Jack (Sam's brother), 23–26, 36, 38, 42, 47, 156–57, 201. 261, 295, 311–12
Altman, Jerry (Sam's father), 15, 21–38, 40–43, 45, 53–54
   affordable housing activism and, 15, 22, 26–30, 32, 34–35, 40–41, 53
   death of and after, 15, 226–28, 233, 248–50, 302
   diamond made of his ashes, 250
   first marriage of, 29
   law school, 35, 37
   move to St. Louis, 37–38, 40

   second marriage of, 33, 35–37, 41–42, 201–2
Altman, Max (Sam's brother), 36–38, 42, 136, 139, 156, 201, 250, 295, 311, 312
Altman, Minnie, 23
Altman, Reba, 23–24
Altman, Richard, 25
Altman, Sam, 1–17, 311–15
   Annie's allegations against, 261–62
   Artix, 68
   belief we are living in a simulation created by, 17
   the "blip" and beyond, 7, 14–15, 276–94, 295–10, 311–15
   blogging, 148, 152
   claim of getting scurvy, 77
   deal with Boost mobile network, 80, 99–105
   death of his father Jerry, 15, 226–28, 233, 248–50, 302
   as dishonest and other complaints, 108–9, 266, 283–84
   dropping out of Stanford, 85–86, 91–93
   early life and family, 15–16
   enter AI, 140–47, 163–66, 169–72
   enter China, 230–31, 267, 314
   enter Covid-19, 246, 248–50
   enter Elon Musk, 16, 147, 170–72, 183, 194, 214–15
   enter Greg Brockman, 172–78, 183–88
   enter Helen Toner, 241–42, 266–67, 276–79, 282–88, 291, 292, 299, 309
   enter Ilya Sutskever, 176–82
   enter Microsoft, 229–33
   enter Mira Murati, 225–26, 242–43, 246–48, 268, 282–93, 304–5, 307–8
   enter Peter Thiel, 1–7, 10, 16–17, 125, 134–48
   enter Saudi Arabia, 231–32

Altman, Sam (*continued*)
  enter Steve Jobs, 61, 85, 87, 90, 110–16
  "good meeting-itis," 107
  ham radios, 32–33, 48
  as head of Y Combinator, 149–63
  Helion Energy nuclear fusion startup, 13, 136, 207, 259, 280, 298
  his cars, 161, 259–60
  his "fashion sense," 49, 115, 126
  his first computer, 44
  his first product launch, 101–3
  his first term sheet (Radiate and NEA), 85–86, 88–93, 95–96
  his public altruism, his private enrichment, 89, 234, 254–61
  his twee all-lowercase writing, 102, 269, 292
  his wide-ranging curiosity, 149, 176
  house in Big Sur, 134, 171, 260
  at the John Burroughs School, 15–16, 45–48, 49–53, 202, 227
  as a kind of doomsday prepper, 154
  Loopt, 16, 96, 101–17, 118–23, 126–28, 133, 138, 150–51, 159, 216, 229–31
  love of cars, 26, 51
  Mac LC II personal computer, 44, 315
  "mercantile spirit," 16, 186
  as a messy eater, 49–50
  "Moore's Law for Everything," 255
  networking talents, 8, 59, 123–24, 274
  at OpenAI's founding, 182–88, 234
  optimism as his brand, 1–4, 15, 17, 30, 101, 103, 128, 136–37, 141, 152, 154, 200
  pink and green polo shirt "douchebag" badge, 115, 126
  political ambitions of, 13, 200, 206, 223, 273–74, 296
  the promises of AI, 254–56
  Radiate (formerly Viendo), 76, 77–82, 85, 88–90, 93–94, 96–101
  from Radiate to Loopt, 103–9
  recurring dream of, 92
  relationship with Lachy Groom, 151–52, 154, 157
  relationship with Matt Krisiloff, 206–7, 275
  relationship with Nathan Watters, 50–51, 201
  relationship with Nick Donahue, 260–61
  relationship with Nick Sivo, 55–56, 60, 64, 85, 89, 91–92, 102–3, 113–14, 128
  relationship with Oliver Mulherin, 275, 288, 295, 311
  religion and spirituality of, 15–17, 24, 31, 43–44, 134, 227
  sexual orientation of, 16, 33, 48–53, 201–3, 295
  "shiny object syndrome," 120–21
  as a Stanford undergrad, 16, 55–61
  travels in Europe, Southeast Asia, and Japan, 134
  in Y Combinator's first Summer Founders Program, 62–66, 75–82
  *see also* OpenAI
Altman, Sol, 23, 24
Altman, Sunny, 23
Amazon, 82, 90, 218, 247
Amazon Web Services, 111, 119, 187, 216
Ambrosia, 258–59
America Online (AOL), 16, 49–50, 76, 162, 229
American Lung Association, 135
American Millinery Company, 31
Amodei, Daniela, 213
Amodei, Dario, 169–70, 178, 184, 209, 213, 226, 241–43, 251–53, 273, 310
Andreessen Horowitz, 155, 158, 161, 257
Andregg, William, 258
angel investing, 70–74, 81, 123, 138, 150–52, 309
Anthropic, 2, 9, 145, 252, 266, 268, 271, 306, 309, 310

# INDEX

Claude the chatbot, 268, 278, 285
Long-Term Benefit Corporation, 277, 299
antisemitism in AI, 270
Anybots, 94
AP Bio exam, 3, 267–68
AP Computer Science, 47
Apple
   App Store, 112
   early personal computers, 44, 86, 117, 195, 315
   iPads, 197
   iPhones, 110–16, 118, 119
   Loopt and, 112–14, 116
   Safari browser, 111
   on the stock exchange, 58
   Vivarium Project, 209–10
   the "walled garden" of, 110
   Worldwide Developers Conference, 111, 114
application programming interfaces (APIs), 125, 245–48, 250–51, 264–65, 269, 279
Arment, Marco, 203
Arnold & Porter law firm, 291
*Ars Technica*, 242
artificial general intelligence (AGI), *see* AI (artificial intelligence)
Artix, 68
Asana workplace management software, 212
Asilomar AI Principles, 208, 211, 213, 233–34
Asimov, Isaac, 47
Assange, Julian, 142
AT&T, 62, 90, 100, 103, 104, 110, 118, 119
Atari, 87, 147, 165, 190
Atkins, Brian and Sabine, 142–43
Atlanta Georgi, 23
*Atlantic, The* (magazine), 46, 297
Atmos home building platform, 260
"Attention Is All You Need" ("the transformer paper"), 218–19, 270
Australia, 99, 266

Authors Guild, 220
Autopilot AI-assisted driving, 225

Baidu, 170
baldness, cure for, 108
Bankman-Fried, Sam, 212, 257, 276
bankruptcies, 31, 87–88, 145, 205
bar exam, AI passing the, 3, 272
Bard conversational model, 271
"bed nets" era of effective altruism, 212
*Beginning of Infinity, The* (Deutsch), 314–15
Bell Labs, 140, 185
Bell, Tom (aka T. O. Morrow), 141–42
Bender, Emily, 252
Bengio, Yoshua, 183–84, 188, 272
Berners-Lee, Tim, 244
Betts-LaCroix, Joe, 257–59
Bezos, Jeff, 229, 260
bias encoded in AI, 252
Biden administration AI policies, 232, 267, 277, 299–300
Biden, Joe, 273, 296–300
Big Island of Hawaii, 227, 250, 311–12
Big Sur, CA, 134, 171, 260
Big Tech companies, 182, 185, 197, 299, 308
Bill and Melinda Gates Foundation, 303
bin Salman, Mohammed, of Saudi Arabia ("MBS"), 231
Bing search engine, 216, 271
Bird, Leah, 28–29, 43
Bitcoin, 142
BitTorrent, 244
Black people and communities, 26–27, 40, 53, 164
   Chicago's first black mayor, 21–22, 33–34, 37
   racism among tech bros, 164–65
   *see also* race and racism
black-pilled, 1
Blackwell, Trevor, 63, 74, 75, 94

"blip, the," 7, 14–15, 276–94, 295–310, 311–15
Blizzard Entertainment, 215
Bloomberg (news site), 67, 207, 276
Bloomberg limited partnership, 234
Bloomberg, Michael, 155
Blumenthal, Richard, 309
BookCorpus dataset, 220
Boom, Steve, 126
Boost mobile network, 80, 99–105
Borgmann, Jolene, 201–2, 226–27
Bostrom, Nick, 5, 163–66, 168–70, 241
  Future of Humanity Institute, 5, 165–66, 241–42, 267
  *Superintelligence: Paths, Dangers, Strategies*, 5, 163–66, 167–68
Botha, Roelof, 88, 93, 96, 124
Bowles, Nellie, 205
Box cloud storage company, 96
Boyden, Ed, 199
*Breakout* (video game), 147
Brennan-Jobs, Lisa, 112–13
Bridgewater Associates hedge fund, 212
Brin, Sergey, 85, 88, 192
Brockman, Anna, 245, 294, 309
Brockman, Greg, 16
  during "the blip," 277–79, 283–84, 286–89, 291–94, 305, 308–9, 310
  in his "code cave," 175–76, 246–47
  as an OpenAI founder, 172–78, 183–88
  role in developing GPT-4, 189–96, 209, 215–23, 242–51, 267–68
Brockman, Ron, 173
Brooklyn's Park Slope, 29–30
Brooks, David, 132
Brown, Jerry, 205, 206
Brown, Willie, 205–5, 256
Brundage, Miles, 241–42, 253
Brunswick, Georgia, 23–25
Buchheit, Paul, 123, 150–51
Bucktown neighborhood of Chicago, 34

Buffet, Warren, 212
BuildZoom digital construction marketplace, 138
bullying, 131, 287
Burger King, 62
Burning Man Festival, 1, 198–99
Burt, Matthew, 92, 93
*Business Insider*, 260, 277

California
  affordable housing crisis in, 197, 205–6
  ballot measure limiting short-term rentals, 262, 274, 292
  Daly City, 162
  Irvine, 99, 105
  Mountain View, 85, 94, 107, 122, 133, 139, 149, 161, 172, 182, 235
  Napa Valley, 185, 190, 260
  Oakland, 195
  YC Startup School at Berkeley and Stanford, 122–23, 194, 234
  *see also* Palo Alto, CA; San Francisco; Silicon Valley
capitalism, 40, 166, 195, 225, 314
"capped profit" model, 229–30, 304
carbon footprint, 252
Carlson, Tucker, 294, 313
Carr, Eric, 120–21
carried interest ("carry"), 73, 158, 281
cars, self-driving, 168, 223–24
cars, tech bros' love of, 26, 51
Castro, Daniel, 300–301
Cement City housing project, 35
Center for AI Policy, 301
Center for Security and Emerging Technology (CSET) at Georgetown, 267, 299, 300
Central Reform Congregation, 15, 43, 227
Centre for the Governance of AI, 267
Centre for the Study of Existential Risk, 168
Chafkin, Max, 131, 205, 258
Chang, Jeff, 198

# INDEX

Character Technologies, Inc., 271
charities/philanthropy, *see* affordable housing; effective altruism (EA)
chatbots, 3, 9, 174, 263, 268–70
  Anthropic's Claude, 268, 278, 285
  antisemitic, 270
  giving a superego to the, 264
  Google's Meena, 270
  Microsoft's Tay, 270
  pedophilia and, 254
  Quora's Poe, 278–79
  "wrappers" around AI, 278–79
ChatGPT, 1, 3, 5, 14–15, 17, 208, 253, 269–73, 276, 278, 285–86, 308, 310
Che, Sally, 25, 49–51, 227, 260–61
Checkr background check automation company, 149–50, 156
Chernobyl nuclear disaster, 135
Chesky, Brian, 157, 236, 263, 270, 274, 277, 289, 292
chess, 67, 151, 145–46, 191, 215–16
Cheung, Adora, 157, 195
Chicago
  Altman family in, 21–38
  Bucktown neighborhood of, 34
  first Black mayor of, 21–22, 33–34, 37
  Lincoln Park neighborhood of, 36
Chicago Toy & Game Group, 209
China, 230–31, 267, 314
*China Syndrome, The* (film), 135
chips, 225, 280–81, 298, 315
  global positioning system (GPS) chips in mobile phones, 57–58, 99
  graphics processing units (GPUs), 176, 181–82, 219, 247, 255
  the US CHIPS and Science Act, 298
Christiano, Paul, 178, 213, 276–77, 299
Chrome web browser, 272
*Chronicle of Philanthropy, The*, 213
Chung, Patrick, 13, 61–62, 74, 78–81, 89–93, 116, 121

Church of Jesus Christ of Latter-day Saints, 55
Churchill, Winston, 256
Cincotta, Gale, 22, 29, 35
Cingular Wireless, 58, 62, 103–4, 110
Cisco, 87, 122, 210
Clark, Chris, 198
Clark, Jack, 207, 241, 253, 266, 273
Clarke, Arthur C., 47
Claude the chatbot, 268, 278, 285
Clayton, a suburb of St. Louis, 23–25, 27, 30, 32–33, 39–54
ClickFacts, 77, 82
Clinton, Bill, 205, 274, 292
Clinton, Hillary, 201
CNBC, 305
Coalition for a Better Acre, 35
Cobbin, Darryl, 102
Coca-Cola Company, The, 102
Cockburn, Chloe, 213
"code caves," 175–76, 246–47
Codex AI software development bot, 262
Coffee Cartel, Clayton, Missouri, 50–51
Cohen, Ryan, 260–61
Cohen-Esrey property management company, 53
"Coherent Extrapolated Volition" (Yudkowsky), 143
Coinbase, 158
"Collison installation," 246
Collison, John, 125, 157, 175
Collison, Patrick, 90, 124–25, 136, 157, 175–76, 190
Comcast, 229
Common Crawl, 243–44
Communication Design Group (CDG), 197–98
compute, 193, 216–19, 221, 224, 243, 245, 282, 298, 306, 314
  Amazon Web Services, 111, 119, 187, 216
  the case of *AI Dungeon*, 247–48, 254–55

compute (*continued*)
  as the currency of the future, 298
  enormous infrastructure costs of running location services, 119
  graphics processing units (GPUs), 176, 181–82, 219, 247, 255
  *see also* energy production
Computer Fraud and Abuse Act, 68
Computer History Museum, Mountain View, CA, 139
computer vision, 168, 182, 184, 247
"Computing Machinery and Intelligence" (Turing), 173–74
Concorde (airplane), 133, 135
Condé Nast, 81, 160–62
Conference on Neural Information Processing Systems (NIPS/NeurIPS), 178, 186, 305
conflicts of interest, 225, 236, 303
Conrad, Parker, 139
Consensus Organizing Institute, 40
conspiracy theories, 253
Continental Bank, 34
Continuity Fund at YC, 159, 194, 232–33, 290
*Contrarian, The* (Chafkin), 205, 258
convertible notes, 94–95, 138
Conway, Ron, 123, 138, 309
Cook, Tim, 229
copyright, 160n, 220, 243, 244
Cornell, 67
Corteau, Kate, 65
Cotra, Ajeya, 277
Country Day School, St. Louis, 27, 45
Coursera course on machine learning, 266
Covid-19, 246, 248–50
criminal justice reform, 213
Crowley, Dennis, 118–19
cryptocurrency, 13, 212, 256
  Bitcoin, 142
  Coinbase, 158
  FTX crypto market, 212, 257
  initial coin offering (ICO), 221–22
  Worldcoin, 13, 256–57

culture, *see* tech bros
Cummings, Dominic, 205
cybersecurity, 68, 106, 140, 154, 200

DAG Ventures, 117
Dahar, Robin, 158–59
Dai, Wei, 142
*Daily Caller*, 204
*Daily News*, 29
Daley, Richard J., 21, 34
DALL-E, 255, 262–63, 268
Daly City, CA, 162
D'Angelo, Adam, 234, 278–79, 284–86, 288, 289, 293
Danzeisen, Matt, 1
De Freitas, Daniel, 270–71
deal flow, 88, 95, 124
Dean, Jeff, 187
"Decoding Intentions: Artificial Intelligence and Costly Signals" (Toner), 285
Deep Blue, 191
"deep learning," 146–47
deepfakes, 241–42, 263
DeepMind, 146–48, 154, 168–69, 171–72, 184, 189–94, 211, 221, 270
Defense Advanced Research Projects Agency (DARPA), 140, 196
"Del Mar divide" in St. Louis, 26, 53
Delicious (website), 75
Deming, Peter, 59–60, 64
democracy, 152–53, 225
Democratic Party, 21, 204, 207–8, 234, 296
Deployment Safety Board (DSB), 279–80, 287
Deshpande, Alok, 58–61, 64, 76, 85, 89, 102, 111, 127
Deshpande, Sheila, 58, 91
Desmond-Hellmann, Sue, 303
Deutsch, David, 314–15
Dewey, John, 45
deworming charities, 211, 213
Diamandis, Peter, 144, 210

diffusion model trained by adding digital "noise," 262–63
Digital Chocolate mobile gaming company, 97
Diller, Barry, 229
Ditton, Andy, 34–35
DJ Kay Slay, 102
DJI drones, 231
DLA Piper lobbying firm, 300
DNNresearch, 182
Doblet phone-charging startup, 149, 177
Dodgeball location-based app, 104, 105, 118
Donahue, Nick, 260–61
*Dota 2* (video game), 215–18, 221–22, 242, 284
dot-com bubble of the late 1990s, 73, 87, 93
Dowling, Steve, 287
Dropbox, 123, 139, 150, 152, 157, 158–59, 247
drop-outs, tech bros as mostly, 62, 88, 91–92, 96, 108, 113, 124, 132, 175, 179, 217–18
Dubai, 274

eBay, 82, 168
economic growth
　democracy and, 152–53, 225
　dot-com bubble of the late 1990s, 73, 87, 93
　financial crisis of 2008, 4, 116–17, 131, 137, 150
　as a kind of spiritual hack, 153
　real wages, 132
Edge web browser, 272
Edison, Thomas, 46
effective altruism, 2, 4, 6 189–214, 224, 265–66, 276, 300
　"bed nets" era of, 212
　deworming charities, 211, 213
　GreaterWrong forum, 301
　Open Philanthropy Project, 213–14, 241, 266–67, 276–77, 299–301
　*see also* AI safety
Effective Ventures, 276
Efficient Market Hypothesis, 144
Eichler, Mike, 30, 40
80,000 Hours, 211–12
Elbaz, Gil, 243
Electronic Arts, 97
Electronic Frontier Foundation (EFF), 59, 98
energy production, 13, 134, 231, 280, 313–14
　nuclear energy, 12–13, 57, 108, 134–36, 154, 177, 205, 230, 259, 280
　solar energy, 28
　*see also* compute
Enigma machine, 139
Enlightenment, 314–15
Enron, 60
esports, 215–16
Evans, Jon, 142
Extropians, 141–42, 144–45, 164–65, 199, 214

Fabolous (DJ), 102
Facebook, 60, 63, 73, 79–80, 96, 101, 116–17, 123, 127, 131, 137, 161, 169, 184, 188, 204, 207–8, 229
fact-checking, 265
Fairchild Semiconductor, 72, 87
fake news, 265
Federal Communications Commission (FCC), 59, 106
Federal Trade Commission, 285
Feldman, Ellen, 173
Fellow Robots robotics company, 210
Fermi's paradox, 170
"few shot" learning, 244
"Feynman method of being a genius," 219
Filan, Daniel, 305
financial crisis of 2008, 4, 116–17, 131, 137, 150
"finding product-market fit," 107, 156
"finding your tribe," 157

Finney, Hal, 142
Firefox web browser, 63
Flexport logistics platform, 138
Florida, 40
"Flowers for Algernon" (Keyes), 140
Foo Camp, 63
*Forbes* (magazine), 257
*Foreign Affairs* (journal), 267
Foresight Institute, a technology think tank, 144
Forstall, Scott, 112, 114, 116
"Founder Mode" (Graham), 263
founders as kings, 6, 60, 65, 68, 70–75, 95
*Founders at Work* (Livingston), 63, 64
Founders Fund venture firm, 2, 6, 132, 139, 147, 226
Foursquare, 118–20, 126
Francis, Peggy, 32, 36
"Free Oceana," 141
Freemasonry, 31
Fridman, Lex, 17, 304
Friend, Tad, 201
*From Zero to One* (Thiel and Masters), 132, 230
frontier in American history, 153
FroSoCo (short for Freshman Sophomore College), 55, 57
FTX crypto market, 212, 257
Furstenberg, Diane von, 229
Future of Humanity Institute, 5, 165–66, 241–42, 267
Future of Life conference, Puerto Rico (2015), 167–70, 207, 211
Future of Life Institute, 145, 168, 208–9, 272

Galef, Julia, 225
Galois, Évariste, 174
game theory, 166, 285
Gates, Bill, 65, 90, 212, 216, 267–68
Gates, Melinda, 212
Gauss, Carl Friedrich, 174
Gawker Media, 137, 204–5
gay marriage rights, 56, 296

Gay Straight Alliance, 52
Gaza, war in, 289
GE (General Electric), 210
Gebbia, Joe, 263
Gebru, Timnit, 252–53, 270–71
Gemini AI model, 307
general artificial intelligence (AGI), *see* AI (artificial intelligence)
generative AI, 1, 3, 9, 219, 221, 270
generative pre-trained transformers (GPTs), 3, 221, *see also various GPTs under* OpenAI
genius, human, 77, 81, 127, 140, 156, 219
"Gentle Seduction, The" (Stiegler), 199–200
George, Henry, 256
Georgia, 23–25, 58, 104, 142
GeoSim 3D modeling company, 209–10
Germany, 274
Gibney, Bruce, 132–33
Gibstine, Connie (Sam's mother), 15, 31–38, 39–50, 53–54, 91, 201–2, 227–28, 248–50, 261, 295, 312
Gil, Elad, 136
Gillette, 62, 71
Ginsberg, Allen, 166
Girard, René, 131
Github Copilot, 262
Gittell, Ross, 35
GiveWell, 212–13, 266
Giving Pledge, 212
Giving What We Can, 211
global positioning system (GPS) chips in mobile phones, 57–58, 99
Gmail, 123, 150, 286–87
Go (game), 191–92, 216–17
"godfathers of AI," 188, 312
Goertzel, Ben, 145
Goetz, Jim, 88
Goldman Sachs, 59, 64, 150, 225
Good Ventures foundation, 212–13
Google
    AdSense, 243
    Alphabet, 194, 271

INDEX

"Attention Is All You Need" ("the transformer paper"), 218–19, 270
Bard conversational model, 271
Chrome web browser, 272
DeepMind acquisition, 146–48, 154, 165, 168–69, 171–72, 184, 189–94, 208, 211, 217, 221, 270
Dodgeball location-based app acquisition, 104, 105, 118, 104
Gemini AI model, 307
Gmail, 123, 150, 286–87
  recent initial public offering (IPO) of, 87
  YouTube acquisition, 93–94
  see also Anthropic
Google assistant, 271
Google Brain, 82, 169, 184, 243, 270
Google Colab Notebook, 247
Google I/O annual developer conference, 307
Google Maps, 59
Google Search, 270
Gordon-Levitt, Joseph, 210, 225
*GotNews* (website), 204
GPTs (generative pre-trained transformers), *see* chatbots; *various GPTs under* OpenAI
Graham, Paul, 3, 13–16, 62–65, 67–76, 81–82, 94, 136, 149, 151, 186, 263
  "A Unified Theory of VC Suckage," 72–73
  "Collison installation," 246
  "Founder Mode," 263
  "How to Start a Startup," 69–70
  painting's influence on, 67–69
  *see also* Y Combinator (YC)
graphical user interfaces, 195
graphics processing units (GPUs), 176, 181–82, 219, 247, 255
Gras, Mike, 249
GreaterWrong forum, 301
Green Dot prepaid debit card company, 126–27, 133
Green-Lowe, Jason, 301–2

Grey, Aubrey de, 144, 259
Groom, Lachy, 151–52, 154, 157
Gross, Daniel, 309
Groupon, 119, 116
*Guardian, The* (newspaper), 220
Gurevich, Mikhail, 77, 82
Gurson, Doktor, 149, 177

*Hacker News* message board, 132, 151
hackers/hacking, 3, 57, 63, 68–70, 160n, 162
Haffner & Gibstine Real Estate, 32
Halcyon Molecular, 257–59
Hall, Ed, 26
*Halo 3* (video game), 95, 109
ham radios, 32–33, 48
Hanson, Robin, 141–42, 144
Harder, Josh, 207
"hardware strategy, the," 226, *see also* compute
Harris, Kamala, 273, 313
Harris, Sylvia, 24–25
*Hartford Courant* (newspaper), 28
Hartford, CT, 28–29
Hartford Institute of Criminal and Social Justice, 28
Hartmann, Frank, 28–29
Harvard Computer Society, 69
Hassabis, Demis, 145–47, 169, 171–72, 192, 209
Hassenfeld, Elie, 212
Hawaii, 227, 250, 260, 295, 311–12
Hawking, Stephen, 169, 211
Hawkins, Trip, 97
HBO's *Silicon Valley*, 101, 258–59
HBO's *Westworld*, 199
heads, frozen, 141
Health Extension Foundation, 258
Helion Energy nuclear fusion startup, 13, 136, 207, 259, 280, 298
Helo telepresence robot, 210
*Her* (film), 307
Herzog, Isaac, 274
Hill, Daniel, 165
Hilton, Jacob, 265

Hinduism, 17
Hinton, Geoff, 178–84, 188, 266, 312–13
hip-hop, 102
Hipmunk travel search company, 162
HIV/AIDS, 33, 43, 49
Hoffman, Reid, 109, 172, 223, 231, 234–35, 277, 296
Hogan, Hulk, 205
Holocaust, attempted analogies to the, 135
Homejoy, 195
Horizon Institute for Public Service, 301
housing, *see* affordable housing
Houston, Drew, 157
"How to Start a Startup" (Graham), 69–70
Howard, James, 110–11
*Howl* (Ginsberg), 166
Huffman, Steve, 69–70, 75, 76–78, 81, 162–63
Hui, Fan, 192
Human Advancement Research Community (HARC), 195, 197–98, 236
humans
　in a democratic society, 135
　human genius, 77, 81, 127, 140, 156, 219
　human-computer interfaces, 210
　learning to truly produce knowledge, 314
　neuroscience and studying the human brain, 145–48, 202
　"only good for hugs or sex," 269
　their brains as the original neural nets, 181
　*see also* chatbots
Hunnewell, H. H., 71
Hurd, William, 266, 278
Hydrazine Capital, 136–38, 139, 156, 161, 207, 260
HyperCard, 209

IBM's Deep Blue, 191
Igleheart, Walker, 27
ImageNet competition, 182, 184
immortality
　Ambrosia, 258–59
　death as a solvable problem, 141, 143, 200, 227, 315
　frozen heads, 141
　Health Extension Foundation, 258
　transhumanist dream of, 170
　young blood transfusions, 258–59
*Inc.* (magazine), 258
*Inception* (film), 210
India, 14, 134, 274–75
Indiana Pacers basketball team, 209
Inflection AI, 277
Infogami, 76
Information Technology and Innovation Foundation, 300
*Information, The* (news site), 311
Infosys, 187, 197, 236
initial coin offering (ICO), 221–22
innovation
　breakthrough technologies, 153, 207–8
　CHIPS and Science Act, 298
　computer vision, 168, 182, 184, 247
　incremental cost of selling an additional widget, 72
　"levers" of change, 154
　moonshots, 13, 16, 136, 194
　"moving fast and breaking things," 237
　nanotechnology, 144, 164
　the need for "adults in the room," 96–98, 217
　the problem of incremental or even fake problems, 132–33, 180
　"productizing" technologies, 245
　quest for a "new Bell Labs," 140, 185
　self-reinforcing cycle of exponential technological progress, 140
　serendipity and, 98
　time horizon of, 153, 174, 185

INDEX

see also software development; tech industry
insider information, 144
InstructGPT, 264
insurance industry, 24, 29, 37, 139, 228, 248, 273
Intel Research Seattle, 110–11
intellectual property, 185, 281
  Common Crawl, 243–44
  copyright, 160n, 220, 243, 244
  open-source technologies, 95, 186, 214, 247, 263
  in the public domain, 196
  Tesla's patents as openly available, 214
  trademark issues, 101
  voice called "Sky," 307–8
Intelligence Age, 314
IntelliGenesis, 145
International, The (global esports championship), 215
Internet Explorer web browser, 272
investing
  angel investing, 70–74, 81, 123, 138, 150–52, 309
  "capped profit" model, 229–30, 304
  conflicts of interest, 225, 236, 303
  deal flow, 88, 95, 124
  Efficient Market Hypothesis, 144
  employee shares, 290
  follow-on investments, 159
  insider information and, 144
  issues of equity, 6, 7, 18, 72, 89–91, 94–95, 138, 158, 172, 185, 190, 222, 233–36, 256, 273–74, 302, 306
  Lehman Brothers collapse, 116
  lessons from the art market's inefficiency and opacity, 68
  limited partners (LPs), 72, 136, 124, 151, 155, 159, 234, 236, 281
  massive valuations of tech companies, 71, 73, 138–40, 152, 290
  preferred stock, 90
  "product-market fit," 107, 157

profiting off carried interest ("carry"), 73, 158, 281
public funds and private investment, 314
Reddit share offerings, 161, 163
Series A investment round, 89, 101, 125, 136, 156, 158–59
university endowments, 72, 155, 159
volatility, 137
see also Sequoia Capital; startups; tech industry; Y Combinator (YC)
IOTA Foundation, 275
iPads, 197
iPhones, 110–16, 118, 119
Irvine, CA, 99, 105
Isaacson, Walter, 171–72
Ishutin, Danil "Dendi," 215–16
Isilon data storage company, 122
Israel, 179, 274
Istanbul, 232
Ive, Jony, 280

Jacobstein, Mark, 60–61, 97–98, 102, 107–9
jailbreaking the iPhone, 112
Japan, 23, 79, 134
Jin, Berber, 11
Jobs, Steve, 61, 85, 87, 90, 110–16
Johansson, Scarlett, 307–8
John Burroughs School, 15–16, 45–46, 53, 202, 227
Johnson, Boris, 205
Johnson, Charles, 204–5
Józefowic, Rafał, 218
Jump, Brian, 46–48, 51
Justin.tv, 82, 157, 293

Kan, Justin, 76, 81–82, 157, 293
Kansas City, 41, 62, 78, 99
Karnofsky, Holden, 212–13, 234, 266–67
Karpathy, Andrej, 184–86, 189–93, 221, 223

Kasparov, Garry, 191
Kay, Alan, 195–98, 209–10
Kennedy, John, 273–74
Kennedy, Len, 105–6
Kennedy, Leroy, 22
Kepchar, Georgeann, 47
Keyes, Daniel, 140
Khashoggi, Jamal, 232
Khosla Ventures, 230, 235, 257
Khosla, Vinod, 230
Kiko online calendar app, 76, 81
Kingma, Durk, 184
Kingsbury, Kathy, 53
Kirtley, David, 259
Kleiner Perkins venture firm, 162, 230
Knapp, Brian, 106, 109, 121
*Knife Fight* (film), 292
Koko the gorilla, 210
Kolpino housing complex in Leningrad, 27n
Kozmo.com Kit-Kat delivery services, 132
Krisiloff brothers (Matt and Scott), 206–7, 296–97
Krizhevsky, Alex, 178, 182, 266
Kurzweil, Ray, 142, 144, 210
Kushner, Josh, 161, 290
Kutcher, Ashton, 138
Kwon, Jason, 290, 293

L5 Society, 144
Ladue suburb of St. Louis, 26, 45
Lagorio-Chafkin, Christine, 161
LaMDA (Language Models for Dialog Application), 270–71
"land value tax," 256
large language models (LLMs), 16–17, 218, 221, 244–46, 252–53, 270–71, 275
Larsen, Thomas, 301
Latino people and communities, 21, 35
Lattice (startup), 311
*Launch Pad: Inside Y Combinator, The* (Stross), 133

Leap Motion, 226
LeCun, Yann, 169, 184, 188
Lee, Jennifer 8., 217
Lee, Kai-Fu, 267
Lee, Wing, 78–80
Legg, Shane, 145–46, 168
Lehane, Chris, 205, 274, 292
Lehman Brothers collapse, 116
Leike, Jan, 264, 284, 305–6, 308
Lemoine, Blake, 271
Lemon, Steve, 119, 121
Leone, Doug, 87, 112
Lessin, Jessica, 185
*LessWrong* (blog), 4, 176, 212, 301
Levchin, Max, 136
"levers" of change, 154
Levie, Aaron, 96
Levy, Carolynn, 86, 94, 232
Levy, Daniel, 309
Levy, Jon, 94, 232–33
LGBTQ people, 43, 50
Li, Fei-Fei, 184
liberalism, 314
libertarianism, 131, 142, 145
Library Genesis (LibGen), 244
*Life 3.0: Being Human in the Age of Artificial Intelligence* (Tegmark), 168
*Life in Seven Songs* podcast, 134, 198, 275
Lightcap, Brad, 293
Lightspeed Grants, 301
LightSurf photo sharing software, 79
limited partners (LPs), 72, 136, 124, 151, 155, 159, 234, 236, 281
Lin, Alfred, 161
Lincoln Park neighborhood of Chicago, 36
Lindsay, Neil, 100
LinkedIn, 223, 234
Lisp programming language, 63, 69, 124, 151
Liu, Min, 96, 107, 114–15
Livingston, Jessica, 63–65, 71–75, 81, 94–95, 122, 151, 157, 187, 232, 235–36

lobbying, 109, 144, 179, 274, 292, 300–301
Local Initiatives Support Corporation (LISC), 34–35, 40
location-based services, 58–59, 106–7, 118–19
　Dodgeball location-based app, 104, 105, 118
　enormous infrastructure costs of, 119
　Foursquare, 118–20, 126
　*see also* mobile phones
Loopt, 16, 96, 101–17, 118–23, 126–28, 133, 138, 150–51, 159, 216, 229–31
Los Alamos National Laboratory, 170
Lourd, Bryan, 307–8
Lowe, Rob, 292
Lowe, Ryan, 265
Lowell, MA, 35
Low-Income Housing Tax Credit program, 22, 35
Lu, Qi, 230–31
Ludacris, 102

MacArthur "genius grant," 81
MacAskill, William, 211–12
Machine Intelligence Research Institute, 210, 301
machine superintelligence, 143, 145, 164, 167, 170, 177, 309, 315, *see also* AI (artificial intelligence)
machine translation, 168, 219, 245
Macron, Emmanuel, 14, 274
Magic: The Gathering card game, 76
Mahoney, Ryan, 312
Mailliard, Page, 85–86, 88–89, 91, 93–95, 122
Makanju, Anna, 274
malaria prevention, 212, 266
"Malicious Use of Artificial Intelligence, The" (Toner), 241–42
"Manhattan Project" for AI, 8, 145, 147, 219
Manson, Richard, 29

Mar-a-Lago, 40
Marciniak, Brian, 97, 98, 105
Marcus, Gary, 268
Mars, colonizing, 147, 170–71
Martin, James, 165
Masayoshi Son, 229
masculinity, *see* tech bros
*Mashable* (news site), 119
Masters, Blake, 132
"math envy," 180
Matheny, Jason, 267, 299
Matrix Partners venture firm, 90
McAdoo, Greg, 86–92, 95, 97–98, 112–16, 119–26, 155, 231
McCain, John, 296–97
McCauley, Matthew, 209
McCauley, Tasha, 209–11, 234, 276, 286, 288, 299
McCaw Cellular, 90
McGrew, Bob, 217–18, 283, 286, 287, 310
McIntyre, Robert, 258
McKinsey, 61–62, 150, 158
Medicare for All, 206
"Meditations on Moloch" (S. Alexander), 165–66
Meena the chatbot, 270
Memamp desktop search company, 76
Mendelsohn, Matt, 45
*Metro* (UK newspaper), 242
Metz, Cade, 145–46, 199
MGX investment fund, 310
Microsoft
　in the AI arms race, 270
　Azure cloud, 216, 237, 272
　and the Biden campaign, 296
　Bill and Melinda Gates Foundation, 303
　Bing search engine, 216, 271
　during the "blip" and beyond, 277, 279–80, 282–83, 288–89, 293–94, 303–4, 310, 313
　Edge web browser, 272
　GitHub Copilot on Visual Studio, 262

Microsoft (*continued*)
  HyperCard, 209
  Inflection AI acquisition, 277
  Internet Explorer web browser, 272
  LinkedIn acquisition, 234
  OpenAI's relationship with, 2, 11, 13, 216–17, 226, 229–30, 234–35, 237, 245, 271–72, 303–4, 310, 313
  QBasic language, 56
  regulation of AI, 299–300
  Tay the chatbot, 270
  valuation of, 298
  Visual Basic language, 56
Millennials, 4, 137, 205
Milner, Yuri, 138, 199
mimetic theory, 131
Minsky, Marvin, 142, 210
MIT (Massachusetts Institute of Technology), 63, 64, 92, 124, 160, 175, 178, 212
*MIT Technology Review*, 258–59, 269
Mithril Capital, 136
mobile phones
  enter the iPhone, 110–16, 118, 119
  global positioning system (GPS) chips in, 57–58, 99
  location-based services, 58–59, 106–7, 118–19
  phone makers as the serfs of the mobile industry, 110
  proprietary software on, 79
  wireless carriers, 58–59, 75, 79, 90, 98, 100–9, 110, 118, 120, 159
mobile virtual network operator (MVNO), 99
Modi, Narendra, 14, 274, 275
Moloch (a god), 165–66
Monongahela Valley of Pennsylvania (Mon Valley), 30, 40
moonshots, 13, 16, 136, 194
Moore, Demi, 138
"Moore's Law for Everything" (Altman), 255
More, Max, 141

Morehead-Cain Scholarship at the University of North Carolina at Chapel Hill, 53
Moritz, Michael, 16, 87–89, 112–13, 117, 186, 206
Morris, Robert Tappan, 63, 68, 74
Moscone Center West in San Francisco, 115
Moskovitz, Dustin, 212–13, 259, 300
Mountain View, CA, 85, 94, 107, 122, 133, 139, 149, 161, 172, 182, 235
"moving fast and breaking things," 237
Mozilla Foundation, 63, 233–35
Mulherin, Oliver, 275, 288, 295, 311
Murati, Mira, 225–26, 242–43, 246–48, 268, 282–93, 304–5, 307–8
Musk, Elon, 2, 5
  as an AI doomer, 2, 5, 163, 167–72, 214–15, 273–74
  children fathered by, 277
  colonization of Mars, 144, 147, 170–71
  DeepMind investment, 147
  departing OpenAI, 216–17, 221–24, 229, 234–36, 277, 304
  fighting aging, 257
  Halcyon Molecular, 257–59
  lawsuit against OpenAI, 14–15, 304, 313
  Neuralink, 193, 234
  at OpenAI's founding and beyond, 170–72, 183, 185–88, 192–94, 198, 213–14, 216–17, 221–24
  in politics, 207, 313
  relationship with Sam Altman, 16, 147, 170–72, 183, 194, 214–15
  SpaceX, 147, 153, 167, 170–71, 187, 194
  Tesla, 153, 167, 194, 223–26, 234, 255
  Twitter, 118, 158, 169, 246–47, 261, 270, 281
  X.com originally, 125
xAI, 15, 277, 310

INDEX

MyMobileMenu (MMM) food ordering system, 75
MySpace, 103, 116

Nadella, Satya, 216, 229–30, 237, 271–72, 273, 288–89, 293–94
Nakamoto, Satoshi, 142
nanotechnology, 144, 164
Napa Valley, 185, 190, 260
NASA, 132–33
NASDAQ, 97
Nathoo, Kirsty, 139
National Center for Missing & Exploited Children, 98
National Institute of Standards and Technology (NIST), 299, 301
National Security Council, 299, 301
"natively multimodal," 307
natural language processing, 174
*Nature* (journal), 148, 192
Nawrocki, Bob, 25–26
NBC Universal, 229
Nectome, 258
Neom, a megacity in Saudi Arabia, 231–32
Netflix, 204, 229
Neuman, John von, 140
neural networks, 147, 164, 176, 179–82, 190–93, 219, 221, 243, 314
Neuralink, 193, 234
NeurIPS (Conference on Neural Information Processing Systems), 178, 186, 305
neuroscience, 145–48, 202
New Enterprise Associates (NEA), 61–62, 74–75, 77, 80–81, 85–86, 89, 91, 92, 117, 121
New Hampshire, 297
*New York* (magazine), 134, 250, 261–62, 281
New York City, 10, 29–30, 43, 67, 86, 101–2
New York Stock Exchange, 163
*New York Times Magazine, The*, 139

*New York Times, The*, 29, 63, 132, 163, 217, 244, 270, 286, 288, 308, 312–13
*New Yorker, The* (magazine), 49, 81, 128, 152–54, 160, 165, 201–2, 259–60
New Zealand, 145, 171
Newhouse, Steve, 162
News Corp, 307
"newscasters," 73
Newsom, Gavin, 205, 309
Newton, Thandiwe, 199
Nextel, 58, 99, 105–6
1910 Genetics, 250
Ng, Andrew, 57, 90, 170, 266
Nicholls, GA, 23
Nobel Prizes in Physics, 219, 312
Noerper, Thelma, 25
Nolan, Jonathan, 199
nonprofit structure of OpenAI, 4–5, 14–16, 28, 172, 184–87, 198, 222–23, 233–34, 236, 266, 275
Nosek, Luke, 147
nuclear energy, 12–13, 57, 108, 134–36, 154, 177, 205, 230, 259, 280
nuclear weapons, 27n, 168, 213
nude celebrity photos leaked by hackers, 162
Nwankwo, Jen, 250

o1 ("Strawberry"), 284, 309
Oakland, CA, 195
Obama, Barack, 21, 172, 207–8, 266
Ohanian, Alexis, 69–70, 75, 81, 162
Oklo fission microreactor startup, 13, 137, 280
Olah, Chris, 178, 184
1billionmasks.com, 250
O'Neill, Megan, 28–29
"On the Dangers of Stochastic Parrots: Can Language Models Be Too Big?" (Bender and Gebru), 252–53
Open Philanthropy Project, 213–14, 241, 266–67, 276–77, 299–301

OpenAI
  the "blip" and beyond, 7, 14–15,
    276–94, 295–310, 311–15
  the case of *AI Dungeon*, 247–48,
    254–55
  charter of, 4–5, 7, 9, 233, 279, 291
  ChatGPT, 1, 3, 5, 14–15, 17, 208,
    253, 269–73, 276, 278, 285–86,
    308, 310
  DALL-E, 255, 262–63, 268
  Deployment Safety Board (DSB) at,
    279–80, 287
  development of application pro-
    gramming interfaces (APIs)s,
    245–48, 250–51, 264–65, 269,
    279
  effective altruism and, 189–214
  Federal Trade Commission investi-
    gation into data practices at, 285
  first commercial product OpenAI
    API, 250–51
  going from "AI training" to "AI
    alignment," 264–65
  GPT-5, 282, 307
  GPT-4, 3, 7–8, 12, 17, 266–69, 272–
    73, 278–80, 284, 287, 307
  GPT-1 and GPT-2, 241–43, 244,
    247, 252, 253
  GPT-3, 242–48, 250–53, 254–55,
    262–65, 272
  o1, 284, 309
  OpenAI Five, 216–17
  pivoting to a for-profit model, 215–
    20, 222–23, 233–37
  the problem of compute power,
    217, 221–22, 225–26
  the recent restructuring of, 295–
    310, 311–15
  research, training, and testing at,
    215–22, 262–64
  Superalignment team, 284, 305–6
  unusual nonprofit structure of, 4–5,
    14–16, 28, 172, 184–87, 198,
    222–23, 233–34, 236, 266, 275
  *see also* Altman, Sam

open-source technologies, 95, 186,
    214, 247, 263
optimism of Sam Altman, the, 1–4,
    15, 17, 30, 101, 103, 128, 136–37,
    141, 152, 154, 200
*Orange County Register, The* (newspa-
    per), 103
Ord, Toby, 211
O'Reilly, Bill, 102
O'Reilly, Tim, 63
*Outline, The* (news site), 206

Pachocki, Jakub, 216, 218, 284, 292,
    305, 308–9
Page, Larry, 85, 88
Paller, Amy, 33
Palm Beach County, FL, 40
Palo Alto, CA, 234
  PF Chang's in Palo Alto, 59–60
  Sand Hill Road, 79, 85, 93, 116–17,
    158, 177
  Tamarine Asian fusion restaurant,
    92
  Xerox's Palo Alto Research Center
    (PARC), 195–97
pandemics, 4, 177, 213, 260
Pao, Ellen, 162, 203
paperclip-making AI, parable of the, 5,
    143–44, 164
Paramount Global, 303
PARC (Xerox's Palo Alto Research
    Center), 195–97
parse trees, 174
patents, 214, *see also* intellectual
    property
Patreon self-streaming monetization
    platform, 138, 247
Pauley, Lucinda, 71
PayPal, 93, 125, 131, 136, 140, 147,
    161, 171
Pelosi, Nancy, 309
Pennsylvania, 28, 30, 40, 135
Pernikoff brothers, 51–52, 64, 65, 92,
    95, 97, 102, 109
Pew Research Center survey, 252

Phillips Academy in Andover, MA, 71–72
Phillips, Dean, 296–98
Phoenix, Joaquin, 307
phone makers, *see* mobile phones
Pichai, Sundar, 263, 271, 273
*Picture of Dorian Gray, The* (Oscar Wilde), 51
Piper, Kelsey, 306
pitchfests, 81, *see also* Y Combinator (YC)
Pitzitz family of Nicholls, GA, 23
Poe the chatbot, 278–79
*Politico* (news site), 299
politics, *see* United States
"polycentric law," 141
Potter, Fred, 111
"product-market fit," 107, 157
Project Gutenberg, 244
Pruitt-Igoe housing complex, St. Louis, 27
*Puck* (magazine), 297–98
Puerto Rico conference on AI safety (2015), 167–70, 207, 211

Qualcomm subsidiary, 120
Quora, 220, 234, 278

race and racism
  antisemitism in chatbots, 270
  first Black mayor of Chicago, 21–22, 33, 37
  Latino people and communities, 21, 35
  racism among tech bros, 164–65
  redlining, racially restrictive covenants, and slum clearance, 22, 26–27, 29, 40
  white flight, 22, 26, 32
  *see also* affordable housing; Black people and communities
Radford, Alec, 218–21, 241–43
Radiate (formerly Viendo), 76, 77–82, 85, 88–90, 93–94, 96–101
Raimondo, Gina, 273

Rain Neuromorphics AI chip startup, 232
Rainert, Alex, 118
Ralston, Geoff, 170, 235–36
Ramsay, Mike, 121–22
RAND Corporation, 267, 299, 300, 301
Rap Genius (later Genius), 133
Rashid Al Maktoum, Maktoum bin Mohammed bin, 274
"rationalism," 143, 176, 210, 212, 225, 264
reality distortion field (RDF), 61
Rechter, Joe, 27
Reconstruction, 26
recurrent neural networks (RNNs), 219
Recurse Center, 176
Reddit, 70, 76, 81, 160–63, 171, 203, 218, 241, 251–52
Reed Hastings, 204
Reform Judaism, 15–16, 24, 44, 31, 43, 227
reinforcement learning, 147, 185, 191–93, 264, 265, 284
*Renewing Cities* (Gittell), 35
rent strikes, 27
Republican Party, 21, 203, 207, 266, 278
RescueTime time-tracking app, 245
*Resident Evil* (video game), 97
Retro Biosciences, 257, 259
Reynolds, Malvina, 162
Rhodes, Elizabeth, 195
"RIP Good Times" deck, 117, 123
Rippling (startup), 250
Rizzo, Frank, 28
Roberts, Steve, 42–43, 53–54, 59
Roble, Jim, 202
Roedinger, Kurt, 46–47, 51–53
*Rolling Stone* article alleged gang rape at the University of Virginia, 204
Roosevelt, Theodore, 46
Rosewood resort, Menlo Park, CA, 177, 183

Ross, Blake, 56, 63
Rostenkowski, Dan, 34–35
Rowghani, Ali, 3, 158–59
RSS-feed format, 63
Rubik's cube–solving robot hand, 218
Rusenko, David, 157
Russell, Stewart, 305–6
Russia, 23, 27n
Russian Hill neighborhood of San Francisco, 86, 260, 293, 315
Rutkowski, Greg, 263
Ryan, Jim, 104

Sacks, David, 139–40
Safari web browser, 111
SAFE (Simple Agreement for Future Equity) convertible notes, 94–95
Safe Superintelligence Inc. (SSI), 309
Saga Ventures, 311
Salkin, Phil, 24
"Sam Altman for President," 150
"Sam Altman Is the Oppenheimer of Our Age," 281–82
Samuels, Lawrence, 40, 42
San Francisco
 Effective Altruism Global conference in San Francisco, 210
 Mission District, 7, 152, 260
 Moscone Center West, 115
 Russian Hill neighborhood, 86, 260, 293, 315
 University of California San Francisco, 202
*San Francisco Chronicle* (newspaper), 206
*San Francisco Standard*'s *Life in Seven Songs* podcast, 134, 198, 275
Sánchez, Pedro, 274
Sandberg, Sheryl, 98, 229
Sanyo flip phones, 80
SAP, 197
Saudi Arabia, megacity in, 231–32
Saudi Aramco, 232
Sauerberg, Bob, 160–63, 171
Saunders, William, 309

SBF (Sam Bankman-Fried), 212, 257, 276
Schmidt, Eric, 192
Schmidt, Steve, 296–97
Schoemehl Jr., Vincent C., 40
Scholz, Olaf, 274
Schulman, John, 184, 189, 226, 245–46, 265, 268–69, 310
Schumer, Chuck, 300
scientific breakthroughs, *see* innovation
sci-fi, 3, 43, 47, 140, 199–200, 214, 220, 307–8
Scott, Kevin, 225, 230, 271, 289, 296
SDKs (software development kits), 110–14, 121
Seasteading Institute, 141–42
Seattle, 31, 76, 110–11, 215, 267, 271
Seeger, Pete, 86, 162
Segways, 75, 210
Seibel, Michael, 194, 232–33
self-driving cars, 168, 223–24
Seligman, Nicole, 303
Selman, Bart, 167–69
semiconductor industry, 72–73, 132, 298
Sentient Networks, 87
sentiment neuron, 218
"Sequence to Sequence" (Sutskever), 219
"sequence to sequence" learning, 178
Sequoia Capital, 16–17, 66, 86–89, 93–97, 116, 100, 106, 112–13, 116–27, 134, 136, 138, 151, 155, 161, 193, 186, 206, 309
 AdMob mobile ad network, 121
 in China, 231
 Green Dot prepaid debit card company, 126–27, 133
 need for "adults in the room," 96–98, 217
 "scout" program at, 123–25, 157
serendipity, enhancing, 98
Series A investment round, 89, 101, 125, 136, 156, 158–59

INDEX

sex/gender, *see* tech bros; "women in tech"
Shazeer, Noam, 270–71
Shear, Emmett, 76, 81–82, 157, 293–94
*Shelley v. Kraemer*, 26
Sikka, Vishal, 197, 236
*Silicon Valley* (TV show), 101, 258–59
Silicon Valley, 2–8, 72–73, 258
  "add a zero" ethos of, 3–4, 272
  Bay 101 cardroom, San Jose, 57
  belief we are living in a computer simulation, 17
  economy of favor-trading, 14
  incestuousness of, 82
  University of California, Berkeley, 94, 111, 122–23, 167, 178, 194, 234, 277, 305
  value system of, 2–3
  "working on the weekends," 175
  zeitgeist of, 4, 6, 137–38
  *see also* Palo Alto, CA; Stanford University; tech bros; tech industry
Simon, Mel, 209
*Sims, The* (video game), 97
Singer, Peter, 211–12
singularity, 140–42, 144–45, 168, 199–200, 305
Singularity Institute for Artificial Intelligence, 142
*Singularity Is Near, The* (Kurzweil), 144
Singularity Summits, 144, 146, 210
Singularity University, an "educational community," 210
Sivo, Nick, 55–56, 60, 64, 85, 89, 91–92, 102–3, 113–14, 128
Skrenta, Rich, 244
"Sky," an AI voice, 307–8
Skype, 5, 144, 168
Slack channels, 245–46, 287, 293–94
*Slashdot* (news site), 63
*Slate Star Codex* (blog), 165–66, 213
Slee, Mark, 96

Slowe, Chris, 76, 81
slum clearance, 26, 34
Smashwords writing platform, 220
Smith, Barbara Ann (Barbie), 52
Smith, Dylan, 96, 289
social contract visionary, 195
social networking, 16, 59, 103, 113, *see also specific social network sites*
Socrates, 265
Sodol, Lee, 192, 216
SoftBank, 229
software development
  "agile" method of, 127
  application programming interfaces (APIs), 125, 245–48, 250–51, 264–65, 269, 279
  Brockman's "code caves," 175–76, 246–47
  hackers/hacking, 3, 57, 63, 68–70, 160n, 162
  Lisp programming language, 63, 69, 124, 151
  object-oriented programming, 195
  open-source technologies, 95, 186, 214, 247, 263
  SDKs (software development kits), 110–14, 121
  "waterfall" method of, 126
  whiteboards, 189–90
Sonsini, Larry, 85
Sony Entertainment, 303
Sony Ericsson phones, 77
SourceCom, 86–87
South by Southwest (SXSW), 118–19
South Carolina, 297
South Korea, 274
Southern District of New York, 290
Southern Pacific Railroad Internal Networking Telephony, *see* Sprint
Soylent meal replacement, 138
space, colonization of, 144, 147, 170–71
SpaceX, 147, 153, 167, 170–71, 187, 194
SPARK Neuro AI dementia detection startup, 275

Springer, Axel, 307
Sprint, 53, 58–59, 62, 78–80, 99–105
St. Louis, 15
   Clayton, a suburb of, 23–25, 27, 30, 32–33, 39–54
   "Del Mar divide" in, 26, 53
   Ladue suburb of, 26, 45
   Mill Creek Valley, 26–27
   Pruitt-Igoe housing complex, 27
*St. Louis Post-Dispatch* (newspaper), 40–41
*Stanford Daily, The* (newspaper), 56, 63
*Stanford Review*, 56, 131
Stanford University
   AI lab at, 90
   annual competition held by, 182
   BASES (Business Association of Stanford Entrepreneurial Students), 60
   courses at, 55, 131
   Education Program for Gifted Youth, 173
   electric vehicle project, 57
   endowment of, 155
   FroSoCo (short for Freshman Sophomore College), 55, 57
   Full Moon on the Quad, 92–93
   generous student leave policy, 91
   program in India, 64
   StartX startup accelerator, 90
   Summer Research College program, 57
   YC Startup School online course at, 122–23, 194, 234
startups
   bankruptcies, 31, 87–88, 145, 205
   discrimination in, 70
   the first venture-funded startup, 72
   founders as kings, 6, 60, 65, 68, 70–75, 87, 95
   normalization of starting a startup, 70, 150
   "RIP Good Times" deck, 117, 123
   Startup School online course, 122–23, 194, 234

unicorns, 150, 311
   *see also* investing; Sequoia Capital; Y Combinator (YC)
StartX, Stanford's startup accelerator, 90
Stiegler, Marc, 140, 199–200
Stone, Zak, 76, 77, 82
"Strawberry" (o1), 284, 309
Streit, Steve, 126–28, 134
Stripe, 4, 16, 123, 125, 136, 139, 150, 152, 157–58, 172, 175–76, 189–90, 246, 251
Stross, Randall, 133
StumbleUpon, 174
Suleyman, Mustafa, 146–47, 277
Summer Founders Program and Angel Day, 62, 64, 68–70, 75–82
Summers, Larry, 294, 303
Sun Microsystems, 230
Sun Valley Resort, 228–29, 263
Sunak, Rishi, 274
Superalignment team, 284, 305–6
superhuman machine intelligence (SMI), 170
superintelligence, machine, 143, 145, 164, 167, 170, 309, 315, *see also* AI (artificial intelligence)
*Superintelligence: Paths, Dangers, Strategies* (Bostrom), 5, 163–66, 167–68
Supreme Court, 26, 52
Sutskever, Ilya, 169, 178–88, 189–96, 209, 211, 217–23, 244–45, 266, 287, 282–94, 304–5, 308–10, 312–13
Swartz, Aaron, 63, 76, 81, 160n
SXSW (South by Southwest), 118–19
Szabo, Nick, 142

Tallinn, Jaan, 5, 144, 168, 300–301
Talve, Susan, 43–44
Tamarine Asian fusion restaurant, Palo Alto, 92
Tan, Garry, 236
Tana, Evan, 96, 102, 121

Tang, Nini, 95–96
taxes, 22, 34–35, 53, 206, 256
Tay the chatbot, 270
Taylor, Bret, 293
*Teach Yourself Visual Basic in 24 Hours*, 56
"teach-ins," 94
tech bros
    cosmic conversations of, 147, 260, 314–15
    hackers/hacking, 3, 57, 63, 68–70, 160n, 162
    love of Burning Man, 1, 198–99
    love of cars, 26, 51
    as mostly college drop-outs, 62, 88, 91–92, 96, 108, 113, 124, 132, 175, 179, 217–18
    racism among, 164–65
    sci-fi, 3, 43, 47, 140, 199–200, 214, 220, 307–8
    space colonization, 144, 147, 170–71
    techno-utopianism among, 205, 256–57
    WhatsApp group of tech CEOs, 289
    wrestling among the, 109
    *see also* cryptocurrency
tech industry
    Big Tech companies, 182, 185, 197, 299, 308
    dot-com bubble of the late 1990s, 73, 87, 93
    lobbying by the, 109, 144, 179, 274, 292, 300–301
    the massive valuations of tech companies, 71, 73, 138–40, 152, 290
    nanotechnology, 144, 164
    semiconductor industry, 72–73, 132, 298
    think tanks in the, 142, 144, 267, 299–300
    "women in tech," 42, 70, 203, 252, 271
    *see also* innovation; software development; startups

*TechCrunch* (news site), 161, 259
technological progress, *see* innovation
techno-utopianism, 205, 256–57
Teespring T-shirt-making startup, 156
Tegmark, Max, 145, 168–69, 208
telepresence, 210
Teller, Sam, 171
Temple Israel, 24, 25
*Terminator* film franchise, 200
Tesla, 153, 167, 194, 223–26, 234, 255
Test of Time Award, 305
TextPayMe, 82
The Game (rapper), 102
"TheFacebook," 60, *see* Facebook
Theranos, 108
Thiel, Peter, 1–7, 10, 16–17, 125, 134–48
    contributions to Trump presidential campaigns, 203–4
    data-mining company, 217
    Facebook and, 131, 204
    Founders Fund venture firm, 2, 6, 132, 139, 147, 226
    founding of the conservative *Stanford Review*, 56, 131
    *From Zero to One* (with Masters), 132, 230
    Halcyon Molecular, 257–59
    Helion Energy nuclear fusion startup, 13, 136, 207, 259, 280, 298
    his "Founders Fund" venture firm, 2, 6, 132, 139, 147, 226
    Inflection AI, 277
    PayPal, 93, 125, 131, 136, 140, 147, 161, 171
    personal beef with Gawker Media, 137, 204–5
    the pessimistic contrarianism of, 4, 131–33, 137–38, 204
    Saudi Arabia and, 231–32
    Thiel Fellowships, 132, 178
think tanks, 142, 144, 267, 299–300
Thole, Craig, 103
Thorpe, Meridith, 46

thought leadership, 89
Three Mile Island nuclear plant, Harrisburg, PA, 135
Thrive Capital, 161, 290, 310
Thrun, Sebastian, 144
*Time* magazine, 1, 112
Time Warner, 229
Tivo, 121
Tokyo, 79
Toner, Helen, 241–42, 266–67, 276–79, 282–88, 291, 292, 299, 309
Tools for Humanity, 256
Torres, Émile, 164
Track Trump website, 204
trademark issues, 101
transcendentalism, 46
"transformer paper, the," 218–19, 270, *see also* AI research/training
transhumanism, 143–45, 165, 166, 170
Trapp, Shel, 22, 35
Traynor, Bill, 35
Tripadvisor, 80
Trump, Donald, 40, 201, 203–4, 208, 296, 313
Tsai, Tommy, 96, 113, 114
Tulyasathien, Charnsin, 110
Tuna, Cari, 212
Turing, Alan, 173–74, 295
Turing Award, 142
Twitch, 82, 157, 292–93
Twitter, 118, 158, 169, 246–47, 261, 270, 281

Uber, 231
"Unified Theory of VC Suckage, A," 72–73
United Arab Emirates (UAE), 281, 298, 310, 313
United Nations Moon Agreement of 1979, 144
United Slate, The (online platform), 206–7
United States
 during the AIDS crisis, 33, 43, 49

Biden administration's AI policies, 232, 267, 277, 299–300
Defense Advanced Research Projects Agency (DARPA), 140, 196
Democratic Party, 21, 204, 207–8, 234, 296
Department of Commerce, 273, 277, 299, 301
Department of Defense, 299
Department of Energy, 196
Department of Homeland Security, 299
Department of Housing and Urban Development (HUD) redevelopment projects, 35
Federal Communications Commission (FCC), 59, 106
Federal Trade Commission, 285
the frontier in American history, 153
Government Accountability Office, 300
during the Great Depression, 24
Intelligence Advanced Research Projects Activity (IARPA), 267
investment blacklists from the, 231
loss of technological mojo since the 1960s, 135
Medicare for All, 206
NASA, 132–33
National Security Council, 299, 301
post–World War II industrial boom, 72
Supreme Court, 26, 52
taxes, 22, 34–35, 53, 206, 256
US Army, 24, 27n, 32
 *see also* US Congress
universal basic income (UBI), 12–14, 194, 205, 256
Universe project, a general AI agent, 191–93
University College London's Gatsby Computational Neuroscience Unit, 145

INDEX

University of California, Berkeley, 94, 111, 122–23, 167, 178, 194, 234, 277, 305
University of Virginia retracted gang rape allegations, 204
Upright Citizens Brigade, 228
US Congress, 301, 309
  CHIPS and Science Act, 298
  Computer Fraud and Abuse Act, 68
  Congressional Internet Caucus, 106
  House of Representatives, 34, 106
  law on location tracking on mobile phones, 58–59
  Senate, 2–3, 5, 273, 296–97, 300, 309–10
user experience, 79, 119, 195, 254
  average revenue per user, 79
  churn rates, 105
  content moderation, 254
  *see also* chatbots
utilitarianism, 2, 143, 211
"utilons," 212
Uygur population in China, 231

Valentine, Don, 87, 112–14
*Valleywag* (blog), 137, 205
"value-lock," 252
Valve, 215
*Venture Beat* (news site), 117
venture capital (VCs), 6–7, 8, 12, 60–62, 66
  "aligning our incentives," 256
  Andreessen Horowitz, 155, 158, 161, 257
  deal flow, 88, 95, 124
  history of, 72–73
  Khosla Ventures, 230, 235, 257
  Kleiner Perkins, 162, 230
  Matrix Partners, 90
  Mithril Capital, 136
  Peter Thiel's Founders Fund, 2, 6, 132, 139, 147, 226
  on Sand Hill Road, Palo Alto, 79, 85, 93, 116–17, 158, 177
  why they suck, 72–75, 77

*see also* investing; Sequoia Capital; startups; Y Combinator (YC)
Verbling video chat language learning service, 138
*Verge, The* (news site), 272
Verizon, 62, 100, 104–5, 110, 119
Viaweb, 68, 70, 73–74
*Vice News Tonight*, 205–6
Victor, Bret, 197
video games
  *AI Dungeon*, 247–48, 254–55
  Atari, 87, 147, 165, 190
  *Breakout*, 147
  *Dota 2*, 215–18, 221–22, 242, 284
  *Halo 3*, 95, 109
  *Resident Evil*, 97
  *The Sims*, 97
  *Warcraft III: Reign of Chaos*, 215
Viendo, 60, 61, 63–65, 74, 76
Viewpoints Research Institute, 198
Vinge, Vernor, 140, 145, 168
viruses, computer, 154, 200
viruses, Covid-19, 248–50
Visual Basic language, 56
Visual Studio development environment, 262
Vivarium Project, 209–10
VotePlz voter registration app, 203–4
"Voting with Your Feet: An Investigative Study of the Relationship Between Place Visit Behavior and Preference" (Potter and Howard), 111
*Vox* (news site), 306

*Wall Street Journal, The*, 7, 10–12, 17, 42, 57, 105, 124, 212, 229–32, 298
"walled garden," 110
Walmart, 126, 133
Walton, Nick, 246–47, 254–55
*Warcraft III: Reign of Chaos* (video game), 215
Washington, Harold, 21–22, 33–34, 37
*Washington Post, The* (newspaper), 232, 301

"waterfall" method of software development, 126
Watters, Nathan, 50–51, 201
*We Are the Nerds* (Lagorio-Chafkin), 161
WebGPT, 265
WebMind stock market trend prediction software, 145
WebText, 241, 244
Weebly, 157
Weiden, David, 89–90, 103–4, 108, 230
Weigend, Andreas, 90
Weil, Elizabeth, 281
Welinder, Peter, 247–48
West, Kanye, 102
Westrup, Evan, 206
*Westworld* (TV show), 199
Wevorce online divorce service, 138
"Where is Ilya?" meme, 304
white flight, 22, 26, 32
whiteboards, 189–90
Whitman, Walt, 46
Wiener, Scott, 310
WikiLeaks, 142
Wikipedia, 244, 252, 265
Wilde, Oscar, 51
Willett Advisors financial management firm, 155
Williams, Walker, 156
WilmerHale law firm, 303
Wilson Sonsini Goodrich & Rosati law firm, 85, 90–91, 93–94, 109, 124, 232
Winer, Lowell, 99–103, 105–6
*Wired*, 81, 138–39, 141–42, 192, 207, 218–20, 242, 251, 254, 310
wireless carriers, 58–59, 75, 79, 90, 98, 100–109, 110, 118, 120, 159, *see also specific telecom companies*
Wolfe, Sam, 24
Wolfram, Steve, 80, 81
"women in tech," 42, 70, 203, 252, 271
Wong, Hannah, 288
Wong, Yishan, 161–62

Wood, Evan Rachel, 199
Woodward, Bob, 296
World Transhumanist Association, 165
World War II, 24, 26, 72, 139, 196
Worldcoin cryptocurrency, 13, 256–57
WorldCom, 60
Worldwide Developers Conference, 111, 114
Wozniak, Steve, 272

X.com (Twitter), 118, 125, 158, 169, 246–47, 261, 270, 281
xAI, 15, 277, 310
Xerox's Palo Alto Research Center (PARC), 195–97
XPRIZE, 144, 210

Y Combinator (YC), 3, 6, 16, 62–63, 65–75, 77, 81–82
  Camp YC the annual glamping getaway, 177
  Condé Nast purchase of, 81, 160–62
  Continuity Fund, 159, 194, 232–33, 290
  Demo Days, 74, 133, 138–39, 155–56, 159, 258
  expansion of the batches (2011–2014), 138–40
  the first Summer Founders Program and Angel Day, 62, 64, 68–70, 75–82
  index fund, 138
  "Manhattan Project" for AI, 8, 145, 147, 219
  as a new Bell Labs, 140, 185
  SAFE convertible notes, 94–95
  scrappy ethos of "ramen profitability," 233
  Startup School online course, 122–23, 194, 234
  unicorns from, 123, 150
  as "woke and bloated," 232

YC China, 230–31
YC Cities, 231
YC Core, 194, 232
YC Group, 194
YC Research, 185, 187, 194–98, 207, 232, 236
YCVC program, 155
Yaeger, Larry, 196, 209
Yahoo!, 68, 80, 87, 101
Yam, Sam, 109, 121, 138
Yanisse, Daniel, 149–50, 156
Yeol, Yoon Suk, 274
Yoder, Shari, 121
Yoon, Diane, 283

YouTube, 89, 93–94, 96–97, 228
Yudkowsky, Eliezer, 1, 4–5, 7, 135, 140–46, 164, 168, 176, 210, 212, 264, 300
Yuen, Phil, 82

Zaremba, Wojciech, 184, 191, 218
Zefer consulting company, 62
Zenefits HR startup, 138–39, 156–57
"zero-shot" responses, 220–21
Zilis, Shivon, 234, 277
Zodiac Aerospace, 225
Zuckerberg, Mark, 6, 96, 98, 137, 212, 260, 276

# BLACK DIAMONDS

*Life and Work in Iowa's Coal Mining Communities, 1895–1925*

DOROTHY SCHWIEDER

*Iowa State University Press*
**AMES**

**For Elmer, David, Diane, and Lyle**

**Dorothy Schwieder** received her B.A. from Dakota Wesleyan University, M.S. from Iowa State University, and Ph.D. from the University of Iowa. She has taught history at Iowa State University since 1969 and is a coauthor with Elmer Schwieder of *A Peculiar People: Iowa's Old Order Amish*. Her most recent publication is as contributing coauthor of *Iowa's Natural Heritage,* published by the Iowa Academy of Science and the Iowa Natural Heritage Foundation.

© 1983 The Iowa State University Press. All rights reserved

Composed and printed by
The Iowa State University Press
Ames, Iowa 50010

No part of this book may be reproduced in any form, by photostat, microfilm, xerography, or any other means, or incorporated into any information retrieval system, electronic or mechanical, without the written permission of the copyright owner.

First edition, 1983

---

Library of Congress Cataloging in Publication Data

Schwieder, Dorothy, 1933–
   Black diamonds.

   Bibliography: p.
   Includes index.
   1. Coal-miners—Iowa—History.  2. Coal mines and mining—Iowa—History.
3. Coal trade—Iowa—History.
I. Title.
HD8039.M62U6682   1983   331.7′622334′09777   82-23412
ISBN 0-8138-0991-6

# CONTENTS

FOREWORD   vii

INTRODUCTION   ix

1. American Coal Mining in the Nineteenth Century   3
2. The Mining Process   27
3. Life above Ground: The Italian-American Family in Iowa   59
4. Life above Ground: The Roles of Italian-American Women   86
5. Populations of Three Iowa Coal Mining Communities   112
6. United Mine Workers in Iowa   126
7. Perspectives   157

APPENDIX   177

NOTES   179

SELECTED BIBLIOGRAPHY   193

INDEX   199

# FOREWORD

IN 1971, Richard Kraemer, manager of Iowa State University Film Production Unit, was going to preview some of the footage that was later incorporated into the award-winning film "The Last Pony Mine." Thinking it would be of interest to historians, he called the Iowa State University History Department to see if anyone there would care to sit in on the viewing. Dorothy Schwieder accepted the invitation. Since that day, she has been engrossed with the history of Iowa coal mining.

Coal production reached its peak in 1918. That year 9,049,806 tons of coal were produced. This is the energy equivalent of over 40 million barrels of oil. It was an economic activity worthy of recognition. Of even greater importance was the impact coal mining had on the people involved, and they were a significant number. In 1910, over 18,000 people were digging coal. They and their families constituted 28 percent of Iowa's nonrural population.

Iowa is so dominated, economically and socially, by its agriculture, that it is not surprising that the history of coal mining has been neglected. Almost too late, Dorothy Schwieder became aware of this neglect.

To capture the intensity of the human experience of life in a coal camp, it is necessary to talk with those who lived it. Their number is dwindling rapidly. It was also a rare opportunity for historian Schwieder—an opportunity to record a significant historical episode by interviewing people who lived it.

It is a fascinating story—one that needed to be told and she tells it well.

Marvin Ross<br>
Iowa State Mine Reclamation Officer

# INTRODUCTION

THROUGHOUT the late nineteenth and early twentieth centuries, thousands of settlers arrived in Iowa, some seeking the state's rich farm land, others seeking the wealth that lay beneath the soil. Like the agriculturists, coal miners came from every part of the nation and from many different foreign countries. This book is about the people who chose to emigrate to Iowa not to become involved in agriculture but rather to earn their living mining the "black diamonds" that lay beneath the earth's surface.

While Iowa's agriculturists have been the subject of literally hundreds of economic and historical studies, Iowa's coal miners have been virtually ignored. With the exception of one general historical study and a dozen or so journal articles and technical studies, most research has centered around the geological aspects of the coal reserves. The lack of historical interest is particularly puzzling given the economic importance of the industry. During the late nineteenth and early twentieth centuries Iowa was primarily an agricultural state. Overall, few industries operated in the state and the ones that did were mostly located along the Mississippi River. Coal mining was the exception to the rule. It constituted a major industrial operation, the largest in the state.

The first written record of coal deposits in Iowa came in 1835 when a United States Geological Survey team descended the Mississippi River between Dubuque and St. Louis. When the survey team reached the Rock River, a man we know only as Featherstonhaugh reported bituminous coal deposits on both sides of the Mississippi River. A year later the federal government sent Lt. Albert Lea to investigate the recently acquired Black Hawk Purchase. In his report Lea noted large coal deposits of excellent quality along the Des Moines River near the Raccoon Fork.[1]

In the 1840s operators opened numerous small mines in eastern and southern Iowa. Samuel Knight in 1840 operated a coal mine at Farmington in Van Buren County, and the company maintained a wagon trade to Keokuk, roughly thirty miles away. With the development of steamboating on the Des Moines River, the mine's output increased rapidly. Operators opened several coal mines at nearby Douds to provide additional coal for the Des Moines steamboat trade that operated between 1837 and 1862. Early Iowa residents used coal for heating their homes as well as for maintaining local industries such as blacksmithing and pottery making. Limited coal mining operations were also carried on in Scott, Muscatine, and Jefferson counties between 1850 and 1900. Most of the mines in eastern Iowa were seasonal in that they either provided fuel for heating homes and businesses or provided fuel for operations such as riverboat trade.[2]

The first major development of coal mining within the state took place in Des Moines. Wesley Redhead, who emigrated from England as a young boy, arrived at Ft. Des Moines in 1851 to work in a clothing store. By 1864 he had become interested in the possibility of developing coal mines in the Des Moines area and along with a dozen associates organized the Des Moines Coal Company. After surviving several disappointing years, Redhead and associates finally reached a four and one-half foot vein of coal in 1873. They named the mine the Black Diamond (later renamed the Pioneer), and within three years the mine employed 150 men and produced two hundred tons per day. Operators opened other mines in and around Des Moines, soon making Polk County one of the largest coal producing counties in the state.[3]

The Whitebreast Fuel Company established the second major coal mining operation in the state when it opened mines in Lucas County in the early 1870s. William Haven, a native of New Hampshire and a graduate of Norwich University, founded the company. Trained as an engineer, Haven worked for some railroads in Illinois before settling in Iowa. The company, capitalized at $150,000, leased a tract of land west of Chariton on Whitebreast Creek and began drilling. The first mine developed rapidly and by 1880 the number of miners reached 360. Equipment in the mine included forty mules, stabled permanently underground. Eventually the company opened mines at Forbush, Keb, Pekay, Flagler, Swan, and Chisholm. The Whitebreast Company continued as a major Iowa coal mining corporation well past the turn of the century. Their last major mines operated at Forbush and Rathbun and continued until shortly before World War I.[4]

Between 1870 and 1885 operators developed coal mines in all

parts of the state's coal fields. In Boone County, for example, local operators opened a mine near Moingona in 1874. The mine found a ready market for its coal through the Chicago and North Western Railroad and prospered until the turn of the century. Two years later T. N. Canfield and C. S. Taylor opened a slope mine west of Boonsboro, marking the first shipping mine in Boone County. Within a short time the mine became the property of the Chicago and North Western Railroad. During the mid-1880s operators opened mines at Incline and Angus. Angus grew rapidly and by 1885 was the largest coal mining town in Iowa, with 3,500 inhabitants. Eventually nine coal companies operated in the Angus area. By 1896 the Iowa Geological Survey reported that Boone County contained more than twenty mines.[5]

Not until Iowa's railroads finished laying tracks across the state in the late 1860s did Iowa operators enjoy a steady demand for coal. In 1867 the Chicago and North Western Railroad completed its line to Council Bluffs with the Rock Island and Burlington railroads following two years later. In 1870 the Illinois Central Railroad reached Sioux City. The completion of the major railroad routes across the state produced two types of demand for coal: the constant demand required by railroads and the seasonal demand required by householders who used coal for fuel. Many Iowa railroads followed the practice of purchasing mineral rights and developing their own mines, popularly referred to as "captive mines." The Consolidation Coal Company at Buxton, owned by the Chicago and North Western Railroad, was the largest captive mine in Iowa.[6]

After 1870, as railroad companies continued to expand their mileage in Iowa, the coal industry grew at a corresponding rate. Between 1870 and 1925 it became increasingly apparent to coal officials that the well-being of the Iowa coal mining industry was closely tied to the policies and practices of Iowa's railroads. By 1895 state mine officials estimated that 342 coal mines operated in the state with a total of 6,863 miners. While the number of mines remained fairly constant over the next thirty years, the number of miners almost doubled. The industry reached its peak in 1917 when wartime demands provided Iowa's coal miners with full employment. By 1925 Iowa's coal mines totaled 354 with 11,241 miners employed.[7]

The presence of the coal industry in Iowa strongly affected the state's economic and social history. Many foreign-born people settled in Iowa because of industrial opportunities provided by coal mining, and their presence contributed to the state's ethnic pluralism. Furthermore, the coal industry accounted for Iowa's second major industry, ranking only beneath agriculture. Iowa coal operators supplied the fuel used to heat the homes and businesses of Iowans; they also provided energy to power the state's limited number of mills and factories. The industry produced the bulk of coal used by Iowa railroads well into the 1920s. Because miners moving into Iowa brought along their zeal for trade unions, the United Mine Workers (UMW) flourished in Iowa. The miners' union stands as the largest trade union in Iowa's history.

This is a study of Iowa's coal mining population between 1895 and 1925. The period encompasses the most prosperous years of the Iowa coal industry. This book will concentrate on the work roles of Iowa coal miners and on the social and economic activities of coal miners' families, particularly the activities of miners' wives. For the most part, coal mining in Iowa was a family centered occupation. The result is that the present work is a study of individuals within an industry rather than a study of the industry per se. The population of three coal mining communities—Beacon, Cincinnati and Seymour—will be analyzed in terms of their social and economic composition in 1895 and in 1915. In addition, the union activities of Iowa miners will be examined.

In the introduction to his book on gold mining in the Black Hills, Joseph Cash acknowledges the importance of corporate influences but insists that "the basic story concerns the workers—individually and collectively. Where they came from, why they came, what they did, how they lived, and how they worked are basic to the study of mining in the West. The corporation, with all its skills and resources, was helpless without them. They dug the gold."[8] The same observation holds for this study. Thousands of men and women extracted Iowa's coal over a generation. This is their story.

I am grateful to many people for assistance in preparing this book. First, although it is impossible to mention individuals, I would like to thank the many former coal miners and their families who graciously consented to be interviewed. Their help was invaluable. I am also grateful to Marvin Ross, state mine inspector, and Richard Kraemer, former director of the Iowa State University Film Production Unit. They share the responsibility for my initial interest in the subject of coal mining. Marvin Ross also aided my work by helping to locate both public and private coal mining materials. John Kolp, University of Iowa, provided much-needed help and support with the quantitative portion of the study. Lyle Risius also assisted with computer work. Greg Sanford, Iowa State University history graduate student, spent many hours coding census data, and Laura Helmers, manuscript typist for the History Department, typed the manuscript and eliminated many technical errors. Malcolm Rohrbough and Linda Kerber, professors of history at the University of Iowa, deserve special credit for the many hours they invested in reading and rereading the manuscript during its dissertation stage. Their insightful comments and their own fastidious scholarship aided immensely throughout all stages of the study. Staff members at the Iowa State Historical Division of Museum and Archives provided helpful assistance, particularly the staff of the Census Division. Most of all I wish to thank my husband, Elmer, for his unlimited patience and encouragement. He accompanied me on many interview trips and offered unending counsel and support during the entire research and writing period.

**BLACK DIAMONDS**

# 1

## AMERICAN COAL MINING IN THE NINETEENTH CENTURY

IN the opening chapter of *Coal and Unionism*, David McDonald and Edward Lynch observe that "the history of the United States might be traced by the development of its coal industry, from an agricultural state to an industrial nation."[1] Their statement provides a useful departure point from which to survey the development of the coal industry in nineteenth century America. During the eighteenth century America was overwhelmingly agricultural; energy needs were simple. Colonial Americans from their earliest settlements utilized the ample stands of timber that grew along the eastern seaboard as their major source of fuel. In the nineteenth century, however, major economic changes took place within the United States that demanded a new and more abundant source of energy. Americans discovered that energy in the form of coal. The first large-scale demand for coal came from the nation's rapidly developing transportation industries: steamboats in the first half of the century and railroads in the last half. Both demanded a steady and uniform supply of coal.

After the Civil War large scale industry developed at a rapid rate; by the 1890s the United States had become the leading manufacturing nation in the world. With adequate coal reserves, a knowledge of coal mining technology, and a sufficient labor force, the nation's coal industry provided an unlimited supply of energy to make possible the burgeoning industrial and transportation sectors of the economy. Responding to the rapid expansion of industrialism, an increasing number of foreign-born people streamed into the cities of the Northeast seeking employment. At the same time, thousands of people moved into the Trans-Missouri West to settle the western portion of the country. At first eastern coal fueled the railroads across the Great Plains; soon, however, local entrepreneurs developed coal fields in Kansas, Colorado, Utah, and Wyoming. By the 1890s all major coal fields in the Middle West, West, and Far West had been opened and were in full production.

The emergence of the coal industry helped to remake the face of America. As the major form of available energy, coal signaled crucial economic and social changes: it signified the change from an agricultural to an industrial nation; the change from a rural society to an urbanized society; and the change to a more ethnically diverse population. For the first two decades of the twentieth century, coal remained the indispensable fuel. After 1920 other forms of energy such as oil, natural gas, and electricity gradually replaced coal. But for the nineteenth century, the dominance of coal as fuel marked the passage of one America and the emergence of another.

The records of early American coal mining activity are scant. It is generally agreed that the French explorers Marquette and Joliet were the first white men to discover coal in what is now the United States when they made their famous exploration down the Mississippi River in 1673. Roughly seventy-five years later the first recorded coal mining in colonial America took place in the vicinity of Richmond, Virginia. Local residents probably used coal for heating in homes and shops and for firing the blacksmith's forge. Within twenty years, coal had come into general use throughout the Richmond area. By 1789 these producers shipped small amounts of coal to the ports of Philadelphia, New York, and Boston.[2]

The areas around Pittsburgh and Wilkes-Barre, Pennsylvania, represented the next coal districts to be explored. Developers first discovered anthracite coal in 1762 around Wilkes-Barre and eventually that area became the most important and largest field in the anthracite industry. Mining began in the Pittsburgh vicinity around 1760. Pittsburgh residents started to use coal for fuel in the early nineteenth century, and local dealers delivered the coal to residents at $1.25 per ton. In 1803 Pittsburgh operators sold their first coal out of the vicinity, and by 1817 they made shipments down the Ohio River to the ports of Maysville, Cincinnati, and Louisville.[3]

The western Maryland counties of Garrett and Allegany marked the third major coal areas to be developed, with miners located around the town of Frostburg, Lanaconing, and Cumberland. By 1820 the Maryland coal trade had become commercially significant. Maryland operators faced the usual problem of shipping coal out by water; they found transportation limited to those periods of high water on the Potomac. In her study of the Maryland coal fields Katherine Harvey describes the coal transports as "usually flatboats eighty feet long, thirteen feet wide, and three feet deep. Each could carry 1,500 to 1,800 bushels of coal. . . ."[4]

Even though these areas gradually expanded their production, the nation's total output of coal did not increase substantially until after 1825 because the main coal deposits were located away from the seaboard area where the majority of people resided. Most coal mined before 1825 was consumed locally. Overall, the demand for coal remained low; most Americans used wood for household heating. As wood became scarce through the midnineteenth century, householders throughout the eastern states gradually shifted to coal. By 1845 steamers on the lower Great Lakes had switched from wood to coal; steamboats on the Ohio River began using coal in the early 1850s. In 1873 Chicago officials noted that residents used little wood because they could purchase coal more cheaply.[5]

Initial coal mining efforts involved only the simplest technology. In the early nineteenth century coal operators looked for seams near the surface or for outcroppings of coal. Once workers located a coal seam, they proceeded to quarry the coal or to follow the drift back into the hillside. Consequently, operators had little need for deep shaft mining before 1860. Because of the soft nature of bituminous coal, most early miners relied on their picks, wedges, and sledges rather than blasting powder. The miners typically cut out a section from under the coal seam and then wedged the coal down with their tools. Early miners used powder only if the coal could not be broken off manually. In the first half of the nineteenth century, coal operators insisted that miners not only dig and load coal, but also deliver it to the mine opening. Miners in the Monongahela River area first used wheelbarrows to move the coal but then began training dogs to assist them. They harnessed the dogs to the front of the small coal cars while they pushed from behind. As the mining operations became more systematic, the miners discontinued the use of dogs and substituted mules that provided greater pulling power.[6]

The continual improvement of transportation facilities throughout the midnineteenth century had a significant influence on the coal industry. Before 1850 coal operators depended primarily on water transportation. After 1850 railroads provided the major means of transporting coal. Ohio coal producers benefited considerably from the development of canal facilities during the 1830s. Developers completed the Hocking River valley branch of the Ohio Canal from Columbus to Nelsonville in 1832, providing a transportation outlet for Hocking valley coal operators. Coal producers in the Tuscarawas valley in Ohio shipped coal to Cleveland following the completion of the Ohio Canal to Massillon in 1834. A few years later operators shipped Tuscarawas valley coal to Chicago and other areas by way of the Great Lakes. In 1844 engineers constructed locks on the Monongahela River enabling coal producers in the Pittsburgh district to ship additional coal. By midcentury the Chesapeake and Ohio Canal between Cumberland and Georgetown, Maryland, was completed providing cheaper and faster transportation facilities for Maryland coal operators. Bituminous coal production for the United States in 1840 totaled 1,103,000 tons, of which 94 percent came from Pennsylvania, Virginia, and Ohio. Mines in Missouri, Iowa, Alabama, and Maryland produced the remaining 6 percent.[7]

As developers built railroads across the Middle West in the fifties, sixties, and seventies, demand for coal greatly increased. In Illinois, for example, miners first dug coal in the 1830s but production did not become significant until 1850. In that year the federal census reported Illinois coal production at 300,000 tons. By 1855 the Illinois Central Railroad had been completed across the state and a short time later local officials began shaft mining at DuQuoin in Perry County. As railroad demand for coal increased, mines opened at Alton, Kingston, Rock Island, Danville, Braceville, and Braidwood. By 1879 Illinois' total coal output reached 2,634,163 tons, with the Belleville field — located near St. Louis — producing 798,810 tons of that total.[8] (See Table 1.1.)

The Civil War gave a strong impetus to the coal industry. Demand for coal remained high throughout the war as Americans util-

Table 1.1. Production of bituminous coal in Middle Western states (in tons)

|          | 1840    | 1850    | 1860      | 1870      |
|----------|---------|---------|-----------|-----------|
| Ohio     | 140,536 | 640,000 | 1,265,600 | 2,527,285 |
| Indiana  | 9,682   | 60,000  | 101,280   | 437,870   |
| Illinois | 16,967  | 300,000 | 728,400   | 2,624,163 |
| Missouri | 9,972   | 100,000 | 280,000   | 621,930   |
| Iowa     | 400     | 15,000  | 41,920    | 263,487   |
| Michigan | ...     | ...     | 2,320     | 28,150    |

Source: *United States Bureau of Mines, Reports.*

ized the mineral in the manufacture of firearms, ammunition, other types of armaments, and naval vessels. The price of coal rose, encouraging operators to open new mines. When many coal miners joined the Union army, a shortage of trained miners occurred, and miners' wages rose "three or four times their former value." One solution to the miners' shortage came with the emigration of many British miners to the United States.[9]

Following the war, production of bituminous coal increased rapidly. In 1860 U.S. production totaled 6,494,200 tons, and by 1876 the output more than doubled to 13,352,400 tons. Pennsylvania operators in 1860 produced the largest amount, with Ohio second and Illinois third. By 1866 all three states more than doubled their 1860 production. After the Civil War the western states of Colorado, Wyoming, and Kansas also began producing coal, although they continued to be the smallest producers for several decades. By the late nineteenth century coal operators and consumers commonly viewed the nation's coal fields as divided into seven major geographical regions: anthracite (eastern Pennsylvania), eastern or Appalachian, middle western, western, southwestern, Rocky Mountain, and Pacific Coast. This list ranks their location from East to West, as well as the commercial value of each field, going from the most profitable to the least profitable.[10]

The American coal industry was strongly influenced by British coal miners throughout much of the nineteenth century. In many ways British miners provided both a social and economic model for their American counterparts. Coal mining had become a well-established industry in Britain by the mideighteenth century. After 1825, when coal mining in the United States became more extensive and systematic, an ever-increasing number of British miners began to emigrate here. By 1850 British miners had become strong advocates of trade unions, a conviction that they continued to support in the United States. British miners also brought along their technological know-how about the methods of mining coal. Although conditions in British mines improved during the 1840s, British miners continued to voice many grievances. With their emigration, British miners carried their fight for shorter hours, better pay, and improved working conditions into the American coal fields.

While small numbers of British miners emigrated to the United

Miner loading coal car in a longwall mine. (Courtesy of Iowa Mines and Minerals Department, Des Moines)

---

States during the early nineteenth century, the number greatly increased in the decade of the 1850s. The discovery of gold in California initially attracted some emigrants; others left Britain because of the failure of the Miners' Association in 1847, brought on by the depression of that year. The United States Census of 1860 reported that 37,523 miners entered the country between 1851 and 1860, although it does not specify country of origin or type of miner. Andrew Roy and Edward Wieck believe that British coal miners constituted the great majority of this group. For many years coal miners identified specific areas with particular nationalities; the Mahoning valley in Ohio attracted mostly Welsh-born miners, while most miners in the Steubenville area came from the north of England.[11]

The British brought to the United States their zeal for trade unions, and soon many became involved in attempts to form miners' unions here. Wieck traces that British influence through the names of the various American unions. Two terms used frequently in American coal fields — association and lodge — were terms common to British

## COAL MINING IN THE NINETEENTH CENTURY

miners. Furthermore, the so-called benevolent association, used by many American mining groups, appeared commonly in the British coal fields. British miners used this term in Britain as a means of protecting themselves against discrimination when feelings against unions ran strong.[12]

British miners also carried technical practices to the United States. The major methods of mining used in Britain—*room and pillar* and *longwall*—became major mining methods utilized in the United States. With the room and pillar method, large columns of coal were left intact to serve as supports; with the longwall method, miners removed all coal. Supports were provided with wooden props and dirt packwalls. Americans made the same distinctions between workers as did the British. British operators referred to men who actually mined coal as *hewers* or *diggers* while they referred to men who laid track or operated motorcars as *mine employees*. Coal operators and miners alike regarded the coal digger or miner as the most skilled worker and as the one who made the most money. Throughout most of the nine-

Underground view of the Riverside Coal Mine illustrating an Iowa coal seam and use of props. (Courtesy of Iowa Mines and Minerals Department, Des Moines)

teenth century, mining officials on both sides of the Atlantic praised the British miner for his skill with a pick. Both British and American operators paid their workers on a tonnage basis; in other words, operators paid for the amount of coal that miners loaded out, rather than for the hours that they worked.[13]

Several institutions in American coal mining communities also followed the British model. Well into the twentieth century many British miners and their families lived in isolated, dreary mining villages. Government officials viewed the miners' isolation as a serious drawback. The isolation minimized the miners' social activities and their children's educational opportunities. Isolation also affected occupational preferences. Because miners' sons had little opportunity to move elsewhere, they usually followed their fathers into the mines. Because of this isolation, British coal operators often provided housing and other community services for their workers. Eventually miners demanded an end to these practices, but without great success. During the nineteenth century British miners organized friendly societies to aid the injured, unemployed, and bereaved. British coal miners frequently attempted to form trade unions as a means of resisting wage reductions, winning insurance benefits, and instigating safer and healthier working conditions. British operators paid miners monthly or every month and one-half. Understandably, the miners complained incessantly that they found it difficult to avoid borrowing money or requesting credit from the company store.[14]

American coal miners shared the stigma of social inferiority carried by their British contemporaries, although perhaps to a lesser degree.[15] In a letter to the editor of the Belleville *Democrat* in 1863, an Illinois coal miner, writing in behalf of the newly formed American Miners Association, noted that the association contained men of talent as well as men of capital and that "at some future period of history of this, our glorious country of adoption, [that fact will] astonish those that look upon the Coal Miner as if he was one of the most degraded of human beings upon the face of the earth."[16] American coal miners had early been characterized as improvident and intemperate. A popular aphorism applied to nineteenth century miners was that they "live hard, drink hard and die hard."[17] Coal operators frequently referred to coal miners as men who were "good spenders" and proud to be known as such. Operators insisted that the habit of "spending freely" had long been considered a virtue in many mining communities and one that miners maintained to insure their popularity. The typical American coal miner, like his British counterpart, lived in an isolated mining village in which the company controlled his life. Consequently, coal miners regarded company housing and company stores as the bane of their existence.[18] Like the British coal miners, American miners faced many hardships in the pursuit of their work. Most coal mining areas were scattered and located in remote areas so there was little opportunity for men to exchange ideas about their work or their grievances. Conditions dictated that miners and their families move often. Along with working in an industry in which small mines experienced a high rate of failure, coal miners had to accommodate themselves to the seasonal nature of mining. Before 1860 the industry's seasonal conditions appeared to dictate most coal mining activity. Given the heavy reliance on water routes for transportation and home heating as the major use for coal, miners had to scramble continually from mine field to mine field to take advantage of the seasonal work. Local observers sometimes referred to coal miners as "birds of passage" because of their need to move several times each year. Coal miners discovered that where operators shipped by water, mining was done in the spring and fall. On the other hand, workers mined coal shipped on the Great Lakes during the summer and dug coal used for heating purposes during the winter. The persistent pressure to move from coal field to coal field led to two difficulties from the miners' point of view. First, it contributed to the view that coal miners as a class were unstable, shiftless, and improvident. Second, the miners found it more difficult to organize themselves into local labor unions.[19]

Even before the Civil War, American coal miners began to organize to protest low wages and poor working conditions. The Monongahela valley miners struck in 1848 to protest a wage reduction. The men had been receiving $0.02 per bushel (80 pounds) but the operators proposed to lower the price to $0.0175. Within a short time the strike failed and the men returned to work at the lower wage. In 1859 the miners again walked off their jobs to force the operators to install scales to weigh the coal. The miners protested that measuring coal by the bushel was highly inaccurate. Eventually the strike affected most of western Pennsylvania; within a few months the miners returned to work, defeated. During the early 1850s miners in the Cumberland field in Maryland struck to protest the lowering of their wages.

They also expressed unhappiness over the coal operators' requirement that they sign a bond forfeiting all wages in the event they engaged in a strike. Moreover, the miners complained that the operators paid them only once a month and then not for work just completed, but for work done during the second preceding month. The company always owed each miner at least one month's wages. Several years later, in 1854, the miners called another suspension to protest a proposed wage reduction of $0.10 per ton.[20]

The men also had many grievances about their working conditions. Before 1860 the average mine employed 25 miners. Operators used steam in only a small number of mines; most workings relied on gin hoists and pumps powered with horses or mules. Ventilation posed one of the greater health and safety problems, and few operators provided adequate ventilating procedures. In the early days most mine operators ventilated their mines with furnaces. Each mine had a single shaft opening that went down to the coal seam and then divided into two sections. One section contained the furnace, while workers used the other to haul miners in and out of the mine and to bring coal to the top. The furnace ventilation system created many problems. The furnace might overheat and ignite the shaft structure, or the furnace tender might fail to keep the fire properly fueled, thus decreasing the air flow through the mine. In 1869 an overheated furnace ignited the wooden shaft structure at Avondale, Pennsylvania, and 109 miners suffocated. Public indignation in the 1870s over several mine disasters such as the Avondale incident combined with the repeated protests of the coal miners led several states to pass laws that provided for two mine openings, minimum standards of ventilation, and the appointment of mine inspectors.[21]

Although the miners demanded safer working conditions, their major grievances were economic. Miners constantly protested two practices that they believed deprived them of their full tonnage. The first practice, the short weighing of coal, reflected the miners' belief that company officials frequently failed to give them credit for all the coal they loaded out. To correct the problem, miners had long suggested that a checkweighman be hired, at the miners' expense, to weigh coal alongside the company weighman. The second practice, excessive dockage, involved the operators' insistence that all coal be screened before being weighed. Operators contended that they could not sell the small chunks of coal and so they would not pay the miners for these. The miners charged that not only was the practice of screening coal unfair in that it deprived them of their full tonnage, but the system was also subject to abuse. First they claimed that the size of the mesh was not uniform; second, they argued that operators placed spreaders on the screens that broke up the coal, forcing even more coal through the screen. The miners proposed that to solve these problems, operators should discard the screens and weigh all the coal loaded. The screening process allowed the operators considerable leeway in wage negotiations. Reduction of miners' wages constituted the major way that operators cut back on expenses yet the operators could also leave the prevailing wage intact but enlarge the size of the screens to decrease the amount of weighable coal.[22]

A third major complaint concerned nonpayment for *dead work*. Dead work covered any work done by the miner other than the actual mining of coal. When he arrived in the morning, the miner frequently had to bail water from his room, lower the floor, remove the ceiling, prop the roof, or lay track. Miners argued that they should be paid for this work because it often took a considerable part of their day; operators insisted that the tonnage rate included payment for such work.[23]

Almost from the beginning of the American coal industry, miners attempted to organize miners' associations. The first local union of coal miners came into existence in the anthracite region of Pennsylvania in 1849. Miner John Bates, an Englishman, organized the local and presented the men's demands for higher wages and better working conditions. Bates soon called the first coal strike in the United States; after several months, however, the men admitted failure and returned to work at the old wage. For the next eleven years there is no evidence that coal miners attempted to organize unions. In 1860 the anthracite miners in Pennsylvania again attempted organization when the miners of the Forrestville Improvement Company formed a local union. Andrew Roy reports that other local unions also organized in the region during the Civil War.[24]

In January 1861 coal miners made their first attempt to form a national union. At that time miners in West Belleville, Illinois, met and organized the Miners' Association. A short time later, they began publication of the *Weekly Miner,* the official publication of the association. Poor economic conditions stemming from the recession of 1860 precipitated the formation of the union. In late 1860 operators in St. Clair County, Illinois, announced a wage reduction of $0.0025 per bushel. A few weeks later operators again announced a similar reduction. The miners reacted to the wage cuts by walking out. Several days later the striking miners organized the Miners' Association to devise a strategy for maintaining the strike. The miners' union first demanded that the operators restore the wage reductions and second, that operators eliminate their method of determining tonnages by measurement. Instead, the miners insisted that the operators install a system of scales so that they could determine tonnage rates by weight. Union members scored an early victory in 1861 when they convinced the Illinois state legislature to pass a bill that required all coal be weighed to determine tonnage. Following passage of the bill, known popularly as the Miners' Bill, operators restored the wage reductions and the miners returned to work. Encouraged union officials began to visit other Illinois coal camps urging men to join the Miners' Association. By the end of March 1861 President Daniel Weaver reported that the association had " 'upwards of five hundred members.' "[25]

The association prospered throughout the Civil War. In September 1862 the St. Clair County miners demanded and received a wage increase of $0.0050 per bushel. A year later the organization began to expand beyond the state of Illinois, taking in members from other states. In 1863 union leaders expressed optimism over the organization's ability to maintain the right of workingmen to organize and to win a wage increase for the miners. Immediately following the end of the war, the association reached its peak membership of 22,000.[26]

In 1865, however, the Miners' Association faced several seemingly insurmountable difficulties. The union suffered its first reverse early in 1865 when operators in the Blossburg District in Pennsylvania succeeded in breaking the union organization there; the breakdown came after a five months' suspension. In September several union officials urged the organization to replace John Hinchcliffe, editor of the *Weekly Miner*. Other officials disagreed and the leadership split over the issue. At the same time, the union was negatively affected by a slowdown in the mining trade after the Civil War.[27]

The dissolution of the American Miners' Association did not mark the end of all union activity. Local unions or lodges continued to exist in many coal mining states. In Pennsylvania, Maryland, Ohio, and Indiana the miners formed local organizations known as the Miners' and Laborers' Benevolent Associations that included surface workers as well as underground miners. Many of these associations began to include economic benefits such as small payments for the relief of injured miners and death benefits for miners' widows and orphans. Although the benefits were not uniform, locals generally allowed $5 for sick members and $40 for funeral expenses. After 1873 increased competition among coal producers led to falling prices; in turn, miners' wages began to decline. Miners began to see more clearly than before that they needed some type of national organization.[28]

In October 1873 the time again seemed propitious to form a national union. John Siney and John James played particularly important roles in drawing coal miners together in Youngstown, Ohio, where they formed the Miners' National Association of the United States of America. President-elect Siney was born in Ireland but grew up in England where he worked as a coal miner. He formerly served as president of the anthracite miners of Pennsylvania and according to a contemporary, Andrew Roy, coal miners throughout the country held him in high regard. James, who served as secretary, was a native of Scotland where he worked with Alexander McDonald, president of the National Union of Miners. James had also been active in the organization of the Miners' Protective Association in Illinois. During their three-day convention, delegates adopted a constitution drawn up by James, who had modeled the document after the constitution of the Miners' National Association of Great Britain.[29]

In his study of coal mining during the nineteenth century, Andrew Roy presented biographical information on twenty-five of the most prominent union organizers and officers in the United States. Of that number, twenty men were born in Britain and at least thirteen had been active in British miners' unions before emigrating to the United States. Roy's work underscores the crucial role British miners played in the formation of American mining unions at the local, state, and national levels. Once organized, men such as Bates, Siney, and James served as presidents of both national and state labor groups. John McBride, an Ohio miner and later president of both the UMW and the American Federation of Labor, pointed out that "foreigners, it will be seen, were the organizers and officers of the first American association of miners." He credited English-born Daniel Weaver with laying the foundation of the American Miners' Association in 1861. Thomas Lloyd, a Welsh-born miner, assisted Weaver with his organizational work. Daniel McLaughlin, an assistant of Alexander McDonald, first suggested the *joint conference,* a meeting between union officials and coal operators, to devise an annual wage scale. In 1885 the National Federation of Miners adopted the idea and issued invitations to coal operators to meet in joint conference later that year.[30]

Following its inception in 1873, the Miners' National Association expanded rapidly. When the group met for its second national convention in October 1874, membership numbered 21,200. Also in 1874 the Schuylkill branch of the Miners' and Laborers' Benevolent Association brought its 1,200 members into the national organization. The union leadership acted on the belief that no strike should be called until every peaceful means to settle a dispute had been exhausted. With that in mind, union officials immediately began visiting coal companies, suggesting that miners and operators develop friendly relations and establish arbitration machinery. Within three years, even in the face of tremendous effort by Siney and James to organize the nation's coal miners, the group faced defeat. The new union was confronted by many unauthorized strikes and by July 1876 its funds were depleted. At that point the officers had no choice but to shut down the national office permanently.[31]

During the next few years the coal miners made several other unsuccessful attempts to form a national union. In 1883 delegates from local unions in Pennsylvania, Ohio, Maryland, and Illinois formed the Amalgamated Association of Miners of the United States. Shortly after this association was formed, however, the membership became involved in the longest and most bitter coal strike of the nineteenth century, the Hocking valley strike in Ohio.[32]

By 1884 Hocking valley had become the most important mining district in Ohio, covering much of Hocking, Athens, and Perry counties. These miners had developed strong locals and a strong state miners' union. In the spring of 1883 all mines in the valley consolidated into two large firms, the Ohio Coal Exchange and the Columbus and Hocking Valley Coal and Iron Company, better known as the Syndicate. During the following year the coal trade throughout the country underwent a depression. The Syndicate attempted to cut miners' wages from $0.70 to $0.60 per ton, arguing that the company could then be more competitive and guarantee the men full-time employment. The miners refused, fearing that this would be the first of several reductions. The company retaliated by posting notices that after June 23, 1884, the miners would receive $0.60 per ton. Refusing to accept the cut, the miners went on strike. Two weeks later the Syndicate announced that the men's wages would be cut $0.20 per ton instead of $0.10. Moreover, the men would be allowed to resume work only if they would sign contracts stating that they would work for $0.50 per ton for a full year. The operators also demanded that the men agree not to join any other strikes nor to join any union for the purpose of increasing their wages. Hearing these terms, the miners became enraged over their loss of freedom and vowed that they would hold out as long as the Syndicate.[33]

The strike proved to be long and bitter; it lasted nine months and involved forty-six mines and more than three thousand miners. With few resources on which to fall back, the miners relied almost totally on contributions of money, food, and clothing from local people and other miners. Many families scarcely knew where they would find their next meal; entire families lived on nothing but apples for several days. Operators evicted many families from their company houses, so the families were reduced to living in tents or moving in with relatives or friends.[34]

The mine operators quickly began sending their agents into the South and back East to hire strikebreakers. They hired blacks as well as Italians, Hungarians, Russians, and Croatians. Altogether, the mining companies employed over one thousand outsiders. To protect the strikebreakers, operators hired Pinkerton detectives to supplement the support of local county sheriffs. The strike finally ended in March 1885, but the end of the struggle also signaled the end of the Amalgamated Association of Miners.[35]

In December 1885 the determined coal miners reassembled to organize a national union. This time the miners succeeded not only in attracting delegates from the major coal producing states of Pennsylvania, Ohio, and Illinois, but also from West Virginia, Iowa, Indiana, and Kansas. The delegates organized the National Federation of Miners and Mine Laborers. Immediately the miners passed a resolution that directed their representatives to seek out coal operators to formulate a wage agreement. After several attempts, coal miners and coal operators came together in Columbus, Ohio, on February 24, 1886, and signed the first interstate wage contract in the United States. This landmark agreement set basic wages for Pennsylvania, Ohio, Indiana, Illinois, Iowa, and West Virginia. The committee also established a joint board of arbitration to deal with disputes.[36]

Following formation of the National Federation, a rival labor group, the Knights of Labor, made one last effort to draw the miners into their union. The Knights, with their philosophy that all laborers should be brought under one banner, had first appealed to the miners in 1875, emphasizing that they, not a separate miners' union, should represent the coal miners. The Knights experienced some success between 1875 and 1880, especially in Pennsylvania and Ohio, but at that point many miners began to feel that the Knights had not moved rapidly enough in their behalf. In 1887 the Knights decided to make a renewed effort. Believing that they had to counteract the influence of the National Federation of Miners, the Knights formed the National Trade Assembly 135, designated as the miners' division. The Knights then demanded that they be allowed equal representation in future miner-operator conferences. Finally the division between the two groups came to an end in January 1890 when delegates from the two organizations met in Columbus to formulate a single national union. There they adopted a constitution that established the UMW. At the time the National Federation (which had recently changed its name to the National Progressive Union of Miners and Mine Laborers) had a membership of about ten thousand while the Knights represented about fifteen thousand miners. For the first time American coal miners had established true union solidarity.[37]

By 1895 the American coal industry had become an integral part of the nation's industrial life. During the previous half-century coal production nearly doubled every ten years. By the turn of the century competition among major coal operators had become intense and overproduction a perennial problem. Coal became the major heating fuel and source of power in both transportation and industry. The economic and social structures that would characterize the industry for the next forty years had been laid down.

Table 1.2. Coal production by states, 1895 (in tons)

| | | | |
|---|---|---|---|
| Pennsylvania | 50,217,228 | IOWA | 1,456,074 |
| Illinois | 17,735,864 | Virginia | 1,368,324 |
| Ohio | 13,355,806 | Indian Terr. | 1,211,185 |
| West Virginia | 11,387,961 | Washington | 1,191,410 |
| Alabama | 5,693,775 | New Mexico | 720,654 |
| Indiana | 3,955,892 | Arkansas | 598,322 |
| Maryland | 3,915,585 | Texas | 484,959 |
| Colorado | 3,675,185 | Utah | 471,863 |
| Kentucky | 3,357,770 | Georgia | 260,998 |
| Kansas | 2,926,870 | Michigan | 112,322 |
| Tennessee | 2,535,644 | California | 75,453 |
| Missouri | 2,372,393 | Oregon | 73,685 |
| Wyoming | 2,246,911 | North Dakota[a] | 42,046 |
| Montana | 1,504,193 | North Carolina | 24,900 |
| Total coal production: | 132,973,550 | | |

Source: *Report of the Secretary of the Interior*, Part 3. Washington, D.C.: Government Printing Office, 1896.
[a] Includes South Dakota.

Miners gather to go underground at the Smoky Hollow Coal Mine, equipped with dinner buckets, teapot lights, and soft hats, circa 1890s. (Courtesy of Iowa State Historical Department, Division of Historical Museum and Archives, Des Moines)

Twenty-eight states mined bituminous coal in 1895. The states of Pennsylvania, Illinois, Ohio, West Virginia, and Alabama produced roughly 72 percent of the total tonnage. Table 1.2 gives the total tonnage produced, by state, in 1895.

Although total coal production in the United States increased substantially throughout the nineteenth century, the basic technique of mining coal underwent little change during the same period. Electric undercutting machines had been introduced into some mines before 1895, but the great majority of miners continued to do their mining by hand. Miners followed the basic procedure of undercutting the coal by hand and then either wedging it down or blasting it loose. Andrew Roy estimated that a good miner could strike an average of forty blows a minute with a pick. Although working conditions varied greatly from mine to mine, Roy believed that a good miner could produce five or six tons in an eight- or nine-hour day if the seam measured five or six feet thick (known as height of seam to noncoal miners). Working a seam of four feet, a miner could produce about three and one-half tons in eight or nine hours.[38]

Mining accidents continued to be common. Falling slate constituted the major cause of individual accidents. Furthermore, major mining disasters continued to increase in frequency. The first well-publicized incident took place at Avondale, Pennsylvania, in 1869. Following that, major mine disasters took place at Braidwood, Illinois, in 1883; Pocahontas, Virginia, in 1884; and the Frick Coal and Coke Company, Pittsburgh, in 1891. By the late 1800s, the majority of mining states had experienced at least one major mining disaster.[39]

During the 1890s state legislatures in coal producing states continued to pass mine safety legislation. Illinois began to require mine officials such as foremen and hoisting engineers, the men who operated the elevators or cages within the mine, to demonstrate their competency through examinations by a state board. In 1899 the Illinois legislature stipulated that two additional inspectors be appointed; in addition, the larger mining counties were required to appoint county inspectors to assist the state inspectors. State officials in Kentucky increased the number of mining supervisors, and the Iowa legislature passed several measures that directed operators to improve mine ventilation and other safety conditions.[40]

By the 1890s the matter of mine safety had become intertwined with the matter of immigration. Because most emigrants from southern and eastern Europe had no mining experience before they emigrated, many mine operators assumed that these men were more likely to be involved in accidents than English-speaking miners. Operators pointed out that since most eastern and southern Europeans were unable to speak or read English, they could not be counted on to read mining procedures or safety regulations. Mine inspectors in particular believed that the miners' lack of English-language capability resulted in carelessness. Mining officials also regarded southern and eastern European miners as less intelligent than English-speaking miners. One corporation executive, asked about the high number of mining accidents, replied: "Yes, but after all, it's not so serious, because most of the men killed are ignorant foreigners who can be easily replaced."[41]

At the same time that miners demanded safer working conditions, they protested what they believed to be restrictive company policies. Operators traditionally followed the practice of the *long pay,* which meant that operators paid their employees once a month. Few men could support their families for a full month without taking some advance on their next month's wages or without obtaining credit at the company store. In the 1890s miners throughout the country argued strongly for semimonthly pay periods. The presence of the company store compounded the difficulty of the long pay periods. Before the Civil War, because coal mines were isolated, company officials established small grocery stores to accommodate their workers. In the 1870s the company practice developed of issuing the men's wages in company scrip rather than in cash. Every payday company officials checked off the miner's house rent and his store charges from the wages he

Company store, 1895, Muchakinock, Iowa.
(Courtesy of Iowa Mines and Minerals Department,
Des Moines)

earned in the mine. On payday some miners discovered that their earnings did not equal their store debt. In that case, company officials checked off the remaining debt on the next payday. For purchases between pay periods, workers turned in scrip at the company store, thus seldom handling any actual cash. Workers apparently accepted the system of scrip so long as the company store was the only outlet, but in the late nineteenth century private merchants began to build stores in and near mining communities. Gradually companies abandoned the practice of issuing wages in scrip but often with the admonition to their workers that they might be fired if they traded elsewhere. Throughout the late nineteenth and early twentieth centuries, coal miners became bitter opponents of the company store. Not only did they feel that it robbed them of freedom of choice in their buying practices, but most believed that company stores charged inflated prices for inferior goods. Increasingly UMW officials argued that to be truly independent workmen, miners must be freed from the credit slavery fostered by the company stores. Critics of company stores pointed out that many companies demonstrated favoritism toward those men who were the "biggest spenders" at the store. Frequently the foremen assigned the best rooms in the mine to those men who charged the most items at the company store rather than to the most able or most thrifty miners.[42]

Company housing added yet another grievance. While this practice was a necessity in many isolated parts of the country, miners believed that companies made high profits on the houses. Even if miners had the option of renting private housing, company officials usually pressured them to live in a company house.[43] Coal miners residing in company housing often discovered that not only must they endure poor housing, but sometimes they found their personal freedom restricted. If the company fired a miner or if the miner were severely injured, he and his family could be evicted immediately from their home. Many companies threatened to evict workers from their homes if they joined unions, thus using company housing as a weapon against unionization. Some companies required miners to sign leases that greatly restricted the miners' freedom of association within their own homes. A mining firm in Fayette County, Pennsylvania, required their employees to sign the following lease:

[The worker agrees] to do no act or thing, nor suffer or cause the same to be done, whereby the public or any person or persons whomsoever, may be invited or allowed to go or trespass upon said premises, or upon said private ways or roads, or upon other grounds of the Lessor, except physicians attending the Lessee and his family; teamsters or draymen moving Lessee and his family belongings into said premises or away from the same; and undertakers with hearse, carriages and drivers, and friends, in case of death of the Lessee or any members of his family.[44]

Upon signing the agreement the miner became, in effect, a captive in his own home. The agreement implicitly allowed for eviction of the family with little or no warning if the family engaged in activity counter to the agreement. In nonunion camps the agreement was clearly intended to keep union organizers out. In some camps, the operators insisted on the right to enter company houses at any time if the operator believed that improper or suspicious persons were within the home. The agreement underscored the linkage between the dwelling and the job and the inability of the renter to control his personal associations.[45]

The length of the miner's working day had long been a point of contention. For many years the men had argued for the eight-hour day but with little success. In the preamble to the constitution of the UMW, delegates succinctly stated that position: "The very nature of our employment, shut out from the sunlight and pure air, working by the aid of an artificial light (and in no instance to exceed one-candle power), would in itself, strongly indicate that, of all men a coal miner has the most righteous claim to an eight-hour day." Moreover, coal miners believed that a shortened work day would bring a greater uniformity to the coal industry. They reasoned that the shorter day would prevent some operators from forcing their men to work longer hours thus making that particular company more competitive. They also hoped that the eight-hour day would act as a check on the tendency of the industry to overproduce.[46]

The seasonal nature of coal mining represented an additional problem within the industry. In nonindustrial states like Iowa, coal producers relied heavily on the consumption of their coal for heating purposes. Mines were closed when warm weather led to a drastic decline in coal sales. In industrialized areas, mines continued to work full time. But because the coal industry had been in a state of overexpansion since the Civil War, supply frequently outdistanced demand. This condition led operators to shut down their mines at various times during the year, particularly during the summer months. Since 1890, when the federal government began to keep labor and production data for coal mines, coal miners have rarely been able to work an eight-hour, six-day week. Between 1890 and 1930, the most days averaged per year was in 1918 when the men averaged only 249 working days. In addition, lack of markets, strikes, shortages of railroad cars, accidents, and breakdowns in equipment also resulted in mine shutdowns. After World War I, mine shutdowns reflected a decreasing demand for coal. This condition stemmed in part from the overproduction capacities of the existing coal mines as well as the increased competition to the coal industry from other fuel producers. In some localities, miners had an opportunity to supplement their miners' wages with short-term work for local farmers, construction companies, or railroad section gangs. In the Southwest and western states (Iowa included), where seasonal mines were commonplace, miners had more opportunity for part-time work. In more isolated areas, however, such as the southern Appalachians, most miners had to survive solely on their mining wages.[47]

Although the coal industry included a high percentage of foreign-born miners from the 1850s on, the ethnic composition changed after 1880. Before the 1880s English, Welsh, and Scottish emigrants constituted the major foreign-born groups. American coal operators and state mining officials considered the Welsh and English to be the most capable miners because many had learned the trade in their native lands. They regarded the English-speaking miners as craftsmen and one would often hear the comment among mine operators that "the old miner was a craftsman and a wonderful craftsman." Operators and officials believed that these men regarded their work "almost as a profession and took pride in it."[48]

After 1880 emigrants from eastern and southern Europe gradually began to dominate the ranks of the foreign-born miners. By 1920, 37 percent of the coal mining population was foreign born. Poland, Italy, Austria, Russia, the British Isles, Slovakia, and Hungary supplied the largest numbers of foreign-born miners. With the advent of a larger number of men from eastern and southern Europe along with greater reliance on mining machines (and less pick work), attitudes toward miners' work changed considerably. Mining officials increasingly began to think that a man needed only a "strong back." As mining machines gradually replaced pick miners in the 1890s, the view persisted that miners needed a strong back but little skill.[49]

By the decade of the nineties, certain basic economic conditions within the coal industry had become too visible to ignore. That the country had vast coal reserves and that operators needed only small investments to open individual mines tended since the Civil War to make coal an overdeveloped industry. By the 1880s and 1890s coal operators faced intense competition, resulting in a continuous search for new markets. In the face of extreme competition, operators often gained a competitive edge through price cutting that in turn led to the lowering of wage rates. Consequently both operators and miners suffered. Morevoer, the industry had always been influenced by seasonal demands that along with overproduction produced irregular operations. By the later 1880s prominent coal mining trade union officials such as John B. Rae, John McBride, David Ross, and William Scaife recognized clearly that their interests would best be served through a strong, unified, national union, capable of presenting a united front to the coal operators.[50]

Moreover, by the 1890s operators and miners alike had come to accept certain realities of coal production. Most important was the sharp variation in the quality and quantity of coal reserves from coal field to coal field. Seams measured from a few inches in width to over twenty feet, with the majority of seams falling between five and ten feet in thickness. This condition meant that there would be considerable difference in the cost of production among the different fields. Operators and miners recognized by the 1890s that wage scales must be devised that would take the thickness and thinness of seams into consideration so that all major coal fields could remain competitive. Coal miners and operators further acknowledged that once miners had organized the UMW, they could meet together to set minimum wage rates for the various districts of the UMW. Given the variation in thickness of seams, quality of coal, and differences in distances to markets, officials would design minimum rates so that each coal district could remain competitive. The first wage rate schedule went into effect in 1891 only to be abandoned the following year. The Depression of 1893 continued to affect the coal industry well into 1895, and the UMW's efforts to hold interstate joint conferences with coal operators in 1895 and 1896 failed. However, in 1898 officials resumed the practice and collective bargaining between UMW officials and coal operators became a permanent feature of the coal industry.[51]

To the 250,000 coal miners scattered through the nation's twenty-eight coal producing states in 1895, conditions must have appeared both good and bad. In a more immediate sense, 1895 was not a good year. The Depression of 1893 brought a disastrous drop in miners' wages; in the Pittsburgh District, for example, miners' wages fell from $0.75 to $0.50 per ton. Mining officials in Pittsburgh reported that miners' children had been seen chasing dogs away from garbage piles so that they might search for food.[52] Overall, miners had made considerable strides by 1895. Through the UMW, miners had convinced the coal operators that both parties would benefit from yearly wage agreements. Although the miners continued to have many strong grievances against the operators, particularly in the economic area, they must have found some solace in the knowledge that at last they had formed a viable national union. Given the number of coal reserves and the increasingly heavy use of coal by all segments of the economy, the long-range prospects for the industry must have appeared bright to coal miners and operators alike.

# 2

## THE MINING PROCESS

THE coal miner going to work in an Iowa mine in the 1890s soon discovered that he had entered an occupation in which he spent most of his waking hours underground, engaged in dirty, dangerous work. Yet his chosen vocation was one of striking contrasts. While the coal miner found his life above ground governed by company policy, he experienced a far different situation underground. Once at his place or work area, the miner's work condition can best be described as isolated and independent. The coal miner perceived himself as a skilled worker and something of an independent contractor. He provided his own tools, miner's light, and blasting powder. He worked at his own pace, established his own procedures, and to a large extent determined the amount of coal that he loaded out. During the course of the working day, the miner exercised a wide range of judgments that affected not only his own safety but also the safety of other mine employees. He needed a thorough knowledge of the composition of the coal seam in order to drill and blast safely and competently. Yet the miner was only one part of an extensive underground network of employees. Underground workers other than miners were known as *company men*; each company man had a work specialty such as track laying, timbering, mule driving, or empty-car coupling. Altogether the company men had one goal: to guarantee that miners continued to dig coal throughout the day and to see that the coal reached the top. Therefore some understanding of the mining process is essential in any analysis of life and work in Iowa's coal mining communities.

The first step in the mining process concerned the type of entry used to reach the coal seam. Local physical conditions largely determined the means of entry as well as the method of underground development. In Iowa operators chose from four designs when planning the entry. Where the coal seam lay horizontal and emerged as an outcropping on a hillside, the entry could be driven directly into the seam, forming a *drift mine*. When the seam lay below ground, but not more than 100 feet or so, a sloping tunnel could be driven downward to intersect with the seam, producing a *slope mine*. When the coal was so deep that a slope would have been excessively long or difficult to build, operators sunk a vertical shaft to the seam, constructing a *shaft mine*. If the coal seam lay close to the surface and the materials above it could be easily removed, workers stripped off the overburden and formed a *strip mine*.[1]

Once the seam was reached, operators determined whether the mine would be developed according to the room and pillar method or

Bringing coal out of the New Gladstone Mine (slope entry), reflecting mining methods of the 1920s. (Courtesy of *Coal Mining & Processing,* May 1970, Copyright © Maclean-Hunter Publishing, Corp., 1970)

the longwall method. Operators utilized the room and pillar system widely throughout the state; the longwall method was used primarily in southern Iowa. Typically the room and pillar method consisted of driving two parallel entries or tunnels from the shaft or slope bottom. These entries averaged eight feet in width, and miners kept the entries as narrow as possible to ensure better roof control. The height of the entry usually measured five feet. Operators designed the entries for permanent use in hauling coal from the work area or room to the shaft bottom where employees loaded coal on the cage and took it topside. Entries were also vital to the ventilation system since air flowed into the mine through one entryway and out of the mine through the other. Between the entryways, miners left pillars of coal ranging from twenty to fifty feet. Operators designated one entry the *haulway* or *haulageway* while they utilized the other only as an entry.[2]

Driving *cross entries* marked the second step in opening the mine. Cross entries were tunnels that turned off the main entries at intervals of about three hundred feet. Because of the continual need for ventilation, the cross entries consisted of two tunnels or entries, one for hauling and the other solely for air flow. *Crosscuts* or breakthroughs at intervals of about fifty feet provided connections between the two cross entries. As entrymen drove the cross entry farther away from the main entry, they sealed off the earlier crosscuts; they then directed the current of air to the farthest crosscut, ventilating the entire length of the cross entry. The ventilating current returned to the main haulageway through the second parallel cross entry. Bricks, stones, boards, and sometimes heavy canvas sealed off the openings. The men mined out the coal between the two sets of cross entries and in the process they created rooms. Specifically, miners began at the first cross entry and worked toward the center of the area while other miners began driving their rooms in from the second cross entry. In this fashion the two sets of men actually worked toward each other, forming rooms that ran parallel to the main entry. Rooms averaged thirty feet in width with a large pillar of coal, typically eight to ten feet, left standing between the rooms for roof support.[3]

The success of longwall mining (sometimes referred to as *advancing longwall*) depended on the pressure or weight of the *overburden,* or the roof, to break down the coal. Other important concerns were width of seam, (known as the height of seam to noncoal miners), softness of coal, and nature of the *bottom,* or the material under the seam. Mining officials continually stressed that thin seams were a major consideration;

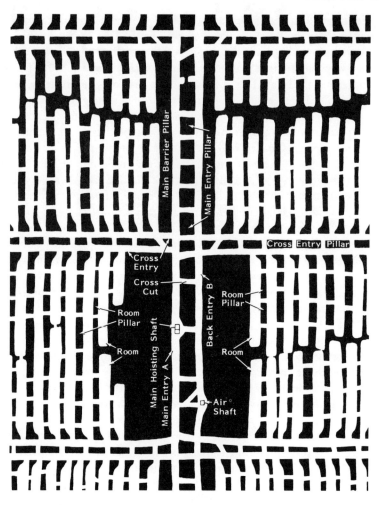

Layout of room and pillar mine.

they regarded seams of around three feet in thickness as ideal while they believed that thicker seams did not work for longwall mining. In the first step of the advancing longwall, the miner undercut the coal seam, removing all dirt and rock from beneath it. Within the next twenty-four hours, the pressure of the overburden broke down the coal seam that had been undercut. The miners then broke up the large chunks with their picks and loaded out the coal.[4]

# THE MINING PROCESS

In the process of continually removing the coal, the face of the seam gradually receded away from the shaft or slope entry. Longwall mining removed the entire seam and miners erected supports to prevent the complete lowering of the mine roof. The first step was to place timbers near the face. While miners continued to remove the coal, dirt and rock were collected from the undercutting and from lowering the floor or bottom (to make room for the railroad tracks). Miners utilized these materials to build *pack walls,* which started about four feet back

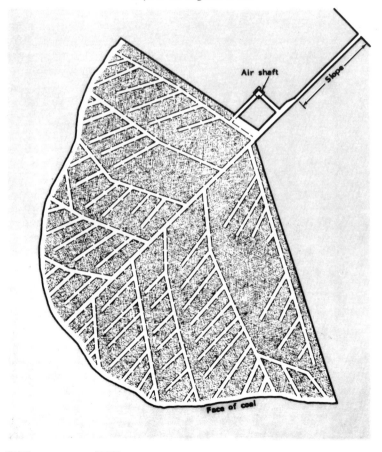

Layout of longwall mine.

from the face. The name pack wall was derived from the process of packing the debris tightly from bottom to roof to prevent the roof from settling behind the miners. The men also built *cribs,* square-shaped wooden supports packed with debris, at points of greater stress such as at the corners where the two entries or tunnels came together. This construction formed entries (in which railroad tracks were laid) with the roof supported by cribs, props, and pack walls on each side of the roadways. The roof continued to settle slowly for several months, but the pack walls eventually prevented further sinking or squeezing. As the seam continued to recede some mines took on a fanlike shape with the main entrance located at the handle end of the fan. In other cases, the mines advanced in an ever-widening circle so that the mined-out areas actually encircled the main shaft.[5]

Physical conditions within the mine determined the longevity of the mining operation and the size of the work force. Regardless of the richness of the vein (measured by absence of rock as well as the presence of a wide seam), the life expectancy of any particular seam rarely exceeded twenty years. The temporary quality of mining became even more pronounced within Iowa's coal fields. Operators here faced "unusually difficult natural conditions," and few mining operations lasted longer than ten years. One Iowa mining engineer explained:

In the first place, the coal beds are irregular in thickness and slope. The beds roll and pitch to such a degree that the average haulageways in the mines resemble in plan and profile a roller coaster speedway. Furthermore, the beds are neither continuous nor contiguous. The coal lies in small disconnected basins. When one of these has been worked out — that is, the mine development carried to a point where the measure "tails" down to a thickness that will not permit of economical production — no other deposit lies adjacent and some miles may be traversed before another may be found.[6]

Iowa operators also encountered difficulties engendered by weather conditions. Most mine roofs were soft slate, making them highly susceptible to temperature changes. In addition, the roof often contained high percentages of lime, which meant that during the summer when moisture formed on the surface, the roof "flaked off" or "weathered," and readily expanded. Later, when the roof contracted, fissures appeared and large slabs of slate eventually broke off. These roof conditions required the use of large quantities of timber to support the roof, a process known as *propping.* Moreover, because of dampness, miners frequently had to replace props. These conditions

# THE MINING PROCESS

Atlas Coal Company, Oskaloosa, showing the top structure of a typical Iowa coal mine. (Courtesy of Iowa Mines and Minerals Department, Des Moines)

meant that upkeep costs on longer haulageways were excessive, so operators consistently tried to keep haulageways short. Frequently operators found it more economical to sink new shafts than to maintain long haulageways. These physical conditions meant that Iowa's mining operations remained small and that Iowa's coal mines had a life span of about half that experienced by many eastern mines.[7]

The great majority of Iowa coal mines worked a single shift from 8 A.M. to 4 P.M. Operators expected miners to be in their workplace at starting time. The men began congregating around the *cage* or elevator by 7:30 A.M., when the engineer started lowering men into the mines, known as the *man trips*. Most cages accommodated 10 men; so with a work force of 150 or 200 men, the man trips might take as long as thirty or even forty-five minutes. Within a short time the first loaded cars appeared on the surface, a process that continued until quitting time. Only a serious accident or a death could disrupt the steady trips of coal to the top. If someone finished his loading before quitting time, he might leave the mine, but he could exit early only by walking up the airshaft.[8]

Before his descent into the mine, however, the miner first faced the problem of transporting himself to the mine. Many families lived

View of top workings including tipple and coal chute at a typical Iowa coal mine in the 1920s. (Courtesy of *Des Moines Register*)

in incorporated communities like Madrid or Albia and the men worked in mines located several miles out of town. Even if one lived in a coal camp, the actual mine entry could be two, three, or five miles away. A coal operator would originally build his camp in close proximity to the mine. However when the original mine was abandoned, the second mine was often developed several miles from the initial site. The distance miners had to travel to work tended to increase as operators were forced by circumstances to build a third and sometimes even a fourth mine. In Buxton, Mine 18 was located eighteen miles from the camp itself.[9] Before the widespread use of automobiles, most miners walked to work. In 1900 one Albia coal miner, along with his nine- and ten-year-old sons walked three miles to the mine and three miles home, fording several creeks along the way. And they made that trek year round.[10] Even after automobiles became common, many miners continued to walk as an economy measure. Sometimes three or four miners formed a car pool. One miner recalled that in an effort to enjoy the luxury of automobile travel but still cut expenses, he and several other miners alternated their means of transportation: they drove the car to work in the morning and walked home that evening; the next morning they walked to work and drove the car home that night.[11] For men working in slope mines, the distance was even greater because they usually walked the entire distance of the slope both to and from their workplace in the mine. Some slopes measured a mile or longer.[12]

In some areas coal operators provided work trains for the miners. In the early 1900s Consolidation Coal Company supplied between eight and ten trains to transport workers to the mines around Buxton. In Colfax, work trains carried miners to three outlying mines as did work trains in the Albia area.[13] Unfortunately for miners who worked the night shift, walking represented the only alternative since no trains operated then. John Corso remembered that "In the summertime it was all right . . . but in the winter, boy, it was cold. You know whenever you come out in 10 or 15 degrees, it's a long walk. I tell you many times I took off my pit pants and it would stand up. They was froze."[14]

When he entered his workplace in the morning, known as a *coal mining room,* the miner found himself in an area resembling a long, narrow tunnel. The inside end, the actual *coal face,* receded a few feet every day as the miner blasted and loaded coal. The width of the seam usually determined the height of the room; frequently men could not stand upright in their rooms. Most rooms averaged 30 feet in width and rarely extended more than 150 feet in length. The walls on either side were part of the seam, but miners left these as support pillars. Narrow tunnels or *breakthroughs* located close to the coal face led to neighboring rooms; these tunnels allowed for ventilation close to the working area. A roadway or railroad track ran into each room to transport coal cars into the room.[15]

Once at his place, the miner first checked the roof. Tapping it with a pick, the miner could tell from the sound whether the slate roof was tight or loose. If it sounded loose, the miner immediately began to prop the roof by placing timbers under the suspected area, topped off with a wedge-shaped cap piece to hold the prop in place. In a room and pillar mine, a company employee known as a *shot firer* traveled through

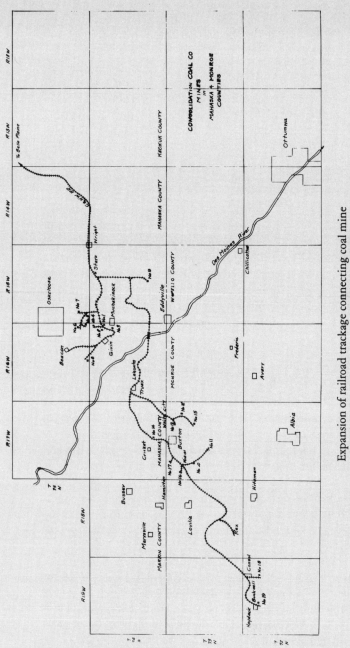

Expansion of railroad trackage connecting coal mine operations in Mahaska and Monroe counties in the early twentieth century. (Courtesy of Iowa State Historical Department, Division of Historical Museum and Archives, Des Moines)

the mine setting off powder charges after all other employees had left work. When the miner entered his room the next morning, he found chunks of coal scattered about. If the explosion had knocked down any props put up the day before, the careful miner replaced these immediately. Once the roof had been secured in the vicinity where he would work, he proceeded to load coal. Occasionally large chunks had to be broken up with a sledge or pick. Apparently new miners quickly picked up the knack of testing the roof. On his first day in the mines, one man commented: "I had never been in a coal mine before and I was nervous. My buddy hit that roof and I knew it was solid. I just lost all fear of the thing falling because it sounded so reassuring."[16]

Loading the coal was the next step in the mining process. Miners quickly learned the tricks to loading the maximum amount of coal into each car. They placed the larger chunks at the open end of the car and around the sides, shoveled the slack coal in the middle, and placed the larger chunks on top. One miner described the correct process as "loading the car good and square and seeing that you had good chunks on each corner, butted against the car so they wouldn't get jerked off."[17] After the loading the miner placed a metal washer marked with his number on the side of the car. This marker enabled the company weighman to credit the coal to the proper account. Some men took considerable care in their loading and consistently loaded out several hundred more pounds than their fellow workers. Certain miners took great pride in "topping the payroll," or being the highest paid miners on payday.[18] Generally each miner developed a distinctive method in loading coal so that often the weighman could tell by merely looking at the load who had done the work and what the tonnage would be. The enterprising miner constantly sought new ways to maximize his tonnage. One miner made it a rule that on his first trip into any mine, he would sit along the side of the coal car, place his elbow on the side of the car, and hold his forearm and fingers up straight. As he rode along, if the roof touched his fingers, he would lower them accordingly so that they would just clear the roof. He knew almost precisely how high he could load his car so that the top pieces would not be knocked off by the low lying roof areas.[19]

Although most miners worked alone, without the help of another miner, it was not unusual for the men to have a helpmate. Iowa coal operators followed the practice of giving miners an extra turn or half-turn if they took a younger person with them into the mine. A *turn* represented the number of coal cars a miner was assigned on any given

The Aubrey brothers at the Munterville Coal Company, who worked as miners and later as operators. Note the carbide lamps, typical lighting devices of the 1920s. (Courtesy of Iowa Mines and Minerals Department, Des Moines)

day. If the company allowed a full turn for the youth, it meant the miner received double his normal number of cars for that day; in effect, he doubled his wages. Moreover, the older the youth, the more coal he could shovel and the more quickly the two workers could load out the cars. In Iowa, miners apparently took their sons into the mines as soon as possible. In 1880 the state legislature prohibited boys under twelve from working in the mines, yet in 1883 the state mine inspector noted that miners continued to violate the law.[20]

Lacking either a son or a nephew, some miners took along friends; occasionally a miner would take along his younger brother. One miner recalled that as a young man, his father had tried several times to leave the mines but had always been urged to go back by his mother. He explained:

[My father] quit a number of times, trying to get out of the mines because there was no future to it and then his mother would say, "there is a new mine opening up. It would be awfully nice if you would take in one of the younger boys." And to help his mother out, and there was something alluring about being back with his friends again, he would start up and take one of the younger boys with him. It would help his mother and father as well as provide a job for himself. That happened several times. . . . It was tough to break away from the coal mines.[21]

Everyone agreed that the younger boys could do little work; they might run errands, pick up tools, or shovel some coal. But because most coal miners were unemployed several months each year, they were unlikely to pass up any opportunity to make a few more dollars. Sometimes miners would pay their sons a small amount for their work, but many acknowledged that after paying the bills, they had nothing left to pay their helpers.

While the short-term economic results of this practice were important for the miner, the long-term results were equally significant for the young apprentice. Taken into the coal mines in their early teenage years, many young men remained in the mines for a lifetime. They learned the mining skills working alongside their fathers, uncles, or older brothers, and by the time they reached seventeen or eighteen, they qualified for a room of their own. Their early entry into the mines also cut short their education. Responses to interviews, however, revealed that the great majority of boys attended school for at least three or four years before going into the mines. This meant that most, if not all, boys could read before beginning their mining careers.

Throughout the day the mule driver picked up the loaded cars and furnished the miners with empty cars. The driver spread his deliveries out so that each miner received an empty car at regular intervals. Miners commonly complained that they always had to wait for empty cars. According to a former mine foreman, the companies spaced the delivery of cars at lengthy intervals to ensure that the men worked the full eight hours. Obviously, the company did not want the men to finish loading all their coal before the shift ended. If they did, it would shut down the operations of the other men such as the mule drivers and the hoisting engineer but leave other company employees such as the timbermen and trackmen working till the end of the shift. In general it did not pay the company to send some employees home early while other men continued to work. Some miners would deliberately choose to stop loading coal early, passing up a final car or two. For the older men, this decision was sometimes a necessity because of fatigue. But other miners, frequently the younger men, gained an advantage because they had an extra car or two to load, thus boosting their tonnage for the day.[22]

Pearson Mine showing typical procedures used in the mines in the 1920s. (Courtesy Iowa Mines and Minerals Department, Des Moines)

---

The use of mules and, to a lesser extent, ponies was indispensable to the mining operation. As a general rule, operators used mules in larger mines and ponies in smaller, local mines. Operators utilized the animals primarily to pull the coal cars from the mine face to the main haulageway where an electric motorcar brought in the empties and took out the full cars. Before operators installed electricity, mules pulled the loaded cars the entire distance to the shaft bottom where workers placed the cars on the cage; on the return trip, the driver brought the empty cars back from the shaft.[23]

Most miners supported the use of mules over ponies. Rarely did miners speak with affection about mine ponies, but they held their mules in high esteem. Mules were regarded as more intelligent than ponies and appeared eager to work. One miner noted that mules could be taught to walk between the ties on the railroad tracks, thus allowing the miners to leave more bottom in their rooms and to eliminate some dead work.[24] Another worker recalled that if his mule caught a hoof under a railroad tie, the mule would wait calmly for the driver to free the hoof. A pony, on the other hand, would frantically try to free the hoof and probably injure himself in the process.[25] Miners sometimes credited mules with an innate ability to sense an impending slate fall. If a mule refused to enter an area, some miners took that as an indication that something was wrong. Several miners remembered that in such instances, slate falls took place a short time later.[26] When mine mules were killed, usually by runaway cars, the occasion was a sad one for the driver.

To expedite the mining operation, larger companies stabled mules underground for long periods of time. In seasonal mines, employees took mules down in September and brought them out in April; in mines that operated full time, companies kept mules underground for many years. To go underground, employees led the mules onto the cage singly and then placed them in a box that confined their movements. Once underground, employees stabled mules in dugout areas adjacent to the workings. Company men fed and watered the mules and had the responsibility for training new mules. In larger

Company employee driving a mule above ground (1920s). (Courtesy of Iowa Mines and Minerals Department, Des Moines).

mines such as those around Buxton, the companies hired full-time mule trainers, while in the smaller mines the individual drivers trained the mules. In smaller slope mines, employees brought up the mules or ponies each night because their movement did not take up valuable time on the cage.[27]

In addition to loading out coal, coal companies required that miners perform *dead work*. Dead work referred to any task not directly related to the miner's primary function, such as cleaning up rock, brushing bottom, or bailing water. One type of dead work that the miner did with regularity was *brushing bottom*. As he continued to remove coal from the face and moved farther into the room, the miner found it necessary to remove a portion of the mine floor. Brushing bottom meant blasting out about twelve to fifteen inches of floor so that the railroad tracks could be lowered by that amount. The result was twofold: the mules that walked along the tracks had sufficient head space; and the brushed bottom lowered the coal car, enabling the miner to scoop and load his coal with less exertion. Before each payday, the foreman passed through the mine to "settle up" with the miners for dead work done since the last payday. It was understood that the company would not pay miners for all dead work, but only for those jobs that took an inordinate amount of the miners' time. Sometimes a miner agreed with the foreman's assessment but not always; if not, a bargaining session ensued. Not always satisfied with the final decision, miners sometimes indicated their unhappiness in other ways. One miner, upset that the foreman would not pay him for the time required to break up and cart out a big rock, enlisted the help of another miner; together they loaded the rock intact on a coal car. The huge rock was carried to the top and rolled down the loading chute onto a customer's truck. The rock was taken to a blacksmith's shop where efforts to break it up with sledges failed. Although the foreman tried to fire the miner over the incident, the miner persisted and finally received pay for the dead work.[28]

In breaking down the coal, Iowa miners practiced several techniques. Before operators adopted undercutting machines, room and pillar miners *shot on the solid*. This meant that the miner did not undercut the seam, but relied totally on the shot to bring down the coal. This technique required deeper holes and larger amounts of powder. After the invention of a successful electric undercutter, however, room and pillar miners had the option of *undercutting* before shooting or continuing to shoot on the solid. Undercutting consisted of removing about a foot of dirt, rock, or other materials present under the coal seam. In some mines, miners used both methods because different parts of a single mine contained different physical conditions. As a general rule, miners used machines if the material under the seam was soft enough to gouge out; with solid material under the seam, miners did not consider undercutting. As late as 1932 the Moran Mine offered two types of pay, depending on the method of breaking down coal. In rooms where miners utilized mining machines, men received $0.83 a ton; in rooms where they did not undercut, the company paid $1.23 a ton.[29]

Before the use of undercutters, longwall miners undercut the coal by hand. Lying on their sides, the men made a three- or four-foot cut under the seam. One miner recalled that his father was so adept at undercutting with a pick that he could lie on his left side and hew the coal with his right hand all along the seam. He could then roll over on his right side and hew the coal equally well with his left arm as he moved back along the same undercut area.[30]

Sometime during his shift, the miner had to prepare his shots. Of all the mining procedures, this required the greatest skill. To carry out the process effectively and safely, the miner had to have a precise knowledge of the composition of the coal seam, the drilling process, and the use of blasting powder. One former mine foreman commented that a miner had to be "almost a mathematician to do his job." The miner first had to determine where the two or three shots should be placed. If he had placed his shots correctly the previous day, the face would be left with a stump or two of coal protruding into the room. Sometimes the protrusions resembled square corners. The miner regarded this condition as ideal for the placement of shots. This condition allowed the miner to drill the holes parallel to the exterior surface, and to provide for a chance, or for a removal, of about three or four feet of coal. One former miner foreman put it this way: "You couldn't make the hole any deeper than the thickness of the coal. If the coal was three feet in thickness, the hole couldn't be any deeper than three feet."[31] The proper placement of shots was crucial to the miner's work. He attempted to place the shot so that the blast would not dislodge any, or very few, props. If the blast was too great, it blew coal all over the room and scattered coal took more time to retrieve and load. The replacement of props also used up valuable time that could be spent loading. Moreover, in setting his shots, the miner not only had to think about the coal to be loosened for the following day but also about the shape of the face after the blasting. Once he decided where to locate the holes, he used an auger to drill the holes into the seam. A tripodlike apparatus secured tightly between the roof and the floor held the auger in place. Following an inspection by the *shot examiner,* an individual hired to certify that shots were properly placed, the miner inserted blasting powder into the holes and placed fuses at the front. He used dirt to seal the openings so that only the fuses extended out of the holes. In effect, the dirt directed the main force of the explosion back into the coal seam. Most miners placed three shots daily, two for loosening coal and one for shooting bottom.[32]

While some miners viewed the room only as a working place, other miners thought of it as a personal possession. Once a man started in a room, the foreman rarely moved him to another working place. He might request transfer to a better room but few miners did so. If a miner was injured or forced to be away from the mine, the foreman held the room for him for a considerable length of time. Because of their independence, miners kept their room in any shape they wished. Foremen continually urged the miners to keep the necks or openings of their rooms well timbered and clear of debris, but what a miner did at the face within his own room was his business. Sometimes the way in which a miner left his room reflected his pride of workmanship. One miner commented: "When you went home in the evening, after your day's work was done, your place looked like a castle, you know, everything was timbered, cleaned up, everything. The track was nice and clean and straight." The same miner, commenting about misplaced shot, lamented: "Sometimes [the next morning] you would find the doggone coal out here and dirt all over. It would just make you sick."[33]

Firing the shots marked the final step in the mining process. In the early days of the industry, miners fired their own shots and did so at will. This meant that many men fired shots two or three times a day. In 1882 the state mine inspector, Park C. Wilson, recommended that the firing of shots be regulated as a way of improving the quality of air in the mines. The inspector reported as follows:

The air is carried in one current around the mine. There was a good current passing but the miners fire the blasts all times in the day. For this reason, the air was loaded with powder smoke. The volume of air was equal to 125 cubic feet per man per minute, but the first man in the air fared better than the last. I recommended to the men to pass a law regulating their firing, but they said they could not make enough by firing twice a day, and for that reason they would rather have the smoke and be allowed to fire at will. . . .[34]

In a later report, the same mine inspector explained why this practice led to a great many accidents. Anxious to begin loading coal, miners would return to their rooms before the smoke cleared. Groping their way through the darkness, they sometimes knocked down props, causing a fall of slate. The concerned inspector urged the men to allow time for the smoke to clear before reentering their rooms. Inexperienced miners inadvertently created an additional hazard because they placed holes improperly, overused powder, and used slack and coal dust for tamping.

The practice of frequent shooting also proved debilitating to the miners' health. Often the rooms and entrances were not cleared of powder smoke until late afternoon. Forced continually to inhale the acrid smoke, miners commonly complained of chronic headaches, nausea, and loss of appetite. Because of the deaths and damage to equipment, miners and operators gradually reached an agreement that the shooting of coal would be permitted only twice a day, at noon and at night.[35]

Nevertheless, accidents continued as a result of improper firing. On January 24, 1902, an explosion at Mine 2 of the Lost Creek Fuel Company was the most serious accident in Iowa mining history. The explosion killed twenty men and injured fourteen. State officials, reacting to the accident, appointed a committee to investigate its causes and to recommend remedial legislation. The committee responded with the following recommendations: that certified shot examiners be employed in all mines where miners shot on the solid; that operators be required to sprinkle entries to reduce coal dust in the air; that miners use only sand, oil, or clay for tamping shots; and that once a shot had blown the tamping, no one should be allowed to recharge it. In 1902 the legislature adopted only the first provision, that where miners shot coal on the solid, a "competent person shall be employed to examine all shots, before they are charged." By 1910 most Iowa coal mines had shot firers, hired at the expense of the miners. Since the law required operators to hire shot examiners, in many mines one man carried out both responsibilities, with operators and miners sharing the cost. At the same time, operators adopted the practice of shooting once a day after after all workers had left the mine. Despite the miners' fears, shooting once a day did not diminish the total output.[36]

In carrying out his responsibilities, the shot firer/examiner walked through the mine twice a day. On the first trip he determined how many shots each miner would set, inspected the holes, and had the final say in approving or condemning a shot. After the shift ended, the engineer notified the shot firer, either by mine telephone or by ringing the mine bell, that people had left the mine. Only then did the shot firer move from room to room, firing the shots. Foremen required each miner to use enough fuse so that the shot firer could be a safe distance away before the blast occurred. Although his job did not require that he return to any room where a shot did not fire, he sometimes did so as a special favor to that particular miner. He would cut the burned fuse off, split it, and relight it. As he made his rounds, he traveled through the openings or breakthroughs located near the face, thus greatly expediting his work.[37]

Opinions varied as to whether or not the shot firer had a dangerous job. Some miners believed that it was the easiest job in the mine because the firer had only "to walk through the mine twice each day." Two former mine foremen remembered that many men liked it because it was "an easy job."[38] Yet other men admitted they did not want the position because they considered it too dangerous. One miner who worked as a shot firer for many years regarded the work as extremely dangerous. Reflecting on that career, he concluded: "Them days I would like to shoot shot, be a shot firer. But now I get a little older, and I realize what I was doin'. Any human that would do that kind of job, or any coal miner that would do that kind of a job, ain't got no respect for his own life. . . ."[39]

Shot firers were not the only mine workers who faced dangers. If a miner's room had a particularly bad ceiling, he might have to surround himself with props. In the course of his work, he might accidentally knock down a prop or two. Miners frequently suffered broken legs or broken arms from falling coal or falling slate. Actually, falling slate represented the major cause of death in the Iowa mines. One miner recalled an accident in which he and a relative were working only a few feet apart. "We were talking and he knocked out a prop. Right away a wedge of slate fell down and killed him dead. The same piece of slate that hit him hit the bill of my mining helmet, and it must have hit my neck too, because the next day I had a stiff neck."[40] One former mine foreman recalled that the two most dangerous times of day were early morning, shortly after the men began their shift, and late afternoon. In the morning many men were anxious to start loading coal immediately and so did not check the roof or their props. Late in the day when many miners found their carbide lamps burning low and with only a short time left to work, perhaps half an hour, some men chose to work with a dimmer light rather than refill their lamp with carbide. Working in the dimmed light increased the possibility of accidents.[41] Driving mules also presented hazards. When the cars began to roll too rapidly, cars sometimes jumped the track. Usually the mule driver attempted to jump clear of the cars, but as one miner pointed out, "There was nothing to give down there; those ribs [walls of haulageway] were solid."[42]

Dust and coal impurities endangered the eyes. Many coal seams contained sulphur, sometimes in the form of a vein or perhaps scattered throughout the seam in small pieces. When a miner's pick hit a piece of sulphur, he described it as being "red hot, just like a meteor." Striking the miner's face or eyes, it could burn badly.[43] The continual presence of coal dust threatened potential eye damage, with miners sometimes losing sight in one or both eyes. One man recalled that his mining career started because his father lost one eye in a mining accident. Following the accident the mine foreman cautioned the miner that he must have a helper for his own safety. The miner took his sixteen-year-old son out of high school and into the mine to satisfy the company order. Several years later when the oldest son got a room of his own, the father took in the younger son for his helper.[44]

Undercutting coal, either by hand or by machine, could also lead to injuries. While undercutting by pick, the miner cut in several feet under the seam and then put up a *sprag* or support as a safety measure. If the coal pressure continued, there was usually a little movement of the sprag to warn the miner of an impending fall, but not always. Before the miner finished the hand cutting, he was literally lying under the overhang of coal. With the machine, a man sometimes got a piece of clothing caught in the revolving blade that would pull in an arm or leg.[45]

The cage itself was a potential source of harm. Occasionally a cable might break or an inexperienced operator would be forced to take over the controls. In some cases the cage fell only a few feet, while in others it crashed to the bottom of the shaft. Some operators, conscious of this ever-present danger, insisted that no two family members ride in the same man trip.[46]

Poor placement of shots and improper packing of holes with powder also increased the miner's vulnerability. If a shot misfired, miners referred to that as a *windy shot* or a *blown out shot*. This resulted when the hole was improperly placed so that instead of the explosion blowing inward and breaking off coal, the force of the explosion blew out of the hole into the room. A windy shot might affect only the room in which it was fired, but it might also send powerful gusts of wind through the entire mine. Shot firers sometimes found themselves knocked down by the shots and sometimes hit by swinging doors. One miner recalled his part in an investigation to locate the windy shot that had killed the shot firer. He examined each of the shots but insisted that there was not a misplaced shot in the mine. He theorized that perhaps several big shots went off at the same time, creating two powerful air currents and the shot firer had been caught in the middle.[47] The most vivid description of a windy shot came from a former shot firer:

A lot of times . . . they call it an explosion. I don't know if you've heard of it or not, an explosion in the mines? Sometimes it's just a wind, but it could take an iron rail and twist it around, or push a car — it'll move anything. Many time I had a hold of a rail, you know, on the track, and I put my hand around it and hang onto it. It throwed me around. That was a bad thing to do, because when it blow like that, the wind push a car, and it'd smash your fingers.[48]

Miners had to be alert to possible complications in the ventilation system as well. If the air flow through the mine was not sufficient, bad air or gases could accumulate. Although methane gas has never been detected in an Iowa coal mine, miners recognized that two other types of gases, carbon dioxide and carbon monoxide, could accumulate. A state mine inspector described carbon dioxide; the miners called it *black damp*:

. . . This gas is accumulated from several causes: The respiration of men and animals, the combustion of the workmen's lights, the decomposition of timber and small coal in the gobs, the explosion of powder . . . and the excrementitious deposits of men and animals; and it also exudes from the roof and floor of the mine. . . . This gas, when mixed with a certain portion of pure air, can be breathed for some time where the miner's lamp cannot be kept burning; but its effect on the miner is such to produce headaches, langour, weakness in the legs, loss of appetite, and general depression. . . .[49]

A long-time miner remembered that with black damp, "You go in there and get tired . . . your arms feel like they awful heavy to lift up, and you get a headache and it sound like it go boom . . . boom . . . boom. . . . You know you don't feel good, so you have to get out."[50]

On the other hand, carbon monoxide or *white damp* — produced by fire, or from exploding blasting powder — was not so easy to detect. Instead of dimming the miners' lamps, it made them burn more brightly. One miner remembered that "With white damp . . . it would work different. White damp, it would get you sleepy. 'Oh,' you say, 'if I only sleep two minutes I feel better.' But if you sit down and close your eyes you're done. That's what white damp is. It'll kill you quick."[51]

While the digging of coal always remained the central focus of any mining operation, the miner's work depended on an extensive support system. The qualities of isolation and independence, so often used to describe the miner's role, were made possible by a wide assortment of company men who maintained the lines of communication between the miner and the outside world. In general, operators and miners referred to employees of the company who did not dig coal as company men. Operators paid them an hourly wage and they worked an eight-hour shift. Operators determined some company positions by age and some by experience. Although the ratio between the miners or diggers and company men varied, the miners typically numbered more than double the number of company men. At the Norwood White Coal Company at Moran in 1931, the company employed 140 diggers along with 60 company men, who worked both above and below ground.[52]

The *trappers* were the youngest company employees. Trappers opened and closed the air doors that directed the air flow through the working portion of the mine. Typically young boys, sometimes as young as nine or ten but more likely twelve or thirteen, sat by a particular door and opened it periodically to allow mule drivers through. The mule driver whistled or signaled in some way that he was approaching; the trapper returned the signal either by giving an all clear sign or indicating the driver should wait. Once through, the trapper immediately closed the door. One man who started his mining career as a trapper, described his job as being like that of a railroad switchman. He claimed that the trapper had to know where all his drivers were at all times and concluded that "sometimes you had as high as four, five drivers that was going up and down your entry . . . and you had to keep them from running together."[53]

The younger workers also coupled empty cars and greased car wheels. Many youths started their mining careers with one of these jobs.

Mule drivers in the late nineteenth century outside Old Number Three, Lockman Coal Camp. Note youth (*left*) who appears to be no more than twelve years old. (Courtesy of Iowa State Historical Department, Division of Historical Museum and Archives, Des Moines)

---

The person coupling empties had to have sufficient strength to pull the cars away from the bottom of the shaft as they came rolling off the cage. The worker ran with the first few empties and placed them a considerable distance away from the cage so that he would have room for about thirty more empties. As the remainder of the cars rolled down the incline from the shaft, they rolled into place. During the winter the wheel grease thickened causing the cars to roll more slowly; the coupler always had to be alert to a possible pileup. As one miner put it: "First thing you knew you would have a terrible bunch of cars in there off the track."[54] For greasing cars, the men dug a little alcove out of the side of the rib, or the wall, near the cage. When the man pulling empties came along, there was a slight lull and the worker in the alcove greased the wheels on that side. He greased the cars half a day on one side and half a day on the other, thus servicing each car completely.[55]

Frequently trappers moved up to driving mules when they reached the age of fifteen or sixteen. Sometimes older men drove mules, but it seemed to be a job that attracted younger workers, rather than middle-aged men. The occupation provided one of the few underground activities in which a worker could show off his physical dexterity as well as his ability to manage the animals. Operators required that each driver be responsible for providing empty cars and collecting loaded cars from a prescribed number of miners; in some mines, drivers handled four to ten miners while in other mines they handled as high as fifteen. Their work involved pulling the loaded cars out of the miners' rooms and taking them to the main haulageway. Loaded or empty, mining personnel referred to the drivers' cars as a "trip." The driver then brought the empty coal cars back from the main haulageway and distributed them at the individual rooms. The rule persisted that a driver "must keep an even turn," which meant that he delivered an equal number of cars to each miner. The company made an exception to the rule if the miner brought along a helper; in that case the miner received an extra turn, or in some cases, an extra half-turn.[56]

Dexterity and quickness were the mule driver's keys to survival. When the mule was moving, the driver "rode the chain" that connected the mule to the first car. The driver stood with one foot on the bumper of the car and the other on the chain; he placed one hand on the mule's rump and the other on the front of the car. Occasionally the entry ceiling measured as low as four feet and the driver had to stay hunched over. If the mine floor was fairly even, the task was easier; many mine floors, however, resembled the surrounding countryside with its rolling terrain. When going down an incline, the driver had to brake his load by placing sprags (pieces of steel or sometimes tree limbs) in the four separate wheels. As the sprag came around and hit the bottom of the box, it locked the wheels. If the hill was not too steep, a miner might get by with locking only one or two wheels. Miners also placed sand on the tracks as a means of braking. If the four locked wheels and the sanded tracks did not slow down the cars, the driver had little choice but to ride out the trip and hope that the cars would not jump the tracks. With a well-trained mule, many drivers did not use mouth bits, but merely relied on the commands of "gee" and "haw." Some miners had their mules so well trained that even though the cars moved rapidly, the driver could unhook the car, holler gee at the mule and as the mule went off the track to the right, the car rolled straight on. One long-time mule driver commented that it was "just like a planned switch."[57] In later years the operators did away with trappers and went to a swinging door to separate sections of the mine. Some miners even trained their mules to nudge the doors open so that the trip could move on through without disruption to the driver.[58]

Two additional company positions necessary to keep the mine operating were *timbermen* and *track layers*. Although the miner timbered or propped his own room, the company hired workers to timber the main entry and the cross entries. The typical mine included entries eight feet wide with timber sets placed about every three to five feet, depending upon the condition of the roof. In rare instances, timbermen found it necessary to form an almost continuous wall of timber sets. Timbermen constructed the sets out of two posts or legs

Mine showing use of ponies underground.
(Courtesy of *Coal Mining & Processing*, May 1970,
Copyright © Maclean-Hunter Publishing, Corp.,
1970)

and a crosspiece or collar. The legs were generally made of round oak, six feet high, with notches hewed at the top. The timbermen also notched the collar so that it fit securely onto the legs. One former timberman commented that "we didn't have any machines to do that, but you got so doggoned use to it that—we did it with hacks and saws—that you could do it blindfolded." The same man, reflecting on his work as a timberman exclaimed: "And you know when you walked down the entry, the timbers were uniform and nice and it really looked fine!"[59]

Following behind the timbermen were the trackmen who put down railroad tracks throughout the mine. Either single or double tracks might be laid through the entries and cross entries, with a single track extending into each miner's room. Every turnoff necessitated building a switching system. In a large mine the haulage system was an intricate one and great care had to be taken to avoid congestion of mule drivers and motormen at key points. Any slowdown either in sending empty cars to the miners or in sending loaded cars to the bottom of the shaft meant a decline in production. Along with laying new track, the trackmen spent a good portion of their time repairing old track.[60]

Additional underground workers were the *motormen, wiremen,* and if the operation were sufficiently large, *brattice men.* The motormen, regarded by many as the most skilled of the underground company men, operated the motorcars that traveled along the main haulageway. Large mines might employ several motormen and each had the responsibility for hauling in a certain portion of the mine. Wiremen worked continually to extend electricity throughout the mine while the brattice men built the doors that directed the air flow through the underground workings.[61]

After the engineer hoisted the coal to the top, the weighing took place. In the nineteenth century, operators hired weighmen who were responsible for weighing the coal and keeping a record of each miner's tonnage. Miners long felt that the company weighman frequently shorted them in the weighing process. Eventually through the UMW, miners gained the right to hire union checkweighmen, at their own expense, to work alongside the company weighman. As the coal cars came onto the scales, both men recorded the weight. The miners believed that the practice safeguarded their interest.

The checkweighman occupied an important position; not only did he weigh coal, but union members expected him to settle certain disputes. One of his responsibilities was to award road coal to the respective miners. As drivers pulled the cars through the mine, some chunks invariably fell off. At night the company dispatched a worker to collect the road coal and the checkweighman then decided which miners should receive credit. The checkweighman knew every digger and how he loaded. If someone's load came up short, the checkweighman usually compensated that miner. A few miners always complained that they did not get their share of road coal, but overall the miners accepted the checkweighman's decisions.[62]

Although mine work was specialized, there existed a high degree of flexibility among employees. Most coal miners had worked in various company capacities before, or even after, their coal digging days. In the event that a trackman or timberman could not work, coal miners could be conscripted by the foreman to leave their rooms and spend the day doing company work. The company paid the miners company wages for that time. Operators particularly valued miners who had served in many capacities because those men could be utilized in many different ways. However some miners hesitated to leave their more lucrative digging to do company work. In the early 1930s one union contract stated that a miner could refuse once to do a company job. The second time the miner refused the foreman's request that he do company work, the foreman could stop his turn. The third time the miner refused the foreman's request, the company had the right to fire the man. One miner recalled that his father became so plagued with continual requests to do company work that he finally refused, telling the foreman: "If you come after me any more, we're going to have trouble." The foreman returned the next day and the miner chased him down the entry with a shovel; thereafter the foreman left him alone.[63]

The mine foreman was responsible for overseeing the entire underground operation. Although the largest mines might also hire a top boss to supervise all operations above ground and a stable boss to handle the underground mules, miners regarded the mine foreman as the top company employee. The typical foreman had previously worked at many different jobs within the mine and had firsthand knowledge of the various work roles under his supervision. He was charged with keeping the mine tonnage up as well as maintaining the mine in good working condition. The safety of all underground employees rested with him. Ideally, the mine foreman traveled through the mine each day, visiting each work place and personally supervising the company men. This frequent visiting was impossible, however, given a common ratio of one hundred underground workers

Slag pile and top structure of an Iowa coal mine.
(Courtesy of Iowa Mines and Minerals Department,
Des Moines)

to one foreman. Typically, the foreman became something of a troubleshooter and spent much of his time supervising special problems such as cave-ins, accidents, water seepages, or special adjustments for dead work. Overall, the foreman usually held a position of respect among the men and wielded considerable economic power, particularly in assigning rooms and setting rates for dead work. Placement in a good working area could make a substantial pay difference to a miner.[64]

Milo Papich's role as foreman points up the interrelationship that existed in any mine between company men and miners. Papich began his mining career at age twelve when he went to work as a trapper. Born in Croatia, Papich had been in the United States only two years when he heard that the mine at Carney needed trappers. He decided on his own that he would like to work in the mine and applied for the job. Within a few years, Papich began coupling empty cars and greasing wheels. At age fifteen, he became a mule driver. After several years as a mule driver — the job he liked best — he tried his hand at timbering, track laying, and finally, digging coal. At age twenty-seven, Papich's foreman asked him if he had ever thought about getting a mine foreman's certificate. When Papich answered no, the mine foreman encouraged him to do so; he even lent Papich his mining books. Papich remembered some of the difficulties:

All the problems were in algebra. All right, I didn't know anything about algebra, I didn't have it when I went to school in 6th grade. So what I did, I asked my brothers-in-law to help me. They'd read it but they couldn't help me because you see with an algebra problem, you have to have a formula to go by and that in mining problems especially. I took those books and I would read and read around them problems, and I would read around those problems and then I turned them problems into long arithmetic. And when I got my certificate, I missed one problem. And that one problem, after awhile, I learned it.[65]

By the time Papich reached his late twenties he became night foreman, a position he held for eight years. He then became assistant mine foreman during the day shift. From there he became what he termed a "main line foreman," filling in for other foremen when they went on vacation; frequently the latter position required that he move from mine to mine. Eventually he became permanent foreman in one of the Scandia Coal Company's mines.[66]

Throughout his career as foreman, Papich's previous mining experience served him well. He recalled that during his younger days as a mule driver, he saw many conditions in the mines he did not like. When he became foreman, he had the opportunity to remedy these. He promptly cleaned and fixed the tracks, kept the switches clean, sumped out water, and kept the mud holes cleaned and *corduroyed* (lined with timber). Consequently the mule drivers working under his supervision enjoyed better working conditions and performed accordingly. A foreman's firsthand knowledge of the different specialties also made him a fair judge of the company men's work. Papich concluded that "a good foreman knew how much work the different company men could do so they would have to make a good showing." Papich reflected that a foreman had to "treat everybody square . . . and that's the whole thing about being a foreman, is you got to treat your fellow man as an equal."[67] A commonplace saying in the mines noted that "you can't drive men in the mines; all you can do is to take care of them." Papich's role seems to bear that out.

Iowa coal miners worked under conditions that were both difficult and dangerous. Regardless of the type of work done in the mine, the workers were continually exposed to falling slate, windy shots, and runaway mules. Mule drivers faced the greatest risk; shot firers also faced constant danger. Even if miners successfully avoided these jobs, they still spent their hours underground in considerable discomfort.

The width of the coal seam determined the height of the room, and in Iowa the coal seams averaged between three and four feet. This meant that miners spent much of their time either bent over or crawling on their hands and knees. Some men in southern Iowa, where coal seams sometimes measured less than three feet, worked lying on their sides. Physical conditions lessened the miners' productivity. If miners happened to have rooms with poor coal seams, they could do little to improve the situation; they simply made less money than their fellow workers. Some men spent considerable time waiting for their assignments to new rooms.

The miners' independence compensated somewhat for the dangers and discomforts inherent in mining. The coal miner was indeed an independent workman. Although he kept specific hours, within that work day he largely determined the amount of coal he loaded out and hence the amount of money that he made. He worked alone in his room, making his own decisions on how he would proceed during the day and the manner in which he would carry out his individual responsibilites. Unlike a factory worker, the coal miner had no supervisor continually "looking over his shoulder." Coal miners also had a certain degree of job security. Although they might be unable to work every day, rarely did operators discharge miners. Most miners worked until the mine closed down or they decided to quit.

The miners did not work alone, however, in carrying out the mining process. In many ways, coal mines took on the look of modern factories in which each phase of the manufacturing process was essential to the production of the final product. In any coal mine, a large supporting network of workmen existed, each carrying out specific responsibilites. The coal miner relied heavily on the efficiency of the company men to keep the general mining operation running smoothly. Even though the mining industry involved a high degree of specialization, most company men learned a wide variety of skills that enabled them to work in several different positions. There was, in other words, a high degree of interchangeability of labor. The final purpose of all workmen, including the miner, was to bring coal to the top in a continuous process throughout the workday. Unfortunately when the process broke down, the miner suffered the most. The company men were paid for the hours spent in the mine; the miner was not.

Iowa coal mines offered possibilities for personal initiative and promotion. As a general rule, coal operators did not bring company men and foremen in from the outside; they chose these men from the group that had started working in the mine as trappers, couplers, or miners. In theory, every man entering the mining profession had the chance to become a mine foreman. The worker who willingly learned several different specialities had additional value for the mine operator. Moreover, he was the man whom the foreman usually tapped to work during a mine slowdown or during the summer when the operator needed only a skeleton crew to maintain the mine.

The Iowa mining industry had at best a haphazard training system for its workers. Company men fared better as their work was more closely supervised, but miners required no special training before they went underground. An informal apprenticeship system did exist through the sponsorship method used by Italians whereby an experienced Italian-American miner supervised an Italian newcomer and through the process of fathers taking sons or other relatives underground. All people involved assumed that somehow the individual taken into the mine would pick up the necessary skills. Yet in practice the training process proceeded unevenly. If the newcomer's supervisor happened to be a careful, thorough worker, the training was good; if the supervisor happened to be a careless, inefficient worker, the training was poor. In either case, after several months the foreman gave the newcomer his own room and he was expected to assume the responsibilities of a full-fledged miner.

# 3

## LIFE ABOVE GROUND: THE ITALIAN-AMERICAN FAMILY IN IOWA

For Iowa's coal mining families in the late nineteenth and early twentieth centuries, coal mining was more than an occupation. It was a way of life. Because of the isolation of coal mining camps, the seasonal characteristics of the industry, the geographic mobility of the miners, and the meager incomes, coal mining families lived a life different from those of other Iowans who lived on farms or in incorporated communities. Because most coal mines were located far from large population centers, operators had to provide housing for the mine workers and their families. The result was that a large percentage of miners and their families lived in company towns where the company provided housing, owned the stores, and generally shaped the communities in which they lived. These conditions posed particularly difficult problems for miners' families. Not only did they have to endure poor housing conditions and frequent moves, but they also faced the problem of surviving on extremely limited incomes.[1]

While these difficulties affected all miners' families regardless of place of origin, it has been possible to document the Italian-American experience in Iowa's coal communities more fully than other ethnic groups. Because the Italian-Americans played a dominant role in Iowa's coal mining industry after 1900, their experience is a more recent one than that of emigrants from northern or western Europe. The Italian-American experience offers the opportunity to combine oral histories with more traditional historical sources. This testimony aids in assessing the emigrants' work experience in Iowa's coal mining fields, and it also helps in the study of cultural and occupational traditions that the emigrants brought from the Old Country. Historians have emphasized the need to study the emigrants' total cultural experience rather than merely the work experience. Herbert Gutman has pointed out in *Work*,

*Culture and Society in Industrializing America* that the new worker was affected not only by the specific job undertaken; rather he was affected by the total new environment. The emigrant discovered that there was an entirely new culture to be absorbed and an old one to be rejected. Throughout the entire emigration process, the family played a crucial role. As Gutman notes: "Tough familial and kin ties made possible the transmission and adaptation of European working class cultural patterns and beliefs to industrializing America."[2]

Chapters Three and Four will deal with the experiences of Italian-Americans in Iowa's coal mining communities. Chapter Three will describe the experiences of the Italian-American families, particularly their European background, emigration process, and initial settlement in the Iowa communities. Moreover, because considerable research has been done recently on the Italian-American experience in urban-industrial settings, it will be possible to contrast the experiences of those families with those of the Italian-American families who settled Iowa's rural-industrial areas. As an additional point of reference, a brief summary will be given of the life-style of Slavic families at Homestead, Pennsylvania, and black families at Buxton, Iowa. Chapter Four will deal with the experiences of Italian-American women in housing, domestic affairs, social activities, and medical treatment. Most of the information for Chapters Three and Four has been obtained from twenty interviews conducted with miners' wives, daughters, and one granddaughter in interviews conducted between 1975 and 1980. Of the twenty women interviewed, five were first generation Italian-American, nine were second generation Italian-American, and two were third generation Italian-American. Two women were of English background but married first generation Italian-American coal miners. All the Italian families emigrated from northern Italy. Although the two black women interviewed were born at Buxton, their parents had lived in Virginia and Missouri. Of the first and second generation Italian-American women, all but two lived in coal camps at least for several years. Additional material on women's roles and coal miners' family lives was acquired through interviews with retired Italian-American coal miners. Father John Gorman, Catholic priest at Granger for over thirty years, provided another perspective and valuable information on Italian-American mining families.[3]

In the late nineteenth century the number of emigrants from southern and eastern Europe greatly increased and these newcomers settled largely in the industrial cities of the Northeast. Italians, along with Croations, Serbians, Slavs, Poles, Lithuanians, and Hungarians went to work in factories, meat packing plants, steel mills, and on construction projects. Between 1870 and 1920, roughly four-fifths of the Italian emigrants settled in cities.[4] Studies of the Italian family have been made in New York, Buffalo, Boston, Chicago, Cleveland, Rochester, Utica, and Kansas City in which the Italian population has been analyzed in regard to work patterns, geographic mobility, social mobility, and family organization and disintegration.

The Italian emigrants' experience in industrial centers differed according to specific areas and particular conditions. But the generalization can be made that most urban Italian emigrants lived in crowded, cramped quarters and worked primarily as unskilled laborers. Italians quickly became known as "birds of passage" because of their desire to remain only temporarily in the United States. Most Italians emigrated thinking that they would save enough money to return home and purchase a small business or buy a small plot of land. Often the emigrants were single; most married men left their wives and children in Italy. By 1880 a steerage ticket between Naples and New York cost merely fifteen dollars, making it possible for the emigrants to cross the ocean many times, visiting and revisiting family and friends in the Old Country. Between 1892 and 1896, for example, for every one hundred Italians coming to the United States, forty-three returned to Italy. Between 1907 and 1911, the rate had risen to seventy-three repatriates for every one hundred Italian emigrants.[5]

Once the married Italian males were located in an urban area and secured employment, they sent for their wives and children. In some cases wives remained in the Old Country for five or six years before receiving money to make the trip to America. The majority of men, however, brought their wives and families to the United States within three and one-half to four years following their own emigration.[6]

Urban housing for Italian emigrants was uniformly bad. In New York City one observer described Italian housing at 5 Jersey Street in the following manner: "Here . . . on lines strung across were thousands of rags hung up to dry; on the ground piled against the board fences rags mixed with bones, bottles and papers; the middle of the yard covered with every imaginable variety of dirt." In some areas of that city, many of the rooms taken by Italians had no windows. Mulberry Street, although colorful, was so congested that even on sunny days the alleys and yards remained dark.[7] One first generation Italian woman vividly recalled her first home in New York, prior to

World War I: "We had a sink in the hall with nothing else, and four families to share it. And one bathroom in the yard where garbage was also thrown. How could a body wash and have a bit of privacy that way? I died a little every time I went there."[8]

Like all emigrants, Italians brought along their Old World traditions of family organization and prescribed male and female roles, fully intending to perpetuate these practices in America. In Italy men believed strongly that women belonged in the home. Although the wife might help her husband in the field, he expected her to "be home-loving, industrious, and obedient to his will. . . ."[9] At certain times of the year the wives and children might migrate to other parts of Italy or to foreign countries to do harvesting and canning. Seasonal emigration to secure work became a way of life among northern Italians for several centuries and because Italians practiced this type of travel so widely one Italian-American scholar declared that over time, Italians developed a "homing instinct not unlike the swallow."[10] Sometimes Italian males took their wives and families along but often they did not. Most northern Italians owned small plots of ground that they returned to farm during the summer.[11] Later, in the early twentieth century, unmarried Italian girls from north central Italy traveled to Marseille, France, to work as domestics during the winter months. They returned home in the spring to help their parents with the agricultural work. At other times young Italian women might work in nearby urban areas or in rural textile mills, but they all contributed to the economic support of their families. Both unmarried sons and daughters turned their earnings over to their fathers or whomever might be head of the household.[12] The point was always "to make enough money to send home." As Joan Scott and Louise Tilly point out in their article entitled "Women's Work and Family in Nineteenth-Century Europe," it is the family's solidarity among peasant societies that "provides the basic framework for mutual aid, control and socialization."[13]

The upbringing of the Italian female in northern Italy left little doubt that her future roles were to be those of wife and mother. Her future husband's roles were equally well defined. He would be the breadwinner and clearly the head of household. The female's roles were prescribed even down to the type of food she would prepare: mainly bread and soups, but little, if any, meat. Even though most people had little money, the typical village family lived in a single family dwelling in which family members could enjoy their privacy. The housewife spent much of her time outside at the public fountain where she col-

View of Riccovolto, a village in northern Italy, from which many Italian-Americans emigrated to Iowa in the early 1900s. (Courtesy of Mary Sertich)

lected water for household use or at the nearest stream where she washed clothes. The burden of household chores was lightened by the sociability of village life, and the housewife carried on these tasks in the company of other village women. Many women even mixed and baked their bread out in the warm sunshine. Women's responsibilities often included such outside work as raising a garden, caring for the animals, and selling surplus garden produce or dairy products. Generally, women worked in the fields at harvest and planting time. In Lombardy, women and children helped feed the silkworms, a seasonal activity.[14] Other women wove extra material, sewed for neighbors, or took in washing. Two first generation Italian-American women in Iowa recalled their work as young teenagers in Italy. Antonia Cerato remembered helping her parents take care of silkworms while Paulina Biondi recalled helping her father operate a grist mill. Both women were only fifteen at the time. Bruna Pieracci recalled that her mother, Filomena Piacentini, worked as a seamstress in Italy and that before her marriage, she walked to neighboring villages with a sewing machine strapped to her back to do the sewing for wealthy families.[15] As Scott and Tilly explain, most of these tasks were extensions of women's "household functions of food provision, animal husbandry and clothing making." Through such work women brought money into the household; their earnings often made the difference between survival and near starvation. Typically women managed the family accounts and as a consequence they gained considerable power within the family.[16]

When they came to the United States, the same general attitudes persisted toward family organization and women's roles: "The Italian ideal was to keep women at home." Italian women and children who settled in the Buffalo area, for example, continued to work as they had in Italy. Buffalo women remained home most of the year, but they did take advantage of part-time work. Every summer many women and children traveled to the Niagara fruit and vegetable areas where they harvested and canned the crops.[17] During the remainder of the year, many Buffalo Italian-American women kept boarders. This was true of Italian wives in other cities. According to Elizabeth Pleck's study on income earnings among married Italian and black women between 1896 and 1911, the percentage of Italian wives who took in boarders varied considerably from time to time. The figures in Table 3.1 were included by Pleck in her study.

In some cities, Italian-American women did factory piecework at home. In New York, one Italian woman explained: "Italians are different from American women, they don't like to work in factories and the men don't want them to do it. They must take the work home especially if they are married." Homework not only satisfied their husbands' demands, but allowed women to continue their traditional roles

Table 3.1. **Percentage of Italian families with income contributed by lodgers**

| Place | Year | Percentage | Number |
| --- | --- | --- | --- |
| Chicago | 1896 | 16 | 1227 |
| Chicago | 1911 | 17 | 219 |
| Buffalo | 1911 | 37 | 115 |
| New York | 1905 | 21 | 2945 |
| New York | 1911 | 20 | 333 |
| Philadelphia | 1911 | 16 | 195 |
| Boston | 1911 | 39 | 210 |
| Cleveland | 1911 | 41 | 111 |

Source: Elizabeth Pleck, "A Mother's Wages: Income Earning Among Married Italian and Black Women, 1896–1911," in *A Heritage of Her Own: Toward a New Social History of American Women,* eds. Nancy F. Cott and Elizabeth Pleck (New York: Simon and Schuster, 1979), p. 372.

of keeping house and caring for children. One area of homework in which Italian families specialized was flower making. In one Italian household, children aged three and four helped their mother and grandmother cut, paste, and assemble the millinery decorations. The four people produced 144 pieces in one hour for which they earned $0.10. Working for twelve hours produced $1.20. In New York City, many Italian women specialized in the "felling and finishing of garments."[18]

Italian-Americans also continued to carry over their Old Country attitudes toward children's work. Italian parents thought of their children as economic assets rather than liabilities and they expected them to go to work at an early age. Social workers and school authorities commonly complained that Italian parents did not value education and that they took their children out of school at the earliest opportunity. This practice forced many youths to enter the job market before they had received any specialized training; in turn, the young Italian-Americans continued their parents' role as unskilled laborers. Because parents believed that daughters should remain under their supervision, Italian girls did not enter the work force, particularly domestic service, in as large numbers as did Polish, German, or Irish immigrant girls. Italian-American parents followed the southern Italian folk proverb that "a good girl does not leave the parental nest before she is married."[19]

Even though Italian-American women spent most of their time at home, the urban environment did offer social and religious outlets. Although Italian-American men tended to be anticlerical, many Italian-American women remained active churchgoers. The church provided a source of comfort in the city as well as continuity with the Old Country. It also made possible a variety of associations for the immigrant family, many of which were social as well as economic. Local lodges provided both sick and death benefits. They also sponsored entertainment such as Saturday night dances.[20] Moreover, urban immigrants were free to use the services of the neighborhood settlement houses. Chicago's Hull House, the nation's best-known settlement, was located in a section known as "Little Italy." At Hull House, mothers learned how to care for sick babies and how to improve their families' diets. Settlement workers provided child care for working mothers along with a place where one could learn English and prepare for naturalization.

Not all southern and eastern Europeans headed toward large urban centers. Margaret Byington's study of Homestead, Pennsylvania, the company town owned by the Carnegie Steel Corporation, provides a view of southern and eastern European women's roles in a single-industry mill town. Although Homestead contained native-born people and English-speaking immigrants as well as many Italians and Slavic people, Byington paid particular attention to the Slavic families.

Life for southern and eastern Europeans in Homestead bore a considerable resemblance to life in a major urban center. Because they were the most recent employees of the company, they received the lowest wages, held the most menial jobs, and were the first to be laid off in a cutback. Their low wages forced the Slavs to live in the cheapest and therefore the most cramped quarters.[21]

Because Homestead was a steel producing plant, wives and daughters were automatically excluded from work at the plant. Instead, Slavic women kept boarders as the major way of supplementing their husbands' wages. In the twenty-one courts (housing areas) studied by Byington, 102 families out of 239 took in boarders. Of the women taking in boarders, three-fourths received an amount at least equal to the family's monthly rent while one-fifth received at least twice their rent money in boarders' fees. From another perspective, over one-half of the women keeping boarders earned one-fourth or more of their husbands wages. Of the total number of families keeping boarders, 62 had 4 boarders or less, 33 kept from 5 to 9 boarders, and 7 families from 10 to 15 boarders.[22]

Yet the additional income was not without cost. Health problems stemmed from the overcrowding, particularly among the young Slavic children. According to birth and death records for Homestead's Second Ward—the most crowded—"one child under two dies for every three born." Physicians reported that "malnutrition due to poor food and overcrowding" caused many of the deaths. Byington noted that many mothers appeared "too poor, too busy, and too ignorant to prepare [proper meals] while the rooms were over-tenanted, and courts too confined to give the fresh air essential for the physical development of children." Pregnant women often preferred the services of a midwife, most of whom were friends and neighbors, rather than physicians. In the view of the company doctor, this preference frequently resulted in incompetent care. Generally, new mothers had to return to work three or four days following the birth of their babies. Byington noted many cases in which the Slavic mothers were in poor health due to overwork and "to lack of proper care during confinement." These conditions combined with poor diet and poor ventilation for the mother, in turn, contributed to the high infant mortality rate.[23]

Although most Homestead families had little money, they did have some leisure opportunities. Six nickelodeons provided the public amusements for young people while adults could patronize the town's fifty saloons. At the same time about fifty lodges existed; some lodges were purely social while others were designed to provide sickness and death benefits as well as social outlets for members. For women, lodges supplied one of their few occasions for outside activity. Church lodges also operated, which provided additional social and economic benefits for both husbands and wives. In addition to the fraternal and church lodges, the Slovaks, Croations, Poles, Hungarians, and Lithuanians established independent lodges that supplied sick and death benefits.[24]

The Catholic Church represented the most significant institution in the lives of the unskilled workers at Homestead. Some families belonged to the Greek Catholic Church; others attended the Roman Catholic Church. By providing lodges for members and parochial schools for the children, churches not only offered spiritual resources but also provided social and educational functions for their parishioners.[25]

For Italian emigrants settling in Iowa, conditions appeared quite different. Missing were the urban centers with tightly packed, steamy tenements and large, mechanized factory operations. Instead, emigrants discovered sprawling coal camps covering the Iowa countryside and small coal mining operations that averaged two hundred employees. Coming from the rural areas of northern Italy, where most emigrants had been agricultural workers, the newcomers moved into rural areas of Iowa where they worked in small industrial operations as coal miners.

Most Italian emigrants relied almost exclusively on assistance from family members or village associates throughout the Iowa immigration and resettlement process. As one Italian-American phrased it: "Family and village ties monitored the entire migration process; they tied thousands of men and women together, irrespective of where they lived at a particular moment; and they provided entry into local communities."[26] Because most Iowa Italians came from poor economic backgrounds, they could seldom leave Italy without sponsors. Typically sponsors were relatives or friends who were already settled in the United

Coal camp housing, Fraser, Iowa, 1905. Larger dwelling (*right*) is the company boardinghouse. Note power plant across the Des Moines River. (Courtesy of Larry Coe, Fraser)

States and who provided the emigrant with money for passage and a place to board. James Battani recalled that Italian-Americans showed tremendous loyalty toward their family and friends in the Old Country. A newly arrived Italian emigrant probably first thought of assisting his immediate family, but he did not forget his extended kin. Battani remembered many cases in which uncles helped their nephews emigrate to Iowa. Battani noted that the father would write from Italy to his brother residing in an Iowa coal camp that "I'm gonna send the boy [his son] over and you help him get started."[27]

Upon his arrival, the newcomer went to work in the mine as an assistant to his sponsor. This was significant because both the emigrant and the sponsor benefited. The practice had long existed in American coal mines of giving a miner double the number of cars ordinarily assigned him on any given day if he took a younger person with him into the mine. Since most miners could load more cars than they were usually assigned, this was an important move. The miner with the assistant could load out double the amount of coal and, in turn, double his wages. The emigrant also benefited because the practice allowed him to repay the sponsor and at the same time learn mining procedure. After five or six months the emigrant could be hired by the coal company as an independent miner.[28]

The second step in the resettling process concerned marriage and the family. Typically the Italian male continued to work until he saved enough money to send for his wife or his fiancée. If he wished to marry an Italian girl, but did not have a specific one in mind, he wrote to family members in Italy, asking their assistance in finding a bride. Some young men returned home to find their own wives. Paulina Biondi recalled that her husband John had come back to his native village in May 1914 looking for a wife. John had emigrated to the United States at age sixteen and six years later returned home to find a marriage partner. Paulina's uncle and aunt introduced the two young people. At first Paulina expressed no interest in the young man, but three months later they were married; two weeks later, they left for the United States.[29] In seeking a wife, the immigrant sometimes enlisted the help of an older Italian-American in his mining camp, known as the social boss. The position of social boss, a practice carried over from the Old Country, was usually held by a highly respected person who could speak and write both English and Italian. He helped the newcomers to write letters, handle legal matters, and apply for naturalization papers. Actually, the person became something of a spokesman for the Italian-American families in the community. In cases where a young man desired to marry a young woman from the Old Country but had no contacts there, the social boss wrote to friends in Italy and arranged for a match.[30]

If the miner already had a wife and family in Italy, reuniting the family took considerably longer. The first step was to save money for the wife's passage and possibly the youngest child or two. Many Italian emigrants followed the practice of placing the older children with relatives in Italy until additional money could be saved for their passage. Once the wife arrived and the family accumulated money for the older children's passage, the family waited until someone from their village was ready to emigrate; they then made arrangements for their children to accompany that adult to the United States.[31]

If the emigrant had not obtained a sponsor, he could count on immediate assistance from fellow Italian-Americans once he arrived at his final destination. For example, Seymour, Iowa, contained numerous Italian-American families. In the early 1900s, Victor and Giacomina Cambruzzi served as the local welcoming committee. The Cambruzzis made arrangements with all railroad conductors to give incoming Italians instructions on how to reach the Cambruzzi home. When the young men arrived, Giacomina Cambruzzi made boarding arrangements for them with local Italian families. Victor Cambruzzi served as the Italian social boss in Seymour and helped the emigrants get jobs at one of the local coal mines.[32]

Once settled in an Iowa mining community, Italian immigrant families faced the challenge of adjusting Old World traditions and values to their new environment. Although the occupation of coal mining imposed considerable hardships on all family members, the type of work as well as the geographical location of the mines provided considerable freedom for the newcomers. The Italian emigrants moved from small villages in northern Italy to small villages in southern Iowa. The major change came in the form of male employment because the male moved out of agricultural work into industrial work. The rural-industrial environment allowed the Italian wife to continue unchanged many of her Old World habits and traditions that she could not have done in an urban setting. Although coal camp houses were small, each family occupied its own house. The Italian male attitude that a wife should not work outside the home was compatible with the Iowa situation; coal camps provided no opportunities for women to work in factories or as domestics. Women could earn money, however, by boarding newly arrived Italian emigrants and many women regularly boarded four or five men. Other women earned money by sewing for outsiders, selling garden produce, or working as midwives. In many cases women realized substantial savings by raising chickens and pigs and milking cows, along with producing large quantities of fruits and vegetables.[33] During the summer months Italian-American women frequently moved some of their activities, such as washing clothes, outdoors. Each coal camp had several wells in the center of the camp. Coming to the wells several times each day for water, the women were able to visit much as they had in the Old Country when they gathered around the village fountain. Because of the Italian dominance in the camps after 1900, Italian-American women were usually surrounded by other Italian-American women. This situation sometimes produced a feeling of comradery and sense of community. Paulina Biondi recalled that during the summer when her twin boys were born, the camp's well went dry. The only operating well was located nearby the boardinghouse. During the day the boardinghouse proprietress put a padlock on the pump to keep the camp women from using it. Paulina decided that with four children, two of whom were infants, she needed water. The woman at the boardinghouse refused to unlock the pump so Paulina decided it was time for action. She went to the coal house, located a big hatchet, and returned to smash the padlock. The proprietress attempted to stop her but could not do so. She returned to her house and along the way spread the word that the women could now get water at the boardinghouse well. Paulina told her story to the mine operator who responded by saying, "Good for Mrs. Biondi!" The order quickly went out that all camp families could obtain water from the boardinghouse well.[34]

While the women worked at these traditional domestic tasks, they were still able to remain in their homes and care for their children. The Italian-American women could carry over their time-honored roles of wife and mother into the Iowa experience. Moreover, the resettlement did not disrupt the equally time-honored practice of the Italian male as breadwinner. Regardless of the meager amount earned by the male, he clearly remained the head of household. The Iowa experience did not cause a breakdown in Italian family structure or in the family's patriarchal organization.

Sunshine Coal Company Number One in the 1930s showing mine workings and camp housing. (Courtesy of Iowa State Historical Department, Division of Historical Museum and Archives, Des Moines)

Once resettled in Iowa, Italian-American families continued to maintain the Old World practice of operating as productive units. The Italian attitude that children were an economic asset and that they should be put to work at an early age was modified somewhat by the local conditions the immigrants faced in Iowa. In 1906 the state restricted work in the mines to boys over the age of fourteen.[35] Nevertheless, some fathers took their sons into the mines before they were fourteen. In cases where father and son worked together, the father collected and retained the paycheck. When the son worked with another adult, the son turned his paycheck over to his father. Second generation Italian-American James Battani recalled that at one time he and six brothers lived at home and all worked in a local coal mine. His parents, like all parents from the Old Country, insisted that their children turn over their paychecks to their mother. He explained:

That was a must. That was a must. With all those [people from the Old Country] that was a must. They insisted. You brought your paycheck home from the mine and you left it on the table for the folks and they would give you back some spending money. And my dad was strict. When I first started, they gave me $2 for two weeks, that was my spending money.

At that time Battani was making between $60 and $70 every two weeks. Battani added, "I don't regret it because my Dad needed the money, that's all there was to it. You know when you have ten kids. . . ."[36] When each son in the Battani family reached twenty-one, the parents allowed him to keep his paycheck; he then paid his mother $10 per week for room and board. The parents also allowed each son to buy an automobile at age twenty-one. Battani added that all Italian-American families followed the same practices in regard to their children's earnings. He admitted that "our family was lucky because somehow or other we did have some extra money." Nine of the Battani children — seven sons and two daughters — played musical instruments and formed a local band. During the summer months they entertained at local picnics where each member earned $5. The parents agreed that the children could keep the band earnings.[37]

The seasonal mobility of Italian families carried over to the Iowa setting. Although the residence of wife and children usually remained the same, the seasonal nature of Iowa coal mining often forced the head of the household to look elsewhere for summer employment. Sometimes there would be prolonged absences from home by the head of the family who worked in another coal mine outside commuting range.

Company store (*first floor*) and miner's hall (*second floor*) at the Highbridge Coal Camp, with miners' houses shown in the background.

---

Many Iowa Italian-Americans went to Chicago to obtain summer employment. Every family interviewed in Seymour had at least one relative, if not more, who had worked in Chicago for at least one summer. Apparently this practice affected both first and second generation families. In Chicago the men did gardening and landscaping for wealthy families, often returning to work for the same family summer after summer. The Italian-American miners in Granger, Madrid, and Rippey followed the same practice. Sometimes wives and children went along and stayed with relatives in Chicago, but more often the men went alone. As the children grew older, fathers took along their oldest sons to help with the work. Most Italian-American men had done agricultural work in Italy and apparently were highly skilled in the practice.[38] One Italian couple, John and Augusta Argenta, traveled to Chicago for summer employment for several years following their marriage. Augusta Argenta had two married sisters in Chicago who housed them and cared for the Argenta's only child. This enabled both Argentas to work. Although Augusta preferred to remain in Chicago where both she and her husband would have an opportunity to work, John Argenta did not like the city so they returned to Seymour each fall.[39] James Battani recalled that he went to Chicago for many summers and was always able to get work with "some Italian firm." The first two summers he worked in a greenhouse; later, he worked on construction. He stayed with relatives in the Chicago area.[40]

Highbridge Coal Camp, Des Moines Township, Dallas County, 1919. Six-room coal camp houses are shown.

---

Many daughters who left home during their late teen years to work in urban areas like Chicago continued to send money home to support the parental household. They often worked as domestics or as nursemaids, thus continuing to use the skills that they had acquired in their own homes. Close ties continued, nonetheless, with their families in Iowa.

The Catholic church as a characteristically influential institution in most Italian-American localities was not carried over into all Iowa coal communities. Seymour never had a Catholic church, nor did Numa construct one until some twenty-five years after the first Italian families settled there. Italian-American women who resided in Dallas County coal camps went to Des Moines to church because local communities did not have a Catholic church for many years. The absence of the church distressed the first generation women. Victoria DeGard stated that her mother, Giacomina Cambruzzi, never forgave her husband for taking her to a place that did not have a church and insisted that at least once a year the family go to Centerville, where the nearest church was located. She then decided that as long as the family was there, they would stay for several masses. The day proved to be an ordeal and costly, too; her husband rented a wagon and horses from the local livery stable to make the trip; the family left before dawn and did not return home until after dark.⁴¹ Many Catholic women kept religious statues and candles in their homes and at least attempted to teach their children the proper prayers. Eugenia Padovan sent her four daughters to the Christian church in Numa because she believed that they should have "at least some religious training."⁴² Paulina Biondi remembered that she had "stayed away from the Church for fifteen years," after she arrived in Iowa. She initially urged her husband to take her to church, but it was difficult because the nearest church was several miles away and they did not have an automobile. She remembered that "little by little I forgot about the church for fifteen long years. When I went back, I'm telling you, it was a blessing." Paulina added that with so many children and so much work to do, she found little time to think about the church.⁴³ Occasionally a Catholic priest from a neighboring community held mass in a private home in Seymour. This same practice was followed in the coal camps in Dallas County.⁴⁴

The absence of the church in the lives of the Italian-American people creates an interesting reservation about the supposedly primary role played by the church in their assimilation into American life. In many studies of immigrant families, the church is assumed to be a basic influence in the immigrant's life; some scholars maintain that immigrants embraced religion even more rigorously in the New World than they did in the Old. One historian concluded that "religion intensified the separateness of the immigrants and transformed them into distinct subcultures."⁴⁵ In Numa and Seymour it is difficult to assess the effect on the original Italian families. Apparently with the second and third generation Italian-Americans the inability to attend church has produced a group of people who feel little loyalty to the church. One second generation woman, Minnie Mores, does not go to church and has told the local priest that she "practiced the ten commandments and that was enough."⁴⁶ Other women interviewed agreed that their religious feelings and affiliations are weak today because they did not attend church regularly during their younger years.

Yet the absence of the Catholic church in many Iowa mining communities was not unusual, especially in light of the Italian-Americans' European background and their American urban experience. According to scholars like Enrico Sartorio, Robert Foerster, and Rudolph Vecoli, Italians traditionally, both in Italy and the United States, attended church only on special religious days and felt little allegiance toward the Catholic church. In his study on Italian immigrants and the Catholic church, Rudolph Vecoli points out that in southern Italy the people had "little sense of reverence" for the church. Nor did the villagers have high regard for the priests; they often regarded the priests as parasites living off the local populace. Priests' education was poor and their parishioners did not perceive them as learned men. While Vecoli describes the southern Italians "nominally" as Roman Catholic, he believes that their religion was more "a folk religion, a fusion of Christian and pre-Christian elements, of animism, polytheism, and sorcery with the sacraments of the Church." Because they greatly feared the supernatural, the Italian peasants were extremely religious. But as Vecoli points out, their beliefs and behavior did not conform to the "doctrines and liturgy of the Church." Once in the United States, they continued to follow their Old World practices. Moreover, the Italian-Americans soon discovered that the American Catholic church discriminated against them as a group and did not consider them worthy of the church's attention. The negative result was that most Italians were married in the church and had their children baptized in the church, but there religious life ended.[47]

Once located within the United States, Italian-Americans found themselves in an alien religious setting. Much to their surprise, the church continually asked them to contribute to its support. Because the Italian government supported the church in Italy, the immigrants were not accustomed to a voluntary financial arrangement. Therefore many Italian-Americans did not contribute to the church. American Catholic officials lamented the absence of Italian-American financial support while pointing out that Italian-Americans continued to expect the presence of a priest in their midst. And the Italian-Americans soon came to the realization that the Irish were the dominant force within the American Catholic church. In turn, the Irish were aghast at the Italians' lack of respect for the clergy, their niggardliness toward the church, and their superstitious ways. As a consequence church officials frequently shunted the Italian-Americans aside or simply forgot them. Some priests regarded them as "little better than pagans and idolators."[48]

If their church attendance was poor in Italy, it was even worse in the United States. In 1918 one Italian scholar estimated that 60 percent of the Italian immigrants had no association with the Catholic church. Although women attended mass more faithfully than men, even they began to fall away after living in the Unites States for several years. The immigrants' stock answer was: "We did not come to America to attend church but to work, make some money, and go back home. When we return to Italy, there we will attend mass." American church officials, however, diagnosed the problem differently. Most ethnic groups brought along their priests to minister to their needs in the United States. The Italians did not, largely because the priests seemed to be indifferent to the fate of the Italians once they arrived here. Therefore the blame was often laid at the doorstep of the Italian clergy. Even when Italian priests did come to the United States they faced many obstacles. Other Catholic officials often treated them as inferiors. Their own parishioners, not accustomed to voluntary church financing, gave little money for their support. Eventually some Italian priests chose to serve other ethnic populations because they were shown more reverence and given better financial support. During the first two decades of the twentieth cenutry, Catholic officials increasingly viewed the fate of the Italian-Americans with alarm. They often referred to the Italian situation as "our biggest Catholic question."[49]

Complicating the situation were the strong anticlerical views of many Italians, both in Italy and in the United States. Many of these people had worked assiduously for the unification of Italy and in the process had come to view the Pope as their chief obstacle. These attitudes predominated among the educated classes. Once in the United States, these people continued to wage their campaign for a united Italy while at the same time they refused to suppress their criticism of the Pope. This element of the Italian-American community evoked even more scorn from Irish church officials.[50]

Northern Italians tended to be the most anticlerical, sometimes repudiating any Italian priest who came into their midst. In a coal mining camp at Chicopee, Kansas, coal miners greeted two Italian priests with a shower of rocks and rotten vegetables. Italian-American stoneworkers at Barre, Vermont, treated Italian priests in similar fashion.[51] The reaction of Italians in Iowa followed somewhat the same pattern. Andrew Zanotti, who emigrated to Iowa at age twelve, held fiercely anti-Catholic views. In 1930, thirty-seven-year-old Zanotti and his family moved to Rippey where they joined an Italian community of about fifty people in a town of four hundred. According to Zanotti's granddaughter, all the Rippey Italian-Americans shared her grandfather's hostility to the Catholic church, citing the absence of a Catholic church there. Andrew Zanotti said many times that the Italians were bitter because the church "bled the people dry in Rome." The Zanottis, like many other Rippey Italian-Americans, belonged to the local Methodist church.[52] Father John Gorman, parish priest at Granger and

William Cuzzens, Sr., and family in front of a typical Buxton residence, 1915. (Courtesy of Iowa Mines and Minerals Department, Des Moines)

manager of the Granger Homestead Projects, also encountered strong anticlericalism among Italian-Americans living in the coal camps around Granger. Father Gorman made repeated overtures to the miners and their families by offering to conduct mass in their homes and to visit them in the event of sickness or death. Over the years Father Gorman succeeded in bringing some of the families back into the church, but others persisted in remaining apart.[53]

While Italian-Americans constituted the major foreign-born populations of most Iowa coal mining communities after 1900, Iowa also included a community which contained many black coal mining families. In 1900 the Consolidation Coal Company founded the coal camp of Buxton. Located in northern Monroe County, the population of Buxton and several outlying coal camps grew quickly to around five thousand. Census data indicate that early in Buxton's history blacks accounted for approximately 54 percent of the town's population; by

1915 blacks constituted less than 50 percent of the population. Consolidation Coal Company, a subsidiary of the Chicago and North Western Railroad, was organized to provide coal for the parent company. Before 1900 the company operated mines near Muchakinock in Mahaska County. But when these mines began to play out the company moved their operations south to Monroe County where they had located new coal reserves. Many Muchakinock residents then moved to Buxton, continuing their employment with Consolidation.

Once relocated in Buxton, Consolidation Coal Company soon earned a reputation as a progressive, equitable employer. The company offered its employees equal wages, substantial and integrated housing, and a wide array of social opportunities. Perhaps the most significant economic aspect of life in Buxton was that Consolidation was a captive mine and thereby sold all its coal to the railroad. As a result, the Buxton mines operated twelve months a year. This full-time employment meant that Buxton miners ranked among the best paid miners in the state. This job security stood in sharp contrast to the fate of the majority of Iowa miners who regularly experienced four months of unemployment each spring and summer. Consolidation's full-time employment, equal hiring and wage policies, and Buxton's general environment led to a widely expressed view by former black residents that Buxton was "the black man's utopia in Iowa."

The houses in Buxton were company owned but were slightly larger than most company homes. They contained five or six rooms, generally with two bedooms on the second floor. Some homes had a small built-on or summer kitchen at the rear of the house while other homes had front or back porches built onto their homes by the residents. According to one former resident, company employees painted all homes a dark red color; the same person remembered that the company maintained the houses adequately. Like most company housing, the homes had no basements.[54] For the most part, white and black families lived side by side in the fifty-odd blocks that composed the town.

The view that emerges from the interviews with former residents is that Buxton was an exceptionally good place to live. In contrast to most Iowa mining camps, Buxton residents had a wide array of activities from which to choose. The Consolidation Coal Company erected two YMCAs, one known as the "Little Y" and the other known as the "Big Y," which apparently served as the social centers of the town. The two "Ys" provided residents with a swimming pool, tennis courts, pool rooms, and roller skating facilities. The "Ys" also sponsored musical concerts and speeches by such notables as George Washington Carver. YMCA officials ran weekly movies and presented minstrel shows. Local residents attended dances at the "Little Y," which also offered them such services as the use of typewriters and sewing machines. Furthermore, the large "Y" contained a well-stocked library and a fully furnished gymnasium.[55]

Yet there were other social activities outside the YMCAs that black women attended. Dorothy Collier remembered her mother and several neighbor women getting together for afternoon quilting parties, combining visiting, quilting, and refreshments. Marjorie Brown, also a former Buxton resident, recalled that her grandmother belonged to the black federated women's club and that her mother belonged to a literary and cultural club. In the early 1900s some publishing companies issued their books in separately bound chapters; when the literary social club selected a book to study, members took turns reviewing the latest chapter of the book. The other club members did handwork while they listened to the review.[56]

The church represented an important institution in the lives of the black families in Buxton. The community supported black Methodist, Baptist, Congregational, and Episcopal churches, as well as a white Methodist and Lutheran church. The black churches had large choirs, and Marjorie Brown's mother, Melvina Lee, a graduate of Leon High School and an accomplished musician, directed the choir and played the pipe organ at St. John's African Methodist Church. These churches sponsored numerous social events during the year, particularly picnics and ice cream socials during the summer.[57]

In addition, black residents belonged to a variety of fraternal lodges. Marjorie Brown's mother belonged to the black chapter of the Eastern Star in Buxton. As a child, Marjorie remembered that she could hardly wait to grow up so she could "wear long evening gowns and go to balls," like her mother. She remembered that her mother and other women attended dances at the YMCA when the streets were so muddy that the only way the women could cross the streets was to wear overshoes and carry their dainty slippers in their evening bags. Black residents also belonged to separate branches of the Benevolent Protective Order of Elks and the Knights of Pythias. All lodges met in the large YMCA.[58]

Every mining camp had its baseball team and Buxton was no exception. The Buxton Wonders, originally an all black team, were the pride of the community; many local observers regarded the Wonders as a professional team. The team, sponsored by the Consolidation Coal Company, traveled extensively throughout the state and hosted visiting teams from Chicago, St. Paul, and Kansas City.[59] Dorothy Collier's father, George Neal, played on the team and she remembered the excitement of attending a ball game. The Consolidation Coal Company erected a grandstand on the edge of town where the Wonders played their games. Dorothy Collier's mother would sometimes rent a surrey from the local livery stable, and the family would drive out to see the game. Dorothy recalled: "That was a big day. On the day they would play the viewing stand would just be loaded. Papa was such a clown, he would cartwheel around to the bases, rather than run."[60] Buxton also had its town band, the Negro Concert Band, which was widely known throughout southern Iowa. At its height the band numbered around fifty members and gave performances in many towns, including several performances at the Iowa State Fair.[61]

Buxton provided a wide array of stores in which local women could do their shopping. The company store, the Monroe Mercantile Store, dominated the business section and boasted over 100 clerks and several different departments such as "ladies' ready to wear" as well as a complete line of groceries. In many ways, the store was a forerunner of a modern day department store. In addition, the business district contained two general stores, several meat markets, bakeries, restaurants, drug stores, and several beauty shops.[62]

One former Buxton resident compared her life in Buxton to "life in an oasis." Reflecting on her childhood there, Marjorie Brown reflected that she knew nothing of the discrimination that lay outside the community. Her parents stressed education and culture, and her family achieved a high standard of living. Her father worked as a hoisting engineer with Consolidation and enjoyed a considerable measure of prestige in the community, as well as a substantial salary. She had never lived in a house that did not have a piano and a carpet on the floor, and every day her mother had a starched white cloth on the dining room table. Since her father came to the table in his work clothes in the morning, her mother put newspapers down so her father would not dirty the cloth. In the evening, however, everyone appeared at the table properly dressed for dinner. Following the death of her mother, when she was ten years old (her father had died one year earlier), Marjorie Brown went to live with her grandparents, also Buxton residents. Within that household as well, education was highly valued and all members were constantly encouraged to improve themselves.⁶³ Every evening as they gathered around the dinner table, her grandfather asked her and her brother, "Well, what did you learn today?" She explained:

Now that didn't mean that you learned to wash dishes; but what did you learn, what did you see, what made it an interesting day? What did you read and what did you learn? And if you said nothing, if you hadn't learned anything, if you sat there dumb, then he would tell you how you were wasting your life. Life meant you should learn. And I thought all the people did that. . . .⁶⁴

For Dorothy Collier, whose father was first a miner and then a tailor (as well as a member of the Buxton Wonders), life was also full of material comforts. She remembered the fine furniture that her family had and that the "front room was used only on Sunday. We wouldn't go in there every day." At mealtime, everyone observed proper manners.

Yes, I remember the table. My brother would always have to pull a chair for Mama, and we had to use the correct silverware. And a blessing was said. . . . And we had to use a napkin. We all had a little ring that you put it in. And always a tablecloth. I could only sit in a certain way. Since I was a little girl, I couldn't sit with my knees crossed. . . . And my father was that way through his entire life. He was very particular about eating at the table; sitting correctly. When you came in the door with a hat on, the first thing he would say, "Did you forget something?"⁶⁵

For the black people, Buxton was indeed a virtual utopia. Marjorie Brown's final comment about life in Buxton sums up well how most Buxton residents felt: "Buxton was something else. You can imagine how we grieved for it."⁶⁶

The Italian-American experience in Iowa's coal mining communities contrasts rather sharply with the experience of Italian-Americans who settled in urban areas of the Northeast. It appears that Iowa Italian-Americans lived more comfortable lives and underwent fewer adjustments than did their urban counterparts. Iowa's rural-industrial setting gave industrial employment to the newcomers but at the same time allowed them to move from a small village environment in northern Italy to a somewhat similar small village environment in Iowa.

The Iowa Italian-Americans seemed to have experienced greater stability in their occupations than did their urban counterparts. While it is true that coal miners frequently needed to travel from one coal camp to another seeking work, many families remained in the same community while the family head changed places of work. The occupation of coal mining continued to be available to Italian-Americans through the 1920s, so they had little need to change occupations. On the other hand, the first and second generations of Italian-Americans who settled in the industrial Northeast often took jobs that were temporary in nature such as railroad construction, building construction, and road building. Both groups experienced seasonal unemployment, and therefore both had to seek temporary work at least part of every year.

The limited size of Iowa's coal companies played a central role in the success of the sponsorship system that enabled Italian-Americans to continue to bring friends and relatives from the Old Country. While many coal miners did not have direct contact with the mine owner, they did have direct contact with the mine foreman. He was easily accessible; when a relative or friend arrived from the Old Country, it was a simple matter to speak to the foreman and arrange for the newcomer to enter the mine. Iowa coal miners did not know a sense of alienation that often accompanied employment with larger firms.

One constant feature of life in both regions was poverty or near poverty. Elizabeth Pleck concludes that the general economic conditions among Italian-Americans in northeastern urban centers between 1896 and 1911 led to a life of "desperate poverty." Although Italian-Americans in Iowa might have been equally hard pressed financially, they had more opportunities to ease the economic pinch. Most coal mining families raised gardens, milked cows, fed chickens, and produced a substantial amount of their food. To earn money, Italian-American women generally kept boarders, and many women contributed to the family income by sewing, nursing, and midwifery.

The experiences of the Slavic families in Homestead form an interesting contrast with the experiences of the Iowa Italian-Americans. Yet in many ways the lives of family members in these two areas show similar patterns. Both groups were always concerned with ways to stretch their meager incomes and both partially solved that problem by taking in boarders. The effect of that action, however, seemed less detrimental to the Italian-Americans than to the Slavs. Homestead families had some social advantages such as the presence of the church and some recreational activities outside the home. But Italian-American women had more opportunities to earn money in the food producing area. No doubt their own families ate more nutritious meals because they raised their own gardens and kept livestock and poultry. Bleak as the coal camps might have been, they offered individual dwellings and a somewhat more sanitary environment than urban tenements. And no observer ever described the Iowa camps as congested.

On the other hand, the experience of black families in Buxton offers little basis for comparison with that of the Italian-Americans. Given Buxton's association with the Chicago and North Western Railroad and the miners' subsequent full-time employment, the standard of living for most Buxton residents was considerably higher than that of other coal camp residents. Few, if any, Buxton miners faced the need to seek outside temporary summer employment. More than any other factor, income provided the main contrast between the life-style of black families in Buxton and those of the Italian-American families. Black families in Buxton did not live in poverty or near poverty. They ate well, dressed well, and lived in comfortable five- or six-room houses built and properly maintained by the Consolidation Coal Company. Furthermore, black families experienced little social or economic isolation and had several black Protestant churches they could attend. The town's many stores provided a full line of goods and services. Of greatest importance, black families could and did attend social functions several times each week at the local YMCAs.

While life in an Iowa coal camp left much to be desired, it offered some advantages to Italian-American families who settled there. Family ties and family relationships were kept intact, much like they had existed in Italy. While coal miners made only meager wages, at least coal mining continued to be an available occupation throughout the 1920s. The general environment, while dreary to many, nevertheless offered the immigrant family a more wholesome and spacious setting than they would have encountered in many eastern cities.

On the other hand, social and religious opportunities for Italian-Americans were almost nonexistent in the Iowa camps. In the cities all family members were able to take part in church and lodge activities, but in the Iowa camps, family activities provided the major social outlet.

Perhaps most important, the emigration system devised by the Iowa Italian-Americans appeared to work. For almost twenty-five years, a steady stream of Italian emigrants flowed into the state, being absorbed quickly and with little difficulty into the economic life of the state. For the most part, this absorption was made easier by strong family and ethnic ties. While each person had to work hard and some had to live on the edge of poverty, the system provided a better life than in the Old Country. Greater economic stability and a promise for future social and economic advancement were implicit in their new environment. At the same time, when compared to immigrant life in urban centers, the Iowa coal communities allowed the immigrants to maintain greater control over their lives in the areas of economics, sanitation, child care, and family stability.

# 4

# LIFE ABOVE GROUND: THE ROLES OF ITALIAN-AMERICAN WOMEN

For the hundreds of Italian-American women who arrived in one of Iowa's coal camps in the early twentieth century, life must have seemed strange indeed. The unfamiliar landscape, the unusual type of housing, the unintelligible language all reinforced the women's knowledge that they had left their homeland far behind. Gone were the familiar mountains and the solid stone buildings. Instead the women gazed over row after row of small, squat wooden houses, many of them set down in the middle of an Iowa cornfield. For many women, the scene was not a welcome one. It had been the husband, not the wife, who determined that the family should emigrate to America. For the Italian women, leaving family and friends behind was difficult to do. Perhaps their main solace was their intention to stay in America only a few years. Then, with their anticipated savings, they would return home to live more comfortably than before.

Once located in an Iowa coal camp, the women soon discovered that their new life would be one of coping with adversities. Coal camp houses were small and usually run-down, and most camps suffered perennial summer water shortages. Winter brought bitterly cold weather and the flimsy company houses provided little protection. Few Italian emigrant women spoke English; all faced a language barrier at one time or another. Most of all, the landscape seemed alien from that of their homeland, northern Italy. Even with these difficulties, most women soon decided that life was better here than in Italy. It didn't take them long to realize that they must seize every opportunity to earn money to insure the economic survival of their families. The first generation Italian-American women revealed a tremendous capacity for hard work in carrying out their domestic responsibilities and at the same time performing services that marked them as important family wage earners. In many ways these women followed much the same routine here as they had followed in the Old Country with the exception that the economic rewards, even though meager, were greater here than in Italy. For second generation Italian-American women, life improved considerably. Although many of these women had to work hard and live on small incomes during their early married years, eventually they experienced some economic security and greater social opportunities.

A pressing problem facing many Italian-American women in Iowa's coal camps was poor housing. The majority of women lived with their families in one of the state's many mining camps where they rented company housing. The remaining families either rented private housing or purchased their own homes. In the latter case, the women experienced a considerably more pleasant existence since most of the private homes were located either on small acreages or on large town lots. Women who lived in company housing quickly discovered that coal operators regarded these facilities as temporary. Since the average life of an Iowa coal mine was only eight to ten years, operators constructed company housing accordingly. Consequently the company houses were small and poorly built. Moreover, operators gave little attention to the actual location of the camps. Observers described the camps as unattractive and depressing, noting that there were few trees, shrubs, or even flowers to break the monotonous landscape. Once constructed, companies resisted spending money on maintenance, and leaky roofs, broken windows, and sagging doors soon became commonplace. Miners themselves felt little motivation to invest time or money in improving homes that they might soon leave behind.[1]

Health and sanitation problems sometimes plagued camp residents. Coal operators frequently selected camp sites hastily with the result that an adequate water supply was not always available. At the same time, company officials did not always give proper consideration to problems of drainage. Heavy rainfall brought seasonal flooding and the many pools of stagnant water provided breeding areas for mosquitos. In other areas workmen failed to encase the wells properly and frequently residents discovered dead rodents and insects in the water. In 1919 the state sanitary inspector concluded in a report to the state commissioner of housing that the overall sanitary conditions in the mining camps were poor. He cited contaminated water supplies (recommending that water from every usable well in every mining camp be analyzed and proper steps be taken to condemn contaminated wells), poorly constructed privies that were "really just shacks," and no systematic scavenger or rubbish collection.²

The mining camps were easily recognized by their long rows of small, square, one-story houses built by the coal operators. The following account given by the state sanitary engineer in 1919 of a four-room mining house at Highbridge typified company housing in Iowa:

> The houses are of the usual mining camp construction—shiplap nailed directly on studding, lathed, and plastered about ⅛ inch thick. The foundations are of brick, and in fairly good condition generally. A few foundations need some repairs. The houses are old, the shiplap has loosened, the window sills are loose, and the doors warped. Many of the roofs leak, the plastering is in bad shape, and the paper has come off in patches. Single flooring only is provided. Consequently, the houses are cold in winter, wet in spring, and not particularly inviting as habitations.³

The monthly rent varied from $8 for a four-room house to $10 for a six-room house, although mining families found the larger houses in short supply.⁴

Living in company housing was generally not a pleasant experience. All of the interviewees remembered that not only were the houses small, but they were poorly insulated. Mary Braida recalled that as an eighteen-year-old bride she moved into a company house in the coal camp of Omnitz in Lucas County that was already occupied by her widowed mother-in-law. Six months later, the young Braida couple moved to the nearby Williamson coal camp. One bitterly cold winter night Mary's husband put his mining boots underneath the heating stove so they would be warm and supple the next morning. He arose the next morning only to discover that although the fire had blazed away all night, his shoes were frozen to the floor. Still, Mary Braida concluded that life was much easier for her than it had been for her mother-in-law when she first came from Italy in 1900 and lived in the Cleveland coal camp. When Mary lived in company housing in the 1920s, the company had electrified the houses although they still contained no indoor plumbing or water. Also, by that time John Braida owned an automobile, "which made a great difference" in their lives.⁵

Angelina Argenta also arrived from Italy around 1900 and took up residence in the Big Jim coal camp, located on the outskirts of Seymour in Wayne County. Much to her distress she discovered that the com pany house her husband rented was "really just a little shack." The house was so poorly constructed that the snow blew in around the windows and covered the bed. The bed itself was in such a dilapidated state that her husband had to tie the springs together with baling wire. When Angelina realized that the shack was to be her home in America, she "cried her heart out for the first year" and kept telling her husband that she wanted to return to Italy.[6]

Along with poor construction and dilapidated furnishings, some women recalled other nuisances. Paulina Biondi remembered that some of the company houses literally crawled with bedbugs: "They were thick, they were thick! I tried to do my best to get rid of them. Many times my fingers were bleeding from so much scrubbing. I said, 'America, America, it's no good.' "[7]

In 1909 Filomena Pieracci arrived from Italy to join her husband, Orlando, in the small coal camp of Berwick. She brought along the couple's only child, a two-year-old daughter named Bruna. Filomena expressed deep disappointment over the flat, drab countryside that was to be her home, and for the remainder of her life she longed to return to her family and friends in her mountain home of Frassinoro. Many years later, Bruna Pieracci, remembering her mother's chronic unhappiness with her life in Iowa, wrote about her family's experiences in several Iowa coal communities. She analyzed the impressions and reactions of the first generation women to their new homes in Iowa:

The mining camps were quite a contrast to the mountain village in the "old country." They were all alike: clusters of small boxlike houses made of wood and painted gray or boxcar red. These immigrants had never seen wooden houses before and they appeared flimsy indeed in comparison to the ancestral homes of stone with walls twenty inches thick. What was even more strange was the absence of trees and shrubbery for this mining camp had been built in the middle of what had been an Iowa cornfield. They were amazed at the flatness of the terrain and among themselves they called it "The Sahara."[8]

During Orlando Pieracci's forty years in the mines, the family lived in three different coal camps. Eventually in 1946 Pieracci purchased a family home in Ankeny.[9]

As the daughter of a coal miner, Bruna Pieracci had her own particular memories about the misery of living in coal camp housing. She recalled that the houses had no porches, closets, or basements. Cheaply built, the houses did little in the winter to shelter the families from the

Michele and Caterina Brugioni, who emigrated from Fiumalbo, Italy, in the early 1900s. (Courtesy of Nellie Fontanini)

---

bitter, cold winds. The heating stove did not heat the house sufficiently and the air was always heavily polluted in the winter because of the sulphur in the coal. Nor did the families find any escape from the steamy heat of Iowa summers. Bruna reported that water always presented problems in the camps. In the winter, pumps froze. Women melted snow and heated the water to pour over the frozen pumps in an effort to thaw them. In the summer, many of the wells dried up. In some camps, because of the water shortage, company employees brought "steam water" from the mine so that the women could wash their clothes. Later the women and children carried the water from the street to their houses to be saved in various containers until the next wash day.[10]

Occasionally first generation women had the chance to live in new company housing, which mitigated somewhat the difficulties of camp living. In 1920, when the Mike Brugioni family moved to the Dallas coal camp, the company had only recently completed the camp; therefore all houses were new. All units had four rooms with double wooden floors but contained neither plumbing nor electricity. The camp had three wells located in the center of the town's main street. Because the camp was built after the passage of the Good Housing Act, enacted by the state legislature in 1919, Dallas mine operators were obligated to lay out the camp in an area with good drainage and an ample water supply. The Brugionis' housekeeping chores were eased considerably because of the clean, sufficient water supply and adequate drainage system.[11]

Even with its many drawbacks, company housing offered some security. When the mines shut down for the summer months or when the miners could work only one or two days a week, families remained in their homes. Many women commented that regardless of the length of unemployment in the mine, operators did not evict the families nor did they raise the rent. However, when the miners struck, operators evicted the participating families immediately.[12]

Some Italian-American women fared considerably better in their housing because they and their husbands purchased homes and avoided camp housing. Living in an incorporated community like Seymour meant that women had an easier time carrying out their domestic tasks because they did not face periodic water shortages, frequent moves, or inadequate shopping facilities. If the families owned their homes, they could continue to make improvements without the fear that they might have to abandon them in a short time. At the same time the women and other family members could plant trees, shrubs, and flowers, thus adding an element of physical beauty to their lives. Many Italian-American families who owned their homes raised fruit trees in addition to large gardens. For them, home ownership frequently improved the quality of life.[13]

The great majority of Italian-American families in Seymour purchased their homes soon after arriving. The John DeGard family was one of the first to arrive in Seymour in the early 1890s. The family resided in the Big Jim coal camp for only a short time and then bought a residence in Seymour. The Victor Cambruzzi family arrived the same year and also purchased a home in Seymour. When Victoria Cambruzzi and coal miner John DeGard, Jr. married in 1924, they first lived in the west part of Seymour and then moved to a house on Main Street. At no time did they live in company housing. The Seymour families were fortunate in that numerous mines were located in that vicinity. When a large mine closed down in Seymour in the early 1920s, miners continued to live there but commuted daily by automobile to one of several mines located near Numa, Promise City, Jerome, and Brazil.[14]

Other Italian-American families in the Seymour area followed the same pattern. In 1913 the David Cerato family rented a house in Numa; there they remained for five years and eventually moved to a small farm two miles outside of Numa (Numa is five miles east of Seymour). The Eugene Ostino family arrived in Seymour in the late 1890s, where they lived for a short time and then moved to Numa and purchased a home. When Eugenia Ostino married John Padovan, a twenty-four-year-old Italian immigrant, they purchased a house in Numa and lived there for over twenty years while John worked in local coal mines.[15] Giovani Argenta came to Seymour in 1902 while George Busetto arrived in 1905. The wives of both men came from Italy, in 1903 and 1906, respectively. The Argenta family resided in the Big Jim camp for a short time and then purchased a home. The Busetto family immediately moved into a home in Seymour.[16] When the two Mores brothers came from Italy in the early 1900s, each married a Cambruzzi daughter, and both families settled permanently in Seymour. Although a small number of Italian-American families in the Seymour area may have continued to live in camps, the general pattern was for the families to purchase a home with a small acreage or at least a large town lot, and for the men to commute to work. Many of the Seymour Italian-American miners worked in seven or eight different mines during their mining careers.[17]

First generation Italian-American women seldom had the time or the occasion for activities outside their homes. Their daily routines were dictated by their circumstances; semi-impoverished conditions meant that the women worked long hours caring for their own families as well as doing a great many outside chores. Moreover, nearly all first generation women took in boarders or did washing, sewing, or baking to earn money.

The work routine of the first generation women in mining communities included planting and tending a garden, canning large amounts of fruits and vegetables, milking a few cows, raising chickens, and maintaining several varieties of fruit trees. Keeping cows enabled the women to make their own butter and cheese. Many families also

Italian-American band, Seymour, Iowa, 1915.
(Courtesy of Edna Padovan)

raised one or two pigs, butchering one and selling the other. Fresh meat such as pork chops had to be eaten quickly. At the same time the women made salami, blood sausage, and headcheese that could be kept for longer periods of time. The single gardening task that the men performed was caring for the grapevines, which most Italian households maintained, and most families made wine once a year.[18]

The women spent many hours each week baking. Most women baked bread two or three times a week and they frequently baked pies, cakes, and cookies. With children in school and men in the mines, women commonly packed a large number of lunches each day. Keeping "the lunch boxes full" was a continual concern.[19] Pauline Biondi recounted that she arose at 4 A.M. during the many years she kept boarders to bake either a pie or cake so "the men would have something fresh and sweet to take in their lunch buckets." Later, her two daughters took turns handling the baking responsibilites.[20]

Each woman made many articles of clothing for her family.

Concetta Moratti lived in the Pershing coal camp for many years where she made all her family's clothing. She also sewed for outsiders, which permitted her to "make a little money on the side." The additional sewing provided extra money so that her daughter, Lola Nizzi, could attend high school and one year of college.[21]

The women also regarded knitting as an important task. Many knit their families' stockings, mittens, caps, and sometimes sweaters. Giacomina Cambruzzi followed the rule that all the children should have two pair of knitted stockings at the beginning of each winter.[22]

An outside chore the first generation women took over was picking up coal. With little cash coming into the household, the women soon recognized that they could save a few dollars each year by picking up the chunks of coal that fell from the loading chute along the railroad tracks or at the mine slag heaps. As the children grew older, this task usually fell on their shoulders. Frequently the families collected enough coal to provide fuel for the entire winter.[23] James Battani, a second generation Italian-American, recalled that for one period of nine and one-half years, his parents had to purchase only one ton of coal. In the Battani family, the children had the responsibility of picking up coal along the tracks and loading chutes and then pulling the coal home in their wagon.[24]

The oldest woman interviewed, Antonia Cerato of Numa, remembered vividly the problems she faced when she first set up housekeeping in Iowa. Antonia was born in Fonzaso, Italy, in 1889. She married David Cerato in 1911 and came from Italy to Numa in 1913. She remembered that in 1913 the fare for her and her small son totaled ninety dollars. On the day she arrived in Numa, her husband had a house rented but had purchased no food. The next day he went to work in the mines. His wife asked him, "What do you have to cook for me?" Fortunately, later that day several merchants came around who spoke Italian and Antonia Cerato purchased what she needed. She remembered saying to the first merchant, "I have nothing here, give me something!" So as she said, "give me this, give me that," he simply marked it down on his pad. The men came every day to take orders from the housewives so Antonia's shopping problems were simplified and language did not present a problem. Later, when she went into Numa to do her shopping, the proprietors of all three grocery stores spoke Italian, so she had no difficulty communicating with them.[25]

In 1918 the Cerato family moved to an eighty-acre farm outside Numa that they eventually purchased. Antonia remembered that she

Rocco Battani, shown in Italian army uniform shortly before emigrating to Iowa from Riccovolto, Italy, 1910. (Courtesy of Mary Sertich)

Camilla Fontanini of Riccovolto, Italy, shortly before her marriage to Rocco Battani. (Courtesy of Mary Sertich)

usually "worked from four in the morning until ten at night to get everything done." She confessed that in looking back over her life, she does not know how she was able to work so hard. The first year on the farm she carried water from a nearby stream for washing and cooking. She carried two buckets at a time, filling double boilers set atop an outside stove. Because of the work involved, she carried water one day and washed the next. On wash day she first had to soften the water; then she boiled the clothes and later scrubbed them on a small washboard. For many years Antonia took in two or three boarders, so the family wash included the boarders' clothes and bedding. The Ceratos had a well drilled on their farm in 1919, which made wash day considerably easier. She also milked twelve cows a day and added to the family income by selling both milk and cream. In addition to peach, pear, and cherry trees, the family raised strawberries and grapes. Sometimes Antonia sold fresh fruit to local residents to increase her earnings.[26]

Almost without exception, first generation women kept boarders, at least for a short time. If the women did not keep boarders, it was usually because they had health problems. Most communities had two types of boarding plans. With the first and most common, the family charged the boarder a set fee, usually $8 per month. The second was called the economy type. Under this system the housewife totaled the weekly bills and then divided the amount among all adult males in the household.[27] Camilla Battani in the Moran coal camp had six children and four boarders. The boarders took over one room of the family's four-room house as their bedroom. The Battani's daughter, Mary Sertich, remembered that her family had a leather couch that opened into a bed in which someone also slept. She explained that "the room was a living room by day and a bedroom by night." Caterina Brugioni kept as many as eight or nine boarders at different times during her husband's coal mining days. Anita Lami, whose family lived in West Des Moines for many years while her father worked in a local coal mine, stated that at one time her mother, Antonia Saiella, had fourteen boarders. In most homes, family members regarded the boarders as part of the family. The boarders were generally young men between the ages of sixteen and eighteen who came from Italy to work in the local mines. Local Italian-American families took the young men in not only to earn a little money, but also to discharge the older women's sense of responsibility and concern for the young, unattached men who were far from their own families. Antonia Cerato kept two or three boarders for many years and charged them $3 for two weeks' lodging. She confessed that she did not like to charge them too much money because the young men were "Italians and the family felt they should help the younger men out."[28] Some women followed the practice of renting two houses from the company. The woman's family lived in one house while she rented the other to eight or ten miners.[29] The women then provided the miners with the housekeeping services of cleaning, cooking, washing clothes, and packing lunches.

    First generation Paulina Biondi recalled that on the day she arrived in Iowa as a seventeen-year-old bride, she immediately took in four boarders. The Biondis lived in two rooms and managed by hanging a burlap curtain between the kitchen where they slept and the bedroom where the four miners slept. Over the next forty years this woman raised a family of five children as well as keeping a minimum of four boarders at all times. During much of that time the family lived in a four-room company house. She speculated that the worst time was when the family lived in a three-room house and had a bed with two men in the kitchen and another bed with two men in the living room. Two men worked at night and two worked in the daytime, which meant that the first two men slept during the day. Commenting on that experience, Paulina Biondi exclaimed: "Yes, that was a mess, a mess, a mess. Let's not talk about it." Reflecting on all the boarders she had kept, Paulina confessed, "It's a wonder they didn't put one in my bed!" Her boarders followed the economy plan in which all male adults in the household shared the expenses equally. Biondi added: "For the woman, she works all these two weeks, serving, washing their pants, ironing, keeping everything clean and she gets a lousy dollar. So, imagine!" Paulina concluded: "If I had my life to start over again, no men around the house, no men around the house. No, out!"[30]

Routine bathing proved an additional hardship for first generation women. Until the late 1920s, Iowa mining companies did not provide shower facilities for the men. When the men came home each evening, the women had to drag out the washtub, set it up in the kitchen, and fill it with warm water. It meant there was little privacy in most mining households. Some families had small sheds behind the house that they used for bathing. Nellie Fontanini remembered that her mother kept

Local residents gather on the steps of the company store at the Carbonado Coal Camp, 1890s. (Courtesy of Iowa State Historical Department, Division of Historical Museum and Archives, Des Moines)

the shed warm all day so that it would be ready when the men arrived home at night. If the household included boarders, the same service had to be provided for them. Each miner had his own tub and all were arranged around the stove and filled with warm water by the time the miners arrived home. In many households the men left their dirty work clothes in the shed. The wife then had the responsibility to light the fire again in the morning so the men could dress in a warm area.[31] After 1927 coal operators agreed to provide showers at the mines. Yet some men still preferred to come home for their baths because if they worked in a large mine which employed two hundred to three hundred men, they often faced a long wait.[32]

The second generation Italian-American women continued many of their mothers' work roles, but in most instances their work loads were lighter. By the early 1930s many mining camp houses had electricity although none had indoor plumbing. Many mining families purchased automobiles during the 1920s, so increasingly families did their shopping in Centerville rather than Seymour or Numa. By the early 1940s most mining camps had disappeared in Iowa, and mining families had moved to the nearest incorporated town where the men commuted to work by automobile.[33]

In general second generation women continued the same domestic tasks of raising large gardens, canning, making butter, and raising fruit trees. Many women continued to sew their families' clothing. For some women, extreme frugality remained a necessity. Victoria DeGard remembered that as a young wife she made her own dresses out of printed feed sacks. When she accumulated three, she could make a dress. If she did not have three prints that matched, she traded with a neighbor. Like the other neighbor women, she also made dish towels and sheets out of the feedsacks. In commenting about this practice she said, "Well, we never thought nothing about it; we just managed."[34]

In managing the household, the women managed the money. In many mining camps the women went to the company store to collect their husband's paychecks. The storekeeper totaled whatever purchases had been made during the two weeks between pay periods and he then subtracted that amount, plus the house rent, from the earnings. During the fall and winter months when the men worked, the women managed fairly well. However, during the months when the mines were closed, credit at the company store was an absolute necessity. A typical expression among the women interviewed was that "you charged all summer and then worked all winter to pay it back." When the men returned to work in the fall, many families could not pay the grocery bill immediately.[35]

Most Italian-American women discovered that managing the families' budgets was made easier by the wage contributions of at least two people within the household. In 1914 the annual income of Iowa coal miners in Beacon, Cincinnati, and Seymour averaged $466.00. That amount averaged out to $38.83 per month. According to interviews, it appears that most first generation Italian-American women took in at least two boarders. Assuming that two boarders represented the average number of boarders, and since the standard boarding rate was $2.00 per week per boarder, the women earned $16.00 a month. Put another way, the women's income equaled almost one-half that of the men's income. Because many women, in addition to keeping boarders, earned money by selling dairy products and garden produce or performing services for other mining families, it is probable that in some households the women's income nearly equaled that of the main breadwinners. It is instructive to note that if the women kept four boarders, which many women did, their earnings from boarding equaled their husbands' earnings. Also of significance was the steadiness of the women's income, while frequently the husbands' wages were nonexistent during the summer months. If the family had children in their teens, probably at least one child worked, thus bringing another income into the household. In some mining households, families enjoyed a total of four or five incomes in addition to boarding fees.[36]

While many former coal camp residents regarded the company store as an oppressive operation, most Italian-American women had different experiences. Bruna Pieracci recalled that in Berwick, the first coal camp where she lived with her parents, the company did not operate a store. In the second and third camps where she resided, Carney and Oralabor, residents had a choice of trading either at company stores or private stores maintained by Italian-American merchants. She noted that in Carney and Oralabor, most of the residents were Italian and they chose to trade at the Italian-American stores. These merchants stocked the foods that Italian-Americans preferred to buy, making their establishments even more popular. Eventually, lack of trade forced the company stores to close in both Carney and Oralabor. Bruna recalled that in Berwick, an Italian baker traveled through the camp daily selling Italian bread and other baked goods. In Seymour first generation Italian-American women always had the option of trading at a company store or at several privately operated stores, one of which was operated by an Italian-American.[37]

In a 1919 report on coal camp housing, Charles Nichols, an Iowa State College engineering professor, explained: "Life in many of the camps is dreary. It is very difficult, except by automobile, to get to town, and there are no attractions or entertainments of any sort for the pleasure of the miners or their families."[38] Professor Nichols's description characterized most mining communities, not just those that were company owned. The isolation of most coal camps meant a lack of social, educational, and occupational opportunities for miners' families. A typical coal camp included a school, miners' hall, pool hall or tavern, company store, and two types of houses: four- or five-room houses for the miners and a small number of commodious, better constructed houses for company officials. While a few mining communities contained modern high schools, most company towns contained only elementary schools. Given the lack of educational and occupational opportunities, miners' sons had little choice but to follow in their fathers' occupational footsteps, and daughters usually remained at home until they married. The next generation then repeated the cycle. Interviews with all women supported the dreariness of camp life, although some women did emphasize the positive roles of family relationships and the fellowship among the mining families.

First generation Italian-American women had little time for social activities. When Antonia Cerato was asked what she did outside the home, she replied: "Nothing, never. There was no time." She did not belong to any activities outside the home during her entire lifetime. She did remember that her husband had taken her to several silent movies in Numa, "many, many years ago."[39] Paulina Biondi responded when asked if she had time for any outside activities: "No, no, no. During the first years there was no time to play cards, I'm telling you. With the washing, and ironing, and mending, and cleaning, cooking, no, no, no time to play cards."[40] Second generation Italian-American women unanimously supported the same view in regard to their mothers. Minnie Mores commented about her mother, Giacomina Cambruzzi: "Her life was just work. She had the boarders to take care of and she fixed their pants and washed, just had a scrub board. Then she baked her own bread. Oh yes, she worked all the time." Later in the

The Battani family, 1928. *Back row*: Frank, Camilla, Anita. *Front row*: Robert, Aldo, Rocco, Clara, Mary. Rocco and Camilla emigrated around 1910 and their children's birthplaces illustrate the high mobility of Iowa's coal mining families: Anita and Frank were born in Carney Coal Camp, Mary and Robert in Highbridge Coal Camp, and Aldo and Clara in Dallas Coal Camp. (Courtesy of Mary Sertich)

---

interview she stated that her mother did not learn English. When asked why, she responded: "Maybe it was because she had too much work to do, kids kept acoming, and just work, and she had boarders to take care of."[41]

Daughters raised in these households experienced the same heavy work routine. Edna Padovan said that as a young girl she always wished for more leisure time. Every day there were many household tasks to do. As a child, she found little time to read or play.[42] As a high school student living near Melcher, Gina Battani had two major responsibilities in addition to helping her mother with household chores. Gina arose at 5:30 A.M. to drive her father to the mine. Arriving home, Gina milked two cows and helped her sister and brother deliver the milk to neighbors. After school, Gina drove to the mine to bring her father home. Then she helped her mother prepare the evening meal. The rule existed in most mining homes that immediately after the men finished bathing, the women served the evening meal. Gina Battani commented that "you had to be pretty sick not to have supper on the table when the men finished their baths." Following supper, Gina milked two cows and delivered the milk before attending any school activities.[43]

Second generation Victoria DeGard remembered participating in few social activities either during her girlhood days or her early married life. She did not recollect going anywhere except to the movies on Saturday night. When she and her husband were courting, their main activity was sitting on the front steps of her parents' home. Some young people had cars and could travel to Centerville, but the DeGards did not have an automobile until after their marriage. Victoria did remember several community celebrations. The first was a special parade and celebration on Columbus Day, and the second was the celebration on the anniversary of the eight-hour day. The miners won the eight-hour day on April 1, 1898, and for many years on April first they held celebrations and parades in smaller towns as well as a major parade in Des Moines. No one expected the miners to work on April first![44]

Several second generation women did leave home to work in other communities before their marriages. Alnora Earls left home at sixteen, after completing one year of high school, to work in Chicago. Several of her girl friends from Seymour had gone to Chicago to work as nursemaids for wealthy families. They wrote to Alnora, telling her that they made "big money, eight to ten dollars a week, and room and board." They also informed her that they had Thursday afternoon off as well as Sunday. When Alnora decided that she did not want to continue going to high school and that she was going to Chicago, it distressed her mother deeply. She said that her mother "cried her heart out" and told her daughter repeatedly, "No, don't do that." Alnora replied that since her father and two brothers were already in Chicago working for the summer, "they could look after me." Alnora recalled that on her day off, she had to "show up where Dad and my two brothers worked." She remained in Chicago for the summer and then returned home. For the next nine years, until she married at age twenty-five, she returned to

Chicago many different times, continuing to work as a nursemaid for various wealthy families. When in Seymour, she stayed at home with her parents and helped her mother with the housework.[45] Augusta Argenta also worked in Chicago before her marriage. She finished the eighth grade at age sixteen and did not want to go on to high school. Her parents agreed that she could go to Chicago, where she had two married sisters. She worked as a live-in domestic and nursemaid but maintained close contact with her sisters. Augusta sent most of her wages home to her parents because "they were so hard up." Augusta eventually married at twenty-two.[46]

All but two second generation Italian-American women remained at home until marriage, and all but one (who married a Croatian coal miner) married first or second generation Italian coal miners. Most married in their early or midtwenties. Typically their husbands were four or five years older. Generally the women graduated from high school and one woman, Lola Nizzi, attended two years at Ottumwa Heights College. She trained to be a teacher and then taught school at the Pershing coal camp. After marriage, she quit teaching because it was "not respectable in that community for a married woman to work outside the home."[47]

Language apparently did not cause a major communications barrier for first generation women. Some women learned sufficient English to communicate with the local merchants but others did not. Victoria DeGard remembered that her mother never learned English but always got along well. When she went to the store, her mother took along samples of items that she wanted and showed them to the clerk. In some mining camps the local merchants came through once a day, allowing the women to place their orders. Usually the clerk was Italian so language posed no problem.[48]

Antonia Cerato did learn English several years after she came to the United States. She recollected vividly the incident that made her realize she must be able to communicate with people other than Italian-Americans. Shortly after she and her husband moved to their farm near Numa, they rented out pasture land. The renter became delinquent in his payments and finally allowed David Cerato to take his buggy as part payment. He also gave Cerato a check for seven dollars. Cerato cashed the check at the local bank only to have the sheriff come to the Cerato house and tell the family that the check had bounced and they must repay the seven dollars to the bank. Antonia Cerato was alone when the sheriff arrived and found that she could not make him understand the details of the transaction. Finally, in desperation, she told him to go to the Numa bank and they would explain the situation to him. At that point she realized that she must learn English if she was going to cope with important problems. She began to pick up the language as her children started school. As they learned English, she learned English. As well as speaking English, Antonia also learned to read English.[49]

Money management sometimes presented problems for first generation women. Antonia Cerato recalled the difficulty she had counting American money early in her marriage. She explained that a friend who lived nearby had helped her count money for purchases. The friend taught her to understand prices by putting everything into ten-cent intervals: an item cost ten cents or twenty cents or thirty cents, or some other amount divisible by ten. Antonia learned quickly and after that she apparently had no difficulty managing the family's finances.[50]

Related to language was the matter of discrimination. Throughout the literature on Iowa's coal mining industry, there is the implicit, and sometimes explicit, belief that coal miners were somehow inferior. Mine operators, state mine inspectors, and local business people often implied that coal miners were more rowdy, intemperate, and slothful than other Iowans, particularly farmers. State officials and businessmen sometimes made exceptions for northern and western Europeans such as the English, Welsh, Irish, and Swedes; but they usually held Italians and Croatians to be in a lower social class. In the interviews with Italian-American women, the effects remained of growing up in an environment that somehow marked them as being different and of an inferior social class. Edna Padovan remembered that she considered herself a foreigner, "because my parents spoke a foreign language. The fact that you were born in this country did not make you an American. I knew it did technically, that I was as much of an American as anybody, and I'm ashamed to admit it, but I was ashamed to admit that I was an Italian." She also recalled that "when I was in college, I felt that people looked down on southern Europeans. If you were German or Norwegian there was not that feeling. People would call Italians wops and dagos and I think I was supersensitive and that bothered me a lot." Edna Padovan felt that after World War II, negative attitudes toward Italian-Americans had been greatly reduced. She did not think that her younger sister was exposed to the same influences that she had been in her teenage years.[51]

Lola Nizzi recalled the same feelings in her elementary school days in Pershing. She remembered that many outside the camp were farmers and Protestants: "We were coal miners and Catholics and Italians to boot—we had three strikes against us. I could sense it; I knew it. I don't know, I had an inferiority complex, I guess." Nellie Fontanini recalled that when her family moved to Granger from the Dallas coal camp, the other children called them "dago or wop. We didn't like it, but we tried to be nice." During their school days, most women felt a sense of deprivation. They admired other girls in school who wore nice clothes and they assumed all the other girls' fathers were rich. Alnora Earls remembered that while some of the farmers' daughters wore a different dress every day, she had one dress that she washed out at night and wore every day of the week.[52]

Even with the bleakness, however, many women remembered the sharing and the fellowship that existed in the mining camps. If anyone became ill, other families responded by bringing in food, helping with the children, and assisting with the housework. According to Father John Gorman there was a great sense of social responsibility among the Italian-Americans in the mining camps. Father Gorman commented: "If someone died, some other family took over the children. Everyone helped everyone else." The foster family usually would not adopt the child but would provide the child with a home until he or she reached working age. Lola Nizzi added that "It was a comforting thought to the people that you knew your family would be taken care of if anything happened to you. You knew that the family would be cared for. They helped each other."[53]

Mining families regularly offered assistance to neighbors and friends that helped ease the frequent tragedies in their lives. In the Dallas coal camp, Mike and Caterina Brugioni often came to the aid of their neighbors in times of distress. When a death occurred in the mine and local officials were reluctant to tell the family, Caterina Brugioni bundled up and headed out over the fields to carry the news herself. Sometimes she helped the family by bathing and preparing the corpse for burial. Mike Brugioni sometimes helped make coffins for infants and babies who died in the community. He and Father Gorman made a number of coffins in the shop of the local Catholic high school, and the women lined the coffins with cotton.[54]

Family and community members usually offered help to miners' widows. When Marie Mores in Seymour was widowed, she moved in with her sister's family, the John DeGards. In the two rooms allotted her in the DeGard home, she took in washing, which she did every day on a scrub board, to support her children. In Dallas County, Italian-American widows sometimes opened their homes to members of the community as a place where they could buy liquor or coffee and play cards. The women remembered that a shot of whiskey during prohibition days cost twenty-five cents. Sometimes other women in the community looked down on this behavior, but according to Flora Betti of Granger, "most women understood that the poor woman had to provide for herself and her children."[55]

Although coal mining families had little money and most people could not afford to travel outside the coal camps for entertainment, some women remembered local Saturday night dances as the main form of community entertainment. These dances were quite popular and participants often referred to them as "kitchen dances" because families held these dances in the kitchen, after all furniture had been pushed back. If the kitchen table still took up too much room, the men took it apart and carried it outside. A fiddler and an accordion player often furnished the music. At the Big Jim mine near Seymour, John Argenta recalled that company officials left one house vacant for the purposes of Saturday night dances. Many nights he remembered playing his accordion at the house with so many people dancing that the house "just rocked back and forth."[56] Sometimes, prospective marriage partners met at such occasions. Mary Peek, a fifteen-year-old miner's daughter at Promise City, worked for her married sister who ran a miners' boardinghouse. One Saturday night a newly arrived Italian immigrant asked Mary to dance. The couple continued to dance together on Saturday nights but could not communicate well because Mary did not speak Italian and Morris did not speak English. Finally one night, Morris spoke three English words very slowly: "I like you." Three weeks later they married. Morris did not have to worry about learning English because Mary quickly learned Italian. In subsequent years she assumed the responsibility of corresponding with his family in Italy.[57]

Italian-American women and their families probably had better medical service than did the residents of Iowa's rural areas or small towns. Most mining companies contracted with a local physician to provide services for mining camp employees and their families for a set monthly fee. In the coal camp of Zookspur, for example, company employees and their families had the option of going to the company doctor, Dr. C. W. Cook, who lived in Madrid (three miles north of Zookspur). Dr. Cook kept a list of patients and any family could sign up for $1.50 a month. For that fee, any member of the family could visit the doctor for an unlimited number of times. Also for that amount, the doctor included prescriptions. If Dr. Cook delivered a baby for a family not on the list, he charged $25.00; if he delivered a baby for a family on the list, he charged $12.50. The family could pay the fee in monthly installments. Following the birth, Dr. Cook came back two or three times to check the mother and baby. The mothers had no prenatal checkups or tests of any kind. A woman went to the doctor, informed him that she was pregnant, and told him when the baby would be born.[58] Most Italian-American women had their babies at home. If a doctor was unavailable, a local midwife usually delivered the baby. Some Italian-American women preferred the services of the midwife since that was the practice in the small villages in northern Italy.[59]

For residents of the early Dallas County coal camps, however, there was no company doctor. Most of the women there had their babies delivered by midwives who were often relatives or friends.[60] James Battani recalled that his mother worked as a midwife in various Iowa coal camps and did not charge for the service. He was certain that she had learned the skill in Iowa.[61]

In Seymour and Numa, a local doctor delivered some of the babies, but many women used midwives. When Giacomina Cambruzzi came to Seymour from Fonzaso, Italy, in 1893, she brought along some knowledge of midwifery. She soon became the neighborhood midwife. In some cases she delivered the baby alone, and in other cases she assisted the local physician. Her daughter, Victoria DeGard, said that the doctor was always pleased when her mother was present because she cared for the baby, while he cared for the mother. Following the birth of the infant, Giacomina Cambruzzi returned to the home twice a day for two weeks, bathing the baby and caring for the mother. She insisted, as did the doctor, that women stay in bed for nine days following the birth. Many women practiced midwifery, at least for a short time. In Dallas County, Antonia Saielli practiced midwifery for many years, teaching the skill to Concitta Moratto. Later, Concitta moved to the Pershing coal camp where she delivered many babies.[62]

In Numa, Antonia Cerato found it too difficult to stay in bed for nine days simply because she had too much work to do. Instead she decided that three days was sufficient to stay in bed after the birth of each child. On one occasion, the birth of their fourth child, she began to experience labor pains and sent her husband to town to call the doctor. He went to town, telephoned the doctor, but stayed in town to talk with a few friends. In the meantime the doctor arrived, delivered the baby, and reached an agreement with Antonia Cerato that the fee for delivering the child should be $7 rather than the usual $5 because he had to travel a greater distance than usual. As the doctor walked out the door, her husband arrived home.[63]

However, for some women the birth experience had lifelong complications. Angelina Argenta had a midwife in attendance for the birth of all her children. Once through the nine-day recuperation period she went back to her usual wash day duties of lifting heavy boilers and buckets of water. Although she experienced no problems with the first two children, soon after the birth of the third child she contracted what her daughter described as "milk fever." Eventually Angelina had a hysterectomy. Her surgeon, however, felt the first operation was not entirely successful so she had surgery for a second time. Afterward she had to wear a corset to "keep her organs in their proper position." Her daughter, Alnora Earls, described her mother as always being sickly.[64]

Filomena Pieracci experienced even greater difficulties over childbearing. After delivering four healthy babies with no complications, she experienced a tubal pregnancy. The situation proved profoundly traumatic for Filomena, and one from which she never fully recovered. Although she did consult the camp doctor about the pregnancy, she had a great fear of doctors. In Italy, physicians resided only in the cities and Filomena Pieracci's village had only the services of a midwife. Furthermore, no one in her family had even been hospitalized. When the camp doctor diagnosed the problem, he sent her to a nearby hospital for surgery. Bruna, her daughter, recollected that other Italian-American women in the camp greatly increased her mother's fears of entering a hospital by telling her that in the Old Country, women sent away to hospitals often did not return. Bruna Pieracci believed that these comments greatly fueled her mother's profound fear of being hospitalized and of undergoing surgery. A physician was able to correct the medical problem, but Mrs. Pieracci experienced such an emotional shock that she never regained her mental health. For the remainder of her life, she required either constant care at home, mostly provided by her eldest daughter Bruna, or institutionalization.[65]

The interview data gathered from first and second generation Italian-American women allow some generalizations to be made about their roles in Iowa mining communities between 1895 and 1925. That all women led difficult lives is apparent, but the experiences of the first generation women stand in a separate category. The foreign-born women knew little except hard work and a life of scrimping and saving to make their limited incomes cover the cost of raising their families. Their work included the full range of domestic duties inside the home as well as many outside chores. Every woman interviewed earned money either by taking in boarders or providing services to community members. But while the income helped ease the economic pinch, it also added greatly to the women's work responsibilities. Perhaps most important, however, the women's incomes appear to have been crucial to their families' economic survival.

The first generation women's adjustment to life in a foreign land is difficult to assess. Certainly all women experienced loneliness for their parents and other family members in Italy. All women, like their husbands, believed that at the time they arrived in Iowa, they would be staying only a few years. They anticipated that after they had saved some money, they would go back home again. A few families did return to Italy but the overwhelming majority did not. Once in America, they soon realized that their lives would be far more comfortable here than in the Old Country. Most did not return to Italy, even to visit. According to their daughters, most of the first generation women expressed no regret over coming to the United States. All came from poor backgrounds in Italy and life there held little promise. They expected to work hard in America so most women expressed little disappointment over the hardships that they encountered.

The presence of many other Italian-American families counterbalanced the loneliness experienced in a new homeland. Over three-fourths of the women interviewed in Seymour, for example, had ancestral ties with the same village in northern Italy. Often brothers, sisters, cousins, and other relatives followed once the first family member emigrated. This practice produced kinship networks as well as an ethnic network. These networks were perpetuated as marriages continued to take place within the Italian-American community. Frequently second generation women married first generation miners. In Dallas County, four of the five second generation women interviewed had married first generation Italian-Americans.

First generation Italian-American women appeared insulated from events outside their communities. Whether families lived in coal camps or in incorporated communities like Seymour, their neighbors were mostly other Italian-Americans. The storekeepers were Italian-

Americans or if not, they hired Italian-Americans as clerks. Some contact with non-Italian-Americans might have come through church attendance, but that was not possible on a regular basis until many of the women had reached their late forties or early fifties. Once their children started school, outside influences began to appear in the home. The children learned English and frequently their mothers learned along with them. Some mothers did not care to learn a new language; their children, of necessity, continued to speak Italian. In some cases, like the Padovan family, the second generation women acted as the interpreters between their mothers who spoke only Italian and their daughters who spoke only English. Clothing styles changed only slightly. The first generation women did not respond to local fashions but continued to wear their longer skirts as well as kerchiefs on their heads. In her study of Italian-Americans in Buffalo, Virginia Yans-McLaughlin noted that the presence of an ethnic community reduced the emigrant's need to participate in American community life. The statement seems equally applicable to the Iowa Italian-Americans.[66]

The greatest opportunity came for the second generation women. Many attended a year or two of high school and several attended college. The daughters lived in many different communities and experienced none of the insulation that surrounded their mothers. Although all married coal miners, their later years were considerably more comfortable, both physically and monetarily, than their mothers' lives had been.

One of the most striking characteristics of the Seymour and Numa Italian families was their permanence. Their practice of purchasing homes clearly disputes the contemporary view that the southern and eastern Europeans moved frequently and hence were less desirable types of citizens and workers. Not only did many Italian-Americans and Croatian-Americans own their own homes, but many maintained small farms or acreages that gave them considerable security in terms of their food needs. This made their experiences similar to the Swedish-Americans in both Boone County and Des Moines.

Throughout the interviews, even though the women related many dreary accounts, they expressed a steady note of optimism. The first generation families had come to America because life in Italy had been too difficult and too hopeless. The emigrants carried the firm conviction that life would be better in America. They were not disappointed; life was better in America even though everyone had to work extremely hard. In many ways, particularly because of the ethnic and family networks in the mining communities, life did not change greatly for the first generation women. They continued to speak Italian, interact with other Italian-Americans, and to perform the same domestic roles they had pursued in Italy. The transition from the Old Country to the new was softened and eased all along the way by family and friends. The Iowa rural-industrial setting seemed to allow a carry-over of many Old World practices. Moreover, when examining the roles of first generation Italian-American women, one must guard against undue pessimism because of their excessive work loads and their lack of social opportunities outside the home. In performing their domestic tasks, the Italian-American women were doing what they regarded as their most important function—acting as wives and mothers. They did this work freely and lovingly and any inability on their part to carry out these time-honored responsibilities would certainly have led to great frustration and unhappiness. Perhaps Paulina Biondi described it best when she stated: "A family without children is just a family with a bunch of boxes for the bees, without any bees."[67] For Paulina Biondi and other Italian-American women, a husband and children fulfilled their lives and gave them their reason for being.

# 5

## POPULATIONS OF THREE IOWA COAL MINING COMMUNITIES

ETWEEN 1895 and 1925, approximately twelve thousand to eighteen thousand coal miners lived in Iowa. Most miners and their families resided in coal camps that dotted the landscape of central and southern Iowa. Bleak and dismal in their outward appearance, these short-lived communities stretched from Wapello County in southeastern Iowa upward to Boone County in the central region of the state. Isolation, single occupational interest, and short-term existence characterized Iowa's coal mining camps. The majority of coal camps were not incorporated; therefore their residents cannot be precisely identified. Miners also lived in incorporated towns that contained numerous businesses other than coal mining. Although incorporated mining communities have been more visible and far more enduring than mining camps, little effort has been made to study systematically their mining populations. This chapter is a study of the populations of the incorporated coal mining communities of Beacon, Cincinnati, and Seymour in 1895 and 1915 with particular emphasis on coal miners. The study is based on the Iowa state censuses of 1895 and 1915. The questions posed are: Where did the residents come from? What work did they do? What were their sexual and racial characteristics? What percentage of people were foreign born? What were their income levels? How many months were they employed each year? Although the main emphasis is on the mining population, comparisons will be made with the other segments of the population to determine social and economic distinctions between the two groups.

The selection of Beacon, Cincinnati, and Seymour as coal mining communities to be investigated for this study was based on several factors. First, the three communities had to be identifiable as coal mining communities. Beacon, Cincinnati, and Seymour were located in the midst of numerous mining operations and all contained sizable num

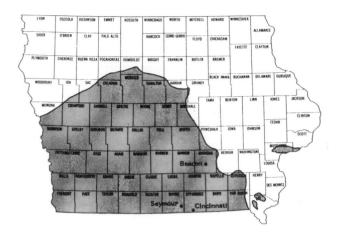

Location of Beacon, Cincinnati, and Seymour
(shaded area is underlain with coal).

---

bers of coal miners. Second, the communities had to be incorporated. People living in the unincorporated coal camp of Zookspur in 1915, for example, were not identified in the state census as Zookspur residents; they were identified only as residents of Des Moines Township (Dallas County), since Zookspur was located in that township. It is essential, therefore, to the researcher that the three communities be incorporated so that their residents could be clearly identified. Third, the communities not only had to be coal mining communities in 1895 but they had to remain active coal mining communities through 1915. The interviewing of many Italian-American coal mining families in Seymour for another part of this study made Seymour a logical choice.[1]

In 1895 Beacon, Cincinnati, and Seymour all appeared to be prosperous coal mining communities with possibilities for further growth. Beacon, the smallest community, contained 971 people; Cincinnati included 1,203, and Seymour, 1,547. The three communities encompassed a wide variety of businesses and professional people as Table 5.1 indicates.[2]

The economy of Beacon was the most strongly influenced by the presence of the coal industry. Excluding coal miners, Beacon contained

Table 5.1. All occupations, 1895

| Occupation | Beacon (%) | Cincinnati (%) | Seymour (%) |
|---|---|---|---|
| Miners | 60.2 | 48.0 | 37.0 |
| Agriculture | 1.0 | 6.8 | 5.4 |
| Professional and semiprofessional | 5.0 | 7.8 | 7.6 |
| Personal services |  | 1.8 | 2.6 |
| Business | 6.0 | 9.7 | 14.8 |
| Transport and communication | 2.3 | 3.8 | 4.8 |
| Labor | 10.0 | 6.8 | 11.4 |
| Artisan | 12.4 | 15.8 | 12.1 |
| Other | 3.1 | 0.3 | 4.3 |
| Total | 100.0 | 100.0 | 100.0 |
| Total number of occupations | 299 | 308 | 462 |

a total of 119 persons who were employed in twenty-two different occupations. Most of these occupations appear to have been mine related, providing services within the mines and maintenance for miners' homes and company buildings. The community contained a small number of craft occupations probably explainable by Beacon's location. Situated two and one-half miles from the much larger community of Oskaloosa, Beacon residents could draw on services available there.

Table 5.1 shows that a wider variety of occupations in both Cincinnati and Seymour could be classified as independent or as unrelated to mining. Located approximately ten miles from a more populous community, Seymour and Cincinnati had a fairly large trade area and accommodated a greater diversity of occupations. The occupational composition of Seymour and Cincinnati is congruent with the analysis of Lewis Atherton in *Main Street on the Middle Border,* which describes the typical small town before the turn of the century as economically self-contained. People traded within their own communities rather than traveling to neighborhood towns to trade. Life was localized both in an economic and a social sense. The result was that each town could support a wide variety of crafts and other businesses.

Women played a minor part in the occupational structure of the three communities. Three occupations appeared clearly identifiable as female occupations: washerwomen, dressmakers, and milliners. The profession of teacher accounted for the largest number of employed women, while women teachers outnumbered men teachers by approximately three to one. Table 5.2 lists all female occupations given in the census of 1895 for Beacon, Cincinnati, and Seymour.[3]

It should be noted, however, that the 1895 census did not list all women's occupations. Interviews with coal miners' wives and daughters

Table 5.2. Female occupations, 1895

| Occupation | Beacon | Cincinnati | Seymour |
|---|---|---|---|
| Teachers | 10 | 6 | 11 |
| Dressmakers | 7 | 1 | 2 |
| Milliners |  | 3 | 8 |
| Washerwomen | 1 |  |  |
| Railroad workers | 1 |  |  |
| Clerks | 1 |  | 3 |
| Typesetters |  | 1 |  |
| Printers |  |  | 1 |
| Hardware store proprietor |  | 1 |  |
| Bookkeepers |  |  | 1 |
| Dentists |  |  | 1 |
| Farmers |  | 1 |  |
| Total | 20 | 13 | 27 |

have revealed that a large number of miners' wives took in boarders as a means of earning money. Miners' wives also earned substantial amounts of money by performing part-time work such as sewing, nursing, and selling excess garden and dairy products. The 1895 census did not include a single woman who listed boarding as an occupation, nor did the census list part-time female occupations such as sewing (other than dressmaker) or selling produce. These omissions should be considered when determining the full range of women's activities in the three communities.

Table 5.3 confirms that coal mining was a young man's occupation. There were few old coal miners in Iowa in 1895. These figures support the view expressed by UMW officials that coal mining was a debilitating occupation. The figures also reiterate the opinion voiced so frequently by British coal miners that because of the difficult nature of coal mining, most men could not work past the age of fifty; actually, many English coal miners did not live beyond their fifties.[4]

Table 5.3. Ages of 499 coal miners, 1895

| Age | Beacon (%) | Cincinnati (%) | Seymour (%) |
|---|---|---|---|
| 1–15 | 2.8 | 0.0 | 0.6 |
| 16–20 | 17.2 | 6.7 | 12.3 |
| 21–30 | 27.2 | 37.8 | 40.4 |
| 31–40 | 16.1 | 36.4 | 31.6 |
| 41–50 | 20.0 | 14.1 | 10.5 |
| 51–60 | 10.0 | 3.3 | 2.9 |
| 61–91 | 6.7 | 1.3 | 1.8 |
| Total | 100.0 | 100.0 | 100.0 |
| Total number | 180 | 148 | 171 |

The census also includes information on children's employment in the mines. As indicated in Table 5.3, only 6 miners out of 499 were under the age of sixteen. This figure is obviously inaccurate. Independent sources have indicated that between 1895 and 1915 many young boys, some as young as nine and ten, worked in the mines. Christina Anderson, for example, remembered that in Albia in the 1890s, two of her brothers, ages nine and ten, worked as trapper boys. She said the practice of hiring boys between the ages of nine and twelve as trapper boys was common.[5] In 1898 District Thirteen official John Reese wrote to the *United Mine Workers Journal:* "We have little boys working in our beloved State for the pittance of fifty cents per day; others getting eighty cents for the same work (trapping) and the same hours of labor." Reese commented that he was not in favor of boys working in the mines but "many of them must do so, and while they do I think it is our duty to see that they get paid for what they do." It should be noted that not a single person, regardless of age, listed his occupation as trapper. The state mine inspector also verified that young boys worked in the mines. In 1880 the state legislature raised the age at which young persons could go into the mines from ten to twelve. Yet three years later the state mine inspector declared that the law was being "continually violated" and urged that parents be required to furnish an affidavit concerning the boys' ages.[6]

Birthplaces for residents of Beacon, Cincinnati, and Seymour parallel places of birth for Iowans in general. Table 5.4 reveals that just under one-half of the total population of the three communities was born in Iowa. The second largest group came from eastern mining states. Out of a total population of 3,721 for Beacon, Cincinnati, and

Table 5.4. Birthplaces of 3,721 residents, 1895

| Location | Beacon (%) | Cincinnati (%) | Seymour (%) |
|---|---|---|---|
| Iowa | 46.5 | 41.0 | 49.7 |
| Eastern coal mining states | 19.2 | 39.2 | 34.0 |
| Other middle western states | 1.6 | 4.1 | 3.8 |
| Southern states | 0.5 | 1.2 | 1.4 |
| Western states | 0.2 | 0.3 | 0.5 |
| Other eastern states | 1.1 | 0.7 | 1.9 |
| British Isles | 22.6 | 10.4 | 4.4 |
| Northern and western Europe | 5.6 | 2.3 | 0.8 |
| Central, southern, and eastern Europe | 2.0 | 0.1 | 3.1 |
| Other | 0.9 | 0.4 | 0.5 |
| Total | 100.0 | 100.0 | 100.0 |
| Total number | 971 | 1203 | 1547 |

Note: No birthplace given for 13 people

Table 5.5. Comparison of birthplaces of 816 coal miner and other heads of households, 1895

| Location | Beacon nonminer (%) | Beacon miner (%) | Cincinnati nonminer (%) | Cincinnati miner (%) | Seymour nonminer (%) | Seymour miner (%) |
|---|---|---|---|---|---|---|
| Iowa | 12.0 | 7.9 | 17.2 | 20.8 | 17.0 | 16.0 |
| Eastern coal mining states | 39.0 | 8.8 | 61.8 | 39.6 | 64.3 | 55.0 |
| Other middle western states | 0.0 | 4.4 | 2.4 | 3.8 | 0.4 | 1.0 |
| Southern states | 1.0 | 0.9 | 2.4 | 0.0 | 3.9 | 0.0 |
| Western states | 0.0 | 0.0 | 0.0 | 0.0 | 0.4 | 0.0 |
| Other eastern states | 3.0 | 0.0 | 3.6 | 0.9 | 4.8 | 3.0 |
| British Isles | 38.0 | 60.5 | 9.6 | 25.4 | 4.8 | 15.0 |
| Northern and western Europe | 4.0 | 13.2 | 1.2 | 11.2 | 3.1 | 1.0 |
| Central, southern, and eastern Europe | 1.0 | 2.7 | 0.0 | 0.7 | 0.4 | 8.0 |
| Other | 2.0 | 1.8 | 1.2 | 0.9 | 0.9 | 1.0 |
| Total | 100.0 | 100.0 | 100.0 | 100.0 | 100.0 | 100.0 |
| Total number | 100 | 114 | 165 | 107 | 230 | 100 |

Seymour in 1895, 247 people were born in Illinois, 244 in Ohio, and 125 in Pennsylvania. The United States Census of 1890 lists the states of Illinois, Ohio, New York, Pennsylvania, and Indiana as contributing the greatest number of people to Iowa. The largest number of foreign-born residents in the three communities came from the British Isles, particularly from England.[7] Table 5.5 compares birthplaces of coal miner heads of households with other heads of households.[8]

The census data show that foreign-born residents in the three communities followed the same clustering patterns evidenced by northern and western Europeans when they settled in Iowa as agriculturists. The Scots dominated in Cincinnati, comprising 41 percent of the total foreign-born population there. In Seymour, the Italians were the dominant ethnic group, constituting 49 percent of the total foreign-born group. Beacon appeared to be a more mixed community with the English-born residents constituting 37 percent of the foreign-born population while the Welsh made up 30 percent.

The three communities had an almost equal number of males and females and illustrates that coal mining communities in Iowa were strikingly different in sexual composition from mining communities in the West. The general conclusion advanced in such studies as Ronald C. Brown's *Hard-Rock Miners: The Intermountain West, 1860-1920* is that western mining communities contained few families or single women. In turn, this condition often led to an increase in lawlessness and violence.[9] In Iowa's coal mining communities, families constituted most household units. Interviews and census data both indicate that when miners moved from one community to another, they usually took their families along. When European emigrants traveled to the United States, with the exception of the Italians and Croatians, they brought along their families. Even in coal camps that were predominately Italian, the family unit still dominated because male newcomers moved into the households of other Italian families.

In 1895 the communities of Beacon, Cincinnati, and Seymour differed considerably from most Iowa mining communities in racial composition. Beacon and Cincinnati had no black residents and Seymour had only three. While figures are not available as to the exact number of blacks and whites in all coal communities, it is clear from census data, interviews, newspapers, and the *United Mine Workers Journal* that by 1895 at least the majority of Iowa's coal communities had some black residents. At the same time, blacks accounted for only a small minority of the total population of coal mining communities. The residents of some communities vehemently opposed the entry of blacks whether they were strikebreakers or regular miners. In at least one community, members of the local UMW refused to honor membership cards of black miners, thus preventing them from transferring into the local. It appears that some communities may have lacked black residents because the all-white community took a strong stand against their entry.[10]

By 1915 the populations of Beacon, Cincinnati, and Seymour changed substantially in terms of numbers and character. Cincinnati and Seymour gained in population while Beacon's population was reduced to about half its 1895 total. Tables 5.6 and 5.7 compare the two sets of census figures and indicate the change in coal mining populations.

Beacon was clearly a declining community by 1915. This is difficult to understand because there was no scarcity of mining operations in the vicinity. In 1915 the Second District Mine Inspector listed twelve mines operating within two miles of Beacon. Eleven of the mines were local,

Table 5.6. Population, 1895 and 1915

| Community | 1895 | 1915 | Change (%) |
|---|---|---|---|
| Beacon | 971 | 483 | −50.3 |
| Cincinnati | 1203 | 1616 | +34.3 |
| Seymour | 1512 | 2134 | +41.1 |

Table 5.7. Coal miners as a percentage of the working population, 1895 and 1915

| Town | 1895 | 1915 |
|---|---|---|
| Beacon | 60 | 32 |
| Cincinnati | 48 | 59 |
| Seymour | 37 | 24 |

which meant that they sold only to area residents and businesses. Generally, local mines were less successful in attracting new miners into the community. One mine, the Bolton-Hoover Company 2, was a shipping mine along the Chicago, Burlington, and Quincy Railroad. At the same time, Beacon was located two miles from Oskaloosa. By 1915 the automobile was available to miners and many men may have begun commuting to work from Oskaloosa. Because 22.4 percent of Beacon's miners were sixty and above (32.5 percent were over age fifty), it appears that some miners decided to remain in Beacon during their later working years and retirement even if their economic opportunities were limited.

On the other hand, the populations of Cincinnati and Seymour expanded substantially, and both communities had extensive coal operations nearby. The Cincinnati area contained four shipping mines and one local mine, while Seymour had two shipping mines and one local mine. The key to Cincinnati's prosperity was the presence of four shipping mines that enabled the community to offer full-time or nearly full-time employment to many coal miners. Each shipping mine had a close affiliation with a particular railroad. Since railroads used coal steadily throughout the year, fewer shipping mines closed down for the summer. If they did close down, the shutdown was shorter than the shutdown for local mines. The Seymour operators shipped on the Chicago, Rock Island, and Pacific Railroad and the Cincinnati operators shipped on the Chicago, Burlington, and Kansas City Railroad. Seymour apparently had strong nonmining economic pursuits of farming and stock raising to provide business for the community and the surrounding areas. Both towns served as central trade areas for numerous small communities located within a five-mile radius.[11]

Table 5.8 confirms that in 1915 the average age of Beacon, Cincinnati, and Seymour coal miners was older than the average age of the coal miners who resided there in 1895. No attempts have been made to check the persistence of coal miners, so it is not known if the same coal miners remained in the three communities and naturally fell into older

Table 5.8. Comparison of ages of coal miners, 1895 and 1915

| Age | Beacon 1895 (%) | Beacon 1915 (%) | Cincinnati 1895 (%) | Cincinnati 1915 (%) | Seymour 1895 (%) | Seymour 1915 (%) |
|---|---|---|---|---|---|---|
| 1-15 | 2.8 | 0.0 | 0.0 | 2.3 | 0.6 | 1.5 |
| 16-20 | 17.2 | 1.7 | 6.8 | 16.7 | 12.3 | 15.5 |
| 21-30 | 27.2 | 19.0 | 37.8 | 25.8 | 40.4 | 26.3 |
| 31-40 | 16.1 | 20.7 | 35.8 | 26.5 | 31.6 | 26.3 |
| 41-50 | 20.0 | 24.1 | 14.2 | 15.7 | 10.5 | 12.4 |
| 51-60 | 10.0 | 12.1 | 3.4 | 10.5 | 2.9 | 14.9 |
| 61-91 | 6.7 | 22.4 | 1.4 | 2.6 | 1.8 | 3.1 |

age categories by 1915. It is known that the number of coal miners in Cincinnati and Seymour increased—Cincinnati from 148 to 306 and Seymour from 171 to 194—while the number of miners in Beacon declined from 180 to 58. Beacon most clearly showed evidence of an aging population. In 1895 only 16.7 percent of Beacon's miners were over age fifty. By 1915 that percentage had more than doubled. Beacon, no doubt, experienced more persistence than Cincinnati or Seymour since the number of Beacon coal miners decreased between 1895 and 1915 and at the same time, the average age of Beacon's miners was older by 1915. Interviews reveal that many Italian-American coal miners in Seymour remained in that community during their entire adult lives and confirm that Italian-American families emigrated from Italy to Seymour into the early 1920s.[12]

Tables 5.9 and 5.10 show that the number of occupations increased substantially in all three communities between 1895 and 1915. The increase appears most striking for Beacon since the population there decreased by roughly one-half while the number of different occupations increased by roughly 50 percent.

Table 5.9. All occupations, 1915

| Occupation | Beacon | Cincinnati | Seymour |
|---|---|---|---|
| Miners | 31.7 | 58.5 | 23.6 |
| Agriculture | 12.6 | 5.0 | 9.5 |
| Professional and semiprofessional | 6.0 | 7.2 | 7.8 |
| Personal services | 0.5 | 1.5 | 4.3 |
| Business | 19.1 | 10.5 | 15.7 |
| Transport and communication | 5.5 | 4.0 | 6.1 |
| Labor | 16.4 | 7.3 | 18.0 |
| Artisans | 7.7 | 5.0 | 12.7 |
| Other | 0.5 | 1.0 | 1.3 |
| Total | 100.0 | 100.0 | 100.0 |
| Total number of occupations | 183 | 523 | 822 |

Table 5.10. Number of different occupations, 1895 and 1915

| Town | 1895 | 1915 |
|---|---|---|
| Beacon | 22 | 33 |
| Cincinnati | 43 | 73 |
| Seymour | 67 | 95 |

Taking the three communities as a whole, there was an increase in every occupational category between 1895 and 1915. The greatest increase came in the business category while the personal services category reflected the second major increase. The category indicating the least increase was that of artisan. Finally, out of all occupations listed, seven were occupations that did not exist in 1895. These were auto livery, auto dealer, auto mechanic, motion picture theater operator, electrician, telegraph operator, and telephone operator. Seymour continued to have the largest number and greatest variety of occupations. This situation seemed consistent with the facts that Seymour had the largest population of the three communities, served a large trade area, and contained the smallest number of coal miners.

The number of women employed in 1915 also reflected substantial increases over 1895 as Table 5.11 indicates.[13]

In comparing occupational census data it is evident that the number of women employed in 1915 increased substantially over the number of women employed in 1895. The category of laborer witnessed the greatest increase, with the occupations of domestics and washerwomen accounting for the greatest number of laborers. These occupational increases were probably based on two conditions: (1) women seeking employment in the early 1900s had little education so they could not obtain higher status, better paying jobs, and (2) the communities involved offered only limited opportunities for female employment. Although both occupations required women to work long hours and re-

Table 5.11. Women's occupations, 1915

| Occupation | Beacon | Cincinnati | Seymour |
|---|---|---|---|
| Agriculture | | | 1 |
| Professional and semiprofessional | 9 | 16 | 28 |
| Personal services | | 2 | 4 |
| Business | 20 | 12 | 23 |
| Transportation and communication | | 5 | 6 |
| Labor | 6 | 4 | 53 |
| Artisan | 1 | 3 | 31 |
| Total | 36 | 42 | 146 |

main standing most of the time, domestics faced the most undesirable working conditions. According to David Katzman's study of domestic workers, the typical domestic worked seven days a week, averaging eleven to twelve hours a day. Moreover, the domestic was closely supervised and was usually isolated from "ordinary social interaction." Washerwomen, by contrast, although their work was described as "backbreaking labor," did have some independence. The typical washerwoman did not work Saturday or Sunday. They could do laundry in their own homes, thus setting their own hours. If washerwomen had children, they could do their work and supervise their children at the same time.[14]

The 1915 census also shows an increase in women's employment in several other categories. Next to laborer, the category of business shows the greatest increase. The 1895 census listed one woman as proprietor of a hardware store while the 1915 census listed thirteen women with business occupations. New jobs or jobs that did not exist in 1895 accounted for only three additional categories: telephone operator, telephone office manager, and motion picture theater manager. The 1915 census identified two women keeping boarders while the 1895 census did not list any. Given the large number of coal miners in each community, it seems likely that many miners' wives kept at least one boarder. In light of this practice, it appears that the number of women employed should be higher in 1915 than the census indicates.

The three communities differed considerably over the birthplaces of miners in 1915 compared to 1895. All communities witnessed a considerable increase in the number of coal miners born in Iowa. This increase appears a logical outcome of the steady development of the Iowa coal industry. As miners moved here in the 1890s and early 1900s, their sons were born in this state, and many sons later went to work in Iowa mines. At the same time, Beacon and Cincinnati saw considerable increases in the number of coal miners born in eastern coal mining states in 1915 over 1895. In Cincinnati, Missouri-born miners accounted for the majority of workers born in the eastern coal mining states. Cincinnati is located only two miles from the Missouri border and this proximity undoubtedly accounted for the prevalence of Missourians. All communities witnessed a decline in the number of miners born in the British Isles. On the other hand, Seymour indicated a substantial increase in the number of miners born in central, southern, and eastern Europe. Table 5.12 compares birthplaces of miners in 1895 and 1915.[15]

The racial composition of Beacon, Cincinnati, and Seymour changed only slightly. In 1895 Seymour had three black residents while

Table 5.12. Comparison of birthplaces of 732 coal miners as heads of households, 1895 and 1915

| Location | Beacon 1895 (%) | Beacon 1915 (%) | Cincinnati 1895 (%) | Cincinnati 1915 (%) | Seymour 1895 (%) | Seymour 1915 (%) |
|---|---|---|---|---|---|---|
| Iowa | 7.9 | 27.1 | 20.8 | 28.3 | 16.0 | 29.0 |
| Eastern coal mining states | 8.8 | 18.8 | 39.6 | 56.2 | 55.0 | 29.7 |
| Other middle western states | 4.4 | 0.0 | 3.8 | 0.5 | 1.0 | 2.4 |
| Southern states | 0.9 | 0.0 | 0.0 | 1.4 | 0.0 | 0.7 |
| Western states | 0.0 | 0.0 | 0.0 | 0.0 | 0.0 | 0.7 |
| Other eastern states | 0.0 | 0.0 | 0.9 | 0.5 | 3.0 | 0.0 |
| British Isles | 60.5 | 39.7 | 25.4 | 9.6 | 15.0 | 13.1 |
| Northern and western Europe | 13.2 | 8.4 | 11.2 | 2.3 | 1.0 | 0.0 |
| Central, southern and eastern Europe | 2.7 | 6.3 | 0.0 | 1.0 | 8.0 | 22.8 |
| Other | 1.8 | 0.0 | 0.9 | 0.5 | 1.0 | 0.7 |

Beacon and Cincinnati had none. Twenty years later, Cincinnati included three blacks while Beacon and Seymour listed none. The general attitude of white miners remained the same; blacks were allowed in some coal communities but they were denied entrance into other communities. The racial situation had changed somewhat in Iowa by 1915 as the town of Buxton had come into existence. The Iowa State Census listed 4,518 residents in Buxton in 1915; blacks accounted for approximately 40 percent of the total population. Buxton was located approximately ten miles from Beacon and over twenty miles from Seymour and Cincinnati.[16]

In 1915 the Iowa state census included two new categories that had particular application to the coal mining population: months unemployed and yearly income. These two categories provide a way to test widely held assumptions gained from interviews and other sources (1) that many miners traditionally faced several months of unemployment each year; and (2) that miners' incomes were extremely low. The census takers asked residents how many months they had been unemployed in 1914. Table 5.13 compares months unemployed for coal miner heads of household with other heads of household.[17]

Table 5.13. Average months unemployed for 535 coal miner and other heads of households, 1914[a]

| | Beacon | | Cincinnati | | Seymour | |
|---|---|---|---|---|---|---|
| | Miner | Nonminer | Miner | Nonminer | Miner | Nonminer |
| Months | 4.2 | 4.7 | 4.5 | 4.9 | 5.1 | 4.9 |

[a] Missing observations = 712

The rate of unemployment among coal miner heads of household varied considerably. The Beacon miners reported the shortest periods of unemployment with 66.7 percent unemployed between two and four months. By contrast, 92.6 percent of the Cincinnati miners reported that they were unemployed between four and six months while 78.3 percent of the Seymour miners were unemployed between four and six months. The high rates of unemployment are surprising inasmuch as most of the mines in and around the three communities were shipping mines. It is not possible to determine from the census data which months out of the year the miners were unemployed. If most of the unemployment came during the spring and summer, it could be assumed that the mine operators failed to get summer contracts with the railroads. If the unemployment came during the winter months, it might have been caused by local strikes. Sometimes mild winter weather brought a higher number of days unemployed.[18]

While Table 5.13 lists months of unemployment for both coal miner and other heads of household, it should be noted that the percent of both groups who were unemployed varied considerably. In the three communities, 58 coal miner heads of household or 14.1 percent of the total number reported no unemployment. By contrast, 472 other heads of household, or 56.4 percent of that total group, reported no unemployment. Overall, the great majority of coal miner heads of household listed some unemployment while slightly over 40 percent of the other heads of household reported some unemployment.

According to information contained in Table 5.13, the unemployed heads of household who were not miners experienced approximately the same rate of unemployment as coal miner heads of household. The unemployed portion of the nonminer group appeared to have jobs closely related to or dependent on the mining industry. Obviously occupations such as draymen and day laborers depended on mine operators for a large part of their work. The remainder of the other heads of household that did not list any unemployment, 56.4 percent, presumably had little or no economic dependence on the coal industry.

The census category of income also provides additional information on the three coal mining populations. Table 5.14 reveals the average annual income for coal miner and other heads of household.[19] This category of income provides one of the sharpest distinctions between miner and other heads of household. (See Appendix Table A.12) Nonminers in all three communities earned more than miners. Taking

Table 5.14. Average annual income for 969 coal miner and other heads of households, 1914[a]

|  | Beacon | | Cincinnati | | Seymour | |
| --- | --- | --- | --- | --- | --- | --- |
|  | Miner | Nonminer | Miner | Nonminer | Miner | Nonminer |
| Income | $461 | $557 | $488 | $749 | $419 | $609 |

[a] Missing observations = 278.

the three communities as a whole, other household heads earned an average of $182 or 27.7 percent more than miners.

Between 1895 and 1915 the populations of Beacon, Cincinnati, and Seymour changed substantially in the areas of size, occupations, and place of birth. Beacon's population declined significantly while the population of Cincinnati and Seymour increased. By 1915 Beacon had few businesses to attract additional residents. Cincinnati appeared to be more closely identified with coal mining in 1915 than in 1895. The number of Cincinnati coal miners increased not only in total number since 1895 but miners as a percent of the total population increased by 6.3 percent. Seymour's economy appeared less affected by coal mining in and around its borders in 1915 than in 1895. The number of different occupations present in the three communities increased dramatically between 1895 and 1915. In 1895 people in the three communities worked in 133 different occupations (Appendix Table A.3); in 1915 that total had risen to 202 (Appendix Table A.4). Women's occupations rose from a total of 12 different occupations in 1895 to 42 different occupations in 1915 (Appendix Table A.5). Changes in birthplaces also appeared in the twenty-year interval (Appendix Table A.6). The most significant change came in the increase of residents born in Iowa. Overall, the number of foreign-born miners declined, particularly those born in the British Isles. Coal miners continued to include a higher percent of foreign-born people than did the other resident populations (Appendix Table A.7). Comparisons of the coal mining and other resident populations in the three communities indicate that there were few major differences in the social composition of the two groups. Yet the two groups did differ most noticeably in the areas of unemployment and income. Consequently, census data from Beacon, Cincinnati, and Seymour substantiate long-held assumptions that coal miners made lower wages and experienced greater unemployment than the other resident populations.[20]

# 6

## UNITED MINE WORKERS IN IOWA

THE UMW was formally organized in 1890, the culmination of a thirty-year effort by American coal miners to establish a national union. Beginning in 1861 with the formation of the American Miners Association and ending in 1885 with the organizing of the National Federation of Miners, coal miners made four unsuccessful attempts to organize nationally, with each organization enjoying only a short-lived dominance. At the same time, coal miners survived a challenge from a rival organization to act as their union representative. The Knights of Labor made several aggressive bids in the 1870s and 1880s to attract coal miners into their organization. After many years of conflict, the National Federation of Miners suggested a merger, and their plan proved successful. In January 1890 the National Federation and the Knights came together and organized the United Mine Workers of America. With the merger, UMW membership stood at almost seventeen thousand.[1]

The UMW constitution called for the general membership to hold annual national conventions to conduct major business; between conventions, a five-member executive board, plus the three national officers, were to handle union business. National officials would work out of Columbus, Ohio, designated as the national UMW headquarters. Once formulated, the executive board moved quickly to organize the nation's miners by dividing the country's coal fields into twenty-one districts. The constitution determined that the national union would supervise the districts but the districts were authorized to elect their own officers. John Rae, UMW president, immediately sent a letter to all mine workers throughout the country outlining union objectives and inviting the miners to join the organization. The executive board appointed three organizers to begin working full time at organizing locals throughout the districts, and the three national officers aided in organizational work. The board also moved to establish an official journal, the *United Mine Workers Journal,* because they believed that the *Journal* would increase membership and educate members in the organization's basic goals. While the UMW consitution listed several major goals, the main issue was the eight-hour day. The executive board voted soon after the union's formation that all union members would go on strike on May 1, 1891, and remain on strike until they won the shorter work day. On February 10, 1891, approximately one year after the founding of the UMW, President Rae proudly announced that the organization was off to an excellent start. The organization had thirty-five thousand members, it had added 116 locals during the first year, organizers had been sent into "every state and territory producing coal," and the first issue of the *Journal* would be issued in April 1891.[2]

The UMW's first effort to organize Iowa's coal miners came shortly after the formation of the national union. W. H. Turner, national vice-president, came to Iowa in late 1890, and following several months' work, called a state convention of Iowa coal miners. The main work of the convention, held at Oskaloosa in February 1891, was to organize a district, known as District Thirteen. District organization paralleled organization at the national level. The district convention called for the election of an executive board member for each subdistrict (District Thirteen was divided into four subdistricts). The constitution determined that the district's executive board, including the district officers, would be the general governing agency of the district. Following its formal establishment, district officers immediately proceeded to set up local unions. Responsibility for organizing locals rested primarily on the shoulders of the district officers with the president devoting almost full time to the effort. Throughout the 1890s the national UMW sent several organizers into Iowa to assist district officers. During the first year, officials successfully established locals in Ottumwa, Oskaloosa, Centerville, Des Moines, Jerome, Carbonado, Laddsdale, Flagler, Mystic, Cincinnati, Oswalt, Angus, Youngstown, and Seymour. The members of each local elected their own local officers.[3]

Along with their organizational work, district officers carried out the day-to-day administration of District Thirteen. The district constitution specified that when problems arose at the local level, local presidents should attend to the difficulties. If they could not do so successfully, they were to contact immediately the district president, who would undertake to handle the problem or assign it to another officer. At the same time the district president was responsible for calling joint conferences. The annual convention provided the time for the membership to come together and handle major district business, including the annual election of district officers. Local members elected delegates to the convention, and each local was assigned a certain number of delegates, based on the number of members within the local. Their constitution required that all nominations for district office be forwarded from local unions to the district secretary-treasurer one month before the convention; subsequently the secretary-treasurer would forward the list of candidates to the locals. The membership of each local would indicate their preference for district officers (and positions on major issues); their delegates would then come to the convention prepared to vote for specific candidates. The convention also provided the miners with an opportunity to present their resolutions of grievances. The entire convention was expected to act on the resolutions, and the majority were turned over to the executive board for final action.[4]

Primary responsibility for district affairs rested on the shoulders of the district president. The president found it necessary to play many roles: administrator, mediator, organizer, promoter, and publicist. He presided over monthly meetings of the executive board. Board minutes indicate that in most cases the board acted in accordance with the president's wishes. The president presided over the annual convention, and was free to present his views to the membership. During a typical year the president issued several circulars, proposing new programs and suggesting solutions for old ones. He initiated the annual joint conference with coal operators and was a central figure in these conferences as well as working frequently throughout the year with coal operators to settle wage and work disputes. District Thirteen's two most influential presidents, John Reese and John White, each displayed different talents; overall, their abilities proved to be complementary. Beginning in 1898, Reese exhibited exceptional organizational ability. When he resigned the district presidency in 1902, he proudly announced that for the first time the district was 100 percent organized. Following in Reese's footsteps, John White's contributions lay primarily in promoting much needed economic programs. During his presidency he persuaded Iowa's miners to establish a defense fund, implement death benefits, and establish a hospitalization program. Of greatest importance, however, White formalized the joint wage agreements with Iowa's coal operators.

The district vice-president and secretary-treasurer also contributed to the growth and vitality of the organization by their responsible leadership. The vice-president aided in organizing locals and assisted the president in solving disputes. The secretary-treasurer had the general responsibility of handling the district's financial matters and keeping all district records, while local officers worked to increase membership, solve local problems, and assist district officers. Strong local presidents produced strong local unions where members abided by the joint agreements and where disputes were settled at the local level. Local union offices provided the training ground for most future district officers. Frequently the president also served in other district offices before assuming the district presidency.

Some district officers were able to continue in union affairs for many years; for some, union work became a full-time career. John White served in UMW positions for almost twenty years, beginning his career as secretary-treasurer of the Pekay Local. When White was thirty, the membership elected him secretary-treasurer of District Thirteen. Two years later the Iowa miners elevated White to the district presidency, a position he held for nine years. White went on to serve one year as international vice-president and six years as international president. Because of his union background, he was able to move into the federal

John White, president, District Thirteen, UMW; later national president, UMW. (Courtesy of Owen McConville, III)

John Reese, 1917, national vice-president, UMW, was earlier vice-president, District Thirteen, UMW. (Courtesy of Owen McConville, III)

bureaucracy, serving as assistant director for the U.S. Fuel Administration. Edwin Perry served in union positions even longer than White. In 1876 the Beacon Local elected Perry its secretary-treasurer. For the next twenty-four years Perry served in every local office. In 1900 District Thirteen elected Perry as vice-president. He served the district as vice-president for one year, president for one year, and as secretary-treasurer for seven years. In 1908 Perry was elected as secretary-treasurer of the international UMW, a position he held until 1916.[5]

The coal miners' overriding concern in 1891 was winning the eight-hour day. For many years coal miners everywhere worked from ten to twelve hours daily. Miners lamented going to work in darkness and returning home in darkness, leaving little time to spend with their families. Their young sons who worked in the mines also labored the same long hours. From its founding, the national UMW was committed to winning the eight-hour day as quickly as possible. In 1891 national officials planned to suspend work until the eight-hour day was won. Shortly before the suspension was to go into effect, however, the national executive board voted to defer action; after surveying district officers, the board agreed that sufficient support for the suspension did not exist among the miners.

Iowa miners believed that the requisite commitment to a shorter work day did exist in District Thirteen. Receiving permission from the national executive board, they called a suspension that went into effect on May 1, and continued for nearly three months. But the Iowa coal operators immediately fought back. In Mahaska County alone, more than three hundred miners were discharged in the spring of 1891 because they had joined the UMW. In Laddsdale, seventy-five men were discharged and ordered out of their houses when they appeared ready to organize. In early June, a coal miner reported from Carbonado that the men were "firm at the wheel . . . notwithstanding the many aggravating surrounds."[6]

Early in the suspension miners began to appeal for aid. On June 11, the district secretary, Julius Fromm, reported to the *Journal* that Iowa operators threatened to "starve the miners of Iowa into submission." He pleaded: "We have women and children living on bread and water and our funds are exhausted, hence we ask you to aid us at once. Don't wait, now is the critical time with us. Aid us at once. . . ."[7] A short time later President Walter Scott of the district wrote to the *Jour-*

*nal* that while traveling through the district, he had observed some sad situations:

> One family, all they had seen for nine days was dry bread and a few boiled beans; three others who had not seen anything but dry bread and water for . . . seven days; another family sold all their furniture down to the bed they lay on and then appealed for aid, and we were compelled to respond with a heavy heart and trembling lips, we can't give you anything.[8]

The first attempt to secure the eight-hour day in Iowa was a failure; within three months, most Iowa miners returned to work under the old conditions. Julius Fromm reported that many men were "victimized," or fired, as a result of joining the union. Although most larger operators refused to grant the concession, Fromm estimated that about a dozen of the state's small operators agreed to implement the shorter work day. However, the great majority of Iowa's miners continued to work ten hours a day.[9]

For the next seven years Iowa miners were unable to make any progress in their fight for shorter hours. Although they did not forget their commitment to the eight-hour day, the Depression of 1893 forced the miners to concentrate on the basic issue of survival. The depression had far-reaching effects on the miners' struggling union in Iowa. In the mid-1890s district officials frequently commented on the "demoralized conditions" that pervaded the Iowa coal fields. During the early months of 1894, scattered reports of the state's economic slowdown began to appear in the *Journal;* in many camps operators reduced the long-time winter price of $1.00 to either $0.75 or $0.80 per ton. A Mystic miner reported to the *Journal* that miners there experienced great difficulties; they worked only four or five days each month. The Mystic merchants had recently denied credit to the men and continued to serve them on a cash basis only.[10]

By spring of 1895 the *Journal* ran a report by Julius Fromm, noting that there had been little work in the state during the previous year. Many mines had actually closed down; as a result, roughly two thousand men left to seek work in other states. He lamented that "what was but a short time ago a healthy organization in District 13 has now wasted away to mere form. . . ." Actually, between March and July 1895 membership in District Thirteen dropped from an estimated twelve thousand to around two thousand.[11]

District officials faced serious organizational problems. In March the district president, John T. Clarkson, resigned. District records do not indicate that Clarkson was replaced. At the same time, the executive board labored diligently to hold its rapidly dwindling membership intact. At their September meeting, however, board members faced the inevitable—given the lack of financial support, District Thirteen could not continue to function. Secretary-Treasurer Fromm reported that only three locals had paid their dues and the income from that source did not exceed thirty dollars. He explained that during the past six months attempts at organizing new locals or reorganizing old ones had met with little success. Fromm estimated that the district did not contain over two hundred paying members. In response, board members drafted a circular to be sent to all members in which they recommended that the board "suspend operations for the time being or until further notice." The board then voted to suspend salaries and expenses after October 1, 1895. As a result, District Thirteen ceased to function for the next two and one-half years.[12]

With the return of economic prosperity in late 1897, Iowa miners began to think about reorganizing District Thirteen. At the urging of the Hiteman miners, 27 delegates from Iowa's mining camps met in Oskaloosa in March 1898 to reorganize the district. The number of miners represented totaled 3,787. Discussion of the eight-hour day dominated the proceedings; delegates reiterated their support for the hours issue, urging district officials to work for its immediate acceptance.[13]

Iowa's coal miners did not have long to wait for the resolution of the hours issue. In October the national UMW convention declared its irrevocable stand for the shorter work day. Soon after, national officials succeeded in gaining the concession from operators in the Central Competitive Coal District, one of the first districts formed by the UMW. It was composed of Illinois, Ohio, Indiana, and western Pennsylvania—the major coal mining regions of the United States—and therefore set the standards for other UMW districts. Known as the Chicago Agreement, that contract became the basis for wage agreements in all other UMW districts. Following the district convention in February 1899, district officials and delegates met with representatives of the newly organized Iowa Coal Operators Association to work out a joint wage agreement for the next twelve months. In early March the joint delegation reached agreement on the wage scale in which the operators' representatives agreed to the eight-hour day. The agreement, negotiated in Centerville, was known as the Centerville Agreement. Unfortunately not all operators accepted the agreement. The failure to win unanimous acceptance forced district officials to conduct separate negotiations in each of the state's four subdistricts.[14]

During the next two months, operators in Subdistricts One, Two and Four signed the Centerville Agreement. Writing to the *Journal* on April 5, John Reese, the district vice-president, appeared ecstatic over the victory in Subdistrict Four, declaring, "This, in our judgment, is the greatest victory ever won in Iowa, in as much as the men have only been organized since April 1st, 1898, and the agreement was made on April 4th. . . ."[15] District leaders particularly relished the victory in Subdistrict Four, because most union activity in the 1890s had been centered in Subdistricts One and Two. Reese himself had earlier visited the Boone District on several occasions and declared that it was "a hard hole to organize." Reese exclaimed that when he arrived in the Boone District on March 29: "There was not an organized miner in the District. Today we can count them by the hundreds."[16]

The last significant eight-hour victory came with the organizing of Subdistrict Three, known as the Des Moines District. Since this subdistrict had been long considered a difficult area, President John Ream and Secretary-Treasurer John Brennan concentrated their efforts there in late spring and early summer. The men had come out on strike on April 1 against a reduction of $0.10 per ton and for the eight-hour day. Within several days, the small mine operators had granted the concessions. The four largest mines in the area—Gibson, Carbondale, Klondike, and Christie—initially refused to grant the concessions. By the end of June, however, the largest operators had agreed to restore the $0.10 reduction and grant the eight-hour day. According to Des Moines District miner M. F. Maher the recent victory represented "the greatest and only complete victory ever won in the Des Moines District. . . ."[17]

While Iowa miners believed the eight-hour day to be the paramount economic issue in the 1890s, they also faced other crucial economic issues. Among miners' top economic priorities were higher wages, acceptance of union checkweighmen, establishment of the joint wage agreement, elimination of coal screens, and elimination of restrictive company store policies. At the same time, miners could not always state their grievances openly. Some operators required that their employees sign agreements that they would not make any complaints about the terms of their employment, such as the screening of coal. Furthermore, many operators wrote agreements that required the miners to bind themselves neither to strike nor to join any organization that sought an advance of wages. If the miners broke the agreement, they forfeited their earned, but unpaid wages. As UMW officials organized locals throughout the state, these practices gradually disappeared. Yet even as late as 1900 some miners complained about operators insisting that they sign such agreements before obtaining employment.[18]

Inadequate wages ranked first on the miners' list of specific economic grievances. Iowa miners shared this problem with miners everywhere. On April 20, 1891, the *Journal* editor reported that in Pennsylvania, Missouri, and Illinois, the miners' daily wage averaged well below $2.00. In Pennsylvania, out of 1,763 coal miners canvassed, the majority earned less than $1.50 per day. In Missouri, 74 percent of the miners had a daily income of less than $1.50 per day. In Illinois, a special commission looking into the strike of 1889 reported that miners in the Spring Valley District averaged $31.63 per month. In Iowa, a miner who recently emigrated from England wrote to the *Journal* that he earned $1.75 a day, a wage that would not support his wife and five children. He added that he paid $4.50 for his house rent along with $0.50 a month for water. These figures supported the national UMW officials' view that throughout the country, coal miners earned less than a living wage.[19]

Wage instability represented another common wage grievance. Iowa miners expressed continual dissatisfaction that operators could arbitrarily manipulate wages at any time. Wage reductions constituted the major reason for strikes throughout the 1890s. Operators claimed that they had to reduce wages from time to time because of declining market prices or declining sales. However, they did not subsequently lower house rents or company store prices. An operator informed his employees of a wage reduction by posting the notice at the entry to the mine, usually a week before the reduction would go into effect. The miners bitterly resented that they not only lacked a voice in negotiating the wage rates, but that they lacked any means of negotiating the matter after wage reductions had gone into effect. Suspending work constituted the miners' only means of protesting the action.

The miners' main solution to wage problems lay in the establishment of an annual district joint wage agreement. Under this arrangement UMW officials and representatives of the Iowa coal operators came together every year to determine wage rates and working conditions for the next twelve months. Through the joint agreements the miners hoped to achieve both higher wages and wage stability. Iowa

Typical scene at a session of the Joint Wage Conference, Indianapolis, 1906. (Courtesy of Owen McConville, III)

miners were following the model provided by the national UMW, which adopted the principle of the joint agreement immediately after they organized in 1890. The formulation and enforcement of the joint agreement constituted a major portion of the UMW's activity, both at the national and the district levels. Once a joint agreement had been drawn up and signed by district officials and at least a majority of coal operators, the participants regarded it as a legal contract. In camps where operators had not signed the agreement, district officials frequently called for suspensions in an effort to bring compliance. From 1893 to 1925, District Thirteen officials and Iowa's coal operators continued to utilize the joint agreement. Both parties soon came to pay tribute to the arrangement, believing that it greatly benefited both labor and management.

The first step in working out a joint wage agreement called for national officials to negotiate a wage agreement with coal operators in the Central Competitive Coal District.[20] National officials believed that the central district agreement would provide both a model and a guide for subsequent joint agreements in the other twenty UMW districts. Moreover, only issues raised in the central district could be raised in subsequent negotiations within the individual districts. Following the adoption of the central district agreement, national officers expected other districts to pursue joint agreements with their local operators. District Thirteen officials attempted to hold a joint conference in the first year of the district's existence but with little success. Two years later, however, district officials and operators successfully negotiated the "Ottumwa Agreement" that provided for a system of separate wage rates for each of Iowa's four subdistricts. Since contracts between operators and the UMW expired on March 31, joint conferences followed the general convention, usually held in late March or early April. Both parties agreed early in the decade that contracts should run from April 1 to March 31. The negotiations sometimes took several months to complete.

While the eight-hour day and the joint wage agreement represented the union's major victories in the 1890s, miners continued to experience many economic grievances. Foremost among these was the failure of operators to pay the miners for all the coal that they loaded out. Operators had long required that all coal pass through a series of screens that separated the smaller chunks, popularly known as nut coal, from the larger pieces. The operators did not pay the miners for loading out nut coal because they insisted that it had little market value. In May 1891 District Thirteen's first president, Walter Scott, calculated the nut coal to be approximately one-third of the miners' total output. He estimated that between 1887 and 1889, the value of nut coal totaled $632,416, for which the Iowa coal miner received nothing. Miners raised the screening question at every district convention held between 1898 and 1916. Apparently, they met strong resistance from the operators but they continued to work toward their goal both through their negotiations with operators as well as through petitioning the state legislature for antiscreening legislation. Faced with repeated failure in both areas, the miners began to advocate uniform screens. With this approach, they hoped to prevent operators from arbitrarily changing screen size to reduce even further the amount of coal loaded out. In 1912 the state legislature passed a uniform screening law. Finally in 1916 the miners achieved victory in their long struggle to abolish coal screens. In that year the international officials succeeded in winning the concession of mine-run coal (or the elimination of screens) from operators in the Central Competitive Coal District. District Thirteen officials quickly secured the same concession from Iowa operators.[21]

Closely related to the matter of screening coal was the position of checkweighman. Shortly after the organization of District Thirteen, miners began to demand the right to appoint checkweighmen at every coal mine. Miners contended that company weighmen frequently did not accurately record the total tonnage when they weighed the miners' cars. The only solution in the miners' view was to have a union checkweighman working alongside the company weighman, verifying the proper weights. Miners first raised the matter of checkweighmen at their 1898 district convention. They proposed that checkweighmen be appointed at every mine, but most operators rejected the proposal; only the operators in Subdistrict Two agreed to permit the union to appoint checkweighmen. Local union officials continued to press for the position with some success. In 1902 John Reese, the district president, noted that most coal mines in the state had the position. Still, the district membership found it necessary to include the demand at every convention between then and 1910.[22]

The presence of the company store represented another longstanding economic grievance. Coal miners were convinced that they paid considerably more for their food, clothing, and mining supplies at the company store than if they traded elsewhere. Miners' wives frequently came to the company store pay office to collect their husbands' checks only to be told that after deductions, there was no money left. One miner's wife remembered that when the family lived in Zookspur, her husband's wages for two weeks totaled ten cents following deductions.[23] In particular, miners resented the company's policy on the purchase of blasting powder. The general rule prevailed throughout the district that miners should purchase their powder at the company store. The men protested that the policy was unfair since they were forced to pay higher prices for powder than if they purchased it elsewhere. The miners' grievance was intensified because the most common method of mining coal in Iowa, shooting on solid, required the heavy use of blasting powder. One worker estimated that each miner used an average of two to three kegs of powder monthly.[24] Throughout the 1890s the powder question remained controversial; operators continued to demand that miners purchase their supply from the company.

Miners also objected strenuously to being paid by the month. They asserted that few families could survive for an entire month without resorting to the use of credit at the company store. Union officials repeatedly argued that miners would become independent workmen only when they were able to break their dependence on the company store. Union officials believed that the first step in that process was implementation of the semimonthly pay. Delegates to every district convention between 1891 and 1910 listed semimonthly pay as a major goal. In 1910 the miners finally succeeded in winning the concession from the coal operators.[25]

During the late 1890s and early 1900s, miners used their union as a vehicle to obtain safer working conditions. By 1900 miners repeatedly emphasized several unsafe conditions within the mines. Their primary concerns were poor ventilation, inadequate supervision, and indiscriminate shot firing. The issue of shot firers differed from other mine safety concerns, however, because the issue was not resolved without loss of life. On January 24, 1901, an explosion at Lost Creek 2 dramatically underscored the need for shot firers. The accident occurred a few minutes before noon as many of the men were preparing to ascend for their noon meal. About one hundred miners were still in the mine. At 11:55 A.M., "those above heard a tremendous report, the earth shook and debris flew high above the shaft." Rescue operations began immediately. The first fifteen men died from the immediate effects of the explosion making identification difficult. In several cases, identification could not be made until articles of clothing were used to confirm identity. Officials immediately speculated that a windy shot had caused the dry dust to ignite and carry the fire to several nearby kegs of powder. The explosion killed twenty miners outright and severely injured nine more. The Lost Creek disaster was the worst accident in Iowa's mining history.[26]

District vice-president John White arrived on the scene soon after the explosion and immediately took charge of the rescue operation. He reported that within an hour or two "the little village was thronged by men volunteering their services. . . ." White reported that "the people of this little village are nearly stupefied as a result of this catastrophe, and it will be some time before they will fully realize what has befallen them." He noted that fifteen of the men had been married leaving "large and helpless families." One of the victims, Michael Fox, Sr., had eleven children; his second son, Michael, Jr., who worked with him in the mine, was also killed. White promised that the district organization would "at once assist in taking care of the immediate wants of these afflicted people." He urged all locals in the district to remember the Lost Creek families.[27]

A short time later, mine officials confirmed that the explosion resulted from a windy shot fired by miner Andrew Pash. The disaster dramatized the need for shot firers and shot examiners. Immediately both operators and miners began to reevaluate shot firing policies. Even before the Lost Creek tragedy, however, agitation had been growing among the miners to establish once-a-day firing. In most mines the men set off shots continually during the eight-hour shift. With the appointment of a shot firer, the shooting would be done only after the men left the mine. A month after the Lost Creek explosion, District Thirteen miners held their annual convention at which the shot firing question dominated the proceedings. Miners came to the convention determined to secure the principle of a shot firer in every mine. The miners were equally determined that the operators should support the cost. After several weeks of "hard fighting," the operators accepted the principle of once-a-day firing, but only if the miners paid the cost of the shot firer. Somewhat earlier the state legislature had passed a law requiring operators to hire shot examiners. But the operators objected

A typical mine rescue team in the 1940s. Iowa coal miners first trained for mine rescue work in the 1920s. (Courtesy of Iowa Mines and Minerals Department, Des Moines)

that they could not bear any further costs. Following the convention, firing was done only once a day. In most mines, operators and miners agreed that one man could serve as both examiner and firer; operators and miners then shared the cost.[28]

After 1905 Iowa's coal miners began to concentrate on conditions that lay outside their world of work. Miners had discovered that higher wages eased some economic burdens but higher wages did not solve the basic social deficiencies that they encountered in their everyday lives. For the next twenty years, coal miners worked through their union to improve the quality of life for themselves and their families, most notably in the areas of death benefits, medical care, housing, and education. Coal miners soon discovered that there were some problem areas over which they had little control. Before 1911 important social and economic changes came primarily in those areas directly controlled by miners themselves. The establishment of death benefits and hospitalization plans rested on fees paid by miners, not on outside funding sources. But changes in areas such as housing and education rested on the agreement and cooperation of outside parties, particularly operators and state legislators. After 1911 Iowa coal miners increasingly turned their attention toward the statehouse in efforts to secure state regulation and assistance. By 1919 the state legislature began to respond to at least some of the miners' needs, particularly aid for education and legislation to upgrade living conditions.

The first social measure brought before the union membership was the matter of death benefits. At the miners' annual convention in 1906, district president White recommended that members establish a death benefit fund. He believed that the district should provide at least $100 for widows or legal heirs of deceased members to be paid out of the district's general fund. White noted that in the previous year, district officers had responded to two separate incidents. In 1905 they had donated a total of $500 to five families who had lost members in the Riverside explosion in Des Moines. District officers also donated $100 to the Shepard family of Hocking who lost two sons in an explosion in the Hocking mine.[29]

White's recommendation touched a prevailing sentiment among the district membership. In times of tragedy, miners' wives had few resources at their disposal. In the event of a miner's death, his widow not only was deprived of her husband's income but also faced burial costs and possible hospital expenses. And she was left with the burden of raising her family alone. John White had experienced firsthand the difficulties caused by a miner's early death; his father had died before White reached school age. In his speeches to the Iowa miners' conventions, White alluded frequently to the hardships his mother had faced as a result of her widowhood. The creation of the death and endowment fund was intended as one small measure of assistance in such situations. The convention followed White's recommendation and created a permanent "death and endowment fund" and stipulated that $10,000 be drawn from the district treasury for the payment of death claims. Benefits would be $100 per miner. If the miner had no heirs, delegates authorized the local union to use the money for funeral expenses and, if sufficient, purchase a tombstone for the deceased.[30]

Three years after the establishment of death benefits, miners began to consider the need for better medical care. John White first raised the issue at the district convention in 1909, but the delegates did not act on the matter. In 1911 Dr. R. P. Miller of the Albia Hospital Association conceived the idea of making hospital services available to miners at low cost, initiating the movement that eventually culminated in the creation of three separate facilities. Dr. Miller began by offering each miner a contract that called for a payment of $0.25 per month so that "in case of accident or surgical operation, of any nature whatever, the miner had the services of the surgeon, nurses, and all the hospital facilities at his command." As miners started joining the program, the hospital expanded its services to cover the miners' entire family. For $0.50 a month the miners' wives and children received comparable care. An Albia resident described the Albia hospital, commonly known as the Albia Miners' Hospital, as an "up-to-date hospital with the latest modern equipment in every department, representing an investment of over $20,000, so that the miner knows he has everything at the best of his service."[31]

Iowa miners responded to the hospitalization plan so enthusiastically that a short time later Dr. T. E. Gutch of Albia started a second miners' hospital. In February 1916 Dr. Gutch expanded his operation to Des Moines where he established a second facility, located at 406 Center Street. Dr. Gutch served as the chief surgeon assisted by Dr. J. M. Griffin and Dr. T. M. Boone of Des Moines. The Des Moines and Albia hospitals followed the same plan of providing medical care to miners and their families for $0.50 per month. In February 1916 hospital authorities estimated that approximately eight thousand miners

Riserville Mine, Albia, 1915. (Courtesy of Sylvia Rosenthal)

in District Thirteen subscribed to services at either the Des Moines or the Albia hospital. District officials prophesied that within the near future "every mine worker and his family in the district will be assured of efficient hospital and surgical care at the lowest cost possible."[32]

Miners also began to introduce cooperative stores in an effort to improve their standard of living. In 1900 district president John Ream urged Iowa miners to consider such establishments, believing that the cooperative store was the only way to eradicate the evil of the company store. During the next five years Iowa miners initiated several cooperative stores but none survived longer than one or two years. With America's entry into World War I, Iowa miners found themselves faced with rapidly rising consumer prices. Although miners were making higher wages than ever before, they believed that they were actually worse off because of the continual rise in the cost of living. Throughout the summer and fall of 1917, many Iowa miners wrote to the *Journal* expressing interest in cooperative buying and selling as the way of "beating the high cost of living."[33]

Nine months after the war began, district officials responded to the growing sentiment for cooperative stores. In early 1918 President

C. J. Lewis presented a resolution to the district's executive board that called for financial assistance to UMW members wishing to open such establishments. In explaining his action, Lewis stated that he had found "a great deal of sentiment for co-operative stores" throughout the district. He noted that UMW members had already opened four stores that appeared to be prospering, but that some members seemed reluctant to initiate such a venture "as it requires considerable capital to stock a store and buy fixtures." He lamented the fact that although the miners had received wage increases, they soon saw these increases eaten up in increased living costs. Because Lewis and other district officials believed that the establishment of cooperative stores would help miners "greatly reduce the cost of living and thereby increase the purchasing value of their earnings," the executive board began a policy of assisting members in starting such stores. When fifty or more members organized a cooperative society in which each had purchased five shares in the society at $5 per share, the executive board authorized the district secretary-treasurer to loan the cooperative society a sum equal to 50 percent of the amount subscribed. By 1920 the district contained five cooperative stores; union officials stated that all five stores were doing well financially.[34]

While the practice of self-help served the miners well in their earlier efforts to improve their standard of living, they soon discovered that the practice had limited impact on their immediate environment. For the most part, solutions to deficient housing, excessive house rents, unsanitary camp conditions, and inadequate schools lay outside the union's control. Housing posed a particular dilemma for the UMW. Camp housing was company housing, and company officials operated under a set of circumstances and attitudes that worked against change. First, operators understood that the life of most coal camps would be short, usually fewer than ten years. Therefore they built the cheapest houses, and once completed, they frequently refused to provide maintenance. Second, operators perceived coal miners as an inferior class of workmen; the operators held that neither the miners nor their families deserved better housing.[35]

There is little evidence before 1910 that coal miners sought the aid of their union to upgrade housing or to lower house rents. In letters written to the *Journal,* Iowa miners occasionally complained about poor housing and high rents, but no one offered any solutions. In 1909, however, miners at Enterprise forced the district leadership to consider the matter of house rents. In September 400 miners walked out of the

Enterprise mine when operators posted notice that house rents would be raised from $1.50 to $2.00 per room. The joint agreement that governed working conditions did not make direct reference to house rents, but it did include the statement that "local conditions not in conflict with the wage agreement itself should not be changed except by mutual consent." The Enterprise operators held that the question of house rents was unrelated to the wage agreement and that they had the right to charge whatever they wished. Local UMW officials immediately called in President White, who, after reviewing the situation, stated that he supported the striking miners. White believed that the rent of $1.50 per room was standard throughout the district, but he felt that rents for the Enterprise houses should be less. He noted that some of the homes were almost twenty years old and the operators had allowed them to fall into a state of disrepair.[36] In presenting the Enterprise matter to the annual district convention, White concluded:

I have an idea as to the cost of erecting the ordinary company house, and I know full well that the rent that is received by the average employer from these houses is indeed excessive. It must be borne in mind that with few exceptions the home is the job. A man must either comply with the terms of the operator in this respect or lose his position, because he will be compelled to give up work when he gives up a house.[37]

The issue was eventually settled when the Enterprise operators agreed to return to the previous rent level.[38]

At the same convention, in response to White's suggestion, delegates thoroughly discussed the house rent question. Convention delegates believed that miners should have some voice in determining house rents; yet they recognized that unless they could write such a position into the joint conference, the operators would totally ignore the miners' right to do so. The delegates requested that the joint conference scale committee take up the house rent matter and prepare a specific rule so that in the case of future rent controversies the miners would have the authority to discuss the issue without recourse to a strike. Although the miners did not succeed in getting the matter of house rents written into the joint agreement, their request did lead to the formation of a camp improvement committee. The following year, operators and miners' representatives attending the joint conference formed the Mine Village Improvement Commission. One miner and one operator from each subdistrict along with the presidents of the district and the operators' association constituted the commission. The joint conference charged the commission with the responsibility of improving the miners' living conditions.[39]

In 1919 the miners' long-time efforts to improve camp conditions began to produce concrete results. In that year the state legislature requested a study of coal-camp housing and sanitary conditions. The legislature intended the study as a basis for establishing minimum requirements governing the construction of new camps. Iowa State College engineering professor Charles Nichols visited over twenty coal camps where he investigated conditions and interviewed camp residents and operators. He submitted his report on August 18, 1919. Nichols listed what he perceived to be the basic causes of poor housing. He first stated that "operators intimate that the miners are as a class inferior morally and mentally to practically all other day laborers, and, therefore, [are] not entitled to the same standard of living." Nichols noted that mine operators assumed that miners wanted the cheapest housing possible. Moreover, operators believed that miners and their families would not take care of company property so "why trust them with anything of value?" Through his interviews with camp residents, Nichols refuted the operators' views. He argued that "talks with many miners, American and foreign alike, confirms the belief that they desire good living quarters for themselves and their families, particularly the children, and that they are willing to pay for what they get." Nichols observed that in some camps operators assisted the miners in building their own homes and these homes were of better quality. Nichols quoted some mine operators as saying that if they were to start another coal camp, there would be no company housing; the company would finance the construction of miners' homes on leased ground, and the miners would pay for the homes in monthly installments.[40]

Nichols's interviews with miners refuted the operators' contention that the miners neither desired nor deserved better housing. Nichols wrote that miners continually expressed dissatisfaction with the "conditions under which they have to live, and particularly as it concerns their families." Nichols concluded:

[The miners] are earning fairly good money in spite of the large amount of idle time; they dress well; they have all they want to eat; many of them own automobiles; and yet, the conditions under which they have to live in order that they may do the only kind of work they know how to do, do not meet their demands.[41]

Several months after receiving Nichols's report, the state legislature passed a "Good Housing" bill that contained a section pertaining to mining camps. The bill stipulated that before anyone could lay out a mining camp with as few as five houses, the person must first "file with the state board of health a plat of the camp, showing in detail the geographical location of same, the character of the houses to be erected, the provisions made for drainage, sewage, outside toilets, and the provisions made to secure water." If the operators convinced the state board of health that the proposed camp would comply with the general provisions of the housing act, a permit for camp construction would be issued. At the same time, the state legislature authorized the state board of health to supervise the existing mining camps.[42]

By 1919 both District Thirteen officials as well as the rank and file were strongly committed to the improvement of Iowa's coal camp schools. UMW officials repeatedly lobbied the state legislature, requesting state aid for camp schools. In that year district officials tried but failed to get the legislature to create a camp school fund based on a tonnage tax.[43] During that campaign, both state education officials and district leaders focused publicity on the poor conditions of the miners' schools. The legislators also heard from people outside the mining industry. In his housing report, Charles Nichols observed that "only a few mining camps have schools extending beyond the Fourth Grade, and these usually run for about four months during the year." He found that most boys finished school at age twelve, but they could not go to work in the mines until they were sixteen. Nichols observed:

There are, therefore, three or four years in the boys' lives at the time when they should be under training, that they are practically turned loose. As a consequence there is no opportunity in the majority of cases for the boys to secure any training which will fit them for useful work. At twelve or thirteen they are too young to strike out for themselves and when they reach sixteen they feel that they are not capable of doing anything else but mining.[44]

Although district officials failed in their bid for a tonnage tax in 1919, they did succeed in getting the legislature to make a two-year appropriation of fifty thousand dollars to upgrade camp school facilities. The school funds became available on July 4, 1919; following that date Vice-President Joe Morris of the district, Association Secretary George Heaps of the Iowa Coal Operators, and officials from the Department of Public Instruction traveled extensively throughout the state, visiting mining schools. According to Morris, the officials soon discovered that the fifty thousand dollars "was entirely inadequate to do more than just remedy the surface and give publicity to the fact that a sum several times [larger] is needed to bring the schools and surroundings up to the conditions where it will be pleasant and profitable for the children to attend school." District officials urged individual miners and their families to contact their state representatives to ask their support for additional aid.[45]

In November 1920 the *Journal* published an account written by F. E. McClennahan, Iowa's state superintendent of schools. McClennahan began his report with the following dramatic statement:

> During the session of the thirty-eight General Assembly a ghost—in the shape of the Mining Camp Schools—rose up to plague and haunt the members of the General Assembly. Stories were told of the horrible conditions that existed in some of these schools. . . .

McClennahan's office immediately began an investigation and discovered that while some camps maintained satisfactory schools, the majority did not. McClennahan cited the camp of Oralabor as an example of the second type in which one teacher "was attempting to teach more than 80 children and handle eight grades of school work." State school officials immediately hired a second teacher and acquired a second school building so that children could be separated into two classrooms. The following summer state officials added a room to one of the buildings and purchased new maps, charts, and supplementary readers.[46]

In 1921 the Department of Public Instruction reported to the Iowa General Assembly on its work in the coal camp schools, and the legislature responded by doubling the appropriations for the next biennium. In August 1921 the state executive council appointed a full-time inspector for the camp schools, Harry D. Kies. In May 1922 Kies announced to the *Journal* that he had devised the following program to improve the schools: (1) establish a school in every camp; (2) encourage better qualified teachers to work in camp schools; (3) improve buildings, equipment, attendance, and instruction; (4) create a better school atmosphere and attempt to make schools the community centers; and (5) equalize taxes between mining camps and surrounding districts.[47]

Between 1895 and 1925, District Thirteen of the UMW initiated several programs and policies that dealt specifically with the needs of Iowa's main ethnic groups. For Italian-Americans and Croatian-

Americans, the district provided joint agreements printed in the Italian and Croatian languages. Efforts at establishing a night school for the foreign-born miners and their children were not successful. After 1900 Italian-Americans, Croatian-Americans, and other southern and eastern Europeans were the primary recipients of improvements in camp housing, sanitation, and education because they constituted the highest number of camp residents. Black families fared less well in their quest for economic and social equality within the coal industry. Throughout the period both district and international leadership stressed the union's policy of equal treatment for all members, but these policies were not always observed at the local level.[48]

In 1905, for the first time, District Thirteen officials as well as national officials expressed concern over the growing number of miners who could not speak or read English. In his report to the national convention in January 1905, national president John Mitchell pointed out that of the nearly 600,000 mine workers in the United States, roughly one-third did not speak the English language. He strongly urged the UMW to circulate pamphlets among the foreign-speaking men, printed in the various foreign languages, acquainting them with the goals of the UMW. He also argued that the joint agreements should be printed in foreign languages. Mitchell cautioned his fellow union members not to write off foreigners as "illiterate in the extreme and not capable of understanding American ideas, American institutions and particularly the object of the American labor unions." Mitchell noted that according to the commissioner of immigration, 53 percent in the most illiterate countries and 97 percent in the most enlightened countries could read and write in their own languages. Mitchell counseled the UMW members to have patience and understanding with the foreign workers because when the foreigners comprehend the mission of the union, "they are ever ready to become members and co-operate with us in carrying into effect the objects of the United Mine Workers of America."[49] In July 1906 District Thirteen's executive board voted to recommend to the next district convention that the joint agreement be printed in Italian and Croatian. Apparently the board decided not to wait for convention approval; three months later they appropriated $287.74 to pay for printing of the agreement and the district constitution in the Austrian and Italian languages. They did not explain the omission of the Croatian language.[50] District officials continued this practice the following year and well into the 1920s.

In 1908 Iowa's state mine inspectors began to respond to the growing number of eastern and southern Europeans working in Iowa's coal mines. Their comments largely echoed those made by inspectors in eastern mining states at the turn of the century. In his 1908 biennial report, Third District inspector Edward Sweeney asserted:

... we are much alarmed over the increased number of accidents occurring in our Iowa mines. The increase is shown as coming to miners while working at the face of their working places. I regret to state that this increase in accidents at the working face indicates a carelessness on the part of the miners themselves, which of late is painfully common. Many of the non-English speaking miners, who have come to us in large numbers during the last few years, are decidedly reckless in their method of workmanship, and evidently incapable of comprehending our mining rules and the meaning of our mining laws.[51]

Two years later, Inspector Sweeney reiterated his warning about the high number of mining accidents. "There is this important fact staring us in the face daily: That we fear to speak of the prime cause of our mine accidents, i.e., we have an army of untrained and impractical miners. The miner is fast falling into the ranks of the unskilled. . . . "[52]

While Italian-Americans could do little about their lack of mining experience before arriving in Iowa, they hurriedly set to work learning the English language. Most immigrants assumed that the operators expected them to begin learning English immediately and they did so. Several former miners stated that most Italian-American men were forced by circumstances to begin learning English as soon as they began work in the mines, with or without the operators' directions.[53] Another miner recalled that at his place of work, the superintendent divided foreign-born miners according to their places of birth: Italian-born miners worked in one section, supervised by a foreman who spoke both English and Italian; Croatian-born miners worked in another section where they were supervised by a foreman who spoke both English and Croatian.[54] It is not known how widespread this policy was nor how long it lasted.

No doubt language played a role in determining union participation. Interviews with Italian-American coal mining families indicate that Italian-American miners belonged to the union and supported it enthusiastically. Interviews generally reveal that as a rule, Italian-Americans did not hold union offices. Bruna Pieracci recalled that one year her father was elected by the Oralabor Local as delegate to the district convention. She remembered that he considered it "a great honor" and talked about his experience for a long time.[55] Second generation Italian-American Judith Milani observed that Italian-

American men did not serve as union officers because they were uncertain of their ability to speak English. She explained that as the men began to learn English, "they understood it in their heads," but that was quite different from being able to articulate their ideas in English.[56] Overall, district records do not indicate that Italian-Americans assumed leadership positions at the local or district level.[57]

In 1918 district officials again turned to the needs of Iowa's foreign-born miners. At the district convention in 1918 delegates voted to support a bill pending before the state legislature that proposed the establishment of an adult night education program. The 1915 Iowa state census listed 273,484 foreign-born persons in Iowa, or 12.3 percent of the total population. Of that number, 37,169 were unable to speak English and 56,307 were of voting age but were not naturalized. Of particular interest to coal miners were the large, foreign-born populations in Appanoose, Wayne, Monroe, Lucas, Wapello, Mahaska, Boone, Polk, Dallas, and Jasper counties, most of whom were engaged in coal mining. Legal department officials estimated that two-thirds of the foreigners in these counties lived in coal camps in which educational facilities were poor if not nonexistent. These officials focused on the common practice of foreign-born miners "to permit their sons to leave school just as soon as they are able to perform some sort of work in the mines." They estimated that there were approximately 4,000 foreign-born miners' sons, the majority of whom were born in the United States, who would "avail themselves of the opportunity for further education in the night schools." Convention delegates voted to support the bill and instructed the legal department to work for the bill's passage, but the bill failed to pass the legislature.[58]

During the 1920s some communities acted independently in establishing night schools. James Battani, a second generation Italian-American, as a teen-ager worked in the mines part-time and went to school part-time. He also attended night school in Woodward for several years and later commented:

[The public schoolteachers] were good enough to teach courses at night. There were a lot of kids in those days, like farm kids that had to work on the farm, or they had to pick the corn in the fall, and we kids had to go to work in the coal mines. We would go to school two days a week and work the other four days in the mine.[59]

Battani also attended summer school. He graduated from high school in 1928 at age nineteen.

The experience of black miners in the UMW differed considerably from that of other ethnic groups. Early in the history of Iowa's coal industry, operators imported black men as strikebreakers. In 1881 operators in both Muchakinock (Mahaska County) and Coalville (Webster County) traveled to the South to hire black men to come North and work in the mines. In Muchakinock the miners went on strike to protest a reduction in wages. Mine operators H.W. and W. A. McNeill responded by hiring a local black businessman, H. A. Armstrong, to go to Virginia and other southern states to recruit blacks to work in the mines.[60] For at least thirty years, Iowa operators continued the practice of bringing in blacks to work the mines during labor disputes.

At the same time a similar situation developed in Webster County. During the winter of 1881, Coalville miners went on strike after the operators of the Ft. Dodge Coal Company refused to grant them a wage increase of $0.15 per ton. The former price had been $1.00 per ton. The miners retaliated by going to work in smaller neighborhood mines. Since all striking miners found other work, the Ft. Dodge operator realized that the men could stay out for an indefinite time. The operator hurriedly traveled to Tennessee and brought back "about seventy-five colored men," offering the blacks only $0.75 per ton. In the meantime, the demand for coal remained high and the white miners' wage rose to $1.50 per ton. When the blacks arrived in Coalville, however, the white miners approached them with an unusual arrangement. The white miners promised that if the blacks would refuse to work, they (the white men) "would furnish them with everything they wanted to live on. . . ." The blacks agreed. The white miners took up collections at all the adjoining mines to support the blacks, who did their part by giving frequent concerts. But with the onset of warm weather the smaller mines began to work only a few days each week, and the white miners were then forced to revise their plans. They decided that they could no longer support the blacks nor did they have the money to send them back to the South. The only solution was for both black and white miners to go to work in the Ft. Dodge mines. Inspector Wilson reported that the strike lasted for three months and throughout the entire period there was no violence.[61]

Following the Muchakinock and Coalville experiences, other coal operators gradually began to hire southern blacks as regular mine workers. Essentially, black workers provided cheap labor in an area where labor was scarce. In August 1891, for example, operators of the Carbonado mines brought in approximately fifty black men to work in the mines, alongside the regular white miners. The operators placed each black man under the supervision of a skilled miner until he could function independently. At the same time, the operators encouraged the blacks to write home and urge relatives and friends to join them in Iowa.[62]

The continuing importation of blacks as strikebreakers throughout the 1890s raised serious problems for the district leadership. In 1891 President Walter Scott declared that the UMW welcomed black men as members and that they could expect equal treatment. Scott's statement reflected the general policy of the district toward black men: The district should be integrated and black members should receive equal treatment with white members. But district officials discovered implementation of that policy at the local level constituted another matter. The use of blacks as strikebreakers tested even the strongest commitments to racial integration of the coal fields. In 1891 President Scott returned from a tour of southern Iowa where he witnessed firsthand the use of black strikebreakers. He noted sorrowfully: "I gave up the best [years] of my life to help free the slave in the South. Today that same black man is used by the money power of this land to make a slave of me and my fellow craftsmen. . . ."[63]

Throughout the 1890s district officials continued to struggle with the issue of black employment. Black miners were apparently hired in some mines but refused employment in others. At the district convention in June 1894, delegates raised the issue of black employment in Iowa mines. Shortly before the convention began, an operator asked for a clarification by delegates of the policy of employing both blacks and whites. The operator's question read as follows: "Heretofore in this area it has been the custom for the miners to strike against certain mines employing colored miners. Will this be continued or will each mine have the privilege of employing anyone regardless of color?" In response, the delegates moved the following resolution: "We accept all colored men into our organization and extend to them the right hand of fellowship."[64] Despite the resolution, racial confrontations continued. The importation of blacks as strikebreakers precipitated some disputes; others developed in camps where blacks had become a regular part of the work force. In 1899 the district convention again issued a statement regarding the employment of black miners:

Be It Resolved, That it is the sense of this convention that, whereas, there has been an injustice practiced upon the colored miners of this State by being refused work in some of the mining camps of said State, we do hereby condemn the above practice and extend the hand of fellowship to all union colored miners of the state of Iowa, and that the same privilege in regard to obtaining work be accorded all creeds, colors and nationalities alike.[65]

The district leadership was not always able to impose its polices on individual locals, however. In the summer of 1911 a racial conflict developed in the Ogden coal camp that sorely tested the district officials' authority. Early in the summer the Ogden miners protested what they believed to be an unfair screening policy. When the company failed to modify the policy, the miners went on strike. Realizing that they had violated the joint agreement, the miners returned to work in two days and announced their decision to quit. The Ogden operator responded by hiring 161 black miners who immediately moved into the Ogden camp housing along with their families. The black men had formerly worked at Buxton and Oralabor where they were union members in good standing. The Ogden camp had previously been all white. Immediately after the black miners arrived, local union officials insisted that the white miners had not quit but rather were out on strike.[66]

For the next fifteen months a bitter struggle ensued between Ogden Local 2433 and District Thirteen officials over whether or not the blacks were strikebreakers or legitimate employees. Because local union officials took the position that the white miners were simply on strike, they insisted that the black miners were scabs. Further, local union officials refused to accept the black miners' UMW membership cards. District officials tried unsuccessfully to settle the controversy by insisting that the Ogden Local honor the black miners' union memberships.[67]

The black miners soon took independent action. In May they hired a highly respected black lawyer from Buxton, George H. Woodson. One month later, Woodson filed a lawsuit in Boone County Court on behalf of the black men. The suit asked the court to force local, state, and national UMW officials and the union membership "to recognize them as union men, [honor] their cards, and to cease calling them scabs." They also requested that the court allow them to create a separate union local at Ogden. Finally, they asked $1,610,000 in damages ($10,000 each).[68] Several months later the black miners pursued a second course of action. They appealed to the international executive board to grant them a charter for a second Ogden local. The international board supported the black miners' right to have their own local, but they left the implementation of the decision to district officials.[69]

Three months later, with the controversy still unsettled, the black miners petitioned the district to have the matter brought before the district convention; they agreed to abide by the convention's decision. Although the black miners were not represented at the convention, Buxton delegates presented their position. Buxton delegate William Brown addressed the convention, stating his belief that the entire Ogden situation was "simply an excuse for running the negro from Ogden." He concluded: "Now gentlemen, fair and square, this is nothing but a race fight from start to finish. You might as well face the issue. You must call a spade a spade. We know this is a white field up there. . . ."[70] After spending ten days debating the Ogden controversy, the convention still had not reached a decision. Finally they decided to turn the matter over to a committee of seven miners.

In April 1912 the Ogden controversy came to an end. The committee first recommended that both black and white miners return to work in the Ogden mines; the Ogden Local rejected the proposal. As a second proposal, the committee recommended that white miners be allowed to return to work and that black miners give up their jobs. In return, the blacks' union memberships would be recognized by other locals where black miners were already employed. The black miners honored their agreement to accept the convention's decision. In addition, they agreed to drop their lawsuit. The district leadership had been unable to exert their authority over Ogden Local 2433. By holding out and refusing to accept any compromise, the Ogden Local had won its case. Blacks were dismissed and the Ogden coal field remained solidly white.[71]

By 1925 District Thirteen's officials experienced some success in dealing with specific needs of Italian-American and black miners. Camp residents, including many Italian-Americans and some blacks, benefited from union efforts to improve housing and general camp facilities. The district's policy of printing joint wage agreements and the UMW constitution in Italian and Croatian no doubt made it far easier for these groups to understand and appreciate union policy. The union's efforts to establish night schools failed, but local communities sometimes filled the void. For the most part, however, Italian-Americans appear to have accommodated themselves to the demands of the industry without union assistance. Italian-Americans joined the union in an enthusiastic manner. Their failure to play leadership roles does not diminish their wholehearted support of the UMW. Overall, Italian-Americans fared better than the industry's black coal miners, but their obstacles were easier to overcome. Italian-Americans were not denied jobs or union membership because of their ethnicity. Blacks on the other hand did experience such discrimination. Black miners sometimes found themselves shut out of mining communities simply because they were black. In some communities, blacks were barred from union membership. Overall, integration of the Iowa coal fields appeared spotty. The Ogden racial controversy clearly revealed weaknesses in the union organization. While local autonomy on certain issues brought strength to the organization, local autonomy had its drawbacks. These were clearly revealed by the district's leadership inability to enforce district policy on the membership of the Ogden Local. By 1925 district officials had made several concerted efforts to eliminate racial prejudice from Iowa's coal fields but with limited success.

Between 1895 and 1925 the UMW was an immensely constructive force in the lives of Iowa's coal mining families. During this period union officials spoke out on every significant issue involving miners and their families. For the most part, the union succeeded in obtaining policies or creating programs desired by its membership. Originally coal miners believed that the purpose of the union was to deal with bread and butter issues. Union officials were able to win shorter hours, higher wages, and greater wage stability. Central to these achievements was the establishment of the joint wage agreement. In their desire for change, however, miners gradually recognized that there were avenues other than the joint agreement open to them. After 1905 Iowa miners worked through their union to achieve a broad range of social and economic programs that applied to life above ground as well as below. These advancements brought greater economic security and an improved quality of life for all family members. Yet progress in housing and education were the most difficult to achieve. In their struggles for change, coal miners learned that the state legislature could be an important ally in the housing and education areas. Considering the growing acceptance nationally of white surpremacist's views, it is easy to understand why the union was least successful in its attempts to eliminate racial prejudice and insure equality for black union members.

Ironically, as District Thirteen came to have greater success in achieving its social goals, the Iowa coal industry entered a period of economic decline. Iowa's coal production reached its peak in 1917;

following that, production gradually decreased. Over the next ten years the number of mines and miners also gradually declined. By 1925 district officials faced the indisputable fact that the economic base for a prosperous Iowa coal mining industry did not exist. In the 1920s Iowa remained an agricultural state with few large industrial operations. The railroads continued to be a major consumer of Iowa coal, but their consumption gradually declined after 1915. Iowans' use of eastern coal also represented a major reason for the decreased production.[72] At the same time, Iowa's particular coal formations did not support large-scale mining investment and production. Given these problems, there appeared to be little that the Iowa miner could do to improve his economic situation, with or without the support of his union. Even in light of the fading strength and size of the district, however, one fact remained certain. After 1925 the lives of Iowa's ever-dwindling coal mining population would be far more pleasant because a strong union organization had existed in District Thirteen during the first quarter of the twentieth century.

# 7

## PERSPECTIVES

THE history of Iowa's coal mining industry is the history of Iowa's coal mining people. It comprises the work habits of the miners and their families, their social concerns, their ethnic experiences, their religious practices, and their economic behavior. These varied experiences constitute the major historical dimensions of the coal industry in Iowa. That history is rich and many sided, touching the lives of literally thousands of native-born and foreign-born people who looked to the Iowa coal industry for their livelihood. Although the coal population encompassed economic and social diversity, it also exhibited certain general characteristics that affected many aspects of the workers' lives. Records of District Thirteen of the UMW, mine inspectors' reports, and contemporary newspapers make it possible to describe and analyze those general characteristics and to study their effects. Other data, particularly census records and oral interviews, permit a look into the actual day-to-day activities of mining families. These records provide social data that enable us to generalize about the mining population. Taken together, these records present a comprehensive social and economic view of the Iowa coal mining population.

The Iowa coal industry exhibited particular characteristics that affected the state's coal mining population in significant ways. The first of these pertained to the nature of mining itself. Because coal mining is an extractive industry, it is a temporary industry in any given location. This characteristic greatly affected Iowa's mining population. The life of an Iowa coal mine rarely exceeded twenty years, and the majority of operations lasted from ten to fifteen years. The tenure of Iowa's coal camps reflected the same impermanence. One state mine inspector estimated that the average life span of an Iowa coal camp was eight years. This had considerable impact on the manner in which many coal mining families lived. Given this short life span, company officials hurriedly selected camp sites, paying little attention to water availability or drainage. They quickly erected houses that generally were small and poorly constructed. Moreover, officials hesitated to invest money in trees, shrubs, sidewalks, or other improvements. The result was that life in an Iowa coal camp was viewed by most residents as a dreary, isolated existence.

Coal camp residents also found themselves the victims of operators' contradictory views about homeowning. Many coal operators believed that the best miners were those men who purchased their homes and became, in the operators' view, "settled" citizens. Yet because companies expected to make money on company housing, they sought to keep the houses continually rented. This made it difficult for miners to purchase homes and to continue working as miners. Because of the operators' attitudes on homeowning, however, the miner who rented company housing continued to be regarded as less responsible by the operator, while at the same time the operator disliked the independence exhibited by the miner who purchased his own home. Thus the miner could not avoid some displeasure by the operator regardless of whether he rented or owned his home.

The temporary nature of mining had a direct impact on geographical mobility within the industry. Coal miners had to be willing to move often. This condition was further complicated by the seasonal nature of many Iowa mining operations. Iowa operators sold primarily to local consumers who used coal for heating homes and businesses. With the onset of warm weather, most operators shut down their mines. Coal mines with railroad contracts usually worked year round, but these operations were not as numerous as the local nonshipping mines. Interviews with mining families indicate that some families moved often from coal camp to coal camp. Nonetheless, many Iowa mining families appeared to be less mobile than mining families in other states. Because of the industry's seasonal nature, many miners developed a second or part-time occupation. Some men purchased small acreages that they farmed during the summer months. Others worked as part-time carpenters, and some men worked as musicians. Several men interviewed worked as railroad section hands and greenhouse employees. One miner even carried on a wallpapering and painting service during the summer. Another miner traveled to Detroit every spring where he worked in the automotive industry for several months.

Every September, however, he faithfully returned to the Iowa coal mines. These part-time jobs enabled many miners to remain in one location for long periods of time regardless of the irregularity of mining work. Some miners turned their part-time work into full-time occupations. In other situations, miners were able to survive one or two years on their part-time work until a second mine opened in the area. As the Iowa coal industry entered a period of irreversible decline in the 1920s and 1930s, a large number of men pursued their part-time jobs on a full-time basis. These part-time activities frequently softened the transition into what otherwise would have been new lines of work. Yet for men who wished to pursue mining on a full-time basis, the Iowa coal industry offered less opportunity than the coal industry in many other states.

Other mining families avoided frequent moves by becoming homeowners and developing subsistence farming operations. Numerous Italian-American families in Wayne and Appanoose Counties purchased their homes shortly after arriving in the state, and they continued to live in their homes for the remainder of their lives. Most of these families, whether they lived in town or on small acreages on the outskirts of town, tended large gardens, raised chickens, ran several pigs, and kept a cow. Many also raised fruit trees. These endeavors enabled them to raise their own food and in some cases produce a surplus of vegetables, fruits, eggs, and dairy products that they sold to neighbors. The Seymour and Numa residents were fortunate in that several mines operated in that vicintiy; when one mine closed down they could seek work in another mine nearby. During the second and third decades of the twentieth century, some mining families acquired automobiles, and commuting ten to fifteen miles to work became commonplace.

An additional characteristic of Iowa's coal industry was its limited size. Although Iowa's coal production ranked fifteenth in the nation in 1900, the state produced only about 1 percent of the nation's coal. Overall, the small size of both the total Iowa industry as well as individual operations had some positive consequences. Only a few Iowa mines employed more than 400 workers. Mines employing 100 to 150 appeared average. The limited size of the Iowa mines allowed workers to know operators, mine superintendents, and mine foremen on a personal level. Moreover, mine operators were directly involved in the Iowa Coal Operators Association, which in turn negotiated directly with representatives of the UMW. Proceedings of District Thirteen's ex ecutive board indicate that most operators had an intimate knowledge of their mining operations; they did not have to rely on secondhand information provided by mine superintendents or mine foremen. Absentee ownership appeared to be relatively unknown in District Thirteen. The alienation suffered by workers in Homestead, Pennsylvania, for example, did not exist in the Iowa mining industry. During interviews, former coal miners occasionally commented that the owner of their mine had been a "good man."

Size had its effect on union organizing activity. District Thirteen was one of the smallest districts in the international organization. Its smaller size meant that district organization required less money and less manpower. Once the organization was completed, wage negotiations were carried out more quickly in Iowa than in most districts because the state contained fewer variations in physical conditions and coal composition. In larger districts a myriad of differentials had to be considered. Iowa had basically two coal fields — the low coal field (Subdistrict Two) and the remainder of the state. Moreover, size probably accounts for the longevity of District Thirteen. During the 1920s and 1930s when many district organizations were falling apart, District Thirteen continued to be a unified, well-disciplined organization.

Limitations in the size of Iowa's coal industry naturally affected miners' lives in other ways. Most Iowa coal camps contained around one hundred houses, although some camps had as few as thirty or forty houses. Generally all houses in a particular camp were built along the same floor plan, and the smallness of the camp size fostered close relations among everyone in camp. Nor did operators find it necessary to place identification numbers on houses, as was the practice in larger company towns in the East. In Iowa, houses were identified by the particular family that lived there, rather than by a number. When water problems developed, most camps were small enough that the problem became the concern of the entire camp, not just one section. Their smaller size undoubtedly helped to promote feelings of fellowship and social consciousness among the residents.

One of the most successful aspects of the Iowa coal miners' experience was their union affiliation. Throughout the entire period from 1895 to 1925, the UMW was an integral part of the Iowa coal miners' existence. From 1891 until 1905, wage and work issues dominated the union's agenda. The union won the eight-hour day, the acceptance of the annual joint wage agreement with the coal operators, and the right to hire shot firers and checkweighmen. However, after 1905 miners began to concentrate on nonwork related issues such as better housing, death benefits, a strike fund, and better schools. In other words, the miners pushed for issues that would improve the quality of life in the mining camps, and they achieved considerable success in all four areas.

Although District Thirteen effectively tackled many social and economic problems, their inactivity on certain fronts is puzzling. There is no evidence that either District Thirteen officials or international officers attempted to solve the fundamental problem of irregular employment. From time to time during the late teens, the *Journal* editor wrote editorials stressing that irregular employment lay at the heart of all the industry's troubles. Occasionally miners and union officials suggested that the men join the Six Hour League in an effort to reduce their workday to six hours. This reform received little support. Contributing to the problem of irregular work was an overabundance of coal miners. The unimpeded entry of thousands of untrained men into the mining profession apparently caused little concern. Actually, the surplus of miners was continually exacerbated by the miners themselves. While lamenting overcrowded conditions and their lack of full-time work, the miners continued to urge friends and relatives to come from the British Isles and from Europe to work in the Iowa coal mines. The continuing emigration brought individual rewards and short-term economic benefits to individual families, but the long-term effects were negative. Newcomers usually went to work in the same mines as their relatives, thus aggravating the already overcrowded conditions. Black miners followed the same tradition. They wrote their relatives in Virginia, Tennessee, and Alabama urging them to "come up to God's country." This continual influx of new workers obviously undermined the economic position of workers already employed. That Iowa miners were not unaware of the impact of surplus workers can be seen in the frequent letters from Iowa miners published in the *Journal* urging miners in other states to stay away from the Iowa coal fields because of overcrowding. Apparently these men did not take the same view toward their relatives and friends arriving from abroad.

Union officials and miners evidenced the same lack of concern over the absence of miners' training programs. Only in the late teens when it was apparent that the industry had suffered a postwar slump, did national leaders begin to speak out in favor of training men before they went underground. But their demands did not result in any training programs. Every day miners went to work knowing that some men in their mine would handle powder and set charges who were inexperienced or untrained for the dangerous work they performed. Not only did the men fail to work consistently for improved safety standards through their union, but there is no evidence that they persistently voiced those concerns. It appears that safety issues were peripheral issues. Union officials first concerned themselves with wage issues. Then, if time permitted, they turned their attention to safety matters. It is instructive to note that Iowa's two major union leaders, John Reese and John White, did not stress safety issues at any time during their tenures as district president. Unless a major disaster took place such as the Lost Creek explosion, wage issues dominated the joint wage conferences.

From another perspective, given the seasonal nature of Iowa coal mining, it is surprising that Iowa's miners exhibited such enthusiastic support for the UMW. Because many men did not rely totally on their mining income for support, it would seem that they might have had less commitment to the union. That many miners had second jobs did not seem to dilute their commitment to the union. Between 1900 and the mid-1920s, international union leaders and *Journal* editors frequently described District Thirteen as being 100 percent organized and having a highly committed membership.

Labor-management relations constituted an important aspect of Iowa's coal mining industry. Each industry exhibits particular characteristics and coal mining was no exception. From any perspective the arrangements between coal operators and coal miners appear unusual. Once a coal miner hired on with a particular company, he entered a conflicting situation. On the one hand, he lost control over some areas of his life such as housing; but, at the same time, he came to enjoy certain job securities. Company officials seldom fired employees and rarely issued reprimands for absenteeism or production slowdowns. Company officials and miners agreed that during a slowdown, all workers would experience a reduction in the number of days worked as opposed to any miners being discharged. It was also possible for men to be legally employed by the company while actually being unemployed. Following the working out of a miner's room, miners sometimes waited several months for assignment to a new room. Under these circumstances, operators allowed miners to remain in company houses whether or not they paid rent. If the mine shut down or went to a reduced work schedule, the miners remained in the company homes. Throughout this period the miners and their families continued to trade at the company store. Apparently there was no limit to the debt that they could incur. The only practice that disrupted these arrangements was if the men went on strike. Also if miners attempted to save money by shopping at a private store, they were frequently fired. Ironically, in some mines, men with the greatest debts at the company store were given the choice room assignments, thus allowing them to mine more coal and pay off their store debt more quickly.

The other side of this unusual economic arrangement involved management policies. Rarely did operators stress increased production or increased efficiency, and they showed little concern over opening new rooms quickly so that miners could get back to work. Sometimes it was the miner who continually pestered the foreman regarding the availability of new rooms. The operators' major concerns were production costs, primarily the cost of labor. The operators' almost lackadaisical attitude toward production is probably explained by the many ways to make a profit within the coal industry. Presumably the operator made the greatest profit from the sale of coal, but he also enjoyed income from company house rentals and company store operations. In some camps, residents paid an additional fee for water and coal for home use. If the company owned the tavern or pool hall, these provided another source of income. The coal operator carried on a multiple business venture, although to the public his major concern appeared to be the production of coal. It was always to the operator's advantage to hire on as many miners as possible, regardless of how many days each man would work. The operator obviously wished to rent all company houses; the presence of families in the houses meant that the volume of business increased at the company store. Generally the income from house rentals and store sales was more regular than profit from the sale of coal. With house rentals and store sales, operators had a captive clientele; workers paid the house rents that operators set and they paid the prescribed prices at the company store. On the other hand, the price of producing coal was subject to negotiation with UMW officials and the selling price of coal depended largely on competition. The renting of houses no doubt produced the greatest profit. In the late 1890s and early 1900s, a coal camp house could be built at a cost of about $150.[1] All sources indicate that the overwhelming majority of operators did little or nothing about maintenance or improvements. Therefore, at the prevailing rents—houses rented at an average of $8 per month—an operator could expect to pay off the cost of a house within eighteen months and the remainder of the time the house was rented it represented, with the exception of taxes, almost pure profit. Following abandonment of the mine, the operator usually sold off the houses for $50 each. He incurred no expenses with these sales because each buyer was responsible for moving his house.[2]

In spite of many drawbacks for the miners, coal mining offered a variety of opportunities for newcomers. For English, Welsh, and Scottish emigrants, many of whom had worked as miners in their homelands, there was the opportunity to continue a familiar occupation. For other men who arrived with no previous mining experience and who did not possess funds to purchase farmland or establish a business, the mines provided a means of livelihood. Entry into and out of the Iowa coal mines was common. Some men represented the third generation of their family to work in the mines; their fathers and grandfathers had worked in mines in Pennsylvania or Ohio before migrating to Iowa. More typical were the men who represented the second generation and who were taken into the mines by their fathers. Some men went into the mines with the attitude that they would work only for a few years and then move into another occupation; in other words, they viewed mining as an interim occupation. A number of miners interviewed had worked as miners before World War II, but after their return from military service, they no longer had any interest in mining. Some men worked several separate "hitches" in the mines, alternating mining with other types of employment. The view did not prevail here, as in England, that mining was a family tradition. In England, a young man became a miner because his father was a miner. Conversely, English miners believed that a young man whose father was not a miner could never become a successful miner.

Men could move into mining with little difficulty because the industry required neither apprenticeship nor previous knowledge of mining. Any able-bodied male, sixteen years or above, would likely be hired on.[3] Sixteen-year-olds would go to work as company employees coupling empties, greasing wheels, or trapping. As their knowledge of mining procedure increased, they would be promoted to jobs requiring greater responsibility. Other young men started working as miners' assistants, learning procedures under the guidance of an experienced miner. It is important to note, however, that the new employees received wages for their work from the first day that they entered the mines. This was particularly important for those emigrants who had no money and who had borrowed for their passage to the United States.

Southern and eastern Europeans usually followed the route of locating a sponsor who provided travel money; they lived with their sponsor who in turn helped them secure a job. For these emigrants, finding immediate work that did not require a skill was imperative. Not only did coal mining offer that type of employment, but it also offered newcomers a considerable degree of freedom. Italian emigrants frequently wished to avoid factory work in which they were closely supervised. Work in the mines, particularly digging coal, offered a chance to work without anyone "looking over their shoulder." While many emigrants from western and northern Europe had first worked in coal mines in Pennsylvania or Ohio, southern and eastern Europeans usually came directly to Iowa. Here the mining industry provided the point of entry into American economic life. For most Italian-Americans, coal mining was a two-generation occupation. By the time the grandsons of the first generation miners reached adulthood, the coal industry had entered a state of decline. In many instances, second generation Italian-Americans did not wish their sons to enter the mines and encouraged them to enter another line of work or attend college.

Interviews and census data provide a more complete understanding of the social composition and economic behavior of Iowa's coal mining population. These sources supply information on places of birth, occupations, work habits, kinship patterns, and household patterns as well as indicating reasons for emigrating and settlement in Iowa. Interviews in particular provide perceptions of how populations not involved in coal mining viewed coal miners and their families. Therefore we know miners came to exhibit particular characteristics that set them apart from other skilled or semiskilled workers of the nineteenth and early twentieth centuries. Some characteristics stemmed from the occupation itself; others were derived from the manner in which mining families lived. Unfortunately most of the characteristics portrayed mining families as inferior to the other resident populations. Their most prominent characteristic related to the type of work miners performed. Coal miners labored below ground in dirty, dangerous, isolated work. This more than any single factor separated coal miners from other occupational groups. Because many coal mining families lived in isolated coal camps, they were separated socially from other families. This separation meant that miners' children frequently received inferior, limited education; following completion of their schooling, they had few occupational opportunities. In many Iowa communities, local residents viewed the coal camps situated a few miles out of town or even on the town's outskirts as totally separated communities where residents lived rather deprived lives of their own choosing. In general, coal miners acquired the reputation of being heavy drinkers, big spenders, and hard fighters. Some coal mining communities carried the labels of "tough towns." Moreover, the occupation of coal mining enlisted large numbers of southern and eastern Europeans. Those not involved in mining as well as coal miners from northern and western Europe viewed the southern and eastern Europeans as different from and inferior to the dominant northern and western European groups that populated most of the state. Moreover, most southern and eastern Europeans were Catholic, making them appear undesirable to a predominately Protestant population. Finally, because coal miners were sometimes required to move to practice their mining trade, they were regarded as transients and therefore less desirable people than more sedentary populations. The results of these characteristics and the subsequent attitudes held by the other resident populations were twofold: First, coal miners were regarded as a "little peculiar"; and second, miners were viewed as inferior and therefore less deserving of higher wages and better living conditions.[4]

Interview and census data indicate that Iowa's coal mining population was family centered and community centered. Most coal miners married and had children. A small percentage of miners lived on acreages but the great majority lived either in incorporated communities or in coal camps. While the coal population is commonly viewed as a distinct population segment, the distinction should be made between populations that lived in coal camps and those that lived in incorporated communities. In the latter, residents had far greater opportunities for education, social activities, religious activity, and employment. Census data from the incorporated communities of Beacon, Cincinnati, and Seymour indicate only slight differences, if any, between coal mining and other resident populations in the areas of sex ratios, religious preference, number of children, total size of households, and total number of employed persons in the households. Substantial differences appeared in the areas of unemployment and income, with coal mining families earning less money and experiencing higher rates of unemployment. In both types of communities miners' children played important economic roles within their families. In Beacon, Cincinnati, and Seymour, miners' sons and daughters were able to work as teachers, store clerks, laborers, coal miners, domestics, seamstresses, delivery boys, teamsters, and bookkeepers. The schooling

Louis Nobile immigrated to Iowa from Italy just before World War I and worked until his early seventies in the mines around Albia. (Courtesy of *Des Moines Register*)

provided in incorporated communities was no doubt superior to that available in mining camps; thus the children were better prepared to qualify for a wide variety of jobs. Families living in mining camps discovered that there were fewer employment opportunities for their children. Generally the young women remained at home until they married; at age sixteen, the young men went into the mines. On the other hand, miners' wives probably had more opportunity to take in boarders in mining camps than in incorporated communities.

Presumably fewer miners' sons followed their fathers into the mines in incorporated communities than in coal camps. In the latter, sons could provide additional income for the family but in other occupations.

Miners' wives played crucial roles in mining households. Their domestic tasks of food preparation, cooking, cleaning, preparing lunches, washing clothes, and child care were essential to the maintenance of the household. At the same time that women performed traditional household tasks, they also served as important wage earners. Interviews indicate that the great majority of Italian-American women kept boarders. They earned an average fee of eight dollars per month per boarder, and many women kept from three to five boarders. Women also earned money by sewing, washing clothes, and selling extra dairy and garden produce. The women's wage earnings were particularly significant because their income tended to be steady. The wives' income no doubt supported their families during the miners' long periods of unemployment.

In the final analysis, the coal industry in Iowa was destined to be a short-term industry. The only consistent, significant demand for Iowa coal came from the railroads; as long as the railroads continued to buy Iowa coal, the coal industry survived. Ironically, railroads provided the major impetus for the development of the state's coal industry in the 1870s; the same railroads presided over the industry's decline in the 1920s and 1930s. Following World War I, Iowa's coal producers were optimistic that coal demands would remain high. As it became increasingly evident, however, that railroads were going to purchase coal out of state, the lack of markets created a serious situation for Iowa operators. In the 1920s coal operators along with other Iowa businessmen mounted a campaign urging Iowans to buy Iowa coal. Operators charged that many Iowans purchased Kentucky coal in the belief that it was superior to Iowa coal, although the operators insisted it was not and that Iowans should support home industries. They also had to contend with the consequences of Iowa's particular coal formations that mitigated against large scale development. Iowa's coal deposits often lay tilted rather than horizontal. These conditions made it difficult for coal companies to develop larger and therefore more profitable operations. Coal sales plummeted even more in the twenties and thirties as many Iowans turned to fuel oil, natural gas, and electricity for heating homes and offices.

After 1925 the number of mines and miners gradually decreased until by the early 1950s only a handful of underground mines continued to operate. One by one operators closed down the coal camps and sold off the houses, and most were moved to nearby towns. Today, coal camp houses can be counted by the dozens in former mining communities such as Boone, Madrid, Des Moines, Seymour, and Centerville, discernible by their square shape, truncated roof, and one-story design. One by one the mining tipples disappeared, leaving large unsightly slag piles to mark where thousands of coal miners once labored underground to eke out a precarious existence.

Coal mining families underwent considerable change as the industry gradually faded away. Many miners and their families moved to neighboring communities like Ames, Des Moines, or Centerville, or to major industrial cities in the East. Even where coal operations remained, mining families ceased to be distinct populations. With the increased use of the automobile, miners abandoned the coal camps and moved into nearby incorporated communities where they frequently purchased homes and merged with the general population. The relocation brought significant changes for mining families. No longer were miners' children stigmatized because of their fathers' occupation. At the same time, they had greater social, educational, and economic opportunities.

The move to nearby communities altered the life-style of miners' wives. Moreover, that relocation coincided with another basic change within the coal industry. During the mid-1920s, the stream of Italian emigrants coming into Iowa coal fields gradually came to an end as did unrestricted immigration into the United States generally. With that change, miners' wives ceased to take in boarders. Moving away from the coal camps brought about two important modifications in the lives of miners' wives. Not only could they enjoy greater social and religious activities in their new locations, but they were liberated from their traditionally heavy work loads.

Although District Thirteen of the UMW continued to be a strong organization into the 1930s, the union faced seemingly insurmountable obstacles as the nation entered the Great Depression. During the latter 1920s and 1930s as operators closed down their mines, a steady stream of locals were abandoned. The average membership in District Thirteen fell from 11,332 in 1925 to 6,804 in 1930. By 1942, the last year that membership figures are available, UMW members in Iowa totaled 4,222. At the same time, miners faced new competition—the development of strip mining in Iowa. At their biennial convention in March 1932, district members passed a resolution requiring their leadership to "protest to the fullest extent to the state legislature against the operation of strip mines in this state." At their biennial convention in 1941, miners again expressed their concern about strip mining. They charged that most employees in strip mines were farmers who worked in the mines during the winter and farmed during the spring and summer. Because the farmers were not totally dependent on mining wages for their livelihood, they expressed little interest in joining the union. In 1945 District Thirteen held its last biennial district convention. Finally, district officials formally disbanded their organization in the early 1950s.[5]

For over fifty years the Iowa coal industry made significant contributions to the economic and social life of the state of Iowa. Between 1895 and 1925, it represented a major Iowa industry, ranking only below agriculture in importance. Consequently, the coal industry accounted for the presence of the largest trade union in the state's history. Its presence changed the ethnic composition of the state, adding an element of pluralism that would not otherwise have been present. Eventually the coal industry faded from the Iowa scene but not before it had altered the state's economic and social composition.

# APPENDIX

THE following twelve tables present social and economic data on Beacon, Cincinnati, and Seymour, Iowa, in 1895 and 1915.

### Table A.1. Religious preference for total populations 1895

|  | Beacon | Cincinnati | Seymour |
|---|---|---|---|
| Catholic | 56 | 39 | 46 |
| Protestant | 387 | 369 | 898 |
| Total | 443 | 408 | 944 |
| Percent of total population in each community | 45.6 | 33.9 | 61.0 |

### Table A.2. Comparisons of religious preference for coal miner and other heads of households, 1895

|  | Beacon | | Cincinnati | | Seymour | |
|---|---|---|---|---|---|---|
|  | miner (%) | nonminer (%) | miner (%) | nonminer (%) | miner (%) | nonminer (%) |
| Catholic | 16.2 | 9.3 | 10.9 | 5.7 | 12.2 | 2.5 |
| Protestant | 83.8 | 90.7 | 89.1 | 94.3 | 87.8 | 97.5 |

### Table A.3. All occupations excluding coal miners, 1895

| Occupation | Beacon | Cincinnati | Seymour |
|---|---|---|---|
| Agricultural | | | |
|   Farmers | 3 | 20 | 8 |
|   Retired farmers | — | 1 | 17 |
|   Total | 3 | 21 | 25 |
| Professional and semiprofessional | | | |
|   Teachers | 10 | 9 | 18 |
|   School principal | | 1 | |
|   Physicians | 3 | 6 | 7 |
|   Clergymen | 2 | 4 | 4 |
|   Dentists | | | 2 |
|   Lawyers | | | 3 |
|   Editor | | | 1 |
|   Pharmacists | — | 2 | — |
|   Total | 15 | 24 | 35 |
| Personal Services | | | |
|   Hotel | | 1 | 3 |
|   Barbers | | 4 | 5 |
|   Restaurant | — | 1 | 4 |
|   Total | | 6 | 12 |

**Table A.3. continued**

| Occupation | Beacon | Cincinnati | Seymour |
|---|---|---|---|
| **Business** | | | |
|   Merchant | 8 | 17 | 25 |
|   Clerks | 8 | 1 | 20 |
|   Insurance agents | 1 | 1 | 2 |
|   Jewelers | | 1 | 2 |
|   Hardware merchants | | 2 | 5 |
|   Lumber dealer | | 1 | |
|   Bankers | | 2 | 5 |
|   Bookkeepers | 1 | 3 | 2 |
|   Coal operators | | 2 | 4 |
|   Building contractor | | | 1 |
|   Traveling salesmen | | | 2 |
|   Total | 18 | 30 | 68 |
| **Transport and communication** | | | |
|   Livery stable | | | 5 |
|   Draymen and teamsters | 6 | 11 | 13 |
|   Station agents | 1 | | 3 |
|   Surveyor | | 1 | |
|   Postmaster | | | 1 |
|   Total | 7 | 12 | 22 |
| **Labor** | | | |
|   Railroaders | 3 | 1 | 5 |
|   General | 21 | 11 | 40 |
|   Domestics | | 4 | 4 |
|   Hoisting engineers | 5 | 3 | 1 |
|   Washerwoman | 1 | | |
|   Mine foremen | | 2 | 1 |
|   Mine superintendent | | 1 | |
|   Policeman | | | 1 |
|   Janitor | | | 1 |
|   Total | 30 | 21 | 53 |
| **Artisans** | | | |
|   Carpenters | 11 | 16 | 14 |
|   Blacksmiths | 7 | 3 | 5 |
|   Brickmakers | | 2 | |
|   Millers | 1 | | 1 |
|   Dressmakers | 7 | 1 | 2 |
|   Butchers | 6 | 1 | 2 |
|   Milliners | | 3 | 8 |
|   Machinists | | 3 | |
|   Bakers | 1 | 1 | 1 |
|   Shoemakers | 3 | 1 | 4 |
|   Harness makers | | 3 | 4 |
|   Tailor | | | 1 |
|   Painters | | 3 | 5 |
|   Plasterers | | 5 | 4 |
|   Printers | | 3 | 2 |
|   Plumber | 1 | | |
|   Photographers | | 3 | 2 |
|   Marble cutter | | 1 | |
|   Well drillers | | | 4 |
|   Total | 37 | 49 | 56 |
| **Other** | | | |
|   Total | 9 | 1 | 20 |
| **Grand total** | 119 | 160 | 291 |

# APPENDIX

## Table A.4. All occupations excluding coal mining, 1915

| Occupation | Beacon | Cincinnati | Seymour |
|---|---|---|---|
| **Agricultural** | | | |
| Farmers | 18 | 5 | 41 |
| Farm laborers | 3 | 16 | 9 |
| Retired farmers | 1 | 4 | 27 |
| Truck farmers | 1 | 1 | 1 |
| Total | 23 | 26 | 78 |
| **Professional and semiprofessional** | | | |
| School superintendents | | 1 | 1 |
| School principal | | 1 | |
| Chiropractor | | | 1 |
| Teachers | 8 | 18 | 26 |
| Physicians | | 5 | 7 |
| Clergymen | | 5 | 5 |
| Dentists | 1 | 1 | 3 |
| Veterinarians | | | 2 |
| Lawyers | | | 3 |
| Editors | | 1 | 2 |
| Pharmacists | | 2 | 3 |
| Baseball player | 1 | | |
| Nurses | 1 | 1 | 4 |
| Musicians | | 1 | 4 |
| Civil engineer | | 1 | |
| Stenographers | | 1 | 2 |
| Government official | | | 1 |
| Total | 11 | 38 | 64 |
| **Personal services** | | | |
| Hotel clerks | | | 2 |
| Saloonkeeper | 1 | | |
| Restaurant operators | | 1 | 6 |
| Hotel proprietors | | 2 | 3 |
| Butler | | 1 | |
| Cooks | | | 7 |
| Barbers | | 3 | 14 |
| Waitress | | 1 | |
| Sewing and boarding | | | 1 |
| Housekeeper | | | 1 |
| Boardinghouse | | | 1 |
| Total | 1 | 8 | 35 |
| **Business** | | | |
| General merchants | 3 | 17 | 35 |
| Clerks | 19 | 11 | 38 |
| Insurance agents | | 3 | 3 |
| Jewelers | | 1 | 1 |
| Hardware merchants | | 1 | 1 |
| Lumber dealers | 1 | 1 | 2 |
| Bankers | | 1 | 1 |
| Bookkeepers | 7 | 4 | 4 |
| Coal operators | 1 | 3 | |
| Grain and livestock dealer | | 1 | 6 |
| Agent | | | 6 |
| Agent, oil company | | 1 | 1 |
| Real estate salesmen | | 2 | 4 |
| Building contractors | 1 | 2 | |
| Traveling salesmen | 3 | 1 | 6 |
| Construction manager | | | 1 |
| Auto livery | | 1 | |

## Table A.4. continued

| Occupation | Beacon | Cincinnati | Seymour |
|---|---|---|---|
| Business (continued) | | | |
| Motion picture theater opearator | | 1 | |
| Nursery dealer | | 1 | |
| Real estate owner | | 1 | |
| Saw mill operator | | 1 | |
| Wholesale produce | | 1 | |
| Electricians | | | 2 |
| Watchmaker and opticians | | | 2 |
| Pool room operator | | | 1 |
| Office managers | | | 3 |
| Laundry operator | | | 1 |
| Monument dealer | | | 1 |
| Manufacturer | | | 1 |
| House mover | | | 1 |
| Decorator | | | 1 |
| Thresher | | | 1 |
| Undertaker | | | 1 |
| Cement and ditching | | | 1 |
| Auctioneer | | | 1 |
| Iceman | | | 1 |
| Auto dealers | | | 2 |
| Total | 35 | 55 | 129 |
| Transport and communication | | | |
| Telegraph operators | | 1 | 4 |
| Draymen and teamsters | 6 | 8 | 16 |
| Hack drivers | | 2 | 3 |
| Station agents | 1 | 3 | 2 |
| Linemen | | | 2 |
| Postmasters | 1 | 1 | 1 |
| Assistant postmasters | 1 | 1 | 1 |
| Mail carriers | 1 | 3 | 4 |
| Telephone operators | 1 | 1 | 7 |
| Bank cashiers | | 1 | 4 |
| Marshal | | | 1 |
| Firemen | | | 3 |
| Cement workers | | | 2 |
| Total | 10 | 21 | 50 |
| Labor | | | |
| Railroaders | | 5 | 19 |
| General | 22 | 14 | 65 |
| Domestics | 4 | 2 | 20 |
| Hoisting engineers | 2 | 3 | 11 |
| Washerwomen | 1 | 1 | 16 |
| Mine foremen | | 7 | 2 |
| Mine superintendents | | 1 | 1 |
| Policeman | | 1 | |
| Janitors | | 1 | 4 |
| Gardeners and laborers | 1 | | 5 |
| Auto mechanics | | 2 | 5 |
| Peddler | | 1 | |
| Total | 30 | 38 | 148 |
| Artisans | | | |
| Masons | | 1 | 3 |
| Carpenters | 5 | 5 | 23 |
| Blacksmiths | 3 | 4 | 5 |
| Dressmakers | 1 | 2 | 20 |

APPENDIX

Table A.4. continued

| Occupation | Beacon | Cincinnati | Seymour |
|---|---|---|---|
| Artisans (continued) | | | |
| Butchers | 1 | | 8 |
| Milliners | | 1 | 5 |
| Machinist | 1 | | |
| Bakers | | 1 | 2 |
| Shoemakers | | 2 | 3 |
| Harness makers | | 3 | 3 |
| Paper-painters | | 1 | 4 |
| Butter maker | | | 1 |
| Painters | | 2 | 2 |
| Plasterers | | 2 | 12 |
| Printers | | 2 | 5 |
| Plumbers | | | 3 |
| Photographers | 1 | | 1 |
| Cordwainer | 1 | | |
| Boilermaker | 1 | | |
| Weavers | | | 4 |
| Total | 14 | 26 | 104 |
| Other | | | |
| Total | 1 | 5 | 10 |
| Grand total | 125 | 217 | 628 |

Table A.5. Women's occupations, 1915

| Occupation | Beacon | Cincinnati | Seymour |
|---|---|---|---|
| Teachers | 8 | 13 | 20 |
| Dressmakers | 1 | 2 | 20 |
| Milliners | | 1 | 4 |
| Washerwomen | 1 | 1 | 16 |
| Clerks | 13 | 2 | 9 |
| Printer | | | 1 |
| Bookkeepers | 6 | 2 | 2 |
| Farmer | | | 1 |
| Domestics | 4 | 2 | 20 |
| Laborers | 1 | | 2 |
| Telephone office manager | 1 | | |
| Nurses | 1 | 1 | 4 |
| Janitors | | | 2 |
| Merchants | | | 5 |
| Cooks | | | 6 |
| Agents | | | 3 |
| Weavers | | | 4 |
| Gardeners and laborers | | | 3 |
| Sewing and boarding | | | 1 |
| Musicians | | 1 | 1 |
| Salesperson | | | 1 |
| Office manager | | | 1 |
| Plasterer | | | 1 |
| Hoisting engineer | | | 1 |
| Telephone operators | | 2 | 6 |
| Stenographers | | 2 | 2 |
| Chiropractor | | | 1 |
| Hotel proprietors | | 1 | 2 |

Table A.5. Women's occupations, 1915 (continued)

| Occupation | Beacon | Cincinnati | Seymour |
|---|---|---|---|
| Peddler | | | 1 |
| Physician | | | 1 |
| Canvasser | | | 1 |
| School superintendent | | | 1 |
| Housekeeper | | | 1 |
| Insurance agent | | 1 | |
| Real estate owners | | 4 | |
| Postmaster | | 1 | |
| Evangelist | | 1 | |
| Boarding house operator | | 1 | |
| Asssistant postmasters | | 2 | |
| Drayman | | 1 | |
| Motion picture theater manager | | 1 | |
| Total | 36 | 42 | 147 |

Table A.6. Birthplaces of 4,229 residents, 1915[a]

| | Beacon | | Cincinnati | | Seymour | |
|---|---|---|---|---|---|---|
| Location | total | (%) | total | (%) | total | (%) |
| Iowa | 307 | 63.6 | 836 | 51.7 | 1235 | 57.8 |
| Eastern coal mining states | 74 | 15.3 | 621 | 38.4 | 631 | 29.6 |
| Other middle western states | 8 | 1.7 | 27 | 1.7 | 66 | 3.1 |
| Southern states | 5 | 1.0 | 23 | 1.5 | 26 | 1.2 |
| Western states | 2 | 0.4 | 6 | 0.4 | 11 | 0.5 |
| Other eastern states | 3 | 0.6 | 9 | 0.6 | 12 | 0.6 |
| British Isles | 53 | 11.0 | 75 | 4.7 | 65 | 3.0 |
| Northern and western Europe | 19 | 3.8 | 9 | 0.5 | 9 | 0.4 |
| Central, southern and eastern Europe | 11 | 2.3 | 6 | 0.4 | 66 | 3.0 |
| Other | 1 | 0.2 | 1 | 0.3 | 12 | 0.5 |
| Total | 483 | | 1613 | | 2133 | |

[a] Four persons did not list birthplace.

Table A.7. Comparison of birthplaces of 1,247 coal miner and other heads of households, 1915

| | Beacon | | Cincinnati | | Seymour | |
|---|---|---|---|---|---|---|
| Location | nonminer (%) | miner (%) | nonminer (%) | miner (%) | nonminer (%) | miner (%) |
| Iowa | 34.5 | 27.1 | 36.0 | 28.3 | 35.9 | 29.0 |
| Eastern coal mining states | 31.0 | 18.8 | 43.6 | 56.2 | 51.5 | 29.7 |
| Other middle western states | 4.6 | 0.0 | 1.3 | 0.5 | 1.9 | 3.4 |
| Southern states | 3.4 | 0.0 | 2.5 | 1.4 | 2.5 | 0.7 |
| Western states | 0.0 | 0.0 | 0.4 | 0.0 | 0.0 | 0.7 |
| Other eastern states | 2.3 | 0.0 | 2.5 | 0.5 | 1.9 | 0.0 |
| British Isles | 16.0 | 39.7 | 11.4 | 9.6 | 2.7 | 13.1 |
| Northern and western Europe | 6.8 | 8.4 | 0.8 | 2.3 | 1.2 | 0.0 |
| Central, southern and eastern Europe | 1.1 | 6.3 | 0.4 | 1.0 | 1.0 | 22.8 |
| Other | 0.0 | 0.0 | 0.8 | 0.5 | 1.4 | 0.7 |

APPENDIX

Table A.8. Sex ratio of 3,721 residents, 1895[a]

|  | Beacon | | Cincinnati | | Seymour | |
|---|---|---|---|---|---|---|
| Sex | no. | (%) | no. | (%) | no. | (%) |
| Male | 499 | 51.3 | 631 | 52.9 | 774 | 50.5 |
| Female | 469 | 48.7 | 567 | 47.1 | 765 | 49.5 |
| Total | 968 | 100.0 | 1198 | 100.0 | 1539 | 100.0 |

[a] Sixteen answers were unintelligible.


Table A.9. Religious preference for 1,870 residents, 1915

|  | Beacon | Cincinnati | Seymour |
|---|---|---|---|
| Catholic | 30 | 23 | 101 |
| Protestant | 242 | 477 | 977 |
| Total | 272 | 500 | 1098 |
| Percent of total population in each community | 56.3 | 30.9 | 51.4 |


Table A.10. Comparisons of religious preference for coal miner and other heads of households, 1915

|  | Beacon | | Cincinnati | | Seymour | |
|---|---|---|---|---|---|---|
|  | miner (%) | nonminer (%) | miner (%) | nonminer (%) | miner (%) | nonminer (%) |
| Catholic | 21.9 | 7.3 | 7.6 | 2.9 | 36.0 | 3.3 |
| Protestant | 78.1 | 92.7 | 92.4 | 97.1 | 64.0 | 96.7 |


Table A.11. Months unemployed for 353 coal miner heads of households, 1914[a]

|  | Beacon | | Cincinnati | | Seymour | |
|---|---|---|---|---|---|---|
| Months | no. | (%) | no. | (%) | no. | (%) |
| 1 | 0 | 0.0 | 2 | 1.0 | 3 | 2.3 |
| 2 | 3 | 14.3 | 3 | 1.5 | 4 | 3.1 |
| 3 | 6 | 28.6 | 6 | 3.0 | 11 | 8.5 |
| 4 | 5 | 23.8 | 95 | 46.8 | 32 | 24.8 |
| 5 | 1 | 4.8 | 64 | 31.5 | 9 | 7.0 |
| 6 | 4 | 19.0 | 29 | 14.3 | 60 | 46.5 |
| 7 | 0 | 0.0 | 0 | 0.0 | 2 | 1.6 |
| 8 | 1 | 4.8 | 3 | 1.5 | 3 | 2.3 |
| 9 | 1 | 4.8 | 1 | 0.5 | 3 | 2.3 |
| 10 | 0 | 0.0 | 0 | 0.0 | 1 | 0.8 |
| 11 | 0 | 0.0 | 0 | 0.0 | 0 | 0.0 |
| 12 | 0 | 0.0 | 0 | 0.0 | 1 | 0.8 |

[a] Missing observations = 58.

Table A.12. Comparison of annual incomes for 969 coal miner and other heads of households, 1914[a]

| Income range (in dollars) | Beacon Miner no. | Beacon Miner (%) | Beacon Nonminer no. | Beacon Nonminer (%) | Cincinnati Miner no. | Cincinnati Miner (%) | Cincinnati Nonminer no. | Cincinnati Nonminer (%) | Seymour Miner no. | Seymour Miner (%) | Seymour Nonminer no. | Seymour Nonminer (%) |
|---|---|---|---|---|---|---|---|---|---|---|---|---|
| 5–100 | 2 | 5.0 | 4 | 7.5 | 0 | 0.0 | 1 | 0.6 | 5 | 3.5 | 29 | 8.0 |
| 101–200 | 3 | 7.5 | 4 | 7.5 | 5 | 2.3 | 9 | 5.8 | 5 | 3.5 | 32 | 8.9 |
| 201–300 | 3 | 7.5 | 5 | 9.4 | 24 | 11.1 | 7 | 4.5 | 26 | 18.3 | 46 | 12.7 |
| 301–400 | 5 | 12.5 | 7 | 13.2 | 56 | 25.8 | 19 | 12.2 | 36 | 25.4 | 34 | 9.4 |
| 401–500 | 17 | 42.5 | 9 | 17.0 | 57 | 26.3 | 28 | 17.9 | 40 | 28.2 | 47 | 13.0 |
| 501–600 | 5 | 12.5 | 11 | 20.8 | 45 | 20.7 | 17 | 10.9 | 25 | 17.6 | 37 | 10.2 |
| 601–700 | 2 | 5.0 | 3 | 5.7 | 16 | 7.4 | 7 | 4.5 | 2 | 1.4 | 18 | 5.0 |
| 701–800 | 2 | 5.0 | 1 | 1.9 | 12 | 5.5 | 16 | 10.3 | 1 | 0.7 | 31 | 8.6 |
| 801–1000 | 1 | 2.5 | 7 | 13.2 | 2 | 0.9 | 25 | 16.0 | 2 | 1.4 | 42 | 11.6 |
| 1001–1500 |  |  | 1 | 1.9 |  |  | 21 | 13.5 |  |  | 35 | 9.7 |
| 1501–2000 |  |  | 0 | 0.0 |  |  | 5 | 3.2 |  |  | 5 | 1.4 |
| 2001–4500 |  |  | 1 | 1.9 |  |  | 1 | 0.6 |  |  | 5 | 1.4 |

[a] Missing observations = 278.

# NOTES

### INTRODUCTION

1. Charles Keyes, "Annotated Bibliography of Iowa Geology and Mining," *Iowa Geological Survey* 22 (1912):119.
2. James H. Lees, "History of Coal Mining in Iowa," *Iowa Geological Survey* 19 (1908):525–33.
3. Ibid., pp. 566–67.
4. Ibid., pp. 550–54; Hubert L. Olin, *Coal Mining in Iowa* (Des Moines: State of Iowa, 1965), pp. 33–35.
5. Lees, "History of Coal Mining," pp. 576–79; N. E. Goldthwait, ed., *History of Boone County*, 2 vols. (Chicago: Pioneer Publishing, 1914), 2:335–44, 398–400; Lees, "History of Coal Mining," p. 579.
6. William J. Petersen, *The Story of Iowa: The Progress of an American State*, 4 vols. (New York: Lewis Historical Publishing, 1952), 1:587.
7. Olin, *Coal Mining*, pp. 80–89; *Census of Iowa, 1895* (Des Moines: State of Iowa, 1896), p. 708; and *Census of Iowa, 1925* (Des Moines: State of Iowa, 1926), p. 220. The Iowa State Census of 1905 is of less value than other state censuses since many parts of the population schedules were destroyed after being badly damaged by water.
8. Joseph H. Cash, *Working the Homestake* (Ames: Iowa State University Press, 1973), p. xi.

### CHAPTER 1

1. David McDonald and Edward Lynch, *Coal and Unionism: A History of the American Coal Miners' Unions* (Silver Springs, Md.: Cornelius Printing, 1939), p. 11.
2. A. T. Shurick, *The Coal Industry* (Boston: Little, Brown, 1924), p. 5.
3. Ibid.; Andrew Roy, *A History of the Coal Miners of the United States* (Columbus, Ohio: J. L. Trauger Printing, 1907), p. 47.
4. Katherine Harvey, *The Best Dressed Miners: Life and Labor in the Maryland Coal Region, 1835–1910* (Ithaca: Cornell University Press, 1969), pp. 5–6.
5. Edward A. Wieck, *The American Miners' Association: A Record of the Origin of Coal Miners' Unions in the United States* (New York: Russell Sage Foundation, 1940), p. 53. It is impossible to state coal production before 1840; neither federal nor state governments kept systematic records of coal production.
6. Ibid., p. 46; Roy, *History of the Coal Miners*, p. 47.
7. Wieck, *American Miners' Association*, p. 54.
8. Roy, *History of the Coal Miners*, p. 60.
9. Ibid., pp. 67–68.
10. Wieck, *American Miners' Association*, pp. 58–59; Shurick, *Coal Industry*, p. 15.

11. Roy, *History of the Coal Miners*, p. 72; Andrew Roy, *Coal Miners* (Cleveland: Robinson, Savage, 1876), p. 311.
12. Wieck, *American Miners' Association*, pp. 78-79.
13. P. E. H. Hair, "The Social History of British Coal Miners in 1800-1845" (Ph.D. diss., Oxford University, 1955), pp. 1-100; Roy, *Coal Miners*, pp. 50-60.
14. G. D. H. Cole and A. W. Filsons, eds., *British Working Class Movements: Select Documents 1789-1875* (London: Macmillan & Co., 1951), pp. 490-95; Great Britain, *Parliamentary Papers* (Commons), "Report on the Truck Commissioners," 1871, 2:17-24; ibid., "Select Committee on the Scarcity and Dearness of Coal," 1873, 5:1-301; ibid., "Royal Commission on the Housing of the Working Class," 1884-1885, 2:341.
15. In a treatise published in 1756 entitled "Vindication of Natural Society," Edmund Burke described English miners as "These unhappy wretches [who] scarce ever see the light of the sun; they are buried in the bowels of the earth, where they work at a severe and dismal task without the least prospect of being delivered from it. . . ." Quoted in Roy, *History of the Coal Miners*, p. 14. Another British observer described them as a "strange, almost half-human stratum of the working class, pugnacious, brutalized by their lives of grimy toil, inhabiting isolated communities which, in leisure time, became dens of drunkenness and savage sport." Quoted in John Burnett, ed., *Autobiographies of British Working Class People 1820-1920* (Bloomington: Indiana University Press, 1974), p. 43.
16. Quoted in Wieck, *American Miners' Association*, p. 256.
17. Quoted in Shurick, *Coal Industry*, p. 311.
18. "Discussion by Readers," *Coal Age* 8 (Nov. 1915):895. Fortunately for the American industry some coal mining practices were not transmitted. The early British practice of utilizing children as young as four or five years old as trappers and utilizing girls and women as pushers (individuals who pushed baskets of coal from the coal seam to the bottom of the shaft) and as errand runners was not adopted in the United States.
19. John McBride, "The Coal Miners," in *The Labor Movement: The Problem of Today*, ed. George E. McNeill (New York: M. W. Hazen, 1888), p. 242.
20. Wieck, *American Miners' Association*, pp. 65-67.
21. Ibid., pp. 73-74.
22. Arthur E. Suffern, *The Coal Miners' Struggle for Industrial Status* (New York: Macmillan, 1926), pp. 15-16.
23. Ibid., p. 16.
24. McDonald and Lynch, *Coal and Unionism*, p. 16; Wieck, *American Miners' Association*, pp. 62-65; Roy, *History of the Coal Miners*, pp. 61-64, 75.
25. Wieck, *American Miners' Association*, pp. 21-30. A few months later, the miners added the word American to the title of their organization.
26. Ibid., pp. 115-17, 178.
27. Ibid., p. 180.
28. Roy, *History of the Coal Miners*, p. 71; McBride, "Coal Miners," p. 250.
29. Roy, *History of the Coal Miners*, pp. 154, 159.
30. Ibid., pp. 42-387; McBride, "Coal Miners," pp. 246-47.
31. McBride, "Coal Miners," p. 250; Roy, *History of the Coal Miners*, pp. 163-64; McDonald and Lynch, *Coal and Unionism*, p. 20.
32. McDonald and Lynch, *Coal and Unionism*, p. 20.
33. Roy, *History of the Coal Miners*, p. 232.
34. Ibid., pp. 333-34.
35. Ibid.
36. McDonald and Lynch, *Coal and Unionism*, p. 21.
37. Ibid., pp. 22-23.

38. Roy, *Coal Miners*, p. 62.
39. William Graebner, *Coal Mining Safety in the Progressive Period: The Political Economy of Reform* (Lexington: University Press of Kentucky, 1976), p. 1.
40. Ibid., p. 124; Roy, *History of the Coal Miners*, pp. 142–43; Willard Jillson, *The Coal Industry in Kentucky* (Frankfort: Kentucky Geological Survey, 1924), p. 32; E. H. Downey, *History of Labor Legislation in Iowa*, Iowa Economic History Series, ed. Benjamin F. Shambaugh (Iowa City: State Historical Society of Iowa, 1910), pp. 45–55.
41. Graebner, *Coal Mining Safety*, pp. 115–17.
42. Suffern, *Coal Miners' Struggle*, p. 16.
43. Ibid.
44. Quoted in Anna Rochester, *Labor and Coal*, Labor and Industry Series (New York: International Publishers, 1931), p. 100.
45. Carter Goodrich, *The Miner's Freedom: A Study of the Working Life in a Changing Industry* (Boston: Marshall Jones, 1925), pp. 99–100.
46. Quoted in McDonald and Lynch, *Coal and Unionism*, p. 27; Suffern, *Coal Miners' Struggle*, p. 17.
47. *A Medical Survey of the Bituminous-Coal Industry: Report of Coal Mines Administration* (Washington, D.C.: U.S. Government Printing Office, 1947), pp. xiv–xv; Edward Hunt, F. G. Tryon, and Joseph Willits, eds., *What the Coal Commission Found: An Authoritative Summary by the Staff*, Human Relations Series, vol. 3 (Baltimore: Williams & Wilkins, 1925).
48. Quoted in Goodrich, *Miner's Freedom*, p. 22.
49. Edward T. Devine, *Coal: Economic Problems of the Mining, Marketing and Consumption of Anthracite and Soft Coal in the United States* (Bloomington, Ill.: American Review Service, 1925), p. 33.
50. Suffern, *Coal Miners' Struggle*, pp. 12, 41; Roy, *History of the Coal Miners*, pp. 270–82. John B. Rae was the master workman of the Knights of Labor, National District 135; John McBride was the president of the Ohio Miners' Union; David Ross was the president of the Illinois Miners' Association; and William Scaife was a labor leader from Illinois.
51. Suffern, *Coal Miners' Struggle*, p. 41, 70–71.
52. Elsie Gluck, *John Mitchell, Miner* (New York: John Day, 1929), p. 25.

## CHAPTER 2

1. Hubert L. Olin, *Coal Mining in Iowa* (Des Moines: State of Iowa, 1966), pp. 80–89.
2. Dean Aubrey, Iowa state mine inspector, interview, Ames, Iowa, Aug. 1972; A. T. Shurick, *The Coal Industry* (Boston: Little, Brown, 1924), p. 62; Olin, *Coal Mining*, p. 80.
3. Shurick, *Coal Industry*, pp. 54, 64; Olin, *Coal Mining*, pp. 80–89.
4. Olin, *Coal Mining*, pp. 80–89.
5. Ibid.; Aubrey interview.
6. Donald Baker, "Operating Conditions Encountered in Iowa," *Coal Age* 14 (Mar. 1921):437.
7. Ibid., pp. 438–39.
8. Milo Papich, former coal miner and mine foreman, interview, Slater, Iowa, May 1979.
9. Alex Erickson, interview, Pershing, Iowa, Sept. 1979.
10. Christina Anderson, interview, Ames, Iowa, Dec. 1978.
11. Ladorrico Sebben, interview, Seymour, Iowa, Oct. 1975.
12. Delno Bingman, interview, Chariton, Iowa, Dec. 1978.

13. James H. Lees, "History of Coal Mining in Iowa," *Iowa Geological Survey* 19 (1908):232; Otto Goodman, interview, Madrid, Iowa, July 1974 and Mar. 1979; Anderson interview.
14. Cynthia Johnson, "A New Life: The Iowa Coal Mines," *Palimpsest* 56 (Mar./Apr. 1975):60; Goodman interview; Anderson interview.
15. Carter Goodrich, *The Miner's Freedom: A Study of the Working Life in a Changing Industry* (Boston: Marshall Jones, 1925), p. 21; Papich interview.
16. John Turner, interview, Madrid, Iowa, July 1974.
17. Papich interview.
18. Goodman interview.
19. Papich interview.
20. *Biennial Report of the State Mine Inspectors, 1882–1883* (Des Moines: State of Iowa, 1883), p. 94. In 1906 the legislature raised the age limit to fourteen. *Laws of Iowa, 1906*, p. 71; *Laws of Iowa, 1880*, p. 199.
21. Turner interview.
22. One authority estimated the time lost by miners waiting for coal cars to be about one and one-half hours per day. See Goodrich, *Miner's Freedom*, p. 31; Papich interview.
23. Aubrey interview.
24. Goodman interview.
25. William Conway, interview, Boone, Iowa, Feb. 1975.
26. Bingman interview.
27. Goodman interview.
28. Ibid.
29. Goodrich, *Miner's Freedom*, p. 24; Turner interview.
30. Gilbert Porter, interview, Chariton, Iowa, Dec. 1978.
31. William Lepovitz, interview, Madrid, Iowa, June, 1979.
32. Goodman interview; Turner interview; Papich interview.
33. Goodrich, *Miner's Freedom*, p. 20; Goodman interview.
34. *Biennial Report of the State Mine Inspectors, 1882–1883*, p. 13.
35. E. H. Downey, *History of Labor Legislation in Iowa*, Iowa Economic History Series, ed. Benjamin F. Shambaugh (Iowa City: State Historical Society of Iowa, 1910), p. 51.
36. *Laws of Iowa, 1902*, p. 63; Downey, *History of Labor Legislation*, p. 52.
37. Papich interview.
38. Lepovitz interview; Papich interview.
39. Johnson, "A New Life," pp. 64.
40. *Slater* (Iowa) *Tri-County Times*, 1 June 1978, p. 3.
41. Papich interview.
42. Turner interview.
43. Papich interview.
44. John Starchovich, interview, Madrid, Iowa, June 1979.
45. Papich interview.
46. Goodman interview.
47. Shurick, *Coal Industry*, p. 89; Goodman interview.
48. Johnson, "A New Life," p. 64.
49. *Biennial Report of the State Mine Inspectors, 1880–1881*, 105.
50. Johnson, "A New Life," p. 61.
51. Ibid.
52. Turner interview.
53. Jesse Frazier, interview, Des Moines, Iowa, Mar. 1979.
54. Turner interview.
55. Goodman interview.
56. James Battani, interview, Woodward, Iowa, Oct. 1979; Papich interview.

57. Goodman interview; Turner interview.
58. Goodman interview.
59. Ibid.
60. Shurick, *Coal Industry*, pp. 114–16.
61. Goodrich, *Miner's Freedom*, pp. 52–53.
62. Turner interview.
63. Ibid.
64. Goodrich, *Miner's Freedom*, p. 32; James Battani interview.
65. Papich interview. Papich was referring to the requirement that all mine foremen hold a certificate of competency issued by the Board of Examiners for state mine inspectors. In 1900 the UMW successfully lobbied the state legislature to pass such legislation. The law made it illegal for anyone to discharge the duties of foreman at any coal mine where the daily output exceeded twenty-five tons unless the person held a competency certificate. Later, hoisting engineers also had to be certified. Miners had to travel to Des Moines to take the examination. Originally the men could take either an oral or written exam but that was later changed to only a written exam. *Biennial Report of the State Mine Inspectors, 1900–1901*, pp. 51–53, 71–72.
66. Papich interview.
67. Ibid.

## CHAPTER 3

1. It is impossible to tell precisely what percentage of Iowa's coal mining families lived in company housing. In 1923 the United States Coal Commission estimated that roughly 20 percent of Iowa's mining families lived in company-owned housing. Comments by state mine inspectors, reports in the UMW *Journal*, and interviews with mining families indicate that the number was much higher in 1895 and that over the next thirty years the number gradually decreased.
2. Herbert Gutmann, *Work, Culture and Society in Industrializing America* (New York: Knopf, 1976), p. 43.
3. See Bibliography for complete list of interviewees.
4. Before 1880, most Italian emigrants came from northern Italy. After 1880, the southern Italian provinces provided the great majority of emigrants to the United States.
5. Thomas Kessner, *The Golden Door: Italian and Jewish Immigrant Mobility in New York City* (New York: Oxford University Press, 1977), pp. 26–28.
6. Virginia Yans-McLaughlin, *Family and Community: Italian Immigrants in Buffalo, 1880–1930* (Ithaca: Cornell University Press, 1977), p. 96.
7. Quoted in Kessner, *Golden Door*, p. 13; Luciano J. Iorizzo, *The Italian-Americans* (New York: Twayne, 1971), p. 5.
8. Quoted in Valentine Rossilli Winsey, "The Italian Immigrant Women Who Arrived in the U.S. before World War I," in *Studies in Italian-American Social History*, ed. Francesco Cordasco (Totowa, N.J.: Roman and Littlefield, 1975), p. 201.
9. Robert Foerster, *The Italian Emigration of Our Times* (Cambridge, Mass.: Harvard University Press, 1919), p. 440.
10. Quoted in Winsey, "Italian Immigrant Women," p. 199.
11. Ibid., p. 199; Bruna Pieracci, interview, Des Moines, Iowa, July 1980.
12. James Battani, interview, Woodward, Iowa, Oct. 1979; Pieracci interview.
13. Joan Scott and Louise Tilly, "Women's Work and Family in Nineteenth Century Europe," *Comparative Studies in Society and History* 18 (1975):43–44, 50–51.
14. Pieracci interview; Winsey, "Italian Immigrant Women," pp. 201–2; Scott and Tilly, "Women's Work and Family," p. 46.
15. Antonia Cerato, interview, Numa, Iowa, Mar. 1978; Paulina Biondi, interview, Madrid, Iowa, Sept. 1979; Pieracci interview.

16. Scott and Tilly, "Women's Work and Family," pp. 45, 48.
17. Yans-McLaughlin, *Family and Community*, p. 53.
18. Kessner, *Golden Door*, pp. 73-74.
19. Ibid., p. 95; Yans-McLaughlin, *Family and Community*, p. 93.
20. Margaret F. Byington, *Homestead: The Households of a Mill Town*, The Pittsburgh Survey, 6 vols. (New York: Russell Sage Foundation, 1910), 4:1-45.
21. Ibid., p. 135.
22. Ibid., pp. 142-43.
23. Ibid., pp. 145, 147.
24. Ibid., 96, 154, 160.
25. Ibid., pp. 158-59.
26. Yans-McLaughlin, *Family and Community*, p. 15.
27. James Battani interview.
28. Victoria DeGard, interview, Seymour, Iowa, Mar. 1978; Mary Sertich, interview, Granger, Iowa, Jan. 1978; James Battani and Papich interviews.
29. Biondi interview.
30. Lola Nizza, interview, Granger, Iowa, Jan. 1978; Nellie Fontanini, interview, Granger, Iowa, Jan. 1978; James Battani and Sertich interviews.
31. Cerato interview and James Battani interview.
32. Minnie Mores, interview, Seymour, Iowa, Mar. 1978; DeGard interview.
33. Cerato interview; Nizzi interview; Sertich interview.
34. Biondi interview.
35. *Laws of Iowa*, 1908, p. 99.
36. James Battani interview.
37. Ibid.
38. Augusta Argenta, interview, Seymour, Iowa, Mar. 1979; Mores interview, Earls interview, and DeGard interview; Gina Zanotti, "The Zanotti Family at Rippey" (Paper, Iowa State University, 1972), p. 10.
39. Augusta Argenta interview.
40. James Battani interview.
41. DeGard interview.
42. Edna Padovan, interview, Centerville, Iowa, Mar. 1978.
43. Biondi interview.
44. Father John Gorman, interview, Elkhart, Iowa, July 1977 and Jan. 1978.
45. Richard Jensen, *The Winning of the Midwest: Social and Political Conflict, 1888-1896* (Chicago: University of Chicago Press, 1971), p. 57.
46. Mores interview.
47. See Rudolph Vecoli, "Peasants and Prelates: Italian Immigrants and the Catholic Church," *Journal of Social History* 2 (Spring 1969):217-68; Henry Charles Sartorio, *Social and Religious Life of Italians in America* (Boston: Christopher Publishing, 1918; reprint ed., Clifton, N.J.: Kelley, 1974), p. 83.
48. Vecoli, "Peasants and Prelates," pp. 223, 260.
49. Sartorio, *Social and Religious Life,* p. 104; Vecoli, "Peasants and Prelates," pp. 230-31, 246-47.
50. Ibid.
51. Ibid., 236.
52. Zanotti, "The Zanotti Family," pp. 6-8.
53. Gorman interview.
54. Dorothy Collier, interview, Des Moines, Iowa, Nov. 1978.
55. Beverly Shiffer, "The Story of Buxton," *Annals of Iowa* 37 (Summer 1964): 344.
56. Collier interview; Marjorie Brown, interview, Waterloo, Iowa, Oct. 1978.
57. Collier interview; Brown interview.
58. Brown interview.

59. Shiffer, "The Story of Buxton," p. 345.
60. Collier interview.
61. Shiffer, "The Story of Buxton," pp. 340–47.
62. Ibid.
63. Brown interview.
64. Ibid.
65. Collier interview.
66. Brown interview.

## CHAPTER 4

1. Charles Nichols, *Housing Conditions: Iowa Coal Mining Camps* (Des Moines: State Printing Office, 1919), pp. 31–34. With the exception of one or two, the Italian-American women interviewed supported the view that camp housing was cheaply built and poorly maintained.
2. Nichols, *Housing Conditions*, pp. 32–35.
3. Ibid., p. 2.
4. Ibid., pp. 32–35.
5. Mary Braida, interview, Chariton, Iowa, Oct. 1975.
6. Alnora Earls, interview, Seymour, Iowa, Mar. 1978.
7. Paulina Biondi interview, Madrid, Iowa, Sept. 1979.
8. Bruna Pieracci, "The Miners," in *The Immigrants Speak: Italian-Americans Tell Their Story*, ed. Salvatore J. La Gumino (New York: Center for Migration Studies, 1979), p. 35.
9. Bruna Pieracci, interview, Des Moines, Iowa, July 1980.
10. Ibid.
11. Nellie Fontanini, interview, Granger, Iowa, Jan. 1978.
12. Mary Sertich, interview, Granger, Iowa, Jan. 1978; Lola Nizzi, interview, Granger, Iowa, Jan. 1978; Fontanini interview.
13. Although some sources like the state mine inspectors state that only a few ethnic groups, particularly the Swedes, purchased their homes rather than renting company housing, this practice was not restricted to just one or two ethnic groups.
14. Victoria DeGard, interview, Seymour, Iowa, Mar. 1978; Minnie Mores, interview, Seymour, Iowa, Mar. 1978.
15. Edna Padovan, interview, Centerville, Iowa, Mar. 1978.
16. Augusta Argenta, interview, Seymour, Iowa, Mar. 1979.
17. DeGard interview; Antonia Cerato, interview, Numa, Iowa, Mar. 1978; Padovan interview; Augusta Argenta interview; Mores interview.
18. Mary Maddalena, interview, Seymour, Iowa, Mar. 1978; DeGard interview; Padovan interview; Mores interview; Nizzi Interview; Augusta Argenta interview; Fontanini interview.
19. DeGard interview; Padovan interview; Augusta Argenta interview; Mores interview; Nizzi interview; Fontanini interview; Biondi interview.
20. Biondi interview.
21. Nizzi interview; DeGard interview.
22. DeGard interview.
23. Alnora Earls, interview, Seymour, Iowa, Mar. 1978; Mores interview; Padovan interview, DeGard interview.
24. Gina Battani, interview, Woodward, Iowa, Oct. 1979.
25. Cerato interview.
26. Ibid.
27. Nizzi interview; Biondi interview.
28. Anita Lami, interview, Granger, Iowa, Jan. 1978; Earls interview; Sertich in-

terview; Fontanini interview; Cerato interview.
29. Milo Papich interview, Slater, Iowa, May 1979.
30. Biondi interview.
31. Gina Battani interview.
32. Fontanini interview.
33. Earls interview; Mores interview; DeGard interview; Augusta Argenta interview; Madallena interview; Braida interview.
34. DeGard interview.
35. Sertich interview; Fontanini interview.
36. Drawn from *Iowa State Census (1915), Manuscript Population Schedules for Beacon, Cincinnati, and Seymour* (Des Moines: Iowa Division of Historical Archives and Museum).
37. Pieracci interview; DeGard interview.
38. Nichols, *Housing Conditions*, p. 32.
39. Cerato interview.
40. Biondi interview.
41. Mores interview.
42. Padovan interview.
43. Gina Battani interview.
44. DeGard interview.
45. Earls interview.
46. Augusta Argenta interview.
47. Nizzi interview.
48. DeGard interview.
49. Cerato interview; Antonia Cerato's family is still perplexed as to how she acquired the skill of reading English. No family member can remember any English-Italian dictionary in the home. She cannot write English, however.
50. Cerato interview.
51. For an example of this attitude see *Biennial Report of the State Mine Inspectors, 1896* (Des Moines: State of Iowa, 1897), pp. 78–79; Padovan interview.
52. Nizzi interview; Fontanini interview; Earls interview.
53. Nizzi interview; Gorman interview.
54. Fontanini interview; Gorman interview.
55. Flora Betti, interview, Granger, Iowa, Jan. 1978; DeGard interview.
56. John Argenta, interview, Seymour, Iowa, Mar. 1978.
57. Maddalena interview.
58. Catherine McClellan, interview, Madrid, Iowa, Apr. 1975.
59. Pieracci interview.
60. Sertich interview; Nizzi interview; Lami interview; Fontanini interview.
61. James Battani interview.
62. Nizzi interview; Fontanini interview; Betti interview; Sertich interview; Lami interview.
63. Cerato interview.
64. Earls interview.
65. Pieracci interview.
66. Virginia Yans-McLaughlin, *Family and Community; Italian Immigrants in Buffalo, 1880*–1930, (Ithaca: Cornell University Press, 1977), p. 132.
67. Biondi interview.

## CHAPTER 5

1. Many towns like Fraser and Angus in Boone County were booming in 1895, but they had almost disappeared by 1915.
2. Cincinnati was founded in 1855 in Appanoose County; Beacon was founded

in 1864 in Mahaska County; and Seymour was founded in 1871 in Wayne County. By 1895 all three communities were clearly identified as coal mining communites. See Appendix Table A.3 for listing of separate occupations.

3. The census takers did not list homemaker as an occupation.

4. Great Britain, Parliament, *Parliamentary Papers* (Commons) "Select Committee on the Scarcity and Dearness of Coal," 1873, 5:8, 15, 31, 100, 192, 229.

5. Christina Anderson, interview, Ames, Iowa, Dec. 1978.

6. *United Mine Workers Journal,* Dec. 22, 1898, p. 1; *Laws of Iowa, 1880,* p. 199; *Biennial Report of the State Mine Inspectors, 1882-1883,* Des Moines: State of Iowa, 1883), p. 94. No doubt the mine inspector was referring to two types of violations: (1) the hiring of trapper boys under the age of twelve and (2) the practice of fathers taking their young sons underground to secure extra turns. With the latter practice, fathers used their sons to run errands, deliver props, and sometimes load coal. Both practices clearly violated the law.

7. U.S., Department of Commerce, Bureau of the Census, *Eleventh Census of the United States, 1890: Population* (Washington, D.C.: Government Printing Office, 1895) 1:560-63. States listed as eastern coal mining states and included in category two on Tables 5.4 and 5.5 are Pennsylvania, Ohio, Indiana, Illinois, Virginia, West Virginia, Kentucky, and Missouri. States listed as other middle western states in category three are Wisconsin, Kansas, Nebraska, Michigan, and Minnesota. States listed as southern states in category four are Tennessee, Maryland, North Carolina, Arkansas, Indian Territory, and Alabama. States listed as western states in category five are Wyoming, Colorado, California, Dakota, Idaho, and Montana. States listed as other eastern states in category six are New York, New Jersey, Massachusetts, Vermont, Delaware, New Hampshire, and Connecticut.

8. Comparisons are made between the coal mining population and the remaining population by selecting for comparison only the heads of household. This is done so that only comparable segments of the population will be utilized. The result is that the two groups compared are: (1) coal miners who are heads of households and (2) all other heads of households.

9. Ronald C. Brown, *Hard-Rock Miners: The Intermountain West, 1860-1920* (College Station, Texas: Texas A & M University Press, 1979), pp. 26-30, 39-40. See Appendix Table A.8 for specific data on sex ratios.

10. Archie Harris, interview, Lovilia, Iowa, Nov. 1974; John DeGard interview, Seymour, Iowa, Oct. 1975; William Ewalt, "The Ogden Coal Strike, 1910-1912," (Paper, Iowa State University, 1974), p. 15.

11. *Biennial Report of the State Mine Inspectors, 1915,* pp. 27-30, 50-55.

12. *Iowa State Censuses (1895* and *1915). Manuscript Population Schedules for Beacon, Cincinnati, and Seymour.* Another possible explanation for the older mining population in 1915 is that of declining birth rates. Beacon's coal miner households averaged 3.3 children in 1895 compared with 2.6 children in 1915; Seymour's coal miner households averaged 3.0 children in 1895 compared to 2.8 in 1915. On the other hand, Cincinnati's coal miner households went from 2.7 children in 1895 to 2.9 children in 1915.

13. See Appendix Table A.5 for specific list of occupations.

14. David M. Katzman, *Seven Days a Week: Women and Domestic Service in Industrializing America* (New York: Oxford University Press, 1978), pp. 21-22, 110-15.

15. See Appendix Table A.7 for further details on birthplaces of miners in 1915.

16. The male-female ratio in the three communities remained approximately the same. The 1915 figures represent a slight change over the male-female ratio in 1895, with the ratios coming closer together. By 1915 Seymour represented the exception as it contained more females than males.

17. See Appendix Table A.11 for specific months unemployed for coal miner heads of households.

18. From the standpoint of union policy, Iowa mines should have enjoyed a higher rate of employment in 1914. In January 1914 the international leadership of the UMW decided that the annual suspensions invoked on March 31 should come to an end. Traditionally, union policy required miners to walk off the job on March 31, the date when the joint agreement expired. The miners did not return to work until a new contract was signed, which sometimes took a full month to negotiate. For years union officials and operators decried the needless loss of work following the contract expiration. In early 1914 union officials and operators agreed that no further suspensions would take place following the contract expiration. As a result, miners continued working after March 31. Apparently this policy change had little effect on the miners in Beacon, Cincinnati, and Seymour.

19. A comparison of coal miner and other heads of households by income categories is included in the Appendix Table A.12.

20. See Appendix Tables A.1, A.2, A.9, and A.10 for religious preference of residents in 1895 and 1915.

## CHAPTER 6

1. Chris Evans, *History of the United Mine Workers of America* (Indianapolis: n.p., 1918?–1920), p. 70.

2. David McDonald and Edward Lynch, *Coal and Unionism: A History of the Coal Miners' Union* (Silver Springs, Md.: Cornelius Printing, 1939), p. 29. John Rae was born in Scotland where he worked as a miner. After coming to the United States, he joined the Knights of Labor; Evans, *History*, p. 67. By 1925 the international UMW claimed 500,000 members located in twenty-six districts. The organization included Canadian locals as well as American locals. See Edward T. Devine, *Coal: Economic Problems of the Mining, Marketing, and Consumption of Anthracite and Soft Coal in the United States* (Bloomington, Ill.: American Review Service, 1925), p. 38.

3. Ben Henry, "A History of District 13" (Paper, Iowa State Historical Department, Division of Museum and Archives, Des Moines), p. 3. It is not possible to know the percentage of miners within a coal camp or mining community that typically joined the union. UMW records state only the number of members; they do not give the total number of employees. During the early 1900s, President John Reese and President John White frequently alluded to the fact that District Thirteen was 100 percent organized. By this statement, Reese and White meant that a local had been established in each camp or coal community. It did not mean that all miners within the camp had joined the union. It is evident from miners' letters to the *Journal* that many camps contained nonunion men. Occasionally union miners wrote that nonunion men were the first to complain when operators lowered prices or reduced the number of working days. It is also not possible to know the exact number of town miners versus coal camp miners. In 1900 towns and cities like What Cheer and Des Moines had twenty-three locals and a total of 3,634 miners. Forty-one camps had a total of 5,408 miners. See *District 13 Executive Board Proceedings, 1895-1925*, 4 reels microfilm, (Des Moines: Iowa State Historical Department, Division of Museum and Archives) 1:34–35. By 1925 the number of coal camp locals had declined so that camps and incorporated towns had roughly the same numbers of locals and members.

4. *District 13 United Mine Workers of America Constitution*, p. 11. Copy deposited at Iowa State Historical Department, Division of Historical Museum and Archives, Des Moines.

5. In 1918 White went into the private sector, accepting an executive position with a mine-related firm; *The United Mine Workers Journal*, Apr. 22, 1909, p. 1. Hereafter cited as *UMWJ*.

6. *UMWJ*, May 7, 1891, p. 4; May 14, 1891, p. 4; June 4, 1891, p. 5.

7. Ibid., June 11, 1891, p. 5.
8. Ibid., July 6, 1891, p. 1.
9. Ibid., July 30, 1891, p. 4.
10. The summer price had been $0.80 cents for many years. See *UMWJ*, Apr. 12, 1894, p. 4; *UMWJ*, Jan. 18, 1894, p. 8. Seven months later Mystic local secretary John Chambers wrote to the *Journal* that "Men are coming in by the score every day." The continual influx of miners created problems for the long-time camp employees. Chambers noted that "We cannot make room for any more of our fellow-slaves, and if they come as they are now, we will have lots of trouble holding our own." See *UMWJ*, Aug. 30, 1894, p. 8.
11. *UMWJ*, Apr. 25, 1895, p. 2; Henry, "A History," p. 4.
12. Henry, "A History," p. 8; Several locals continued to function even though the district organization ceased to exist.
13. *UMWJ*, Mar. 16, 1899, p. 1.
14. Arthur E. Suffern, *The Coal Miners' Struggle for Industrial Status* (New York: Macmillan, 1926), pp. 46-72; *UMWJ*, Mar. 30, 1899, p. 1. In 1891 union officials divided District Thirteen into four subdistricts. Subdistrict One contained Putnam, Missouri, Wayne, and Appanoose counties; Subdistrict Two contained Marian, Mahaska, and Keokuk counties; Subdistrict Three contained Jasper, Polk, and Dallas counties; and Subdistrict Four contained Boone, Green, and Webster counties.
15. *UMWJ*, Apr. 13, 1899, p. 2. Reese should have stated that Subdistrict Four was reorganized in three days. Subdistrict Four was organized in 1891 and became inactive in 1895 along with the other subdistricts. Therefore it was a matter of reactivating locals in Subdistrict Four.
16. *UMWJ*, Apr. 13, 1899, p. 2.
17. Ibid., July 6, 1899, p. 2; July 20, 1899, p. 7.
18. Ibid., May 14, 1891, p. 4.
19. Ibid., Apr. 30, 1891, p. 3; June 4, 1891, p. 5; June 11, 1891, p. 1.
20. Suffern, *Coal Miners' Struggle*, p. 64.
21. *UMWJ*, May 14, 1891, p. 4; Dec. 16, 1915, p. 8; Mar. 28, 1912, p. 6; Dec. 19, 1912, p. 6; Mar. 30, 1916, p. 9; July 27, 1916, p. 11. In 1905 the national UMW took in several Canadian locals. From then on it used the name International United Mine Workers of America.
22. *UMWJ*, Sept. 25, 1902, p. 3. Although there is no way to be certain that all mines had checkweighmen by 1910, the fact that miners stopped raising the issue at their annual convention appears to indicate that they had accomplished their goal.
23. Catherine McClellan, interview, Madrid, Iowa, Apr. 1975.
24. *UMWJ*, Dec. 12, 1901, p. 3. In the fall of 1894 a typical dispute arose over the powder issue. At the coal camp of Keb, two union men purchased their powder at a private store for $1.50 per keg, $0.75 less than at the company store. The company immediately discharged the men.
25. In 1905 the district leadership achieved another major goal — the establishment of a defense fund. At the 1905 convention President John White urged the delegates to consider the matter. The delegates responded by voting to assess all members 1 percent of their gross earnings for the establishment of a permanent defense fund. See *District 13 Executive Board Proceedings*, 2:38. Hereafter cited as *District 13 EBP*. Also see *UMWJ*, Apr. 20, 1905, p. 1. District Thirteen became one of the first districts to take such action. But the following year the defense fund amounted to $210,000. Three years later the defense fund totaled nearly $500,000. See *District 13 EBP*, 3:17; *UMWJ*, Mar. 8, 1906, p. 5; ibid., Mar. 23, 1909, p. 1.
26. *UMWJ*, Jan. 30, 1902, p. 8.
27. Ibid.
28. Ibid., Feb. 13, 1902, p. 7. E. H. Downey, *History of Labor Legislation in Iowa*, Iowa Economic History Series, ed. Benjamin Shambaugh (Iowa City: State Historical

Society of Iowa, 1910), pp. 50–53. Miners had little success in securing other safety measures. Miners believed that lack of mine supervision represented a potential hazard. In 1886 the state had three mine inspectors to supervise 350 mines. In 1899 district officials suggested that the governor divide the state into four inspection districts and assign an inspector to each. The legislature did not heed the miners' advice, however, and the state continued to employ only three inspectors. See *UMWJ*, Sept. 7, 1899, p. 3, and *District 13 EBP*, 1:151.

29. *UMWJ*, Mar. 8, 1906, p. 5.

30. *District 13 EBP*, 2:81–82. One year later, John White announced that the death and endowment fund was working satisfactorily. The district paid death benefits of $100 in a total of ninety cases. Of that number, "forty-five men died from miscellaneous diseases; fourteen died from accidents outside the mine; eight died from tuberculosis; and twenty-three died from accidents in the mine." See *District 13 EBP*, 2:5. Three years later, again at White's urging, convention delegates voted to provide death benefits of $50 for each miner's wife. See *District 13 EBP*, 2:1175–181. By 1918 the miners raised the benefits to $200 for each UMW member and $100 for each wife. Miners also included $100 for the miner's mother if she depended upon her son for support. See *UMWJ*, Mar. 31, 1919, p. 12.

31. *UMWJ*, Aug. 5, 1909, p. 2; Dec. 30, 1915, p. 9; Jan. 20, 1916, p. 21.

32. Ibid., Feb. 17, 1916, p. 9.

33. Ibid., May 25, 1905, p. 4; June 28, 1917, p. 11.

34. *District 13 EBP*, 3:32–44. The executive board imposed the following conditions on all participating members: the store must be organized and conducted under the Rockdale Plan; the district executive board must have the right to audit the business, and the society's officers must take out a two-year promissary note, payable to the district, in the amount loaned to the society. Although there is no total record of cooperative stores started after January 1918, in April 1920 the following stores were in operation: Des Moines Co-Operative Mercantile Association; Monroe County Co-Operative Association; Boone Co-Operative Store; Colfax Co-Operative Store; and Melcher Miners' Store. See *District 13 EBP*, 23–24 and *UMWJ*, Apr. 4, 1920, p. 3.

35. *Biennial Report of the State Mine Inspectors, 1914–1915* (Des Moines: State of Iowa, 1916), pp. 41–43; Charles Nichols, *Housing Conditions: Iowa Coal Mining Camps* (Des Moines: State Printing Office, 1919), p. 31.

36. *UMWJ*, Sept. 23, 1909, p. 1.

37. *District 13 EBP*, 2:53–60.

38. Ibid.

39. Ibid., 2:60–70.

40. Nichols, *Housing Conditions*, p. 31.

41. Ibid., p. 32.

42. *Laws of Iowa, 1919*, p. 150. The law is very brief. It states simply that the state board of health will supervise the existing camps. It does not specify in what manner. In 1922 the U.S. Coal Commission estimated that approximately 20 percent of Iowa's coal mining families lived in company housing. The commission concluded that about one-half of the nation's bituminous coal miners lived in company houses. See *UMWJ*, Mar. 15, 1924, p. 3. My own view is that more than 20 percent of Iowa's coal miners lived in company housing in 1922. I included the 20 percent as an estimate. Presumably the people who did not live in coal camps owned their own homes.

43. *UMWJ*, Jan. 11, 1920, p. 14.

44. Nichols, *Housing Conditions*, p. 32.

45. *UMWJ*, Jan. 11, 1920, p. 14.

46. Ibid., Jan. 11, 1920, p. 15.

47. Ibid., Jan. 1, 1922, p. 16.

48. It is impossible to know the precise percentage of foreign-born families in each mining camp. The only way to determine the percentage would be to record the places

of birth of all coal miners in Iowa during the census years. According to state mine inspectors' reports, the UMW executive board, and interviews with Italian-Americans, increasingly after 1900 southern and eastern Europeans dominated the coal camps. During the late teens and twenties these sources state that many camps were composed almost entirely of Italian- and Croatian-Americans, with Italian-Americans predominating.

49. *UMWJ*, Jan. 19, 1905, p. 7.
50. *District 13 EBP*, 2:146.
51. *Biennial Report of the State Mine Inspectors, 1908*, p. 98. Second District Inspector R. T. Rhys included similar comments in his 1908 report, p. 59.
52. *Biennial Report of the State Mine Inspectors, 1910*, p. 72. Although it is difficult to determine whether or not southern and eastern Europeans were less careful than other miners, it should be noted that mine inspectors had been complaining about miners' carelessness since the office of mine inspector was initiated in 1880. In 1883 Inspector Park Wilson noted that "Nine-tenths of the accidents happening in and around the mines, if properly traced to the cause, would prove to be carelessness. We often hear the remark from miners, 'I know my roof is bad, and I intend to prop it as soon as I get my coal loaded'; and perhaps before he gets his coal loaded, there is a fall of roof, and he is badly hurt, or perhaps killed. . . ." See *Biennial Report of the State Mine Inspectors, 1882-1883*, pp. 10-11. Later in the same report Wilson noted that ". . . the miners seemed to prefer to take a chance losing their lives than lose a car of coal." Ibid., p. 21.
53. Interviewees like Milo Papich, James Battani, and Armand and Silvio Milani agreed that men did begin to learn English at once because of their mining work while the women felt little pressure to learn English.
54. Milo Papich, interview, Slater, Iowa, May 1979.
55. Bruna Pieracci, interview, Des Moines, Iowa, Aug. 1980.
56. Judith Milani, interview, Madrid, Iowa, Feb. 1981.
57. It is more difficult to know if Italian-Americans served in local offices because local records have not survived. The only information on the Italian-Americans' local union participation emerges from interviews.
58. *District 13 EBP*, 3:21-22. District Thirteen created the legal department in 1914 so that the district would have the benefit of a full-time legal staff. See *UMWJ*, Apr. 23, 1914, p. 6.
59. James Battani, interview, Woodward, Iowa, Oct. 1979.
60. Hubert L. Olin, *Coal Mining in Iowa* (Des Moines: State of Iowa, 1965), pp. 49-50; *Biennial Report of the State Mine Inspectors, 1880-1881*, pp. 120-21.
61. Ibid., p. 121.
62. Olin, *Coal Mining*, p. 49; *UMWJ*, Aug. 27, 1891, p. 3.
63. *UMWJ*, July 23, 1891, p. 2. District Thirteen was following a racial policy established in the previous decade. In 1883 Iowa miners formed the short-lived Amalgamated Association of Miners of the State of Iowa. At their organizational meeting delegates voted "to welcome within our fold all colored men who espouse and advocate our cause." See *Biennial Report of the State Mine Inspectors, 1882-1883*, p. 83.
64. *District 13 EBP*, 1:6-7. Unfortunately, the District Thirteen executive board proceedings do not give the name or address of the operator who asked the question about black employment. Interviews with white miners have supported the view that some camps were integrated and some were not. Alex Erickson, who worked as a miner in Buxton and Pershing for almost forty years, stated that Bussey and Pershing contained no blacks at any time because the white miners would not allow blacks to come in. See Alex Erickson, interview, Pershing, Iowa, Sept. 1979. Italian-Americans in Seymour have made similar statements. James Ray of Madrid also expressed this view. See James Ray, interview, Madrid, Iowa, Nov. 1975.
65. *UMWJ*, Sept. 7, 1899, p. 3.

66. *District 13 EBP,* 2:1–40.
67. Ibid.
68. C. W. Webb, G. W. Montague, M. J. Bradford, Committee, and Certain United Mine workers of America of Ogden, Iowa v. the United Mine Workers of America. Case 11092 filed June 22, 1911, District Court of Iowa, Boone County. Until a decision was reached on the major issues, the black miners asked Judge R. W. Wright to issue a temporary injunction against the UMW. Judge Wright complied on June 20, 1911, by issuing an injunction restraining the Ogden Local from expelling black miners from the union and from refusing their transfer cards and dues. One week later, Judge Wright modified the injunction to restrain the Ogden Local from depriving the black miners of their union membership and from interfering with their work at the mine. The judge no longer required the local to give the black miners transfer cards or to receive their dues. See Order for Temporary Injunction, June 28, 1911, Case 11092; Order Modifying the Temporary Injunction, Aug. 5, 1911, Case 11092.
69. *District 13 EBP,* 2:1–40. District officials did not take action on the matter.
70. Ibid.
71. It was then possible for all miners, both black and white, to return to work as the operator had recently opened a second mine. William Ewalt, "The Ogden Coal Strike, 1910–1912," (Paper, Iowa State University, 1975).
72. "Iowa," *Coal Age* 18 (May 1925):736.

## CHAPTER 7

1. It is presumed that the operator originally purchased sufficient land for the coal camp as well as for the coal mine itself.
2. Further calculations indicate that if an operator developed a camp consisting of 150 houses, each built at a cost of $150 and rented at $8 per month, the operator could realize an income of $144,000 in ten years. With 150 houses, the cost of building the homes would total $22,500 leaving the operator a profit of $121,500. Once the operator had sold all 150 houses, he realized another $7,500 profit. While it is recognized that the operator would pay yearly taxes, it appears that housing represented a considerable source of profit for the operator. Comments by operators present the view that the only reason for company housing is to "accommodate" the worker. Overall, operators could realize a return five and one-half times their original investment over a ten-year period. Miners continually charged that rents were excessive. No doubt they were referring to the operators' rate of return on their investments rather than the actual rent of the houses. It is impossible to estimate operators' profits on company store sales because no business records of these business operations are known to exist.
3. In 1906 the state legislature raised the legal age at which boys could be employed in the mines to fourteen. See *Laws of Iowa, 1906,* p. 71.
4. At times, the negative attitudes toward miners applied to northern and western Europeans as well as to eastern and southern Europeans. At other times, however, the term "foreigner" applied only to those people who did not speak English. Under this definition, English-, Welsh-, and Scottish-born miners clearly were not regarded as foreigners, but Swedish-born miners also appeared to be grouped in the nonforeign category even though they did not speak English.
5. *District 13 Executive Board Proceedings,* 4:50, 21.

# SELECTED BIBLIOGRAPHY

### STATE DOCUMENTS

*Biennial Report of the State Mine Inspectors, 1880-1925.* Des Moines: State of Iowa.
*Executive Board Proceedings, 1891-1925.* District 13, United Mine Workers of America. Des Moines: Iowa Division of Historical Archives and Museum Microfilms.
*Iowa State Census (1895). Manuscript Population Schedules for Beacon, Cincinnati, and Seymour.* Des Moines: Iowa Division of Historical Archives and Museum Microfilms.
*Iowa State Census (1915). Manuscript Population Schedules for Beacon, Cincinnati, and Seymour.* Des Moines: Iowa Division of Historical Archives and Museum Microfilms.
*Laws of Iowa, 1880-1925.* Des Moines: State Printer, 1880-1925.
McFarland, W. M., comp. *Census of Iowa for the Year 1895.* Des Moines: Iowa State Printing Office, 1896.
Nichols, Charles. *Housing Conditions: Iowa Coal Camps.* Des Moines: State Printing Office, 1919.
Webb, C. W.; Montague, G. W.; Bradford, M. J., Committee and Certain United Mine Workers of America of Ogden, Iowa v. The United Mine Workers of America. Case 11092, Filed June 22, 1911, District Court of Iowa, Boone County.

### FEDERAL DOCUMENTS

Sheridan, Frank J. "Italian, Slavic and Hungarian Unskilled Immigrant Laborers in the United States." *Labor Bureau Bulletin* 15 (1907): 403-86.
U.S., Department of Commerce, Bureau of the Census. *Eleventh Census of the United States, 1890: Population,* vol. 2, pt. 1. Washington, D.C.: Government Printing Office, 1895.
U.S., Department of the Interior, Coal Mines Administration. *A Medical Survey of the Bituminous-Coal Industry: Report of the Coal Mines Administration.* Washington, D.C.: Government Printing Office, 1947.

### BRITISH DOCUMENTS

Great Britain. Parliament. *Parliamentary Papers* (Commons), vol. 2, "Report on the Truck Commissioners," 1871.

Great Britain. Parliament. *Parliamentary Papers* (Commons), vol. 5, "Select Committee on the Scarcity and Dearness of Coal," 1873.
Great Britain. Parliament. *Parliamentary Papers* (Commons), vol. 2, "Royal Commission on the Housing of the Working Class," 1884–1885.

### NEWSPAPERS AND JOURNALS

Slater *Tri-County Times*, 1978.
*United Mine Workers Journal*, 1891–1925.

### BOOKS

Allen, James B. *The Company Town in the American West*. Norman: University of Oklahoma Press, 1977.
Aurand, Harold W. *From the Molly Maguires to the United Mine Workers*. Philadelphia: Temple University Press, 1971.
Brown, Ronald C. *Hard-Rock Miners: The Intermountain West, 1860–1910*. College Station: Texas A & M University Press, 1979.
Burnett, John, ed. *Autobiographies of British Working Class People 1820–1920*. Bloomington: Indiana University Press, 1974.
Byington, Margaret F. *Homestead: The Households of a Mill Town*. The Pittsburgh Survey. 6 vols. New York. Russell Sage Foundation, 1910.
Cash, Joseph H. *Working the Homestake*. Ames: Iowa State University Press, 1973.
Cole, G. D. H., and Filsons, A. W., eds. *British Working Class Movements: Select Documents 1789–1875*. London: Macmillan & Co., 1951.
Devine, Edward T. *Coal: Economic Problems of the Mining, Marketing, and Consumption of Anthracite and Soft Coal in the United States*. Bloomington, Ill.: American Review Service, 1925.
Downey, E. H. *History of Labor Legislation in Iowa*. Iowa Economic History Series. Edited by Benjamin F. Shambaugh. Iowa City: State Historical Society of Iowa, 1910.
Dubovsky, Melvyn, and Van Tine, Warren. *John L. Lewis: A Biography*. New York: Quadrangle/New York Times, 1977.
Evans, Chris. *History of the United Mine Workers of America*. 2 vols. Indianapolis: n.p., 1918?–1920.
Finley, Joseph E. *The Corrupt Kingdom: The Rise and Fall of the United Mine Workers*. New York: Simon and Schuster, 1972.
Foerster, Robert. *The Italian Emigration of Our Times*. Cambridge, Mass.: Harvard University Press, 1919.
Gluck, Elsie. *John Mitchell, Miner*. New York: John Day, 1929.
Goldthwait, N. E., ed. *History of Boone County*. 2 vols. Chicago: Pioneer Publishing, 1914.
Goodrich, Carter. *The Miner's Freedom: A Study of the Working Life in a Changing Industry*. Boston: Marshall Jones, 1925.

# SELECTED BIBLIOGRAPHY

Graebner, William. *Coal Mining Safety in the Progressive Period: The Political Economy of Reform.* Lexington: University Press of Kentucky, 1976.
Gutmann, Herbert. *Work, Culture and Society in Industrializing America.* New York: Knopf, 1976.
Harl, Neil; Achterhof, John; Anderson, Paul; and Wiese, Karen. *Coal Mine Maps for Eight Iowa Counties.* Iowa State University: Agriculture and Home Economics Experiment Station, 1977.
Harvey, Katherine. *The Best Dressed Miners: Life and Labor in the Maryland Coal Region, 1835-1910.* Ithaca: Cornell University Press, 1969.
Hunt, Edward; Tryon, F. G.; and Willits, Joseph, eds. *What the Coal Commission Found: An Authoritative Summary by the Staff.* Human Relations Series. Vol. 3. Baltimore: Williams & Wilkins, 1925.
Iorizzo. Luciano J. *The Italian-Americans.* New York: Twayne, 1971.
Jensen, Richard. *The Winning of the Midwest: Social and Political Conflict, 1888-1896.* Chicago: University of Chicago Press, 1971.
Jillson, Willard. *The Coal Industry in Kentucky.* Frankfort: Kentucky Geological Survey, 1924.
Katzman, David M. *Seven Days a Week: Women and Domestic Service in Industrializing America.* New York: Oxford University Press, 1978.
Kessner, Thomas. *The Golden Door: Italian and Jewish Immigrant Mobility in New York City.* New York: Oxford University Press, 1977.
Lunt, Richard D. *Law and Order vs. the Miners: West Virginia, 1907-1933.* Hamden, Conn.: Archon, 1979.
McBride, John. "The Coal Miners." In *The Labor Movement: The Problems of Today,* edited by George E. McNeill, pp. 241-67. New York: M. W. Hazen, 1888.
McDonald, David, and Lynch, Edward. *Coal and Unionism: A History of the Coal Miners' Union.* Silver Springs, Md.: Cornelius Printing, 1939.
Morris, Homer. *The Plight of the Bituminous Coal Miner.* Philadelphia: University of Pennsylvania Press, 1934.
Nelli, Humbert. *Italians in Chicago 1880-1930: A Study in Ethnic Mobility.* New York: Oxford University Press, 1970.
Olin, Hubert L. *Coal Mining in Iowa.* Des Moines: State of Iowa, 1965.
Petersen, William J. *The Story of Iowa: The Progress of an American State.* 4 vols. New York: Lewis Historical Publishing, 1952.
Pieracci, Bruna. "The Miners." In *The Immigrants Speak: Italian-Americans Tell Their Story,* edited by Salvator J. La Gumino, pp. 33-47. New York: Center for Migration Studies, 1979.
Rochester, Anna. *Labor and Coal.* Labor and Industry Series. New York: International Publishers, 1931.
Rolle, Andrew. *The Immigrant Upraised.* Norman: University of Oklahoma Press, 1968.
Roy, Andrew. *The Coal Miners.* Cleveland: Robinson, Savage, 1876.
―――. *A History of the Coal Miners of the United States.* Columbus, Ohio: J. L. Tauger Printing, 1907.
Sartorio, Henry Charles. *Social and Religious Life of Italians in America.* 1918.

Reprint. Clifton, N.J.: Kelley, 1974.
Shurick, A. T. *The Coal Industry.* Boston: Little, Brown, 1924.
Suffern, Arthur E. *The Coal Miners' Struggle for Industrial Status.* New York: Macmillan, 1926.
Wieck, Edward A. *The American Miners' Association: A Record of the Origin of Coal Miners' Unions in the United States.* New York: Russell Sage Foundation, 1940.
Winsey, Valentine Rossilli. "The Italian Immigrant Women Who Arrived in the U.S. before World War I." In *Studies in Italian-American Social History,* edited by Francesco Cordasco, pp. 199-210. Totowa, N.J.: Roman and Littlefield, 1975.
Yans-McLaughlin, Virgina. *Family and Community: Italian Immigrants in Buffalo, 1880-1930.* Ithaca: Cornell University Press, 1977.

## ARTICLES

Baker, Donald. "Operating Conditions Encountered in Iowa." *Coal Age* 14 (Mar. 1921):437-41.
Bergman, Leola. "The Negro in Iowa." *Iowa Journal of History and Politics* 46 (Jan. 1948):3-90.
Hoffman, Phil. "The Lost Creek Disaster." *The Palimpsest* 26 (Jan. 1945): 21-27.
_____. "The Powder House Explosion." *The Palimpsest* 26 (Aug. 1945): 247-56.
"Iowa." *Coal Age* 18 (May 1925):736.
Johnson, Cynthia. "A New Life: The Iowa Coal Mines." *The Palimpsest* 56 (Mar./Apr. 1975):56-64.
Keyes, Charles. "Annotated Bibliography of Iowa Geology and Mining." *Iowa Geological Survey* 22 (1912):89-121.
Lees, James H. "History of Coal Mining in Iowa." *Iowa Geological Survey* 22 (1913):535-88.
Rutland, Robert. "The Mining Camps of Iowa: Faded Source of Hawkeye History." *Iowa Journal of History* 54 (Jan. 1956):35-42.
Rye, Stephen. "Buxton: Black Metropolis of Iowa." *Annals of Iowa* 41 (Spring 1972):939-57.
Scott, Joan W., and Tilly, Louise A. "Women's Work and the Family in Nineteenth-Century Europe." *Comparative Studies in Society and History* 18 (1975):36-64.
Shiffer, Beverly. "The Story of Buxton." *Annals of Iowa* 37 (Summer 1964):339-47.
Swisher, Jacob. "Mining in Iowa." *Iowa Journal of History* 43 (Oct. 1945): 305-56.
_____. "The Rise and Fall of Buxton." *The Palimpsest.* 26 (June 1945): 179-92.
Vecoli, Rudolph J. "Prelates and Peasants: Italian Immigrants and the Catholic Church." *Journal of Social History* 2 (Spring 1969):217-68.

# SELECTED BIBLIOGRAPHY

## DISSERTATIONS AND UNPUBLISHED PAPERS

Ewalt, William. "The Ogden Coal Strike, 1910–1912." Paper, Iowa State University, 1974.
Hair, P. E. H. "The Social History of British Coal Miners in 1800–1845." Ph.D. diss., Oxford University, 1955.
Henry, Ben. "A History of District 13." Paper, Iowa Division of Historical Archives and Museum, Des Moines.
Zanotti, Gina. "The Zanotti Family at Rippey." Paper, Iowa State University, 1972.

## INTERVIEWS

Anderson, Christina. Ames, Iowa, 15 Dec. 1978.
Aquilani, Merico. Ogden, Iowa, 7 Feb. 1976.
Argenta, Augusta. Seymour, Iowa, 24 Apr. 1978.
Argenta, John. Seymour, Iowa, 24 Apr. 1978.
Aubrey, Dean. Ames, Iowa, 2 Aug. 1972.
Battani, Gina. Woodward, Iowa, 24 Oct. 1979.
Battani, James. Woodward, Iowa, 24 Oct. 1979.
Betti, Flora. Granger, Iowa, 10 Jan. 1978.
Bingman, Delno. Chariton, Iowa, 28 Dec. 1978.
Biondi, Paulina. Madrid, Iowa, 22 Sept. 1979.
Blasovitch, Rudy. Seymour, Iowa, 4 Oct. 1975.
Braida, Mary. Chariton, Iowa, 20 Oct. 1975.
Brown, Marjorie. Waterloo, Iowa, 15 Oct. 1978.
Cathcart, John. Seymour, Iowa, 4 Oct. 1975.
Cerato, Antonia. Numa, Iowa, 25 Mar. 1978.
Collier, Dorothy. Des Moines, Iowa, 14 Nov. 1978.
Conway, William. Boone, Iowa, 12 Feb. 1975.
DeGard, John. Seymour, Iowa, 4 Oct. 1975.
DeGard, Victoria. Seymour, Iowa, 1 Mar. 1978.
Earls, Alnora. Seymour, Iowa, 24 Mar. 1978.
Erickson, Alex. Pershing, Iowa, 15 Sept. 1979.
Fontanini, Nellie. Granger, Iowa, 10 Jan. 1978.
Fox, Charles. Seymour, Iowa, 4 Oct. 1975.
Frazier, Jesse. Des Moines, Iowa, 10 Mar. 1979.
Goodman, Mary. Madrid, Iowa, 31 July 1974 and 17 Mar. 1979.
Goodman, Otto. Madrid, Iowa, 31 July 1974 and 17 Mar. 1979.
Harris, Archie. Lovilia, Iowa, 17 Nov. 1974.
Gorman, John. Elkhart, Iowa, 12 July 1977 and 10 Jan. 1978.
Lami, Anita. Granger, Iowa, 10 Jan. 1978.
Lepovitz, William. Madrid, Iowa, 15 June 1979.
McClellan, Catherine. Madrid, Iowa, 4 Apr. 1975.
Maddalena, Mary. Seymour, Iowa, 1 Mar. 1978.
Milani, Armand. Madrid, Iowa, 17 Feb. 1981.

Milani, Judith. Madrid, Iowa, 17 Feb. 1981.
Milani, Silvio. Madrid, Iowa, 17 Feb. 1981.
Mores, Minnie. Seymour, Iowa, 24 Mar. 1978.
Nizzi, Lola. Granger, Iowa, 10 Jan. 1978.
Padovan, Edna. Centerville, Iowa, 1 Mar. 1978.
Padovan, Eugenia. Centerville, Iowa, 1 Mar. 1978.
Papich, Milo. Slater, Iowa, 8 May 1979.
Pieracci, Bruna. Des Moines, Iowa, 1 Aug. 1980.
Porter, Gilbert. Chariton, Iowa, 28 Dec. 1978.
Ray, James. Madrid, Iowa, 25 Nov. 1975.
Sebben, Ladorrico. Seymour, Iowa, 4 Oct. 1975.
Sebben, Tony. Seymour, Iowa, 4 Oct. 1975.
Sertich, Mary. Granger, Iowa, 10 Jan. 1978.
Starchovich, John. Madrid, Iowa, 1 June 1979.
Turner, John. Madrid, Iowa, 31 July 1974.
Voyce, Bert. Seymour, Iowa, 4 Oct. 1975.

# INDEX

Albia, Iowa, 34-35, 116, 141-42
  Miners' Hospital, 141
Allegany County, Maryland, 4
Amalgamated Association of Miners of the United States, 16-17
American Federation of Labor, 15
American Miners Association, 10, 126
Anderson, Christina, 116
Angus, xi, 127
Argenta, Angelina, 88-89, 108
Argenta, Augusta, 73
Argenta, Giovani, 92
Argenta, John, 73, 106
Armstrong, Hobe, 151
Atherton, Lewis, 114

Bates, John, 13, 15
Battani, Camilla, 96
Battani, Gina, 101-2
Battani, James, 68, 72-73, 94, 107, 150
Beacon, Iowa, xii, 112-13, 166
  birthplaces of residents, 116-17, 122
  employment of children, 116
  income, 1915, 123-25
  miners' ages, 115
  occupations, general, 114, 120-21
  population, 113, 118
  racial composition, 118, 122-23
  sex ratios, 117-18
  unemployment, 1915, 123-25
  women's occupations, 114-15, 121-22
Berwick coal camp, 89, 99
Betti, Flora, 106
Big Jim coal camp, 88, 91-92, 106
Biondi, John, 69
Biondi, Paulina, 69-71, 75, 89, 93, 96-97, 100, 111
Black damp, 48
Black Diamond Mine (later Pioneer), x
Blacks in Iowa, 118, 148, 161. *See also* Buxton
  in Buxton, 60, 84
  racial discrimination, 151-55
  social activities of women, 80-81
  UMW, treatment in, 151
Bolton-Hoover Company, 2, 119
Boone, T. M., 141

Boonsboro, xi
Braida, John, 88
Braida, Mary, 88
Brennan, John, 133
British coal miners
  and American miners, 7
  and American unions, 15
  communities of, 10
  and immigration to U.S., 7-8
  and mining methods, 9
  and trade unions, 8
Brown, Marjorie Lee, 80-82
Brown, Ronald C., 117
Brown, William, 154
Brugioni, Caterina, 96, 105
Brugioni, Michele, 91, 105
Buffalo, New York, 64, 110
Burlington Northern Railroad (earlier Chicago, Burlington, and Quincy), 6, 119
Busetto, George, 92
Buxton, Iowa, xi, 34-35, 42, 60, 84, 123, 153
  churches, 80
  coal mines, 79
  founding of, 78-79
  houses, 79
  lodges, 80
  and Monroe Mercantile Company, 81
  women's clubs, 80
  Wonders, 80-81
  YMCAs, 79-80
Byington, Margaret, 66

Cambruzzi, Giacomina, 69, 74, 94, 100
Cambruzzi, Victor, 69, 91
Carbonado, 127, 130, 151
Carnegie Steel Corporation, 66
Carney coal camp, 55, 99
Cash, Joseph, xii
Catholic Church, 67, 75
  and discrimination against Italian-Americans, 76
  and Italian-Americans, 76
  and reunification of Italy, 77
Catholic Church, Iowa
  absence of church, 75
  and Italian-American communities, 74-78

# INDEX

Centerville, Iowa, 74, 98, 102, 127, 132–33, 169
Central Competitive Coal District, 132, 135–36
Cerato, Antonia, 63, 94–95, 96, 100, 103–4, 107–8
Cerato, David, 92, 94, 103
Checkweighman, 53–54, 133, 137
Chicago, Illinois, 61, 65, 73, 74, 81, 102–3
Chicago and North Western Railroad, xi, 79, 84
Chicago, Burlington, and Kansas City Railroad, 119
Chicago, Burlington, and Quincy Railroad, 119
Chicago, Rock Island, and Pacific Railroad, 119
Cincinnati, Iowa, xii, 113, 127, 166
 birthplaces of residents, 116–17, 122
 employment of children, 116
 income, 1915, 123–25
 miners' ages, 115
 occupations, general, 114, 120–21
 population, 113, 118
 racial composition, 118, 122–23
 sex ratios, 117–18
 unemployment, 1915, 123–25
 women's occupations, 114–15, 121–22
Civil War, 3, 6–7, 23, 25
Clarkson, John T., 131–32
Cleveland coal camp, 61, 88
Coal camps, Iowa, 87. *See also* Italian-American families, Iowa
 attitudes of operators, 145
 descriptions of, 89, 100, 158
 house costs and rents, 163–64
 isolation and lack of social activities, 100
 longevity of, 157–58
 medical treatment of residents, 106–8
 size, 160
 water shortages, 90
Coal miners
 early hardships, 11
 family characteristics, 166–67
 foreign-born element, 24
 social inferiority and isolation of, 10, 104, 165–66
 wages, 26, 134
Coal miners, Iowa
 economic grievances, 133–35
 geographic mobility, 158
 independence, 27, 57
 location, 34
 mining hazards, 45–49
 night education program, 150
 part-time work, 158–59
 safety concerns, 138
 social inferiority, 165
 transportation to mines, 34–35
 work schedules, 33
 work trains, 35
Coal miners, Italian-American, 147–49, 165
 English language capabilities, 148
 as home owners, 159
 UMW, participation in, 149–50
Coal miners' labor unions, 11–13. *See also* American Miners Association; Miners' National Association of the United States of America; National Federation of Miners and Mine Laborers; United Mine Workers of America
Coal mining, Iowa
 earliest mining operations, ix, x, xi
 longevity of mines, 32
 number of mines in 1895 and 1925, xi
 physical conditions, 32
 seasonal nature of, 23
 training, lack of, 58
Coal mining, U.S.
 development in Middle West, 6
 earliest operations and technology, 4–5
 earliest uses of coal, 4–5
 immigrants, role of, 24
 major geographical regions, 7
 mining methods, 19–20
 part-time work, 24
 physical variations in mines, 25
 seasonal nature of, 23
 and water routes, 5–6
Coal mining accidents and disasters, 12, 20
 Lost Creek, 45, 138
Coal mining methods
 longwall, 9, 29–32
 mine entry, types of, 27–28
 pick mining, 19–20
 room and pillar, 9, 29
 undercutting machines, 19
Coal mining procedures
 child labor, 38–39
 deadwork, 42
 loading coal, 35
 miners' helpers, 37–38
 mules, use of, 40–41
 preparing shots, 43
 safety precautions, 35
 shooting on solid, 42–43
 undercutting, 43
Coal mining safety legislation, 20
Coal operators. *See also* Iowa Coal Operators Association
 and company housing, 158
 and labor-management, 162–63

INDEX

Coal screening, 12-13, 133, 136
Coalville coal camp, 151
Colfax, Iowa, 35
Collier, Dorothy, 80-82
Company housing
  description of, 87-91
  grievances about, 22-23
Company men, 27, 57
  brattice men, 53
  couplers, 49-50
  foremen, 42, 54-58
  hoisting engineer, 53
  motormen, 53
  mule drivers, 39, 51, 56
  mule trainers, 42
  ratio to miners, 49
  shot examiners, 44, 46
  shot firers, 35, 46, 56
  timbermen, 52-53
  track layers, 52-53
  trappers, 49, 52
  undercutters, 47
  wiremen, 53
Company store, 21, 133
  credit, 98-100
  grievances about, 137
  inflated prices, 137
  scrip, 22
Consolidation Coal Company, xi, 35, 78-79, 81, 84
Cook, C. W., 107
Corso, John, 35
Croatian-American miners, Iowa, 104, 148-49, 154

Dallas coal camp, 91, 105
Dead work, 13, 42
Death benefits, 140
DeGard, John, Jr., 91, 105
DeGard, John, Sr., 91
DeGard, Victoria, 74, 98, 102, 103, 107
Department of Public Instruction, 146-47
Depression of 1893, 131
Des Moines, Iowa, x, 74, 102, 127, 141-42, 169
District Thirteen, UMW, 157, 159, 169
  accomplishments, 160-70
  adult night school, 150
  camp school improvements, 146-47
  cooperative stores, 142-43
  death benefits, 140-41, 169-70
  disbanding in 1895, 132
  and eight-hour day, 130-31
  and English-language capabilities, 148-49
  and housing improvements, 143-44
  and irregular employment, 161
  medical care and hospitalization plan, 140-41
  and mine safety, 138
  organization of, 127
  and racial discrimination issues, 152-55
  reorganization of, 132
  and shot firer/shot examiner, 139
  and strike breaking, 151-52
  structure and functions, 127-29
  subdistricts, 127
  and training programs, 161-62, 164
  union organizing, 160

Earls, Alnora, 102-3, 105
Eight-hour day, issue of, 23, 127, 130, 132. *See also* District Thirteen, UMW; United Mine Workers of America
English immigrants, 24, 104, 164
Enterprise, Iowa, 143-44

Foerster, Robert, 75
Fontanini, Nellie, 97, 105
Fonzaso, Italy, 94, 107
Ft. Dodge Coal Company, 151
Frassinoro, Italy, 89
Fromm, Julius, 130-32

Garrett County, Maryland, 4
Good Housing Act, 91, 146
Gorman, Rev. John, 60, 77-78, 105
Granger, Iowa, 73, 77, 105-6
Griffin, J. M., 141
Gutch, T. E., 141
Gutman, Herbert, 59-60

Haven, William, x
Heaps, George, 146
Hinchcliffe, John, 14
Hocking valley strike, 16-17
Homestead, Pennsylvania, 60, 66
Hull House, 65

Illinois Central Railroad, xi, 6
Illinois coal industry, 6, 13-14, 16-17, 19-20, 117, 132, 134
Indiana coal industry, 14, 17, 117, 132
Iowa Coal Operators Association, 132, 159

Iowa Geological Survey, ix
Irish-American immigrants, 104
Italian-American coal miners, Iowa
  English-language capabilities, 149
  union participation, 149
Italian-American families, Iowa
  children's work and wages, 72
  and company housing, 87–91
  ethnic discrimination, 104–5
  and health and sanitation of, 87
  and home ownership, 91–92
  and leisure time, 100–102, 106
  medical treatment of, 106
  midwives, use of, 107
  poverty, 83
  seasonal mobility of, 72–73
Italian-Americans, Iowa, 60, 67, 82–83, 104
  assistance of family, 68
  immigration patterns, 59
  and marriage, 69
  role of sponsors, 67–68
  and rural industrial environment, 70–71
  seasonal mobility of, 72–73
  social boss, role of, 69
Italian-Americans, Northeast, 82–83, 165
  children's education, 65
  immigration of, 61
  housing of, 61–62
  religious activity of, 65
  social activity of, 65
  studies of, 61
  and traditions from Europe, 62
  women's employment, 64–65
Italian-American women, Iowa
  and boarders, 95–97
  and childbirth, 107
  and child care, 71
  and daughters' activities, 101–2
  domestic routines of, 92–95
  earnings of, 99
  language difficulties of, 103–4
  and marriage, 103
  second generation, 98–99, 110
  and social activities, 100–102
  as wage earners, 70–71
Italy, northern, 62–64

James, John, 15, 16
Joint conference, 15
Joint wage agreement, District Thirteen, 132–36, 144. *See also* Joint conference
  printed in foreign languages, 148

Kansas, 17, 77
Katzman, David, 122

Kentucky, 20, 168
Kies, Harry D., 147
Knight, Samuel, ix
Knights of Labor, 17–18, 126–27

Laddsdale, 127, 130
Lami, Anita, 96
Lea, Albert, ix
Lewis, C. J., 143
Lloyd, Thomas, 15
Lost Creek Fuel Company, explosion at, 138–39

McBride, John, 15, 25
McClennahan, F. E., 147
McDonald, Alexander, 15
McLaughlin, Daniel, 15
McNeill, H. W., 151
McNeill, W. A., 151
Madrid, Iowa, 34, 73, 169
Maher, M. F., 133
Maryland coal industry, 5, 11, 14, 16
Milani, Judith, 149
Miller, R. P., 141
Miners' and Laborers' Benevolent Association, 14, 16
Miners' Association, 13–14
Miners' National Association of the United States of America, 15–16
Miners' Protective Association, 15
Mine Village Improvement Commission, 144
Mitchell, John, 148
Moingona, xi
Moran coal camp, 43, 49, 96
Moratti, Concitta, 94, 107
Mores, Marie, 105–6
Mores, Minnie, 75, 100–101
Morris, Joe, 146

National Federation of Miners and Mine Laborers, 15, 17–18, 126
Neal, George, 81
Nichols, Charles, 100, 145–46
Nizzi, Lola, 94, 103, 105
Numa, 74–75, 92, 94, 98, 104, 107, 110, 159

Ogden coal camp, 153–55
Ohio coal industry, 8, 11, 14, 17, 19, 117, 132, 165
Olmitz coal camp, 88
Oralabor, Iowa, 99, 147, 153
Oral histories, use of, 59–60
Ostino, Eugene, 92

# INDEX

Padovan, Edna, 101, 104
Padovan, Eugenia, 75, 92
Padovan, John, 92
Papich, Milo, 55–56
Pash, Andrew, 138
Peek, Mary, 106
Pennsylvania coal industry, 11–14, 16–17, 18, 20, 22, 26, 117, 132, 134, 165
Perry, Edwin, 130
Pershing coal camp, 94, 104, 107
Piacentini, Filomena, 63
Pieracci, Bruna, 63, 89, 99, 108, 149
Pieracci, Filomena, 89, 108
Pieracci, Orlando, 89
Pleck, Elizabeth, 64, 83

Rae, John B., 25, 126–27
Ream, John, 133, 142
Redhead, Wesley, x
Reese, John, 116, 128, 133, 137, 162
Richmond, Virginia, 4
Ross, David, 25
Roy, Andrew, 8, 15, 19

Saiella, Antonia, 96, 107
Sartorio, Enrico, 75
Scaife, William, 25
Scandia Coal Company, 56
Scott, Joan, 62–63
Scott, Walter, 130–31, 136, 152
Scottish immigrants, 24, 164
Sertich, Mary, 96
Seymour, Iowa, xii, 69–70, 73, 75, 88, 91–92, 98–99, 105–7, 109–10, 113, 127, 159, 166, 169
  birthplaces of residents, 116–17, 122
  employment of children, 116
  income, 1915, 123–35
  miners' ages, 115
  occupations, general, 114, 120–21
  population, 113, 118
  racial composition, 118, 122–23
  sex ratios, 117–18
  unemployment, 1915, 123–35
  women's occupations, 114–15, 121–22
Siney, John, 15–16
Six Hour League, 161
Slavic immigrants, 60
  and Catholic Church, Homestead, Pennsylvania, 67
  Homestead, Pennsylvania, settlement of, 66–67, 83–84
  women's roles, 66–67
Southern and eastern European immigrants, 20, 24, 165–66
State legislature, Iowa, 38, 116, 140

adult night education program, 150
Good Housing Act, 91, 146
study of coal camps, 1919, 145–46
Strike breaking, 17, 151–52
Strip mining, 170
Swedish-American immigration, 104, 110
Sweeney, Edward, 149

Tilly, Louise, 62–63
Turner, W. H., 127

United Mine Workers Journal, 116, 118, 127, 130–31, 133–34, 142–43, 147, 161–62
United Mine Workers of America (UMW), xii, 15, 18, 22–23, 25–26, 53, 118, 143, 151, 163
  and eight-hour day, 130
  and English-language capabilities, 148
  and foreign-born populations, 147–48
  and joint wage agreement, 134–36
  membership of, 127
  organization of, 126
  structure of, 126–29
U.S. Geological Survey, ix

Vecoli, Rudolph, 75–76

Weaver, Daniel, 14–15
Welsh immigrants, 24, 104, 164
West Virginia coal industry, 17, 19
White, John, 128–30, 138, 140–41, 144, 162
Whitebreast Fuel Company, x
White damp, 49
Wieck, Edward, 8
Wilkes-Barre, Pennsylvania, 4
Williamson coal camp, 88
Wilson, Park C., 44
Women, 168. See also Coal camps, Iowa; Italian-American women, Iowa; Italy, northern; Buxton
Woodson, George, 153
Woodward, 150

Yans-McLaughlin, Virginia, 110

Zanotti, Andrew, 77
Zookspur, 106–7, 112–13, 137